BIOETHICS

BIOETHICS

PASSING THE BOARDS, PROVIDING PATIENT CARE, AND BEYOND

Jeffrey W. Bulger

Oxford University Press is a department of the University of Oxford. It furthers
the University's objective of excellence in research, scholarship, and education
by publishing worldwide. Oxford is a registered trade mark of Oxford University
Press in the UK and certain other countries.

Published in the United States of America by Oxford University Press
198 Madison Avenue, New York, NY 10016, United States of America.

© Jeffrey Bulger 2024

All rights reserved. No part of this publication may be reproduced, stored in
a retrieval system, or transmitted, in any form or by any means, without the
prior permission in writing of Oxford University Press, or as expressly permitted
by law, by license, or under terms agreed with the appropriate reproduction
rights organization. Inquiries concerning reproduction outside the scope of the
above should be sent to the Rights Department, Oxford University Press, at the
address above.

You must not circulate this work in any other form
and you must impose this same condition on any acquirer.

Library of Congress Cataloging-in-Publication Data
Names: Bulger, Jeffrey W., author.
Title: Bioethics : passing the boards, providing patient care, and beyond / Jeffrey W. Bulger.
Description: New York, NY : Oxford University Press, [2024] |
Includes bibliographical references and index. |
Identifiers: LCCN 2024024750 | ISBN 9780197772195 (pb) |
ISBN 9780197772218 (epub) | ISBN 9780197772201 (other ebook) |
ISBN 9780197772232 (other ebook) | ISBN 9780197772225 (other ebook)
Subjects: LCSH: Medical ethics.
Classification: LCC R724 .B844 2024 | DDC 174.2—dc23/eng/20240628
LC record available at https://lccn.loc.gov/2024024750

This material is not intended to be, and should not be considered, a substitute for medical or other professional
advice. Treatment for the conditions described in this material is highly dependent on the individual circumstances.
And, while this material is designed to offer accurate information with respect to the subject matter covered and to be
current as of the time it was written, research and knowledge about medical and health issues is constantly evolving
and dose schedules for medications are being revised continually, with new side effects recognized and accounted
for regularly. Readers must therefore always check the product information and clinical procedures with the most
up-to-date published product information and data sheets provided by the manufacturers and the most recent codes
of conduct and safety regulation. The publisher and the authors make no representations or warranties to readers,
express or implied, as to the accuracy or completeness of this material. Without limiting the foregoing, the publisher
and the authors make no representations or warranties as to the accuracy or efficacy of the drug dosages mentioned in
the material. The authors and the publisher do not accept, and expressly disclaim, any responsibility for any liability,
loss, or risk that may be claimed or incurred as a consequence of the use and/or application of any of the contents of
this material.

DOI: 10.1093/med/9780197772195.001.0001

Printed by Marquis Book Printing, Canada

CONTENTS

About the Author	xxiii

SECTION 1
FOUNDATIONAL TOPICS

1. The Four Principles — 3
    Tripartite Process — 4
        Think — 4
        Assess — 4
        Conclude — 4
    The Patient: The Axis of Healthcare — 5
    The Practitioner: The Guardian of Ethical Integrity — 5
    Public Policy: The Bedrock of Ethical Standards — 6
    Navigating the Ethical Labyrinth — 6
    Review Questions — 6
    Clinical Vignettes — 7
    Web Links — 9

2. The Belmont Report—Abridged — 13
    Think — 13
    Assess — 13
    Conclude — 15
    Review Questions — 15
    Clinical Vignettes — 16
    Web Links — 18

3. The Hippocratic Oath — 21
    Think — 21
    Assess — 21
    Conclude — 22
    Review Questions — 22
    Web Links — 25

4. The Declaration of Geneva — 27
    Think — 27
    Assess — 27
    Conclude — 28
    Review Questions — 28
    Web Links — 30

SECTION 2
60 TOPICS

5. Abortion — 35
    Frozen Betrayal: A Battle of Convictions and the Perilous
        Consequences of Healthcare Negligence — 35
    Think — 36

		Assess	37
		Patient: Autonomy	37
		Practitioner: Beneficence and Nonmaleficence	37
		Public Policy: Justice	37
		Conclude	38
		Review Questions	38
		Clinical Vignettes	39
		Reflection Vignettes	41
		Web Links	42
6.		Abuse: Child, Elder, and Intimate Partner	43
		Whispered Cruelty: A Wake-Up Call for Mandatory Reporting in Healthcare	43
		Think	44
		Assess	44
		Patient: Autonomy	44
		Practitioner: Beneficence and Nonmaleficence	44
		Public Policy: Justice	45
		Conclude	45
		Review Questions	46
		Clinical Vignettes	47
		Reflection Vignettes	49
		Web Links	49
7.		Addressing Oneself and Patient	51
		Deceptive Prescriptions: The Importance of Proper Introductions and Titles in the Healthcare Profession	51
		Think	52
		Assess	52
		Patient: Autonomy	52
		Practitioner: Beneficence and Nonmaleficence	53
		Public Policy: Justice	53
		Conclude	54
		Review Questions	54
		Clinical Vignettes	55
		Reflection Vignettes	57
		Web Links	57
8.		Advance Directives—Planning	59
		Documented Wishes: The Importance of Advance Directives and Durable Power of Attorney for Healthcare	59
		Think	60
		Assess	60
		Patient: Autonomy	60
		Practitioner: Beneficence and Nonmaleficence	60
		Public Policy: Justice	61
		Conclude	61
		Review Questions	61
		Clinical Vignettes	62
		Reflection Vignettes	64
		Web Links	65

9.	Assisted Suicide	67
	A Dance with Morality: The Conundrum of Assisted Suicide in Healthcare	67
	Think	68
	Assess	68
	Patient: Autonomy	68
	Practitioner: Beneficence and Nonmaleficence	68
	Public Policy: Justice	69
	Conclude	69
	Review Questions	69
	Clinical Vignettes	70
	Reflection Vignettes	72
	Web Links	72
10.	Capital Punishment—Executions	75
	Life and Death in the Gray Zone: Probing the Healthcare Profession's Involvement in Capital Punishment	75
	Think	76
	Assess	76
	Patient: Autonomy	76
	Practitioner: Beneficence and Nonmaleficence	76
	Public Policy: Justice	77
	Conclude	77
	Review Questions	77
	Clinical Vignettes	78
	Reflections Vignettes	80
	Web Links	81
11.	Chaperones and Personal Privacy	83
	Privacy Betrayal: Unraveling the Intersection of Chaperones and Personal Privacy in the Healthcare System	83
	Think	84
	Assess	84
	Patient: Autonomy	84
	Practitioner: Beneficence and Nonmaleficence	84
	Public Policy: Justice	84
	Conclude	84
	Review Questions	85
	Clinical Vignettes	85
	Reflection Vignettes	87
	Web Links	88
12.	Confidentiality	89
	Whispered Betrayals: Navigating the Treacherous Waters of Confidentiality Breaches	89
	Think	90
	Assess	90
	Patient: Autonomy	90
	Practitioner: Beneficence and Nonmaleficence	90
	Public Policy: Justice	90
	Conclude	91
	Review Questions	91
	Clinical Vignettes	92

| | Reflection Vignettes | 94 |
| | Web Links | 95 |

13. Conflict of Interest — 97
Veiled Ties: A Dance with Transparency in the Labyrinth of Conflicts of Interest — 97
- Think — 98
- Assess — 98
  - Patient: Autonomy — 98
  - Practitioner: Beneficence and Nonmaleficence — 98
  - Public Policy: Justice — 98
- Conclude — 98
- Review Questions — 99
- Clinical Vignettes — 99
- Reflection Vignettes — 101
- Web Links — 102

14. Contraception — 103
Veiled Agendas: The Ethical Divide in the Contraception Conundrum — 103
- Think — 104
- Assess — 104
  - Patient: Autonomy — 104
  - Practitioner: Beneficence and Nonmaleficence — 104
  - Public Policy: Justice — 105
- Conclude — 105
- Review Questions — 105
- Clinical Vignettes — 106
- Reflection Vignettes — 108
- Web Links — 108

15. Death with Dignity and Assisted Dying — 111
The Veil of Serenity: A Dance with Ethical Shadows of Assisted Dying — 111
- Think — 112
- Assess — 112
  - Patient: Autonomy — 112
  - Practitioner: Beneficence and Nonmaleficence — 112
  - Public Policy: Justice — 113
- Conclude — 113
- Review Questions — 114
- Clinical Vignettes — 114
- Reflection Vignettes — 116
- Web Links — 117

16. Disagreements: Attending Versus Resident — 119
An Unruly Accord: The Crucial Dance of Collaboration and Respect in Quelling Tumults Between Attending Physicians and Residents — 119
- Think — 120
- Assess — 120
  - Patient: Autonomy — 120
  - Practitioner: Beneficence and Nonmaleficence — 120
  - Public Policy: Justice — 121
- Conclude — 121
- Review Questions — 121
- Clinical Vignettes — 122

	Reflection Vignettes	125
	Web Links	125
17.	Do Not Resuscitate Order, Do Not Attempt Resuscitation Order, and Allow Natural Death Order	127
	Vanished Directive: Navigating the Digital Labyrinth and Upholding Dignity at EchoLife Decision Hospital	127
	Think	128
	Assess	128
	Patient: Autonomy	128
	Practitioner: Beneficence and Nonmaleficence	128
	Public Policy: Justice	128
	Conclude	129
	Review Questions	129
	Clinical Vignettes	129
	Reflection Vignettes	132
	Web Links	133
18.	The Doctrine of Double Effect	135
	Dilemma of Dual Outcomes: Navigating the Razor-Edge Path of Ethical Consequences in Oncology	135
	Think	136
	Assess	136
	Patient: Autonomy	136
	Practitioner: Beneficence and Nonmaleficence	137
	Public Policy: Justice	137
	Weighing and Balancing	137
	Conclude	137
	Review Questions	138
	Clinical Vignettes	138
	Reflection Vignettes	140
	Web Links	141
19.	Email and Electronic Communication	143
	Breach of Trust: The High-Stakes Fallout of a Digital Misstep in Healthcare	143
	Think	144
	Assess	144
	Patient: Autonomy	144
	Practitioner: Beneficence and Nonmaleficence	144
	Public Policy: Justice	145
	Conclude	145
	Review Questions	145
	Clinical Vignettes	146
	Reflection Vignettes	148
	Web Links	149
20.	Errors	151
	A Winter Night's Wake-Up Call: Transparency, Accountability, and the Cold Truth in Surgical Errors	151
	Think	152
	Assess	152
	Patient: Autonomy	152
	Practitioner: Beneficence and Nonmaleficence	152
	Public Policy: Justice	152

	Conclude	153
	Review Questions	153
	Clinical Vignettes	154
	Reflection Vignettes	156
	Web Links	157
21.	Euthanasia	159
	Edge of Existence: An Unveiling of the Ethical Quandary in Mercy's Closure	159
	Think	160
	Assess	160
	Patient: Autonomy	160
	Practitioner: Beneficence and Nonmaleficence	160
	Public Policy: Justice	161
	Conclude	161
	Review Questions	161
	Clinical Vignettes	162
	Reflection Vignettes	164
	Web Links	165
22.	Financial Disclosures	167
	The Tempest of Hidden Agendas: Navigating the Murky Waters of Financial Disclosures in Healthcare Ethics	167
	Think	168
	Assess	168
	Patient: Autonomy	168
	Practitioner: Beneficence and Nonmaleficence	168
	Public Policy: Justice	168
	Conclude	169
	Review Questions	169
	Clinical Vignettes	170
	Reflection Vignettes	172
	Web Links	172
23.	Futile Treatment	175
	On the Brink of Desperation: Navigating the Ethical Labyrinth of Futile Treatments in Medicine	175
	Think	176
	Assess	176
	Patient: Autonomy	176
	Practitioner: Beneficence and Nonmaleficence	176
	Public Policy: Justice	177
	Conclude	177
	Review Questions	177
	Clinical Vignettes	178
	Reflection Vignettes	180
	Web Links	181
24.	Genetic Information Nondiscrimination Act	183
	Genetic Secrets: Genetic Discrimination and the Pursuit of Justice	183
	Think	184
	Assess	184
	Patient: Autonomy	184
	Practitioner: Beneficence and Nonmaleficence	184
	Public Policy: Justice	184

	Conclude	185
	Review Questions	185
	Clinical Vignettes	185
	Reflection Vignettes	187
	Web Links	188
25.	Gifts	189
	Ethical Dilemma: The Case of the Mysterious Gifts and the Importance of Healthcare Ethics	189
	Think	190
	Assess	190
	Patient: Autonomy	190
	Practitioner: Beneficence and Nonmaleficence	190
	Public Policy: Justice	190
	Conclude	190
	Review Questions	191
	Clinical Vignettes	191
	Reflection Vignettes	193
	Web Links	194
26.	Gunshot Wounds	195
	Code Name Confidential Gunshots: The Delicate Balancing Act of Mandatory Reporting and Patient Confidentiality	195
	Think	196
	Assess	196
	Patient: Autonomy	196
	Practitioner: Beneficence and Nonmaleficence	196
	Public Policy: Justice	196
	Conclude	197
	Review Questions	197
	Clinical Vignettes	198
	Reflection Vignettes	200
	Web Links	200
27.	Hospice	203
	The Ethereal Specter of Mercy: Hospice Care Walking the Fine Line of Ethical Imperatives	203
	Think	204
	Assess	204
	Patient: Autonomy	204
	Practitioner: Beneficence and Nonmaleficence	204
	Public Policy: Justice	204
	Conclude	204
	Review Questions	205
	Clinical Vignettes	205
	Reflection Vignettes	208
	Web Links	209
28.	Immunization—Vaccine Hesitancy	211
	The Serpent in Eden: A Tale of Immunological Defiance	211
	Think	212
	Assess	212
	Patient: Autonomy	212
	Practitioner: Beneficence and Nonmaleficence	212
	Public Policy: Justice	213

	Conclude	213
	Review Questions	213
	Clinical Vignettes	214
	Reflection Vignettes	217
	Web Links	218
29.	**Impaired Driver**	**219**
	The Reaper's Edict: A Deadly Consequence of Upholding Healthcare Responsibility	219
	Think	220
	Assess	220
	Patient: Autonomy	220
	Practitioner: Beneficence and Nonmaleficence	221
	Public Policy: Justice	221
	Conclude	221
	Review Questions	221
	Clinical Vignettes	222
	Reflection Vignettes	224
	Web Links	225
30.	**Impaired Practitioner**	**227**
	Falling Façade: The Puzzling Mystery of an Ailing Practitioner	227
	Think	228
	Assess	229
	Patient: Autonomy	229
	Practitioner: Beneficence and Nonmaleficence	229
	Public Policy: Justice	229
	Conclude	229
	Review Questions	230
	Clinical Vignettes	230
	Reflection Vignettes	233
	Web Links	233
31.	**Interrogation**	**235**
	Shadows of Coercion: The Repercussions of Unlawful Interrogation	235
	Think	236
	Assess	236
	Patient: Autonomy	236
	Practitioner: Beneficence and Nonmaleficence	236
	Public Policy: Justice	237
	Conclude	237
	Review Questions	237
	Clinical Vignettes	238
	Reflection Vignettes	240
	Web Links	241
32.	**Lifelong Learning**	**243**
	Academic Assassin: The Importance of Lifelong Learning in Healthcare	243
	Think	244
	Assess	244
	Patient: Autonomy	244
	Practitioner: Beneficence and Nonmaleficence	244
	Public Policy: Justice	245
	Conclude	245

	Review Questions	246
	Clinical Vignettes	246
	Reflection Vignettes	249
	Web Links	249
33.	Malpractice	251
	Oversight Redemption	251
	Think	252
	Assess	252
	Patient: Autonomy	252
	Practitioner: Beneficence and Nonmaleficence	252
	Public Policy: Justice	253
	Conclude	253
	Review Questions	253
	Clinical Vignettes	254
	Reflection Vignettes	257
	Web Links	257
34.	Media Communications	259
	An Unintended Breach: An Intriguing Cautionary Tale for the Healthcare Fraternity	259
	Think	260
	Assess	260
	Patient: Autonomy	260
	Practitioner: Beneficence and Nonmaleficence	261
	Public Policy: Justice	261
	Conclude	261
	Review Questions	261
	Clinical Vignettes	262
	Reflection Vignettes	265
	Web Links	265
35.	Healthcare Records	267
	Data Intrusion: A Cautionary Tale of Trust Shattered and Privacy Breached in Healthcare	267
	Think	268
	Assess	268
	Patient: Autonomy	268
	Practitioner: Beneficence and Nonmaleficence	268
	Public Policy: Justice	269
	Conclude	269
	Review Questions	269
	Clinical Vignettes	270
	Reflection Vignettes	273
	Web Links	273
36.	Minor Patients	275
	Conflicting Principles: Struggling Between Parental Rights and Patient Autonomy in the Heart of a Pediatric Mystery	275
	Think	276
	Assess	276
	Patient: Autonomy	276
	Practitioner: Beneficence and Nonmaleficence	276
	Public Policy: Justice	277

	Conclude	277
	Review Questions	277
	Clinical Vignettes	278
	Reflection Vignettes	281
	Web Links	281
37.	Nurses and Allied Health Professionals	283
	Deadly Authority: The Unraveling of a Healthcare Tyrant	283
	Think	284
	Assess	284
	Patient: Autonomy	284
	Practitioner: Beneficence and Nonmaleficence	284
	Public Policy: Justice	285
	Conclude	285
	Review Questions	285
	Clinical Vignettes	286
	Reflection Vignettes	289
	Web Links	289
38.	Organ Donations	291
	Organ Deception: The Shadows of LifeSavers Transplant Clinic	291
	Think	292
	Plasma Donations	292
	Sperm Donations	292
	Eggs Donations	292
	Assess	292
	Patient: Autonomy	292
	Practitioner: Beneficence and Nonmaleficence	293
	Public Policy: Justice	293
	Conclude	293
	Review Questions	294
	Clinical Vignettes	295
	Reflection Vignettes	297
	Web Links	297
39.	Palliative Care	301
	Palliative Betrayal: The Importance of Palliative Care and Medication Management in Healthcare	301
	Think	302
	Assess	302
	Patient: Autonomy	302
	Practitioner: Beneficence and Nonmaleficence	303
	Public Policy: Justice	303
	Conclude	303
	Review Questions	303
	Clinical Vignettes	304
	Reflection Vignettes	306
	Web Links	306
40.	Patient-Practitioner Relationship	309
	Abandonment Affliction: The Case of the Abandoned Patient	309
	Think	310
	Assess	310
	Patient: Autonomy	310

	Practitioner: Beneficence and Nonmaleficence	310
	Public Policy: Justice	310
	Conclude	311
	Review Questions	311
	Clinical Vignettes	312
	Reflection Vignettes	315
	Web Links	315
41.	Practitioner Disagreements	317
	Disputed Prescriptions: The Clash of Convictions in a Healthcare Center	317
	Think	318
	Assess	318
	Patient: Autonomy	318
	Practitioner: Beneficence and Nonmaleficence	318
	Public Policy: Justice	319
	Conclude	319
	Review Questions	320
	Clinical Vignettes	320
	Reflection Vignettes	323
	Web Links	323
42.	Precision Medicine	325
	Genetic Orphans: Uncovering the Unseen Betrayal in Precision Medicine	325
	Think	326
	Assess	326
	Patient: Autonomy	326
	Practitioner: Beneficence and Nonmaleficence	327
	Public Policy: Justice	328
	Conclude	328
	Review Questions	328
	Clinical Vignettes	329
	Reflection Vignettes	332
	Web Links	332
43.	Pregnant Patients	335
	Ethical Dance: Navigating Autonomy, Healthcare, and Law in a Pregnant Patient's Journey	335
	Think	336
	Assess	336
	Patient: Autonomy	336
	Practitioner: Beneficence and Nonmaleficence	337
	Public Policy: Justice	337
	Conclude	338
	Review Questions	338
	Clinical Vignettes	339
	Reflection Vignettes	342
	Web Links	342
44.	Product Sales	345
	Compromised Care: The Mysterious Case of the Conflicted Practitioner and a Community's Reckoning	345
	Think	346
	Assess	346
	Patient: Autonomy	346

		Practitioner: Beneficence and Nonmaleficence	346
		Public Policy: Justice	347
	Conclude		347
	Review Questions		348
	Clinical Vignettes		348
	Reflection Vignettes		351
	Web Links		351

45. Racial Concordance — 353
  Concordance Conundrum: Racial Concordance and the Ethics of
     Patient Autonomy in Medicine — 353
  Think — 354
  Assess — 354
     Patient: Autonomy — 354
     Practitioner: Beneficence and Nonmaleficence — 355
     Public Policy: Justice — 355
  Conclude — 355
  Review Questions — 356
  Clinical Vignettes — 356
  Reflection Vignettes — 358
  Web Links — 359

46. Referrals and Fee Splitting — 361
  Referral Ruse: A Tangled Web of Deception and Redemption in
     Healthcare Practice — 361
  Think — 362
  Assess — 362
     Patient: Autonomy — 362
     Practitioner: Beneficence and Nonmaleficence — 363
     Public Policy: Justice — 363
  Conclude — 363
  Review Questions — 364
  Clinical Vignettes — 365
  Reflection Vignettes — 367
  Web Links — 368

47. Reportable Infections and Illnesses — 369
  Contagion Conflict: Balancing Patient Confidentiality and Public
     Health in Reporting Infectious Diseases — 369
  Think — 370
  Assess — 370
     Patient: Autonomy — 370
     Practitioner: Beneficence and Nonmaleficence — 370
     Public Policy: Justice — 371
  Conclude — 371
  Review Questions — 371
  Clinical Vignettes — 372
  Reflection Vignettes — 375
  Web Links — 375

48. Research and Clinical Equipoise — 377
  Equipoise Enigma: The Case of Clinical Equipoise — 377
  Think — 378
  Assess — 379

	Patient: Autonomy	379
	Practitioner: Beneficence and Nonmaleficence	379
	Public Policy: Justice	379
	Conclude	380
	Review Questions	380
	Clinical Vignettes	381
	Reflection Vignettes	383
	Web Links	384
49.	Research on Human Subjects	385
	Consent Crisis: The Unraveling of Practitioner Novak's Unethical Research	385
	Think	386
	Assess	387
	Human Subject: Autonomy	387
	Researcher: Beneficence and Nonmaleficence	387
	Public Policy: Justice	387
	Conclude	387
	Review Questions	388
	Clinical Vignettes	389
	Reflection Vignettes	392
	Web Links	392
50.	Self-Treatment and Family Treatment	395
	Treatment Tragedy: The Line Between Compassion and Objectivity at Ethical Boundaries Hospital	395
	Think	396
	Assess	396
	Patient: Autonomy	396
	Practitioner: Beneficence and Nonmaleficence	396
	Public Policy: Justice	397
	Conclude	397
	Review Questions	398
	Clinical Vignettes	398
	Reflection Vignettes	400
	Web Links	401
51.	Sexual Boundaries	403
	Seductive Malpractice: The Case of the Unethical Practitioner at Sacred Boundaries Hospital	403
	Think	404
	Assess	404
	Patient: Autonomy	404
	Practitioner: Beneficence and Nonmaleficence	404
	Public Policy: Justice	405
	Conclude	405
	Review Questions	405
	Clinical Vignettes	406
	Reflection Vignettes	409
	Web Links	409
52.	Sexually Transmitted Infections	411
	Infection Intrigue: The LotusBloom STI Clinic Dilemma	411
	Think	412
	Assess	412

	Patient: Autonomy	412
	Practitioner: Beneficence and Nonmaleficence	412
	Public Policy: Justice	413
	Conclude	414
	Review Questions	414
	Clinical Vignettes	415
	Reflection Vignettes	417
	Web Links	418
53.	Social Media Boundaries	421
	Privacy Peril: The MediaManners Hospital Dilemma	421
	Think	422
	Assess	423
	Patient: Autonomy	423
	Practitioner: Beneficence and Nonmaleficence	423
	Public Policy: Justice	423
	Conclude	424
	Review Questions	424
	Clinical Vignettes	425
	Reflection Vignettes	426
	Web Links	427
54.	Sterilization	429
	Sabotaged Sterility: A Battle Between Personal Beliefs and Professional Ethics	429
	Think	430
	Assess	430
	Patient: Autonomy	430
	Practitioner: Beneficence and Nonmaleficence	431
	Public Policy: Justice	431
	Conclude	432
	Review Questions	432
	Clinical Vignettes	433
	Reflection Vignettes	436
	Web Links	436
55.	Strikes—Unionization	439
	Bargaining Dilemma: The Power and Limits of Collective Bargaining in Healthcare	439
	Think	440
	Assess	440
	Patient: Autonomy	440
	Practitioner: Beneficence and Nonmaleficence	440
	Public Policy: Justice	440
	Conclude	441
	Review Questions	441
	Clinical Vignettes	442
	Reflection Vignettes	445
	Web Links	445
56.	Structural Injustice	447
	Injustice Unveiled: Championing Equality at Justice Care Hospital	447
	Think	448
	Assess	448

	Patient: Autonomy	448
	Practitioner: Beneficence and Nonmaleficence	449
	Public Policy: Justice	449
	Conclude	449
	Review Questions	450
	Clinical Vignettes	450
	Reflection Vignettes	453
	Web Links	453
57.	Student Patient Care	455
	Identity Confusion: A Lesson in Honesty at PatientTrust Healthcare Center	455
	Think	456
	Assess	456
	Patient: Autonomy	456
	Practitioner: Beneficence and Nonmaleficence	456
	Public Policy: Justice	457
	Conclude	457
	Review Questions	457
	Clinical Vignettes	458
	Reflection Vignettes	461
	Web Links	462
58.	Surrogate Decision-Making	463
	Surrogate Choices: Surrogate Decision-Making in End-of-Life Care	463
	Think	464
	Assess	464
	Patient: Autonomy	464
	Practitioner: Beneficence and Nonmaleficence	464
	Public Policy: Justice	465
	Conclude	465
	Review Questions	465
	Clinical Vignettes	466
	Reflection Vignettes	469
	Web Links	470
59.	Telemedicine	473
	Digital Deception: The Telemedicine Mystery	473
	Think	474
	Assess	474
	Patient: Autonomy	474
	Practitioner: Beneficence and Nonmaleficence	474
	Public Policy: Justice	474
	Conclude	476
	Review Questions	476
	Clinical Vignettes	477
	Reflection Vignettes	480
	Web Links	480
60.	Terminal and Palliative Sedation	483
	Sedation Secrets: Terminal/Palliative Sedation in End-of-Life Care	483
	Think	484
	Assess	484
	Patient: Autonomy	484

	Practitioner: Beneficence and Nonmaleficence	485
	Public Policy: Justice	485
	Conclude	485
	Review Questions	486
	Clinical Vignettes	486
	Reflection Vignettes	489
	Web Links	489
61.	Testimonials and Quackery	491
	Testimonial Deception: The Seduction of Quackery and the Perilous Road of False Promises	491
	Think	492
	Assess	492
	Patient: Autonomy	492
	Practitioner: Beneficence and Nonmaleficence	492
	Public Policy: Justice	492
	Conclude	493
	Review Questions	493
	Clinical Vignettes	494
	Reflection Vignettes	496
	Web Links	497
62.	Torture	499
	Torture's Betrayal: The Prohibition of Torture and Healthcare Providers' Responsibilities	499
	Think	500
	Assess	500
	Patient: Autonomy	500
	Practitioner: Beneficence and Nonmaleficence	500
	Public Policy: Justice	500
	Conclude	501
	Review Questions	501
	Clinical Vignettes	502
	Reflection Vignettes	505
	Web Links	505
63.	Triage	507
	Triage Dilemma: Balancing Needs and Limited Resources	507
	Think	508
	Assess	508
	Patient: Autonomy	508
	Practitioner: Beneficence and Nonmaleficence	508
	Public Policy: Justice	508
	Conclude	508
	Review Questions	509
	Clinical Vignettes	510
	Reflection Vignettes	512
	Web Links	513

64.	Withholding and Withdrawing Treatment	515
	The Right to Choose: A Story of Autonomy and Care	515
	Think	516
	Assess	516
	Patient: Autonomy	516
	Practitioner: Beneficence and Nonmaleficence	516
	Public Policy: Justice	517
	Conclude	517
	Review Questions	517
	Clinical Vignettes	518
	Reflection Vignettes	521
	Web Links	522

APPENDIX: STEP EXAMS   523
GLOSSARY   525

ABOUT THE AUTHOR

Dr. Jeffrey W. Bulger stands at the forefront of biomedical ethics, blending a rich academic foundation with real-world application. With a PhD in Philosophy from the University of Tennessee, his expertise spans healthcare ethics, clinical ethics, and the philosophy of science, underpinned by a deep commitment to human rights and healthcare education.

A revered educator, Dr. Bulger has molded minds at various academic levels, from introductory courses to graduate seminars. His pedagogical excellence shines through the success of his students, 50 of whom have presented peer-reviewed papers at the prestigious National Conference of Undergraduate Research. Dr. Bulger's teaching is not just about imparting knowledge; it is about sparking transformation, a fact celebrated by his induction into the Master Teacher Guild.

In service, Dr. Bulger's contributions are profound. His work on institutional governance committees and hospital ethics boards underscores a career dedicated to ethical decision-making and policy development. His leadership roles in clinical and academic settings have significantly advanced patient care, curriculum development, and interdisciplinary collaboration.

Dr. Bulger's research addresses critical questions at the intersection of ethics and informed consent, shaping the discourse on patient rights and public policy. His scholarly work includes authoritative texts on biomedical ethics, alongside numerous papers and online courses that bridge theoretical ethics with clinical practice.

Beyond academia, Dr. Bulger is a champion for diversity and inclusion, advocating for equitable treatment across socioeconomic and cultural communities. His approach to education and ethics is inclusive, recognizing the diverse needs and perspectives of all individuals.

A lifelong learner and educator, Dr. Bulger's journey is one of continuous growth and unwavering dedication to excellence in teaching, scholarship, and service. His work not only educates but also inspires, fostering a more ethical, informed, and compassionate society.

SECTION 1

FOUNDATIONAL TOPICS

THE FOUR PRINCIPLES

1

As a member of the medical profession, I solemnly pledge to dedicate my life to the service of humanity.
—Declaration of Geneva

DELVE INTO THE INTRIGUING WORLD OF BIOETHICS with a twist, where each ethical dilemma is introduced with a dramatic moral mystery story. Far from the usual healthcare case studies, these tales are crafted not just to educate but to captivate. This innovative resource presents 60 riveting topics designed to educate and captivate healthcare students and practitioners. Each issue focuses on high-yield content crucial for boards, licensing exams, and continuing healthcare education for students and practitioners. The book is an essential reference for clinical practice, ethics consultations, and academic writings. It explores the complex ethical challenges faced by today's healthcare professionals, guiding readers in striking the delicate balance between the patient's reasonable goals, values, and priorities, professional codes of conduct, and the societal expectations.

While the ethical issues discussed may spark varied opinions rooted in personal, religious, or political beliefs, the book draws from moral positions issued by various healthcare professional organizations, offering a framework for behavioral expectations within the healing community. By immersing oneself into the moral mystery stories, the ethical analysis, and applying the acquired knowledge to review questions and the national board-style clinical vignettes, healthcare professionals will gain the necessary tools and skills to adhere to ethical, professional, and legal standards of care. Engage in this thrilling exploration of bioethics and emerge well-equipped to navigate the challenges of making life-altering decisions.

Healthcare ethics has become a high priority within healthcare education and practice as the field has evolved from a paternalistic hierarchy—the practitioner knows best—to a shared decision-making process between the patient and the practitioner. The focus is now

> the maximization of the patient's best interests as determined by the patient's reasonable goals, values, and priorities within the medical healthcare standards of care framework.

This book presents the methodology and the ethical principles necessary for evaluating 60 moral topics that practitioners frequently encounter in the art of healing. Every topic covered focuses on maximizing the patient's best interests from three perspectives: patient, practitioner, and public policy. The goal is to gain the knowledge, tools, and skills to develop new patient insights, practitioner discernments, and public policy understandings necessary for maximizing the patient's best interests.

The structure used for each topic is as follows:

Think
Assess
    Patient: Autonomy
    Practitioner: Beneficence and nonmaleficence
    Public policy: Justice
Conclude

TRIPARTITE PROCESS

When assessing ethical issues, the practitioner should methodically think, assess, and conclude (Table 1.1).

Table 1.1 Tripartite Process

Think	Ascertain precisely what the ethical issue is.
Assess	Specify and balance the four principles of bioethics.
Conclude	Determine an answer appropriate to and consistent with the standards of care.

Think

It is about more than having answers but asking the right questions at this stage. It is a reflective process, urging practitioners to delve deep, challenge their assumptions, and understand the crux of the ethical concern. It is an exercise that is often more difficult than it sounds, considering the multifaceted nature of most ethical dilemmas in healthcare.

Assess

The assessment phase is analytical. The four principles of bioethics play a central role here. But they are not static, one-size-fits-all principles. The book emphasizes the dynamic interplay between them, advocating for a specified and balanced approach. For instance, although patient autonomy (informed consent) is paramount, it does not exist in a vacuum. It interacts, sometimes even clashes, with other principles such as beneficence (do good) or justice (be fair). It is in these intersections that most ethical dilemmas reside. Therefore, it is paramount to carefully specify how each moral principle relates to the specific topic and determine the weight of each relevant ethical principle in the decision-making process, both being unique for that specific topic. The four principles of bioethics used are autonomy (informed consent), beneficence (do good), nonmaleficence (do no harm), and justice (be fair) (Table 1.2).

Table 1.2 Principles of Biomedical Ethics

Principle	Responsible	Application
Autonomy	Patient–practitioner	Informed consent
Beneficence	Practitioner	Do good
Nonmaleficence	Practitioner	Do no harm
Justice	Public policy	Be fair

1. Patient: Autonomy (informed consent)
2. Practitioner: Beneficence (do good), nonmaleficence (do no harm)
3. Public policy: Justice (be fair)
4. Specification: Determine how the four principles apply to a particular circumstance.
5. Balancing: Determine the relative weight of each specified principle for the particular circumstance encountered.

Conclude

Drawing conclusions in bioethics is not about finding an absolute answer but a contextually appropriate one. The book equips practitioners with the tools to draw conclusions that resonate with the broader healthcare standards, ensuring decisions made are not in isolation but anchored in a wider ethical and professional framework, making them both ethically sound and professionally defensible. However, because specific state laws vary so much from state to state and date to date, this text primarily focuses on the national standards of care. Standards of care are peer-reviewed, evidence-based care recognized by the healthcare community and established by the healthcare profession's statements on behavioral expectations.

THE PATIENT: THE AXIS OF HEALTHCARE

Every facet of healthcare, whether a surgical intervention or a simple diagnostic test, revolves around the patient. In this intricate dance of healthcare, the patient is the lead. They have the inherent right to be informed and make healthcare decisions. The principle of autonomy (informed consent) underpins this right, ensuring patients can make choices aligned with their unique life narratives, cultural backgrounds, and personal beliefs. But autonomy is not a carte blanche; it comes with responsibilities. For patients to make truly informed decisions, they must understand their condition, the potential interventions, and the implications of each choice (Table 1.3).

Table 1.3 Informed Consent

Diagnosis and prognosis	Patient demonstrates knowledge about their diagnosis and prognosis.
Options	Patient demonstrates knowledge of the various treatment options, including no treatment.
Purpose	Patient demonstrates knowledge of the purpose of the various treatment options.
Benefits and risks	Patient is able to weigh and balance the benefits and risks of the various treatment options, including no treatment, using the patient's own goals, values, and priorities.
Rational	Patient can provide logical reasons for their treatment or no treatment authorization that are in line with the patient's reasonable goals, values, and priorities.
Consent	Patient can communicate the authorization for treatment option.

However, the practitioner is not a mere bystander in this process; informed consent is a shared decision-making process. Practitioners play a pivotal role in ensuring the patient's autonomy is both informed and respected. It requires practitioners to not only have clinical acumen but also possess the ability to communicate effectively. They must distill complex healthcare jargon into digestible information, ensuring patients are truly informed.

The era in which patients were passive recipients of care is long gone. Today, they are informed, active participants, advocating for their rights and demanding a say in their healthcare decisions. This empowerment is a double-edged sword. While it promotes autonomy, it also presents challenges, especially when patients' wishes diverge from established standards of care or when cultural or personal beliefs clash with scientific evidence.

THE PRACTITIONER: THE GUARDIAN OF ETHICAL INTEGRITY

While practitioners must respect patient autonomy, they also have their own set of ethical obligations. The principles of beneficence (do good) and nonmaleficence (do no harm) guide them. Simply stated, while they strive to do good (beneficence), they must ensure they do no harm (nonmaleficence). The practitioner has the professional obligation to provide patient-centered care that maximizes the patient's best interests, as determined by the patient's reasonable goals, values, and priorities, and without violating the principles of nonmaleficence (do no harm), beneficence (do good), and the standards of care.

But these principles are not always black and white in a world of ever-evolving healthcare technologies and treatments. In a landscape where harm is not always clear-cut, where the line between intervention and inaction is blurred, making the right choice becomes a monumental task. Balancing their professional obligations with their personal values, all while respecting patient autonomy (informed consent), demands a nuanced understanding of ethics.

PUBLIC POLICY: THE BEDROCK OF ETHICAL STANDARDS

Beyond the immediate patient–practitioner relationship lies a vast landscape shaped by public policy. From overarching healthcare legislation to specific regulations governing healthcare research, public policy sets the parameters for ethical healthcare practice. Central to this is the principle of justice, ensuring that healthcare is of the highest standard and accessible to all, irrespective of socioeconomic or cultural background. Public policy has the social responsibility and authority to create legislation for protecting the rights of patients and subjects and minimizing healthcare disparities by promoting the equitable distribution of benefits and burdens and actualizing the social and moral principle of justice (be fair).

In the United States, research involving human subjects is regulated by the federal Common Rule (45 Code of Federal Regulations 46 [45CFR46]), which embodies the ethical principles outlined in the 1979 Belmont Report. These bioethical principles, essential for addressing moral issues in medicine, have been integrated into the standards of care not just because the principles are federally enforced throughout the entire country and the boundaries between research and the practice of healthcare are sometimes blurred but also primarily because the principles are easy enough for all parties to understand and they are sufficient enough to address most moral issues related to medicine.

Such policies ensure that as science and medicine advance, they do so with a profound respect for human dignity. Public policy shapes the contours of the ethical landscape, setting boundaries and ensuring checks and balances. However, as with any broad-based solution, public policy sometimes struggles to cater to individual nuances, leading to tensions between standardized care and personalized treatments.

NAVIGATING THE ETHICAL LABYRINTH

Bioethics: Passing the Boards, Providing Patient Care, and Beyond is not just a guide; it is an invitation—an invitation to engage, to question, and to understand. It encourages readers not to consume knowledge passively but to participate actively in the ever-evolving healthcare ethics discourse.

By integrating moral mystery stories with rigorous ethical analysis, the book achieves a rare feat: making bioethics both intellectually stimulating and emotionally resonant. As readers turn the pages, they are not just learning about ethical principles; they are experiencing their implications, feeling their weight, and understanding their significance.

In a world in which medicine and morals are increasingly intertwined, this book serves as a beacon, guiding all those who navigate the challenging terrains of healthcare, ensuring that the human touch is never lost, even in the face of the most daunting ethical dilemmas.

* * *

REVIEW QUESTIONS

1. The three categories of moral responsibility used in the book Bioethics are
    A. Hospital institutions, insurance companies, and political leaders
    B. Patient, practitioner, and public policy
    C. Consequentialism, deontology, and political
    D. Right, wrong, and relative

2. Standards of care are
    A. Evidence-based healthcare practice
    B. Healthcare practice that is peer-reviewed in the healthcare literature
    C. Healthcare practice that is generally recognized by the healthcare community
    D. All of the above

3. To meet all the informed consent requirements, the practitioner must effectively communicate and educate the patient about their diagnosis and prognosis, treatment options, treatment purposes, benefits and risks, and answer their questions.
   A. True
   B. False

4. Informed consent is solely a patient's autonomous responsibility, not a shared decision-making process.
   A. True
   B. False

5. Autonomy (informed consent), beneficence (do good), and nonmaleficence (do no harm) should be exercised only within the boundaries of standards of care.
   A. True
   B. False

6. All research conducted on human subjects is carefully regulated by federal law Common Rule 45CFR46, the legal implementation of the 1979 Belmont Report.
   A. True
   B. False

7. The principles of bioethics were adopted by the healthcare profession as part of the standards of care because
   A. The principles are sufficient to address most moral issues related to medicine
   B. The principles are already federally enforced throughout the entire country
   C. The boundaries between research and the practice of healthcare are blurred
   D. All of the above

Answers: 1B, 2D, 3A, 4B, 5A, 6A, 7D

* * *

CLINICAL VIGNETTES

1. A researcher is studying the efficacy of a new medication for treating depression. The study involves administering the medication to participants and comparing their symptoms to a control group who will receive a placebo. Which of the following best describes the regulations that apply to this study?
   A. The study is not regulated by any laws.
   B. The study is governed primarily by state laws regarding human subject research.
   C. The study is governed by the Common Rule 45CFR46, the legal implementation of the 1979 Belmont Report.
   D. The study is governed by international laws regarding human subjects research.

   Explanation: The Common Rule 45CFR46 is a federal law that governs all research conducted on human subjects in the United States. It is the legal implementation of the 1979 Belmont Report, which outlines the ethical principles for conducting human subjects research, such as respect for a person's autonomy (informed consent), beneficence (do good), nonmaleficence (do no harm), and justice (be fair). The regulation ensures that human subjects are protected from harm and their rights are respected throughout the research process. *Answer*: C

2. Ms. Heather Patel, a 27-year-old architect, is being treated by a practitioner and raises a concern about the ethical principles guiding the healthcare profession. The practitioner

explains that the healthcare profession adopted the principles of biomedical ethics as part of the standards of care because

- A. The principles are sufficient to address most moral issues related to medicine. This means that the principles provide a comprehensive framework for addressing ethical dilemmas in the healthcare field.
- B. The principles are already federally enforced throughout the entire country. This means that the principles have been adopted as law by the government and are mandatory for all healthcare providers to follow.
- C. The boundaries between research and the practice of healthcare are blurred. This means that the principles guide ethical conduct in both research and patient care, as the two are often interrelated.
- D. All of the above

Explanation: The healthcare profession adopted the principles of biomedical ethics as part of the standards of care for several reasons. First, the principles are sufficient to address most moral issues related to medicine and provide a comprehensive framework for addressing ethical dilemmas in healthcare. Second, the principles are already federally enforced throughout the country, meaning they have been adopted as law by the government and are mandatory for all healthcare providers. Last, the boundaries between research and the practice of healthcare are blurred, and the principles guide ethical conduct in research and patient care, as the two are often interrelated. *Answer*: D

3. Mr. Armando Esparza, a 47-year-old artist, presents with symptoms of a healthcare condition. The practitioner must determine the most appropriate course of treatment based on the current standards of care. What is the best definition of "medical healthcare standards of care"?
   - A. Healthcare practice that is evidence-based
   - B. Healthcare practice that is peer-reviewed in the healthcare literature
   - C. Healthcare practice that is generally recognized by the healthcare community
   - D. All of the above

Explanation: "Medical healthcare standards of care" refers to the best practices and guidelines recognized and followed by the healthcare community in diagnosing and treating a healthcare condition. This includes evidence-based treatment options, meaning they are supported by scientific evidence and research, and practices that are peer-reviewed in the healthcare literature, meaning they have been evaluated and approved by experts in the field. In addition, standards of care are generally recognized by the healthcare community, meaning they are widely accepted as the most appropriate and effective methods for treating a particular condition. By following the standards of care, the practitioner can ensure that the patient receives the most appropriate and effective treatment for their condition. *Answer*: D

4. Ms. Eliza Brown, a 52-year-old firefighter, has been diagnosed with a chronic healthcare condition and has come to the clinic seeking information about treatment options. To meet all informed consent requirements, the practitioner must effectively communicate and educate the patient regarding their diagnosis, prognosis, treatment options, purposes, and benefits and risks and answer any patient's questions. Which of the following actions by the practitioner would best meet the informed consent requirements in this situation?
   - A. Simply providing the patient with a written informational brochure about the condition and treatment options
   - B. Quickly reviewing the diagnosis and treatment options without allowing the patient to ask questions
   - C. Spending the necessary amount of time to discuss the diagnosis, prognosis, treatment options, and benefits and risks and answer the patient's questions

D. Telling the patient that they do not need to know the details about their condition or treatment options, as the practitioner will make all decisions for them

Explanation: Spending the necessary amount of time to discuss the diagnosis, prognosis, treatment options, and benefits and risks and answer the patient's questions ensures that the patient fully understands their condition, treatment options, and any potential risks and benefits and can make an informed decision about their care. Simply providing a brochure does not meet the requirement for effective communication and education, quickly reviewing the information without allowing questions does not allow for a thorough understanding, and not providing information at all goes against the informed consent requirements.
Answer: C

5. Mr. Horacio Gonzalez, a 22-year-old construction worker, is scheduled for a routine healthcare procedure. Gonzalez has been engaged in a collaborative process with the practitioner to understand the risks and benefits of the procedure, and they have ultimately made a shared decision to undergo it. Which of the following best describes the individual's process in agreeing to the procedure?
   A. The patient was not involved in the decision-making process.
   B. The decision was made solely by the practitioner.
   C. The patient and their practitioner engaged in a shared decision-making process.
   D. The decision was made through coercion.

Explanation: Informed consent is a collaborative process in which the individual is actively involved in decision-making about their healthcare treatment. This involves discussing the risks and benefits of the procedure, as well as alternative options, with their practitioner. In this case, the individual was engaged in a shared decision-making process with their practitioner, indicating that they actively participated in the decision to undergo the procedure.
Answer: C

* * *

WEB LINKS

1. "The Belmont Report." Office of the Secretary, The National Commission for the Protection of Human Subjects of Biomedical and Behavioral Research. April 18, 1979. https://www.hhs.gov/ohrp/regulations-and-policy/belmont-report/read-the-belmont-report/index.html#xbasic
   Outlines three fundamental ethical principles relevant to research involving human subjects—respect for persons, beneficence, and justice—aimed at guiding the resolution of ethical problems in such research.
2. "Principles of Clinical Ethics and Their Application to Practice." Basil Varkey. 2021. https://www.ncbi.nlm.nih.gov/pmc/articles/PMC7923912
   This article reviews the four main ethical principles—beneficence, nonmaleficence, autonomy, and justice—and their application in clinical practice.
3. "In Whose Interests? The Best Interests Principle Under Ethical Scrutiny." Susan Bailey. 2001. https://www.sciencedirect.com/science/article/pii/S1036731405800591
   This article discusses the best interests principle and its application in intensive care units.
4. "Beneficence, Interests, and Wellbeing in Medicine: What It Means to Provide Benefit to Patients." Johan Bester. 2020. https://pubmed.ncbi.nlm.nih.gov/32105204
   This article discusses the principle of beneficence and its role in promoting and protecting the patient's well-being.
5. "Patient Rights and Ethics." Jacob P. Olejarczyk and Michael Young. 2022. https://www.ncbi.nlm.nih.gov/books/NBK538279
   This article discusses patient rights and ethics, emphasizing the importance of beneficence and autonomy in patient care.

6. "The Best Interests Standard for Incompetent or Incapacitated Persons of All Ages." Loretta Kopelman. 2007. https://www.cambridge.org/core/journals/journal-of-law-medicine-and-ethics/article/abs/best-interests-standard-for-incompetent-or-incapacitated-persons-of-all-ages/E94339297C33220E658F3F8ECE391195
   This article discusses the best interests standard for incompetent or incapacitated persons.
7. "A Bioethical Framework to Guide the Decision-Making Process in the Care of Seriously Ill Patients." Daniel Neves Forte, Kawai, Fernando, and Cláudio Cohen. 2018. https://bmcmedethics.biomedcentral.com/articles/10.1186/s12910-018-0317-y
   This article presents a bioethical framework to guide the decision-making process in the care of seriously ill patients.
8. "Revisiting the Best Interest Standard: Uses and Misuses." Douglas S. Diekema. 2011. https://pubmed.ncbi.nlm.nih.gov/21837884
   This article revisits the best interest standard and its uses and misuse in healthcare.
9. "Bioethics and Health Informatics in Nursing." Nevada State College. https://online.nsc.edu/nursing/rn-to-bsn/bioethics-and-health-informatics-in-healthcare
   This article discusses the principles of bioethics and health informatics in nursing.
10. "What Is Bioethics?" Michigan State University. https://bioethics.msu.edu/what-is-bioethics
    This article overviews bioethics and its role in healthcare.
11. "Clarification of Ethical Principle of Beneficence in Nursing Care: An Integrative Review." Rozita Cheraghi, Leila Valizadeh, Vahid Zamanzadeh, Hadi Hassankhani, and Anahita Jafarzadeh. 2023. https://bmcnurs.biomedcentral.com/articles/10.1186/s12912-023-01246-4
    This article clarifies the ethical principle of beneficence in nursing care.
12. "Clinical Ethics and Law." Lisa V. Brock and Anna Mastroianni. https://depts.washington.edu/bhdept/ethics-medicine/bioethics-topics/detail/56
    This article discusses clinical ethics and law's overlap and unique parameters.
13. "The Theorisation of 'Best Interests' in Bioethical Accounts of Decision-Making." Giles Berchley. 2021. https://bmcmedethics.biomedcentral.com/articles/10.1186/s12910-021-00636-0
    This article discusses the theory underlying the concept of best interests in healthcare policy and practice.
14. "Book Review: *Principles of Biomedical Ethics*, 5th edn." Soren Holm. 2002. https://jme.bmj.com/content/28/5/332.2
    This article reviews the classic work in the field of bioethics, *Principles of Biomedical Ethics* by Beauchamp and Childress, which introduced the four principles of respect for autonomy, nonmaleficence, beneficence, and justice.
15. "The Four Principles of Biomedical Ethics." Healthcare Ethics and Law. https://www.healthcareethicsandlaw.co.uk/intro-healthcare-ethics-law/principlesofbiomedethics
    This post discusses the four principles of biomedical ethics: autonomy, nonmaleficence, beneficence, and justice.
16. "Principlism's Balancing Act: Why the Principles of Biomedical Ethics Need a Theory of the Good." Matthew Shea. 2020. https://pubmed.ncbi.nlm.nih.gov/32726809
    This article discusses the bioethical theory of principlism, centered on the four moral principles of beneficence, nonmaleficence, respect for autonomy, and justice.
17. *Principles of Biomedical Ethics*, 8th ed. Tom L. Beauchamp and James F. Childress. 2019. https://global.oup.com/ushe/product/principles-of-biomedical-ethics-9780190640873
    This page provides an overview of the book *Principles of Biomedical Ethics* by Tom L. Beauchamp and James F. Childress, which thoroughly develops and advocates for the four principles at the core of moral reasoning in healthcare.
18. "Applying the Four-Principle Approach." John-Stewart Gordon, Oliver Rauprich, and Jochen Vollmann. 2011. https://bioethics.jhu.edu/wp-content/uploads/2021/10/Gordon-et-al-Applying-the-Four-Principle-Approach.pdf

This document discusses the application of the four principles—autonomy, nonmaleficence, beneficence, and justice—to different cases in biomedical ethics.

19. "Principles of Bioethics." Thomas R. McCormick. https://depts.washington.edu/bhdept/ethics-medicine/bioethics-topics/articles/principles-bioethics
    This page provides a brief overview of the four commonly accepted principles of healthcare ethics, excerpted from Beauchamp and Childress (2008).
20. "Applying the Four Principles." R. Macklin. 2003. https://jme.bmj.com/content/29/5/275
    This article discusses applying the four principles of biomedical ethics to a case involving a Jehovah's Witness patient.

THE BELMONT REPORT—ABRIDGED

2

As a member of the medical profession, the health and well-being of my patient will be my first consideration.
—Declaration of Geneva

* * *

THINK

Principlism is an ethical decision-making approach used in the healthcare profession. It was first formalized in the Belmont Report and put into federal law for all research on human subjects by 45 Code of Federal Regulations 46 (45CFR46), also known as the Common Rule. The four principles of bioethics are autonomy (informed consent), beneficence (do good), nonmaleficence (do no harm), and justice (be fair).

The principles of biomedical ethics are considered comprehensive enough to address a large portion of healthcare decision-making. The principles are regarded as common ground in that they are socially agreed-upon moral principles that are derivable from, consistent with, or at least not in conflict with the majority of ethical, theological, and social approaches toward moral decision-making.

ASSESS

THE BELMONT REPORT—ABRIDGED AND EDITED

HEALTHCARE PRACTICE VERSUS RESEARCH

The term *practice* refers to interventions designed solely to enhance the well-being of an individual patient or client and that have a reasonable expectation of success. The purpose of healthcare or behavioral practice is to provide diagnosis, preventive treatment, or therapy to particular individuals. When a practitioner departs significantly from the standards of care, the innovation does not, in and of itself, constitute research. The fact that a procedure is "experimental"—in the sense of new, untested, or different—does not automatically place it in the category of research.

By contrast, the term *research* designates an activity designed to test a hypothesis, permit conclusions to be drawn, and thereby develop or contribute to generalizable knowledge (expressed, for example, in theories, principles, and statements of relationships). Research is usually described in a formal protocol that sets forth an objective and a set of procedures designed to reach that objective.

BASIC ETHICAL PRINCIPLES

The expression "basic ethical principles" refers to general judgments that serve as a primary justification for the many particular ethical prescriptions and evaluations of human actions. The basic principles that are particularly relevant to the ethics of research involving human subjects and the standards of care are autonomy (informed consent), beneficence (do good), nonmaleficence (do no harm), and justice (be fair) (Table 2.1).

Patient: Autonomy (Informed Consent)
An autonomous human subject or patient is an individual capable of deliberation about their reasonable goals, values, and priorities and acting under the direction of such deliberation. Autonomy (informed consent) demands that subjects and patients are voluntary and provide informed consent.

Practitioner: Beneficence (Do Good) and Nonmaleficence (Do No Harm)
Persons are treated in an ethical manner not only by respecting their autonomy (informed consent) but also by the legal and professional responsibility of nonmaleficence (do no harm) and beneficence (do good).

The Hippocratic maxim of nonmaleficence (do no harm) has long been a fundamental principle of bioethics. Furthermore, the Hippocratic Oath requires beneficence (do good) and the practitioner's responsibility to promote their patient's best interests per the researcher's or practitioner's best healthcare judgment using the standards of care.

Public Policy: Justice (Be Fair)
Justice addresses "fairness in distribution" or "what is deserved." An injustice occurs when some benefit to which a person is entitled is denied without good reason or when some burden is imposed unduly.

APPLICATIONS
Patient Autonomy (Informed Consent)
Autonomy requires that human subjects and patients, to the degree that they are capable, be given the opportunity to choose what shall or shall not happen to them.

A. Information: Most research and healthcare practice codes establish specific items for disclosure to ensure that human subjects and patients are given sufficient information.
   1. Procedures
   2. Purposes
   3. Risks and benefits
   4. Alternatives
   5. Answer questions
   6. Withdrawal at any time
B. Comprehension: Because the human subject's and patient's ability to understand is a function of intelligence, rationality, maturity, and language, it is necessary to adapt the presentation of the information to the human subject's and patient's capacities.
C. Voluntariness: An agreement to participate in research or clinical treatment constitutes informed consent only if voluntarily given, free of coercion and undue influence.

PRACTITIONER: BENEFICENCE (DO GOOD) AND NONMALEFICENCE (DO NO HARM)
Research and clinical practice must be justified based on a favorable risk–benefit assessment maximizing the subject's and patient's best interests as determined by the subject's and patient's reasonable goals, values, and priorities.

Public Policy: Justice (Be Fair)
The principle of justice (be fair) gives rise to moral requirements of fair procedures and outcomes in selecting human research subjects and outcomes (see Table 2.1).

Table 2.1 The Belmont Report: Principles and Applications	
Principles	**Applications**
Respect for persons	Informed consent
Subjects/patients who are autonomous: Informed consent	Provide information of
	Procedures
Subjects/patients with diminished autonomy: Protected	Purpose
	Risks and benefits
Autonomy (informed consent)	Alternatives
	Answer questions
	Withdraw any time
	Comprehension
	Voluntariness
Beneficence	**Risks and benefits**
Practitioner must maximize benefits	Nature and scope of risks and benefits
Practitioner must do no harm	Assessment of risks and benefits
Beneficence (do good), nonmaleficence (do no harm)	
Justice	**Subject selection**
Public policy: Fair distribution	Fair procedures: Selection and outcomes
Justice (be fair)	

Source: The National Commission for the Protection of Human Subjects of Biomedical and Behavioral Research.

CONCLUDE

The Belmont Report has significantly shaped the ethical landscape of healthcare decision-making by introducing the principlism framework. The four principles—autonomy (informed consent), beneficence (do good), nonmaleficence (do no harm), and justice (be fair)—provide a comprehensive and widely accepted foundation for moral decision-making in both healthcare practice and research. By adhering to these principles, healthcare professionals can ensure ethical conduct, protect patient rights, and promote a morally sound healthcare environment.

* * *

REVIEW QUESTIONS

1. The Belmont Report was put into federal law, 45CFR46, known as the Common Rule.
   A. True
   B. False

2. The principles of bioethics are regarded as common ground in that they are socially agreed-upon moral principles that are derivable from, consistent with, or at least not in conflict with the majority of ethical, theological, and social approaches toward moral decision-making.
   A. True
   B. False

3. If a practitioner performs a procedure on a patient that is "experimental" in the sense of providing a treatment that is new, untested, or different, then that treatment is automatically in the category of research.
   A. True
   B. False

4. The Hippocratic maxim of autonomy (informed consent) has long been a fundamental bioethics principle.
   A. True
   B. False

5. When disclosing the diagnosis, prognosis, treatment options, and risks and benefits of the options and answering patient questions, the practitioner must adapt the presentation of the information per the subject's or patient's capacities.
   A. True
   B. False

6. Coercion is defined as what a patient considers a credible threat.
   A. True
   B. False

7. A patient subjected to coercion or manipulation cannot be considered free, invalidating informed consent.
   A. True
   B. False

Answers: 1A, 2A, 3B, 4B, 5A, 6A, 7A

* * *

CLINICAL VIGNETTES

1. Ms. Phoebe Thompson, a 37-year-old social worker, seeks to participate in a research study involving human subjects. Before enrolling in the study, the patient is informed about the principles outlined in the Belmont Report, which have been put into federal law as 45CFR46, also known as the Common Rule. What is the purpose of the Belmont Report and the Common Rule?
   A. To ensure the safety of human research participants
   B. To ensure that human research participants provide informed consent
   C. To ensure the validity of research findings
   D. Both A and B

   Explanation: Both the Belmont Report and the Common Rule (45CFR46) are designed to safeguard human research participants. Option A is correct because these guidelines are centered around protecting research participants' welfare, guided by beneficence (maximizing benefits and minimizing harm) and nonmaleficence (doing no harm). Option B is accurate because it stresses obtaining informed consent; operationalizing the principle of autonomy; and ensuring participants know procedures, purposes, risks, benefits, alternatives, and their right to withdraw. Option C, however, needs to be primarily addressed in these regulations. Although research validity is important, the main focus of the Belmont Report and Common Rule is the ethical treatment of subjects, not the outcomes of the research. Therefore, the answer is D, "Both A and B," which aligns with these rules' protective and informed consent purposes. *Answer*: D

2. A healthcare team is discussing the care of Mr. Gage Kissinger, a 62-year-old locomotive engineer admitted to the hospital with a serious illness. The patient is unconscious and unable to make any decisions regarding their healthcare, and the patient's next of kin is unavailable to provide guidance or decisions. The healthcare team is trying to reach a consensus on the best course of treatment and care. Which of the following principles of bioethics is the healthcare team most likely considering as the common ground in their decision-making process, in that they are socially agreed-upon moral principles that are

derivable from, consistent with, or at least not in conflict with the majority of ethical, theological, and social approaches toward moral decision-making?
A. Autonomy
B. Beneficence
C. Nonmaleficence
D. Justice

Explanation: The principles of bioethics are considered common ground in that they reflect socially agreed-upon moral principles that are derivable from, consistent with, or at least not in conflict with the majority of ethical, theological, and social approaches toward moral decision-making. In this scenario, the healthcare team primarily focuses on providing "the best course of treatment and care" to the patient, aligning with the beneficence principle. Beneficence requires healthcare professionals to act in the patient's best interest and provide appropriate care and treatment. The principle of nonmaleficence, which requires healthcare professionals not to cause harm to their patients, is also relevant in this scenario because the team must balance the need to provide care with the need to avoid causing harm. The principle of autonomy, which allows patients to make decisions about their healthcare, does not apply in this situation because the patient is unconscious and the next of kin is unavailable to provide guidance or decisions. The principle of justice, which requires healthcare professionals to treat all patients fairly, is not the focus of the decision-making process in this specific scenario. *Answer*: B

3. Ms. Xena Flores, a 58-year-old chef, presents with a rare form of cancer that has not responded to conventional treatments. The practitioner informs the patient that a new and untested treatment option is available but is considered "experimental." The practitioner further explains that research involves a systematic investigation, including clinical trials, to gain new knowledge or establish a treatment's safety and efficacy. From what has been stated by the practitioner, is the experimental treatment research?
A. Yes
B. No
C. Indeterminable

Explanation: The experimental treatment of a rare form of cancer may or may not be considered research. According to the practitioner, research involves a systematic investigation, including clinical trials, to gain new knowledge or establish a treatment's safety and efficacy. Thus, whether the experimental treatment is considered research depends on the specific details and characteristics of the treatment itself and how the treatment is conducted, and none of that has been disclosed. Therefore, it is indeterminable with the information provided. *Answer*: C

4. Ms. Trinity Patel, a 24-year-old engineer, is being treated for a chronic illness and is considering enrolling in a clinical trial to test a new treatment option. The patient expresses concern about the potential risks and benefits of the trial. The practitioner assures the patient that the ethical committees have thoroughly vetted and approved the trial. Although the Hippocratic Oath failed to mention any concept of autonomy or informed consent, the principle of informed consent and respect for persons in healthcare research and treatment has been reinforced by several historical events, including the Belmont Report (1979), the Nuremberg Trials (post-World War II), and the Declaration of Geneva (1948). Which historical event was most instrumental in establishing informed consent as a principle in biomedical ethics?
A. The Hippocratic Oath
B. The Nuremberg Trials
C. The Declaration of Geneva
D. The Belmont Report

Explanation: The Belmont Report, issued in 1979, is considered the most instrumental historical event in establishing informed consent and respect for persons as principles in biomedical ethics and practice. The report emphasized the importance of these principles and helped establish them as cornerstones of modern biomedical ethics, filling the gap left by the Hippocratic Oath, which failed to mention any concept of autonomy or informed consent. The Nuremberg Trials and the Declaration of Geneva also reinforced the principle of informed consent, but the Belmont Report is considered the key event that established it as a principle in biomedical ethics and enforced by federal law. *Answer*: D

5. Mr. Curtis Parker, a 66-year-old lawyer, comes to the clinic complaining of fatigue, headaches, and blurred vision. After a thorough examination and laboratory tests, the practitioner determines that the patient has been diagnosed with a serious healthcare condition. The practitioner must now communicate the patient's diagnosis, prognosis, treatment options, and risks and benefits. Which of the following is the most appropriate approach for the practitioner to take when communicating this information to the patient?
   A. Present all of the information in a detailed and technical manner, assuming the patient has a high level of healthcare knowledge
   B. Adapt the discussion in accordance with the patient's intelligence, rationality, maturity, and language
   C. Refuse to discuss the diagnosis, prognosis, treatment options, and risks and benefits of the options with the patient
   D. Provide the information in a rushed and cursory manner without allowing the patient to ask questions

Explanation: The most important aspect of communicating healthcare information to patients is ensuring they can fully understand it. This requires the practitioner to adapt their approach based on the patient's abilities and needs, including intelligence, rationality, maturity, and language. By adjusting the discussion to the patient's needs, the practitioner can ensure that the patient can fully comprehend the diagnosis, prognosis, treatment options, and risks and benefits of the options, allowing them to make informed decisions about their health. *Answer*: B

* * *

WEB LINKS
1. "The Belmont Report at 40: Reckoning with Time." Eli Y. Adashi, LeRoy Walters, and Jerry Menikoff. 2018. https://www.ajph.org/doi/full/10.2105/AJPH.2018.304580
   This article comprehensively reviews the Belmont Report, its historical context, and its enduring influence on the ethical conduct of research involving human subjects.
2. "The Belmont Report." U.S. Department of Health and Human Services. https://www.hhs.gov/ohrp/regulations-and-policy/belmont-report/index.html
   This is the official government website for the Belmont Report, which identifies basic ethical principles and guidelines that address ethical issues arising from the conduct of research with human subjects.
3. "Belmont Report." Wikipedia. https://en.wikipedia.org/wiki/Belmont_Report
   This Wikipedia page summarizes the Belmont Report and its three core principles: respect for persons, beneficence, and justice.
4. "Belmont Report." National Institutes of Health, Office of NIH History and Stetten Museum. https://history.nih.gov/display/history/Belmont+Report
   This NIH History and Stetten Museum page provides a detailed overview of the Belmont Report and its applications in research ethics.
5. "The Belmont Report: The Triple Crown of Research Ethics." Vickie A. Miracle. 2016. https://pubmed.ncbi.nlm.nih.gov/27258959

This article discusses the Belmont Report as a critical document for those involved in research and its applicability to clinical practice.

6. "The Belmont Report. Ethical Principles and Guidelines for the Protection of Human Subjects of Research." U.S. Department of Health, Education, and Welfare, & National Commission for the Protection of Human Subjects of Biomedical and Behavioral Research. 2014. https://pubmed.ncbi.nlm.nih.gov/25951677

   This article provides a summary of the Belmont Report and its basic ethical principles identified in the course of its deliberations.

7. "The History of the Belmont Report." Teachers College Institutional Review Board. https://www.tc.columbia.edu/institutional-review-board/irb-blog/2020/the-history-of-the-belmont-report

   This blog post discusses the history of the Belmont Report and its impact on biomedical and behavioral human subjects research.

8. "The Belmont Report: What Is It and How Does It Relate to Today's Clinical Trials?" Cancer Support Community. https://www.cancersupportcommunity.org/blog/2019/10/belmont-report-what-it-and-how-does-it-relate

   This blog post discusses the Belmont Report and its relevance to today's human subject research.

9. "Belmont Report: Overview and Principles?" [Video]. Study.com. https://study.com/academy/lesson/the-belmont-reports-impact-importance.html

   This lesson overviews the Belmont Report and its impact and importance.

THE HIPPOCRATIC OATH

3

As a member of the medical profession, I will respect the autonomy and dignity of my patient.
—Declaration of Geneva

* * *

THINK

The Hippocratic Oath is one of the earliest and most significant statements on bioethics. Assumed to have been written by Hippocrates, the Oath is closely aligned with the Pythagorean school's teachings on holistic well-being. Diverging from wider Greek thought, Pythagoras emphasized the interconnectedness of mind, body, and cosmos. For him, true health was a harmonious balance between the physical, mental, and spiritual dimensions, transcending the mere absence of illness. His beliefs, advocating for practices such as dietary recommendations and the therapeutic role of music, laid foundational ideas for holistic health. Estimates of the Oath's actual date of origin vary from the sixth century BCE to the first century CE.

The Hippocratic Oath exemplifies the importance of being a clinical preceptor and a lifelong learner, along with being respectful of professional boundaries, confidentiality, and the moral principles of beneficence (do good), nonmaleficence (do no harm), and justice (be fair). Yet contrary to popular belief, few healthcare schools use the original Hippocratic Oath for graduation because it is considered archaic, paternalistic, and controversial.

ASSESS

THE HIPPOCRATIC OATH

[1] I swear by Apollo the physician [practitioner], and Asclepius, and Hygieia and Panacea and all the gods and goddesses as my witnesses, that, according to my ability and judgement, I will keep this Oath and this contract:

[2] To hold him who taught me this art equally dear to me as my parents, to be a partner in life with him, and to fulfill his needs when required; to look upon his offspring as equals to my own siblings, and to teach them this art, if they shall wish to learn it, without fee or contract; and that by the set rules, lectures, and every other mode of instruction, I will impart a knowledge of the art to my own sons, and those of my teachers, and to students bound by this contract and having sworn this Oath to the law of medicine, but to no others.

[3] I will use those dietary regimens which will benefit my patients according to my greatest ability and judgement, and I will do no harm or injustice to them.

[4] I will not give a lethal drug to anyone if I am asked, nor will I advise such a plan; and similarly I will not give a woman a pessary to cause an abortion.

[5] In purity and according to divine law will I carry out my life and my art.

[6] I will not use the knife, even upon those suffering from stones, but I will leave this to those who are trained in this craft.

[7] Into whatever homes I go, I will enter them for the benefit of the sick, avoiding any voluntary act of impropriety or corruption, including the seduction of women or men, whether they are free men or slaves.

[8] Whatever I see or hear in the lives of my patients, whether in connection with my professional practice or not, which ought not to be spoken of outside, I will keep secret, as considering all such things to be private.

> [9] So long as I maintain this Oath faithfully and without corruption, may it be granted to me to partake of life fully and the practice of my art, gaining the respect of all men for all time. However, should I transgress this Oath and violate it, may the opposite be my fate.
>
> *Translated by Michael North, National Library of Medicine, 2002*

CONCLUDE

Although the original Hippocratic Oath is no longer widely used in modern healthcare education, its influence on the ethical principles guiding the healthcare profession remains significant. The Oath provides a foundation for understanding the importance of upholding ethical standards in patient care, respecting professional boundaries, and maintaining patient confidentiality. As a historical document, the Hippocratic Oath serves as a reminder of the core values that have shaped the healthcare profession and continues to inspire the development of contemporary bioethics.

* * *

REVIEW QUESTIONS

1. The Hippocratic Oath states that a practitioner should teach the art of healing only to their own sons and not to the children of their teachers.
   A. True
   B. False

   Explanation: The text states that the practitioner swears to impart knowledge of the art to their own sons, their teachers' children, and students bound by the contract and having sworn the Hippocratic Oath to the law of medicine [2]. Therefore, the statement is false because the practitioner is willing to teach not only their own children but also the children of their teachers and other students. *Answer*: B

2. The Hippocratic Oath states, [3] "I will do no harm . . . to them." Which of the four principles of biomedical ethics is best reflected in this statement?
   A. Autonomy (informed consent)
   B. Practitioner: Beneficence (do good)
   C. Practitioner: Nonmaleficence (do no harm)
   D. Public policy: Justice (be fair)

   Explanation: The Hippocratic Oath's statement "I will do no harm" is best reflected in the principle of nonmaleficence, one of the four principles of biomedical ethics. Nonmaleficence refers to the obligation of healthcare professionals to avoid causing harm or injury to their patients. This principle is at the heart of the Hippocratic Oath, a set of ethical guidelines for practitioners, and has been considered the foundation of bioethics for centuries. By pledging to do no harm, healthcare professionals are committing to prioritize the well-being and safety of their patients above all else, which is a fundamental aspect of ethical healthcare practice. *Answer*: C

3. The Hippocratic Oath states, [3] "I will do no . . . injustice to them." Which of the four principles of biomedical ethics is best reflected in this statement?
   A. Autonomy (informed consent)
   B. Practitioner: Beneficence (do good)

C. Practitioner: Nonmaleficence (do no harm)
D. Public policy: Justice (be fair)

Explanation: The statement "I will do no . . . injustice to them" in the Hippocratic Oath is best reflected in the principle of justice, one of the four principles of biomedical ethics. Justice refers to the obligation of healthcare professionals to distribute benefits and burdens fairly among their patients. This principle is reflected in the Hippocratic Oath, which requires practitioners to be fair in their treatment of patients and to avoid engaging in any behavior that could be considered unjust or unfair. By pledging to do no injustice, healthcare professionals are committing to treat all patients equally and impartially, a fundamental aspect of ethical healthcare practice. The principle of justice is a crucial component of biomedical ethics because it helps ensure that patients receive the care and treatment they need, regardless of their social status or background. *Answer*: D

4. According to the Hippocratic Oath, what is the practitioner's stance on administering lethal drugs or advising such plans?
   A. They will provide them if asked.
   B. They will consider it on a case-by-case basis.
   C. They will not give lethal drugs or advise such plans.
   D. They will consult with other practitioners before deciding.

Explanation: In the Hippocratic Oath, the practitioner swears not to give anyone a lethal drug if asked or advises such a plan [4]. This indicates a clear stance against administering lethal drugs or advising others to do so. *Answer*: C

5. The Hippocratic Oath states, [4] "I will not give a lethal drug to anyone if I am asked, nor will I advise such a plan." Which of the four principles of biomedical ethics is best reflected in this statement?
   A. Autonomy (informed consent)
   B. Practitioner: Beneficence (do good)
   C. Practitioner: Nonmaleficence (do no harm)
   D. Public policy: Justice (be fair)

Explanation: The statement "I will not give a lethal drug to anyone if I am asked, nor will I advise such a plan" in the Hippocratic Oath best reflects the principle of nonmaleficence, one of the four principles of biomedical ethics. Nonmaleficence refers to the obligation of healthcare professionals to avoid causing harm or injury to their patients. This principle is emphasized in the Hippocratic Oath, which requires practitioners to refrain from providing or advising on treatments that could cause harm or result in death. By pledging not to give a lethal drug, healthcare professionals are committing to prioritize the well-being and safety of their patients above all else, which is a fundamental aspect of ethical healthcare practice. Nonmaleficence is essential to healthcare because it helps ensure patients receive care that promotes their health and well-being rather than causing harm or injury. *Answer*: C

6. According to the Hippocratic Oath, a practitioner swears not to use a knife even on patients suffering from stones, leaving this task to those trained in the craft.
   A. True
   B. False

Explanation: The text states that the practitioner swears not to use the knife, even upon those suffering from stones, and to leave this task to those trained in the craft [6]. Thus, the statement is true because the practitioner explicitly promises not to perform such procedures. *Answer*: A

7. The Hippocratic Oath states, [7] "Into whatever homes I go, I will enter them for the benefit of the sick." Which of the four principles of biomedical ethics is best reflected in this statement?
   A. Autonomy (informed consent)
   B. Practitioner: Beneficence (do good)
   C. Practitioner: Nonmaleficence (do no harm)
   D. Public policy: Justice (be fair)

   Explanation: The statement "Into whatever homes I go, I will enter them for the benefit of the sick" in the Hippocratic Oath is best reflected in the principle of beneficence, one of the four principles of biomedical ethics. Beneficence refers to the obligation of healthcare professionals to promote well-being and do good for their patients. This principle is at the forefront of the Hippocratic Oath, which requires practitioners to enter the homes of the sick to provide care and treatment that will benefit their health. By pledging to enter homes for the benefit of the sick, healthcare professionals are committing to prioritize the well-being and health of their patients, which is a fundamental aspect of ethical healthcare practice. The principle of beneficence is essential to healthcare because it helps ensure that patients receive care that promotes their health and well-being rather than causing harm or injury. *Answer*: B

8. The Hippocratic Oath allows a practitioner to reveal information about a patient's life if it is related to their professional practice.
   A. True
   B. False

   Explanation: According to the Hippocratic Oath, the practitioner swears to keep secret whatever they see or hear in the lives of their patients, whether in connection with their professional practice or not [8]. This demonstrates a commitment to patient confidentiality in all aspects of the patient's life. *Answer*: B

9. What is the practitioner's obligation regarding patient confidentiality according to the Hippocratic Oath?
   A. To keep patient information confidential only when it is related to their professional practice
   B. To share patient information with other practitioners for the benefit of the patient
   C. To keep all patient information secret, regardless of its connection to their professional practice
   D. To report any illegal activities they learn about from their patients

   Explanation: The text states that the practitioner swears to keep whatever they see or hear in the lives of their patients secret, whether it is in connection with their professional practice or not, considering all such things to be private [8]. This demonstrates a strong commitment to patient confidentiality in all aspects of the patient's life. *Answer*: C

10. The Hippocratic Oath states, [8] "Whatever I see or hear in the lives of my patients, whether in connection with my professional practice or not, which ought not to be spoken of outside, I will keep secret, as considering all such things to be private." Which of the four principles of biomedical ethics is best reflected in this statement?
    A. Autonomy (informed consent)
    B. Practitioner: Beneficence (do good)
    C. Practitioner: Nonmaleficence (do no harm)
    D. Public policy: Justice (be fair)

    Explanation: The statement "Whatever I see or hear in the lives of my patients, whether in connection with my professional practice or not, which ought not to be spoken of outside, I will keep secret, as considering all such things to be private" in the Hippocratic Oath is best reflected in the principle of confidentiality, which is a key aspect of the principle of nonmaleficence. Nonmaleficence refers to the obligation of healthcare professionals to avoid causing harm or injury to their patients. Confidentiality is an important aspect of this principle because it helps protect patients' sensitive and personal information, thereby avoiding harm or injury resulting from unauthorized disclosure. By pledging to keep secrets and consider all information to be private, healthcare professionals are committing to protect the privacy and confidentiality of their patients, which is a fundamental aspect of ethical practice. Confidentiality is essential to healthcare because it helps build trust between healthcare professionals and their patients and ensures that patients receive care in a safe and secure environment. *Answer*: C

* * *

WEB LINKS

1. "Hippocratic Oath." Wikipedia. https://en.wikipedia.org/wiki/Hippocratic_Oath
   This Wikipedia page provides a comprehensive overview of the Hippocratic Oath, an oath of ethics historically taken by practitioners, and its significance in Greek healthcare texts.
2. "Hippocratic Oath: Definition, Summary, and Facts." Britannica. https://www.britannica.com/topic/Hippocratic-oath
   This Britannica entry provides a summary and history of the Hippocratic Oath, an ethical code attributed to the ancient Greek practitioner Hippocrates.
3. "The Hippocratic Oath: The Original and Revised Version." Practo. https://doctors.practo.com/the-hippocratic-oath-the-original-and-revised-version/
   This article discusses the original and modern versions of the Hippocratic Oath, the oldest and most widely known treatise on bioethics.
4. "Greek Medicine: The Hippocratic Oath." National Library of Medicine. https://www.nlm.nih.gov/hmd/greek/greek_oath.html
   This page from the National Library of Medicine provides a detailed look at the Hippocratic Oath and its requirements for new practitioners.
5. "The Hippocratic Oath Today." Peter Tyson. 2001. https://www.pbs.org/wgbh/nova/article/hippocratic-oath-today
   This article discusses the Hippocratic Oath, one of the oldest binding documents in history, in the modern context.
6. "First, Do No Harm." Robert Shmerling. 2020. https://www.health.harvard.edu/blog/first-do-no-harm-201510138421
   This Harvard Health Blog post discusses the Hippocratic Oath and its famous dictum, "First, do no harm."
7. "The Myth of the Hippocratic Oath." Harvard Health Blog. 2015. https://www.health.harvard.edu/blog/the-myth-of-the-hippocratic-oath-201511258447
   This article discusses the myth and reality of the Hippocratic Oath, named after the ancient Greek practitioner Hippocrates.
8. "What Is the 'Hippocratic Oath,' and Who Was Hippocrates?" Live Science. https://www.livescience.com/62515-hippocrates.html
   This Live Science article provides a brief history of the Hippocratic Oath and its creator, Hippocrates.
9. "The History of the Hippocratic Oath." Laura McPherson. 2015. https://absn.northeastern.edu/blog/the-history-of-the-hippocratic-oath
   This article provides a history of the Hippocratic Oath, one of the oldest and most widely known codes of ethics.
10. "Modern Hippocratic Oath." UCLA David Geffen School of Medicine. 2018. https://medschool.ucla.edu/blog/modern-hippocratic-oath-holds-underlying-values-medicine-digital-world
    This article discusses the modern interpretation of the Hippocratic Oath and its role in today's digital world of medicine.

THE DECLARATION OF GENEVA

4

As a member of the medical profession, I will respect the secrets that are confided in me, even after the patient has died.
—Declaration of Geneva

* * *

THINK

The Declaration of Geneva, or a modified version, is used at many healthcare school graduations. Like the Hippocratic Oath, the Declaration attempts to exemplify the importance of being a clinical preceptor and a lifelong learner, along with being respectful of professional boundaries, confidentiality, and the moral principles of beneficence (do good), nonmaleficence (do no harm), and justice (be fair). In addition, the Declaration adds the principle of autonomy (informed consent) so as not to be paternalistic. It also avoids some archaic and controversial topics found in the Hippocratic Oath.

ASSESS

> **DECLARATION OF GENEVA**
>
> As a member of the medical profession:
>
> [1] I solemnly pledge to dedicate my life to the service of humanity;
> [2] The health and well-being of my patient will be my first consideration;
> [3] I will respect the autonomy and dignity of my patient;
> [4] I will maintain the utmost respect for human life;
> [5] I will not permit considerations of age, disease or disability, creed, ethnic origin, gender, nationality, political affiliation, race, sexual orientation, social standing or any other factor to intervene between my duty and my patient;
> [6] I will respect the secrets that are confided in me, even after the patient has died;
> [7] I will practice my profession with conscience and dignity and in accordance with good medical practice;
> [8] I will foster the honor and noble traditions of the medical profession;
> [9] I will give to my teachers, colleagues, and students the respect and gratitude that is their due;
> [10] I will share my medical knowledge for the benefit of the patient and the advancement of healthcare;
> [11] I will attend to my own health, well-being, and abilities in order to provide care of the highest standard;
> [12] I will not use my medical knowledge to violate human rights and civil liberties, even under threat;
> [13] I make these promises solemnly, freely, and upon my honor.
>
> Adopted by the 2nd General Assembly of the World Medical Association, Geneva, Switzerland, September 1948, and amended by the 22nd World Medical Assembly, Sydney, Australia, August 1968, and the 35th World Medical Assembly, Venice, Italy, October 1983, and the 46th WMA General Assembly, Stockholm, Sweden, September 1994, and editorially revised by the 170th WMA Council Session, Divonne-les-Bains, France, May 2005, and the 173rd WMA Council Session, Divonne-les-Bains, France, May 2006, and amended by the 68th WMA General Assembly, Chicago, United States, October 2017.

CONCLUDE

The Declaration of Geneva captures the core principles that guide healthcare professionals in their practice. By pledging to prioritize patient well-being, uphold confidentiality, respect colleagues, and continue learning, healthcare professionals demonstrate their commitment to ethical conduct. The Declaration serves as a reminder of the values and principles that underpin the healthcare profession, inspiring generations of practitioners to provide compassionate, patient-centered care while respecting human rights and civil liberties.

* * *

REVIEW QUESTIONS

1. According to the Declaration of Geneva, which of the following aspects should be a practitioner's first consideration?
   A. Their own well-being
   B. Advancement of healthcare
   C. The health and well-being of their patient
   D. Respecting the noble traditions of the healthcare profession

   Explanation: The text states that a practitioner's first consideration should be the health and well-being of their patient [2]. Although other aspects such as personal well-being, advancement of healthcare, and respecting traditions are mentioned in the pledge, the practitioner's primary focus is the patient's health and well-being. *Answer*: C

2. The Declaration of Geneva states, [2] "The health and well-being of my patient will be my first consideration." Which of the four principles of biomedical ethics is best reflected in this statement?
   A. Autonomy (informed consent)
   B. Practitioner: Beneficence (do good)
   C. Practitioner: Nonmaleficence (do no harm)
   D. Public policy: Justice (be fair)

   Explanation: The statement "The health and well-being of my patient will be my first consideration" in the Declaration of Geneva is best reflected in the principle of beneficence, which is one of the four principles of biomedical ethics. Beneficence refers to the obligation of healthcare professionals to promote the well-being and do good for their patients. This principle is reflected in the Declaration of Geneva, which requires practitioners to prioritize the health and well-being of their patients above all else. By pledging to make the health and well-being of their patients their first consideration, healthcare professionals are committing to providing care and treatment that promotes the health and well-being of their patients, which is a fundamental aspect of ethical healthcare practice. The principle of beneficence is essential to healthcare because it helps ensure that patients receive care that promotes their health and well-being rather than causing harm or injury. *Answer*: B

3. The Declaration of Geneva states, [3] "I will respect the autonomy and dignity of my patient." Which of the four principles of biomedical ethics is best reflected in this statement?
   A. Autonomy (informed consent)
   B. Practitioner: Beneficence (do good)
   C. Practitioner: Nonmaleficence (do no harm)
   D. Public policy: Justice (be fair)

   Explanation: The statement "I will respect the autonomy and dignity of my patient" in the Declaration of Geneva best reflects the principle of autonomy, one of the four principles of biomedical ethics. Autonomy refers to the right of individuals to make decisions about their own lives and to have control over the healthcare treatments they receive. This principle is reflected in the Declaration of Geneva, which requires practitioners to respect the autonomy and dignity of their patients. By pledging to respect the autonomy and dignity of their patients, healthcare professionals are committing to allow their patients to make informed decisions about their care and to exercise control over the healthcare treatments they receive, which is a fundamental aspect of ethical healthcare practice. The principle of autonomy is essential to the practice of healthcare because it helps ensure that patients can make informed decisions about their care and receive treatment consistent with their values and beliefs. *Answer*: A

4. The Declaration of Geneva states, [4] "I will maintain the utmost respect for human life." Which of the four principles of biomedical ethics is best reflected in this statement?
   A. Autonomy (informed consent)
   B. Practitioner: Beneficence (do good)
   C. Practitioner: Nonmaleficence (do no harm)
   D. Public policy: Justice (be fair)

   Explanation: The statement "I will maintain the utmost respect for human life" in the Declaration of Geneva is best reflected in beneficence, one of the four principles of biomedical ethics. Beneficence refers to the obligation of healthcare professionals to promote the well-being and do good for their patients. This principle is reflected in the Declaration of Geneva, which requires practitioners to maintain the utmost respect for human life. By pledging to respect human life, healthcare professionals are committing to promoting their patients' health and well-being and avoiding causing harm or injury, a fundamental aspect of ethical healthcare practice. Beneficence is essential to healthcare because it helps ensure patients receive care and treatment that promotes their health and well-being rather than causing harm or injury. By upholding the principle of beneficence, healthcare professionals can provide care and treatment that respect the dignity and value of human life. *Answer*: B

5. What does the Declaration of Geneva say about respecting the secrets of a patient?
   A. Respect secrets only while the patient is alive.
   B. Respect secrets even after the patient has died.
   C. Do not respect secrets.
   D. Respect secrets only if they are relevant to the treatment.

   Explanation: The text states [6] "I will respect the secrets that are confided in me, even after the patient has died." This commitment emphasizes the importance of maintaining patient confidentiality. *Answer*: B

6. The Declaration of Geneva states, [6] "I will respect the secrets that are confided in me." Which of the four principles of biomedical ethics is best reflected in this statement?
   A. Autonomy (informed consent)
   B. Practitioner: Beneficence (do good)
   C. Practitioner: Nonmaleficence (do no harm)
   D. Public policy: Justice (be fair)

   Explanation: The statement "I will respect the secrets that are confided in me," from the Declaration of Geneva, aligns most closely with the principle of nonmaleficence or "do no harm." This principle emphasizes the prevention of harm, which in this context involves protecting patient confidentiality to prevent potential damage such as embarrassment, stigma, or discrimination. Although beneficence, or "do good," can be seen as relevant, it primarily pertains to proactive actions to improve patient well-being rather than avoiding harm. As such, in the case of respecting secrets, nonmaleficence's commitment to not cause harm is more directly applicable than the active principle of beneficence. Therefore, the best-fit principle for this statement is nonmaleficence. *Answer*: C

7. Practitioners should respect and be grateful to their teachers, colleagues, and students.
   A. True
   B. False

   Explanation: The text states [9] "I will give to my teachers, colleagues, and students the respect and gratitude that is their due." This highlights the importance of mutual respect and appreciation in healthcare. *Answer*: A

8. A practitioner should consider their own well-being and abilities to provide the highest standard of care.
   A. True
   B. False

   Explanation: The text states [11] "I will attend to my own health, well-being, and abilities in order to provide care of the highest standard." This means that taking care of oneself is essential to care for patients effectively. *Answer*: A

9. Under extreme circumstances, practitioners can use their healthcare knowledge to violate human rights and civil liberties.
   A. True
   B. False

   Explanation: The text explicitly states [12] "I will not use my healthcare knowledge to violate human rights and civil liberties, even under threat." This commitment emphasizes the ethical responsibility of a practitioner to uphold human rights. *Answer*: B

10. The Declaration of Geneva states, [12] "I will not use my healthcare knowledge to violate human rights and civil liberties, even under threat." Which of the four principles of biomedical ethics is best reflected in this statement?
    A. Autonomy (informed consent)
    B. Practitioner: Beneficence (do good)
    C. Practitioner: Nonmaleficence (do no harm)
    D. Public policy: Justice (be fair)

    Explanation: The statement "I will not use my healthcare knowledge to violate human rights and civil liberties, even under threat" in the Declaration of Geneva best reflects the principle of justice, one of the four principles of biomedical ethics. Justice refers to the obligation of healthcare professionals to distribute benefits and burdens fairly among their patients and to avoid engaging in behavior that could be considered unjust or unfair. This principle is reflected in the Declaration of Geneva, which requires practitioners not to use their healthcare knowledge to violate human rights and civil liberties, even under threat. By pledging not to engage in behavior that violates human rights and civil liberties, healthcare professionals are committing to treat all patients equally and impartially, a fundamental aspect of ethical healthcare practice. The principle of justice is a crucial component of biomedical ethics because it helps ensure that patients receive the care and treatment they need, regardless of their social status or background. By upholding the principle of justice, healthcare professionals can provide care and treatment that respect the dignity and value of human life. *Answer*: D

* * *

WEB LINKS

1. "Declaration of Geneva." World Medical Association. https://www.wma.net/what-we-do/medical-ethics/declaration-of-geneva
   This source provides an overview of the Declaration of Geneva, its history, and its significance in bioethics.

2. "Declaration of Geneva." Wikipedia. https://en.wikipedia.org/wiki/Declaration_of_Geneva
   This Wikipedia entry provides a comprehensive overview of the Declaration of Geneva, including its purpose, history, and revisions.
3. "The Revised Declaration of Geneva: A Modern-Day Physician's Pledge." Ramin Walter Parsa-Parsi. 2017. https://jamanetwork.com/journals/jama/fullarticle/2658261
   This article discusses the 2017 revision of the World Medical Association's Declaration of Geneva.
4. "WMA Declaration of Geneva." World Medical Association. https://www.wma.net/policies-post/wma-declaration-of-geneva
   This source provides the full text of the Declaration of Geneva, along with its history and amendments.
5. "Ethical Principles for Medical Research and Practice." *JAMA Network*. https://sites.jamanetwork.com/research-ethics
   This source discusses the Declaration of Geneva and its role in outlining the professional duties of practitioners.
6. "Declaration of the Rights of the Child." Wikipedia. https://en.wikipedia.org/wiki/Declaration_of_the_Rights_of_the_Child
   This Wikipedia entry discusses the Declaration of the Rights of the Child, also known as the Geneva Declaration of the Rights of the Child.
7. "Declaration of Geneva." Encyclopedia.com. https://www.encyclopedia.com/science/encyclopedias-almanacs-transcripts-and-maps/declaration-geneva
   This encyclopedia entry provides a detailed overview of the Declaration of Geneva, including its adoption and amendments.
8. "The Revised Declaration of Geneva: A Modern-Day Physician's Pledge." Ramin Walter Parsa-Parsi. 2017. https://pubmed.ncbi.nlm.nih.gov/29049507
   This source briefly overviews "The Revised Declaration of Geneva: A Modern-Day Practitioner's Pledge."
9. "History of Child Rights." UNICEF. https://www.unicef.org/child-rights-convention/history-child-rights
   This source discusses the history of child rights, including the Declaration of Geneva's articulation of the rights owed to children.
10. "Geneva Declaration of the Rights of the Child, 1924." Humanium. https://www.humanium.org/en/geneva-declaration
    This source provides an overview of the 1924 Geneva Declaration of the Rights of the Child, summarizing the fundamental needs of children outlined in the Declaration.

SECTION 2

60 TOPICS

ABORTION

5

It is no part of a practitioner's business to use manipulation or compulsion with patients.
—Aristotle

* * *

MYSTERY STORY

FROZEN BETRAYAL: A BATTLE OF CONVICTIONS AND THE PERILOUS CONSEQUENCES OF HEALTHCARE NEGLIGENCE

The January frost hung heavy in the air as a chilling emergency call crackled through the quiet hum of the local police department. The frantic voice belonged to a woman who had been hurriedly transported to the Blossom Women's Choice Hospital, her life ebbing away due to an alarming hemorrhage. Confused and petrified, the woman was pregnant, but she had no idea how the river of blood had started nor the depth of her injuries.

As the hospital's healthcare team raced against time to restore her delicate equilibrium, a horrifying realization dawned upon them—she had been subjected to an illegal abortion. Her life was hanging by a precarious thread, her body bearing the gruesome testament of her botched procedure.

The law enforcement machinery jolted into swift action, instigating an urgent search for the person behind the clandestine abortion. The pursuit was rendered complex by recent legislative shifts overturning abortion laws, catapulting the power to regulate abortions back into the hands of state legislators rather than a private decision between the patient and practitioner. The investigating team found themselves threading a labyrinth of legalities, gathering crucial evidence to ensure the culprits faced justice.

As the intricate layers of the case unfurled, a disturbing pattern emerged. The woman had approached a licensed practitioner for a legal abortion. However, due to personal beliefs, the practitioner refused her request. Instead, the pregnant woman had been thrust into the clutches of a person devoid of any license or expertise to perform abortions—a catastrophic choice that led to her tragic condition.

The team's investigation revealed another unsettling facet. The practitioner, in a glaring disregard for the principles of beneficence (do good) and nonmaleficence (do no harm), had failed to provide the woman with comprehensive counseling. The woman had been left uninformed about her alternatives and the inherent dangers associated with resorting to unlicensed healthcare practitioners. This abandonment of professional responsibility culminated in a tragedy.

The relentless pursuit of the investigation team bore fruit as both the unauthorized abortionist and the neglectful practitioner were finally held accountable. The case illuminated the critical importance of proper patient counseling and referrals, particularly in the volatile landscape of abortion laws, which are constantly changing in accordance with the political landscape.

The case served as a stark reminder to all those at Blossom Women's Choice Hospital of the sacred responsibility of healthcare practitioners to embody the principles of beneficence (do good) and nonmaleficence (do no harm), irrespective of their personal ideologies. Amid the perpetual debate surrounding abortion's legality, practitioners must remain unwavering in their commitment to patient welfare, enabling patients to make informed treatment decisions through comprehensive guidance and compassionate care.

* * *

THINK

On January 22, 1973, in the *Roe v. Wade* ruling, the Supreme Court of the United States (SCOTUS) affirmed that access to safe and legal abortions was a fundamental constitutional right.

Fundamental rights are rights that have a high degree of protection from government encroachment, usually identified in the Constitution or found under the Due Process Clause. The Due Process Clause of the 14th Amendment is the source of an array of constitutional rights not explicitly listed in the Constitution.

It was argued in *Roe v. Wade* that the Due Process Clause provided a fundamental "right to privacy" that protected a pregnant woman's liberty to choose whether or not to have an abortion. Because access to safe and legal abortions was determined to be a fundamental constitutional right, every state was mandated to have at least one abortion clinic. States that had only one abortion clinic were Kentucky, Mississippi, Missouri, North Dakota, South Dakota, and West Virginia.

On June 29, 1992, in the *Planned Parenthood v. Casey* ruling, SCOTUS upheld the constitutional right to have an abortion and rejected the necessity for a waiting period, spousal notification, and parental consent for minors. However, the Supreme Court also overturned the *Roe* trimester framework, replacing it with fetal viability, allowing states to implement first-trimester restrictions.

On June 24, 2022, in the *Dobbs v. Jackson* ruling, the Supreme Court majority opinion overruled *Roe* and *Casey*, declaring,

> The Constitution makes no reference to abortion, and no such right is implicitly protected by any constitutional provision.... The inescapable conclusion is that a right to abortion is not deeply rooted in the Nation's history and traditions.... Our job, is to interpret the law, apply longstanding principles of stare decisis [precedence].... *Roe* and *Casey* must be overruled, and the authority to regulate abortion must be returned to the people and their elected representatives.

However, the minority opinion dissented, arguing,

> Today, the Court . . . says that from the very moment of fertilization . . . a State can force [a woman] to bring a pregnancy to term. The Court does not act "neutrally" when it leaves everything up to the States.... Withdrawing a woman's right to choose whether to continue a pregnancy does not mean that no choice is being made. It means that a majority of today's Court has [transferred] this choice from women and given it to the States.... We dissent.

And on the same day, the American Medical Association (AMA) in a press release stated

> The American Medical Association is deeply disturbed by the U.S. Supreme Court's decision to overturn nearly a half-century of precedent protecting patients' right to critical reproductive healthcare—representing an egregious allowance of government intrusion into the medical examination room, a direct attack on the practice of medicine and the patient–physician relationship, and a brazen violation of patients' rights to evidence-based reproductive health services.

On June 29, 2022, the Association of Bioethics Program Directors, comprising the leadership of nearly 100 academic bioethics programs at healthcare centers and universities throughout North America, wrote and approved the "Bioethics Guidance for the Post-Dobbes Landscape," affirming that practitioners are to

- counsel their prenatal patients about all available options within the standards of care and where such care is legally available;
- support economic equity for prenatal and postpartum patients, neonates, and children; and
- protect the patient's right to confidentiality and privacy and the legality of patient referral practices.

ASSESS

Patient: Autonomy

A patient's right to provide informed consent is a central tenet of the patient–practitioner relationship. An adult pregnant patient with decisional capacity has the right to provide an informed consent decision for whether or not a practitioner is authorized to perform an abortion in states where it is legal.

For minors, most states have some parental involvement laws, such as parental notification or parental consent. However, if parental consent is required and not attainable, the minor can file for a judicial bypass. If the minor does not want to inform the parents about their pregnancy, then the practitioner needs to determine why that is the case, along with encouraging the benefits of discussing the minor's pregnancy with their parents, all while ensuring that the minor's patient–practitioner confidentiality and privacy will not be abrogated unless there is a healthcare emergency or required by law.

Practitioner: Beneficence and Nonmaleficence

Practitioners are morally obligated to promote the patient's best interests using the patient's reasonable goals, values, and priorities. The moral principles of beneficence (do good) and nonmaleficence (do no harm) come from this professional mandate. Depending on the circumstances, an abortion can satisfy one or both moral principles, professionally obligating the practitioner to make available or perform a safe and legal abortion.

If under justifiable conditions, a practitioner is unwilling to perform a legal abortion due to personal convictions, then the practitioner must, under the professional principles of beneficence (do good) and nonmaleficence (do no harm), refer the patient, in a nonjudgmental way, to another qualified practitioner in a timely manner. The American College of Obstetricians and Gynecologists (ACOG) recommends that providers with moral or religious objections to abortion should practice with or in proximity to individuals who do not share their views and that referral processes are in place to provide access to the service that the practitioner does not wish to provide.

ACOG states,

> Physicians and other healthcare providers have the duty to refer patients in a timely manner to other providers if they do not feel that they can in conscience provide the standard reproductive services that patients request. In resource-poor areas, access to safe and legal reproductive services should be maintained. Providers with moral or religious objections should either practice in proximity to individuals who do not share their views or ensure that referral processes are in place. In an emergency in which referral is impossible or might negatively have an impact on a patient's physical or mental health, providers have an obligation to provide medically indicated and requested care.

As stated, there are exceptions to this conscientious objector provision, such as during a healthcare emergency. If, during a healthcare emergency, it is determined that an abortive procedure is what the standards of care would provide, and no other practitioners are available to perform the abortion, then failure to provide the procedure would be a violation of the professional principles of beneficence (do good), nonmaleficence (do no harm), and the public policy principle of justice (be fair).

Public Policy: Justice

Although access to safe and legal abortions is no longer a fundamental constitutional right, that does not mean abortion is unconstitutional. Rather, it means that the federal government (by Congress) and states (through legislation) now have the authority to regulate abortion. As a result, abortion laws will vary dramatically from state to state and from date to date, with differing regulations regarding who, when, where, how, and even if patients can have access to abortions.

Socially, there is very little consensus on the fair distribution of abortion access because there is very little consensus as to what the status of a fetus is or ought to be. The public is deeply divided, and even with public discussions, there has been very little progress toward reaching a social consensus or compromise.

CONCLUDE

Abortion laws vary dramatically from state to state and are subject to the rapid change of partisan politics. However, the moral and professional responsibility for safe and appropriate healthcare referrals and for encouraging pregnant minor patients to discuss their pregnancy with their parents is a professional and socially recognized responsibility of practitioners.

Abortion laws continue to be a complex and ever-changing issue in the United States, with significant variation across states and subject to the influence of partisan politics. Despite this, practitioners must maintain their commitment to promoting patients' best interests, providing or referring for legal abortion services when necessary, and supporting minors in discussing their pregnancies with their parents. Upholding these professional responsibilities is essential in ensuring safe and appropriate patient care, regardless of the political climate.

* * *

REVIEW QUESTIONS

1. On January 22, 1973, in the *Roe v. Wade* ruling, the Supreme Court of the United States (SCOTUS) affirmed that access to safe and legal abortions was a fundamental constitutional right.
   A. True
   B. False

2. On June 29, 1992, in the *Planned Parenthood v. Casey* ruling, SCOTUS upheld the constitutional right to abortion and rejected the necessity for a waiting period, spousal notice, and parental consent for minors.
   A. True
   B. False

3. On June 24, 2022, in the *Dobbs v. Jackson Women's Health Organization*, SCOTUS concluded, "We, therefore, hold that the Constitution does not confer a right to abortion. *Roe* and *Casey* must be overruled, and the authority to regulate abortion must be returned to the people and their elected representatives."
   A. True
   B. False

4. For a minor who does not want to inform their parents about their pregnancy, it is essential that the practitioner
   A. Encourage the minor to discuss their pregnancy with their parent(s)
   B. Ensure the patient's privacy will not be abrogated
   C. Inquire as to why they do not want to inform their parents
   D. All of the above

5. Practitioners who are unwilling to perform a legal abortion due to personal convictions must
   A. Refer the patient to a qualified practitioner
   B. Refer patients in a timely manner
   C. Both A and B

6. In an emergency in which standards of care require an abortion, and there is no other available practitioner to perform the procedure, the practitioner must provide the abortion even if the practitioner is a conscientious objector to it.
   A. True
   B. False

Answers: 1A, 2A, 3A, 4D, 5C, 6A

* * *

CLINICAL VIGNETTES

1. On January 22, 1973, Ms. Kristen Anderson, a 22-year-old teacher, presented to your clinic seeking information on their reproductive options. The patient was unmarried and sexually active and had a positive pregnancy test. The patient was concerned about their ability to access safe and legal abortion services and heard conflicting information about the laws surrounding abortion. According to the Supreme Court of the United States (SCOTUS) ruling in *Roe v. Wade* on January 22, 1973, what determined the legal status of access to safe and legal abortion?
   A. Access to safe and legal abortion was restricted to certain circumstances, such as rape or incest.
   B. Access to safe and legal abortion was not a constitutional right, and it was illegal in many states.
   C. Access to safe and legal abortion was a fundamental constitutional right.

   Explanation: The SCOTUS ruling in *Roe v. Wade* on January 22, 1973, determined that access to safe and legal abortion was a fundamental constitutional right. This means that women have the right to decide about their reproductive health, including the decision to have an abortion, without undue interference from the government or other third parties. The ruling established a framework for balancing pregnant women's rights and the state's interests in protecting potential life. This ruling has been the subject of ongoing political and legal controversy in the United States. *Answer*: C

2. On June 29, 1992, Ms. Diana Hendrickson, a 17-year-old waitress, presented to your clinic seeking information on their reproductive options. The patient was unmarried and sexually active and had a positive pregnancy test. The patient was concerned about their ability to access safe and legal abortion services and heard conflicting information about the laws surrounding abortion. According to the Supreme Court of the United States (SCOTUS) ruling in *Planned Parenthood v. Casey* on June 29, 1992, what determined the legal status of access to safe and legal abortion?
   A. The constitutional right to have an abortion was overturned.
   B. Restrictions of a waiting period before having an abortion were upheld.
   C. Restrictions of spousal notification and parental consent for minors before having an abortion were upheld.
   D. The constitutional right to have an abortion was upheld; rejected the necessity of a waiting period, spousal notification, and parental consent for minors; and overturned the *Roe* trimester framework, replacing it with fetus viability.

   Explanation: The SCOTUS ruling in *Planned Parenthood v. Casey* on June 29, 1992, upheld the constitutional right to abortion but rejected the trimester framework established by *Roe v. Wade* and replaced it with the concept of fetal viability. The ruling stated that the state is interested in protecting the potential life of the fetus from the point of viability, generally approximately 24 weeks of pregnancy, and can regulate or prohibit abortion after that point. However, before viability, the state cannot impose an undue burden on women's right to access safe and legal abortion services. The ruling also struck down the requirement for spousal notification and a waiting period before obtaining an abortion. Still, it upheld the requirement for parental consent for minors, providing a judicial bypass option. The ruling reaffirmed the fundamental constitutional right to access safe and legal abortion but allowed greater state regulation of the procedure than under the previous *Roe* framework. *Answer*: D

3. Ms. Hazel Reyes, a 25-year-old delivery driver, presents to your clinic to discuss their reproductive options. The patient is unmarried and sexually active and has a positive pregnancy test. The patient is concerned about their ability to access safe and legal abortion services and has heard conflicting information about the laws surrounding abortion. According to the Supreme Court of the United States (SCOTUS) ruling in *Dobbs v. Jackson* on June 24, 2022, what determined the legal status of access to safe and legal abortion?
   A. The right to have an abortion was declared unconstitutional.
   B. The right to have an abortion was declared to be a constitutional right.
   C. The case did not affect the constitutionality of abortion.
   D. *Roe* and *Casey* were overruled, and the authority to regulate abortion was transferred from the patient–practitioner decision to elected state representatives.

   Explanation: The SCOTUS ruling in *Dobbs v. Jackson* on June 24, 2022, overruled the previous rulings in *Roe v. Wade* and *Planned Parenthood v. Casey* and transferred the authority to regulate abortion from the patient–practitioner decision to elected state representatives. The ruling upheld a Mississippi law prohibiting abortions after 15 weeks of pregnancy, with only narrow exceptions for healthcare emergencies or fetal abnormalities. It effectively allowed states to ban abortions before fetal viability, typically approximately 24 weeks of pregnancy. The ruling rejected the viability standard established in *Roe* and *Casey*, stating that the Constitution did not support it and that the state's interest in protecting potential life begins at conception. The ruling did not declare the right to have an abortion unconstitutional, but it allows states to regulate or prohibit abortions as they see fit, subject to certain limits. This ruling has been highly controversial and seen as a significant setback for reproductive rights in the United States. *Answer*: D

4. Ms. Kaylee Hernandez, a 16-year-old high school student, presents to your clinic seeking information about their reproductive options. The patient is pregnant and does not want to inform their parents about the pregnancy. The patient has heard about the option of a judicial bypass but is unsure about the process. What should the practitioner do to inform the minor's parents about the pregnancy and protect the patient's confidentiality and privacy?
   A. Inform the minor's parents about the pregnancy without considering the reasons for the minor patient wanting to keep it confidential
   B. Encourage the benefits of discussing the minor's pregnancy with their parents and ensure that the minor's patient–practitioner confidentiality and privacy are not abrogated unless there is a healthcare emergency or required by law
   C. Deny the minor access to information and services regarding their pregnancy and reproductive options

   Explanation: When a minor patient desires to keep their pregnancy confidential from their parents, the practitioner should respect their wishes and protect their patient–practitioner confidentiality and privacy. The practitioner should provide the minor with information about their reproductive options, including the option of a judicial bypass, which allows the minor to obtain an abortion without parental involvement. The practitioner should encourage the minor to discuss their pregnancy with their parents because this may be in the best interest of the minor's health and well-being. Still, ultimately, the decision should be left up to the minor. The practitioner should also ensure the minor is fully informed about their reproductive options' risks and benefits and provide appropriate referrals for further care, such as counseling or other support services. It is important to remember that minors can access reproductive healthcare services and make decisions about their own bodies, even if they are not legally adults. *Answer*: B

5. Ms. Yvonne Kim, a 35-year-old accountant, presents to the emergency room with severe abdominal pain and bleeding. After examination and assessment, it is determined that the patient requires an abortion due to a healthcare emergency per the standards of care. However, the practitioner has personal conviction against performing abortions and is the only practitioner on duty and available. What does the practitioner do?
   A. Refuse to perform the procedure due to personal convictions, even though it is the standards of care in a healthcare emergency
   B. Provide the procedure to save the patient's life, as it is the standards of care in a healthcare emergency and failure to do so would violate professional principles and public policy
   C. Delay the procedure until another practitioner becomes available, even though the patient's life may be at risk

   Explanation: In this scenario, the practitioner has a professional and ethical duty to provide the necessary healthcare treatment to save the patient's life, which, in this case, requires an abortion due to a healthcare emergency. Refusing the procedure due to personal convictions would violate the practitioner's professional obligations. It could risk the patient's life, which is unacceptable in the healthcare profession. Delaying the procedure until another practitioner becomes available could also put the patient's life at risk and would not align with the standards of care. Therefore, the practitioner should provide the necessary procedure to save the patient's life, even if it goes against their personal convictions, as this aligns with professional principles and public policy. However, it is important for the practitioner to receive support and counseling to help process any emotional or moral conflicts they may experience due to providing care that goes against their personal convictions. *Answer*: B

REFLECTION VIGNETTES

1. Ms. Alice Johnson, a 29-year-old patient, presents to practitioner Rachel Brown at the local clinic for an abortion. Ms. Johnson is 8 weeks pregnant and has decided to terminate the pregnancy. She is accompanied by her partner, Mr. Thomas Smith, who is also the biological father of the fetus. The state in which they reside allows abortions to be performed until the end of the first trimester. During the intake process, Ms. Johnson expressed her reasons for wanting an abortion, citing personal and financial concerns and being in the early stages of her career. She also states that she is not emotionally or financially ready to become a mother. However, Mr. Smith disagrees with her decision and argues that he has every right to participate in the fate of the fetus. He wants Ms. Johnson to keep the pregnancy and raise the child with him.

   What is the ethical issue(s)?

   How should the ethical issue(s) be addressed?

2. Samantha, a 16-year-old patient, presents to Practitioner Patel at the local clinic for an abortion. Samantha is 4 weeks pregnant and has decided that she is not ready to become a mother. The state in which she resides allows abortions until the end of the first trimester. Samantha confides in the practitioner that her parents are strict Catholics opposed to abortions. She fears they will not support her decision and may try to prevent her from terminating the pregnancy. Samantha requests that her parents not be notified and that her healthcare records be kept confidential.

   What is the ethical issue(s)?

   How should the ethical issue(s) be addressed?

* * *

WEB LINKS

1. *Code of Medical Ethics.* American Medical Association. https://code-medical-ethics.ama-assn.org

   The AMA *Code of Medical Ethics*, established by the American Medical Association, is a comprehensive guide for healthcare practitioners. It addresses issues and challenges, promotes adherence to standards of care, and is continuously updated to reflect contemporary practices and challenges.

2. "The Limits of Conscientious Refusal in Reproductive Medicine." American College of Obstetricians and Gynecologists. 2007. https://www.acog.org/clinical/clinical-guidance/committee-opinion/articles/2007/11/the-limits-of-conscientious-refusal-in-reproductive-medicine

   This document discusses the ethical boundaries and obligations of healthcare providers in balancing their personal beliefs with professional responsibilities in reproductive healthcare.

3. "How a Bioethicist and Doctor Sees Abortion." Alvin Powell. 2022. https://news.harvard.edu/gazette/story/2022/05/how-a-bioethicist-and-doctor-sees-abortion

   This source provides insights from a bioethicist and doctor on abortion.

4. "For Doctors, Abortion Restrictions Create an 'Impossible Choice' When Providing Care." Selena Simmons-Duffin. 2022. https://www.npr.org/sections/health-shots/2022/06/24/1107316711/doctors-ethical-bind-abortion

   This article discusses the ethical dilemmas practitioners face due to abortion restrictions.

5. "An Ethical Issue: Nurses' Conscientious Objection Regarding Induced Abortion in South Korea." Chung Mee Ko, Chin Kang Koh, and Ye Sol Lee. 2020. https://bmcmedethics.biomedcentral.com/articles/10.1186/s12910-020-00552-9

   This paper discusses the ethical issues related to nurses' conscientious objection regarding induced abortion.

6. "What Ethical Health Care Looks Like When Abortion Is Criminalized." Isaac Chotiner. 2022. https://www.newyorker.com/news/q-and-a/ethical-health-care-after-roe

   This article discusses the ethical implications of healthcare when abortion is criminalized.

7. "Know Your Rights: Reproductive Health Care." U.S. Department of Health and Human Services. 2022. https://www.hhs.gov/about/news/2022/06/25/know-your-rights-reproductive-health-care.html

   This government source provides information on patient rights in the context of reproductive healthcare, including abortion.

8. "Legislating Abortion Care." Jody Steinauer and Carolyn Sufrin. 2014. https://journalofethics.ama-assn.org/article/legislating-abortion-care/2014-04

   This article discusses the legislation of abortion care and its implications for bioethics.

9. "Patient Rights and Ethics." Jacob P. Olejarczyk and Michael Young. 2022. https://www.ncbi.nlm.nih.gov/books/NBK538279

   This source provides a comprehensive overview of patient rights and ethics in healthcare, including abortion.

ABUSE

6

CHILD, ELDER, AND INTIMATE PARTNER

Make a habit of two things: to help [beneficence] or at least to do no harm [nonmaleficence].
—Hippocrates

* * *

MYSTERY STORY

WHISPERED CRUELTY: A WAKE-UP CALL FOR MANDATORY REPORTING IN HEALTHCARE

In the usually bustling pediatric wing of the Phoenix Rising Support Hospital, Dr. Emily Peterson noticed an eerie silence hanging heavy around her young patient, 10-year-old Jacob. This routine checkup was marred by an unsettling discovery—a constellation of bruises marring the canvas of the child's arms and legs. Dr. Peterson found herself confronting a wall of evasive glances and trembling silence when she questioned Jacob about the injuries. Her instincts screaming at the chilling possibility of home abuse, she took the courageous step of reporting her suspicions to Child Protective Services (CPS).

But the next day, the hospital's usual rhythm was shattered by a grim revelation. Jacob, the boy whose fearful silence had just begun to raise alarm bells, was dead. As the cause of his untimely demise lurked in the shadows, an icy suspicion of foul play gripped Dr. Peterson.

The police were swiftly called in to untangle the mysteries surrounding Jacob's abrupt death. Dr. Peterson found herself under their questioning, sharing her chilling observations and the actions she had taken with CPS. The seed of suspicion she had sown began to take a dark shape as the investigation delved deeper.

The inquiries unearthed a disturbing history—Jacob's father had a violent past, marred by domestic abuse and a prior arrest for assaulting his wife. The police began to suspect that the man had orchestrated a horrific act, snuffing out his son's life and masking the atrocity as an accident.

Then came a revelation that sent shock waves through the community. Jacob's father had been under the care of a healthcare provider for several years, but his abusive history had been quietly swept under the rug. The provider, viewing their role as a healer and not an intruder, had balked at the idea of meddling in the family's private affairs.

This dangerous silence had proven lethal, highlighting the grave implications of failing to report abuse. The tragic case underscored the importance of mandatory reporting laws in safeguarding the vulnerable from unseen threats lurking in their safe havens.

As the final chapter of the investigation unfolded, Jacob's father found himself facing the consequences of his actions, shackled in the cold grip of justice. Dr. Peterson, once a silent observer, was now hailed a hero for her courage in shining a light on the suspected abuse. The echo of her actions served as a sobering reminder to healthcare providers—their duty extended beyond their clinics, reaching into the realm of safeguarding their patients from harm.

Jacob's tragic story marked a pivotal moment for the Phoenix Rising Support Hospital community, illuminating the role mandatory reporting laws play in shielding the vulnerable. It stood as a stark reminder that silence, although comforting, can sometimes harbor

> a deadly secret. The duties of healthcare practitioners extend beyond treatment; they are also the guardians of the vulnerable, tasked with the responsibility to speak up, to break the silence, and to protect those unable to protect themselves.

* * *

THINK

Practitioners are mandatory reporters of child and elder abuse because minors and elders are considered vulnerable populations. There is no discretion as to whether or not to report abuse. By law, child and elder abuse must be reported. This social policy has been implemented so that protective services can immediately interview, judge, and intervene to protect and prevent further harm to the victim. Although the practitioner has no authority to remove a child from parental custody, CPS does have that authority. If the practitioner is sincerely and honestly reporting the suspected child or elder abuse to protective services, then there will be no legal liability, even if it turns out that there was no abuse.

Confidentiality and privacy are two central components of the patient–practitioner relationship. Mandatory reporting laws are not justifiable breaches of that trust; rather, mandated reporting laws are part of the patient–practitioner social contract. This means that if a practitioner does not report a child or elder abuse incident, that would violate the patient-centered, patient–practitioner social contract.

ASSESS

Patient: Autonomy

Informed consent is the practical application of the moral principle of autonomy, which means self-rule and is implemented practically with informed consent. Competent adults have the legal, professional, and moral right to provide informed consent for the authorization of the practitioner to provide treatment to the patient.

Children are a vulnerable population because they cannot give informed consent or make treatment decisions due to not being competent and not having legal decision-making authority. The elderly, in contrast, are legally competent but may have compromised decisional capacity because of aging and their dependency on those who provide residence and care. Because minors lack competency, and the elderly may have diminished decisional capacity, and because both children and the elderly depend on others for their protection and care, both children and the elderly are legally categorized as vulnerable populations.

Practitioners are mandated by law, profession, and morally to report to CPS or Adult Protective Services (APS), respectively, if there is any reasonable cause to suspect that a child or elderly person has been abused or neglected, no matter what the family members or other caregivers may say in their own defense. In addition, even if the elderly person with decisional capacity objects to the report and does not provide informed consent, the practitioner is still mandated to report the potential elder abuse to APS. This is not a breach of patient–practitioner confidentiality because child and elder abuse reporting is part of the patient–practitioner social contract by law and standards of care.

However, intimate partners, such as a spouse, with decisional capacity are not considered a vulnerable population. As a result, practitioners do not have the authority and legal protection to go against the victim's lack of consent. Practitioners must get the expressed consent of the victim and document the consent authorization to report in the healthcare records before reporting to the police or any other authority, as to do otherwise would be a breach of confidentiality and privacy of the patient–practitioner relationship and a violation of autonomy (informed consent).

Practitioner: Beneficence and Nonmaleficence

Based on the professional principles of beneficence (do good) and nonmaleficence (do no harm), the practitioner has the mandatory professional obligation to report child abuse to CPS and elder abuse to APS. This is based on the notion that reporting would be consistent with what a reasonable person would consider being in their best interests. Therefore, failure to report child or elder abuse violates the patient–practitioner relationship because mandated reporting laws are part of that social contract.

In contrast, the practitioner dealing with a possible intimate partner abuse victim must use the victim's reasonable goals, values, and priorities to determine what would be in the victim's best interests. However, because the victim is generally the best person to know and assess what would be in their best interests, the victim's autonomous decision will generally be the standard for whether or not to report.

If the practitioner judges that it would be best to report and the intimate partner victim does not consent or give authorization for the practitioner to report, then the practitioner should

- professionally enquire as to why and how the patient has arrived at their decision;
- encourage the victim to report the abuse; and
- make sure the patient has a safe place to retreat.

However, if it is determined that the intimate partner victim is being coerced into not consenting to the reporting of the abuse, then that denial is not a freely determined autonomous refusal, and an exception can be made based on professional principles of beneficence (do good) and nonmaleficence (do no harm). When abuse is reported, patient confidentiality and privacy must be protected as much as possible by disclosing the minimal amount of information necessary.

Every practitioner needs to know about and be able to direct the abused patient to community and private healthcare resources and safe locations to help avert any further harm caused by violence and abuse.

Public Policy: Justice

The role of government in healthcare is to implement justice (fair distribution of benefits and burdens) concerning healthcare disparities, including protecting vulnerable populations.

In 1974, Congress passed the Child Abuse Prevention and Treatment Act, which required all states "to prevent, identify and treat child abuse and neglect." State governments meet this requirement with CPS, also known in some states as the Department of Children and Family Services. APS provides social services for abused, neglected, or exploited older adults and adults with significant disabilities.

The Belmont Report, formalized by The National Commission for the Protection of Human Subjects, was implemented into federal law by the Department of Health, Education, and Welfare in 1979 as Common Rule 45 Code of Federal Regulations 46 (45CFR46). The Belmont Report defines five categories of vulnerable human populations that require increased protection:

1. Children
2. Pregnant females
3. Prisoners
4. Mentally disabled
5. Other vulnerable groups

Common Rule 45CFR46 also provides federal protections for confidentiality and privacy concerns that are part of the patient–practitioner protected health information and legislates that whenever there are audio and video recordings, extra precautions must also be in place to make sure confidentiality and privacy are not breached.

Mandatory reporting is an attempt of the government to ensure that federal and state services are available for interviewing, judging, and intervening to protect vulnerable populations that are found to be at risk by the very people whose duty is to protect, nurture, and provide care.

CONCLUDE

Practitioners who sincerely and honestly suspect elder or child abuse must always report the incident to CPS or APS regardless of others' explanations in their defense. Responses such as "Encourage the victim to report" or "Encourage discussion" will always be incorrect. In contrast, with an intimate partner or spousal abuse, the correct response would be "Encourage the victim to report."

In summary, practitioners must be vigilant in recognizing and reporting suspected abuse cases involving children, the elderly, and intimate partners. Mandatory reporting is a crucial aspect of the patient–practitioner social contract, aiming to protect vulnerable populations and uphold the principles of beneficence (do good) and nonmaleficence (do no harm). For cases involving intimate partners with decisional capacity, practitioners should encourage the victim to report the abuse and should provide resources for support. Ultimately, healthcare professionals play a crucial role in safeguarding the well-being of those who cannot protect themselves. (For more information on mandatory reporting, see Chapter 26).

* * *

REVIEW QUESTIONS

1. The practitioner has the authority to remove a child from parental custody if the practitioner sincerely and honestly believes that the child is being abused.
   A. True
   B. False

2. Before reporting to Child or Adult Protective Services, it is legally, professionally, and ethically mandatory for practitioners to inquire with family members and caregivers to get an explanation.
   A. True
   B. False

3. Mandatory reporting requires that the practitioner reports intimate partner abuse regardless of whether or not the victim consents.
   A. True
   B. False

4. Failure to report child or elder abuse violates the patient–practitioner relationship because mandated reporting laws are part of that social contract.
   A. True
   B. False

5. In 1974, Congress passed the Child Abuse Prevention and Treatment Act, which requires all states to prevent, identify, and treat child abuse and neglect.
   A. True
   B. False

6. If a practitioner sincerely and honestly suspects elder or child abuse, then the correct answer will be which of the following:
   A. Encourage the victim to report.
   B. Report to Child or Adult Protective Services.

7. If a practitioner sincerely and honestly suspects intimate partner abuse, then the correct answer will be which of the following:
   A. Encourage the victim to report.
   B. Report to Adult Protective Services.
   C. Encourage discussion.

Answers: 1B, 2B, 3B, 4A, 5A, 6B, 7A

* * *

CLINICAL VIGNETTES

1. Aurora Cooper, an 8-year-old elementary student, walks into a clinic, and the practitioner notices bruises and signs of abuse on the patient's body. The practitioner suspects that the patient is a victim of child abuse. The practitioner is mandated by law to report it to CPS. According to the laws and regulations surrounding mandatory child abuse reporting, what is the practitioner's role and responsibility in this situation?
   A. The practitioner has the authority to remove the child from parental custody.
   B. The practitioner is not required to report the suspected abuse because it may turn out to be false.
   C. The practitioner must report the suspected abuse to CPS, but only if they are certain it is true.
   D. The practitioner must report the suspected abuse to CPS and is protected from legal liability if they are sincerely and honestly reporting the abuse.

   Explanation: The practitioner's role and responsibility in reporting suspected child abuse are governed by laws and regulations surrounding mandatory reporting. According to these laws, the practitioner's duty is to report the suspected abuse to CPS, and they are protected from legal liability if they sincerely and honestly report the abuse. However, practitioners do not have the authority to remove the child from parental custody, nor are they allowed to choose not to report the suspected abuse, even if they are uncertain of its validity. Failure to report any suspicion of child abuse to CPS can result in legal consequences. *Answer*: D

2. A practitioner is conducting a routine checkup on Ms. Lila Hensley, an 82-year-old retired sales clerk who lives with their adult daughter. During the examination, the practitioner notices bruises on the patient's arms that are inconsistent with the explanation given by the daughter. The daughter asserts that the bruises resulted from a fall and that the patient is "clumsy." What should the practitioner do in this situation?
   A. Report the suspected elder abuse to APS, even if the patient objects and the daughter provides a defense for the bruises
   B. Ignore the bruises and not report the suspected abuse because the patient and their daughter have provided an explanation for them
   C. Ask the patient for permission to report the suspected abuse to APS

   Explanation: In this situation, the practitioner has a legal and ethical obligation to report any suspicion of elder abuse to APS, even if the patient objects and the daughter provides a defense for the bruises. Failure to report the suspected abuse can harm the patient. The patient's objection or the daughter's explanation for the bruises does not absolve the practitioner of their responsibility to report the suspected abuse. Asking the patient for permission to report the suspected abuse is unnecessary because the practitioner has a mandatory reporting requirement and does not need the patient's consent to make the report. Ignoring the bruises and not reporting the suspected abuse is not a responsible or ethical course of action. *Answer*: A

3. Ms. Greta Green, a 26-year-old musician, is seeking healthcare treatment for an injury sustained during a physical altercation with their intimate partner. The patient does not want to involve the police or any other authority and refuses to provide informed consent for the practitioner to report the incident. What is the practitioner's most appropriate course of action?
   A. Report the incident to the police without the patient's consent.
   B. Refuse to treat the patient and refer them to another practitioner.
   C. Without consent, documented in the healthcare records, the practitioner does not have the authority to report intimate partner abuse.
   D. Without consent, the practitioner is not allowed to council or assure the patient that they have a safe place to go if needed.

Explanation: In this scenario, the practitioner's most appropriate course of action is to respect the patient's wishes and not report the incident to the police or any other authority, as the patient has not provided informed consent. However, the practitioner should document the patient's intimate partner violence report and refusal to involve the police or other authorities in the healthcare records. The practitioner should also provide the patient with information on available resources for victims of intimate partner violence, including counseling and advocacy services. The practitioner should not refuse to treat the patient and refer them to another practitioner, as this may put the patient at further risk of harm. The practitioner should inform the patient about available resources, including safe places to go, and offer support and referrals to appropriate services.. *Answer*: C

4. Mr. Tony Roberts, a 79-year-old retiree, comes to you with bruises and other signs of physical abuse. After a thorough evaluation, you suspect the patient is a victim of elder abuse. The elderly victim refuses to provide informed consent for you to report the abuse. How does mandatory reporting of elder abuse impact informed consent in the patient–practitioner relationship?
   A. It goes against informed consent because the practitioner is required to report the abuse without the patient's consent.
   B. It does not impact informed consent because mandatory reporting is part of the patient–practitioner public policy social contract, which includes compliance with the law and the professional principles of beneficence (do good) and nonmaleficence (do no harm).
   C. It is not relevant because informed consent only applies to healthcare treatment and not reporting of abuse.
   D. It undermines informed consent because reporting elder abuse goes against the principle of confidentiality in the patient–practitioner relationship.

   Explanation: The mandatory reporting of elder abuse does not necessarily impact informed consent in the patient–practitioner relationship because it is part of the patient–practitioner public policy social contract, which includes compliance to the law and the professional principles of beneficence (do good) and nonmaleficence (do no harm). Although the patient can refuse informed consent for reporting the abuse, the practitioner has a legal and ethical obligation to report suspected elder abuse to the appropriate authorities. This obligation is based on protecting vulnerable individuals from harm and preventing further abuse or neglect. In cases in which the patient is unable or unwilling to provide informed consent, the practitioner should follow the legal and ethical requirements for mandatory reporting of elder abuse and ensure that the patient is aware of their rights and available resources. Reporting elder abuse does not necessarily go against the principle of confidentiality, as reporting is typically done in a confidential manner and with respect for the patient's privacy. *Answer*: B

5. Eli Adams, a 9-year-old elementary student, comes to the clinic with multiple bruises and other signs of physical abuse. After a thorough evaluation, you suspect the patient is a victim of child abuse. What are the main purposes of the Child Abuse Prevention and Treatment Act (CAPTA) and the role of CPS?
   A. To provide social services for older adults and adults with significant disabilities
   B. To support research on child abuse and neglect
   C. To prevent, identify, and treat child abuse and neglect and to ensure that states have the necessary resources to fulfill this purpose
   D. To investigate and prosecute criminal cases of child abuse and neglect

   Explanation: The main purposes of CAPTA are to prevent, identify, and treat child abuse and neglect and to ensure that states have the necessary resources to fulfill this purpose. The role of CPS is to investigate reports of suspected child abuse and neglect and to provide social services and support to families in need. CPS works with law enforcement and other agencies to ensure the safety and well-being of children who have been victims of abuse or neglect and may remove children from their homes to protect them from further harm. CAPTA also provides funding for research on child abuse and neglect and supports the developing of effective prevention and intervention programs. The ultimate goal of CAPTA and CPS is to ensure that all children are safe and protected from abuse and neglect. *Answer*: C

REFLECTION VIGNETTES

1. Mr. Frederic Holmes, an 80-year-old patient, presents to practitioner Jessica Chen with a fractured wrist and oval-shaped bruises on the patient's arms. The patient's adult offspring, who provides housing and care for the patient, reports that the patient fell. The patient had been admitted to the hospital 6 months earlier with a dislocated shoulder, also attributed to a fall. Upon examining the patient's injuries, the practitioner suspects that the patient's bruises may have been caused by physical abuse. Practitioner Chen initiates a conversation with the patient in private, expressing concern about the possibility of abuse and asking about the patient's living situation. The patient confides that the injuries were not the result of a fall but, rather, were caused by the patient's offspring. The patient is adamant about not reporting the situation, as the patient depends on their offspring for housing, food, and care.

    What is the ethical issue(s)?

    How should the ethical issue(s) be addressed?

2. Mr. Dylan Eastman, a 30-year-old patient, presents to practitioner Maya Singh, a primary care practitioner, for a routine physical checkup. He is accompanied by his male partner, Mr. Edwin Crawford. The practitioner noticed several new and old bruises on Mr. Eastman's body during the examination. The practitioner initiates a private conversation with Mr. Eastman, expressing concern about the bruises and asking about their cause. In the private conversation, Mr. Eastman reveals that the bruises were caused by his partner, Mr. Crawford. He explains that he is financially dependent on his partner for housing, food, and care and is afraid of reporting the situation.

    What is the ethical issue(s)?

    How should the ethical issue(s) be addressed?

* * *

WEB LINKS

1. *Code of Medical Ethics*. American Medical Association. https://code-medical-ethics.ama-assn.org
    The AMA *Code of Medical Ethics*, established by the American Medical Association, is a comprehensive guide for healthcare practitioners. It addresses issues and challenges, promotes adherence to standards of care, and is continuously updated to reflect contemporary practices and challenges.
2. "Mandatory Reporting Laws." Richard Thomas and Monique Reeves. https://www.ncbi.nlm.nih.gov/books/NBK560690
    This source provides a comprehensive overview of mandatory reporting laws in the United States, particularly for those with contact with vulnerable populations.
3. "Competency: Five Basic Categories." Jeffrey Bulger. http://platospress.com/page-6-13_5comp.html
    This website details five methods to evaluate patient competency in healthcare decisions, focusing on the decision's existence, outcome, rationality, understanding of risks and benefits, and the level of understanding needed.

4. "Mandatory Reporting of Injuries Inflicted by Intimate Partner Violence." Carolyn J. Sachs, 2007. https://journalofethics.ama-assn.org/article/mandatory-reporting-injuries-inflicted-intimate-partner-violence/2007-12
This article discusses the mandatory reporting of injuries inflicted by intimate partner violence, with a focus on the responsibilities of healthcare workers.
5. "Abuse: Child Elderly and Intimate Partner." Quizlet. https://quizlet.com/308427019/abuse-child-elderly-and-intimate-partner-flash-cards
This source provides definitions and possible signs of child abuse, elder abuse, and intimate partner violence, as well as guidelines for providers if abuse is suspected.
6. "Understanding a Nurse's Role as a Mandated Reporter." Keith Carlson. https://nursejournal.org/resources/understanding-nurses-role-as-a-mandated-reporter
This article discusses the role of nurses as mandated reporters, including their legal obligation to report suspected or known instances of abuse.
7. "Screening and Detection of Elder Abuse: Research Opportunities and lessons learned from emergency geriatric care, intimate partner violence, and child abuse." Scott Beach, Christopher Carpenter, Tony Rosen, Phyllis Sharps, & Richard Gelles. 2016. https://www.ncbi.nlm.nih.gov/pmc/articles/PMC7339956
This source provides an overview of elder abuse screening and detection methods and challenges in these areas.
8. "Policy Finder: Violence and Abuse." American Medical Association. https://policysearch.ama-assn.org/policyfinder/detail/Violence%20and%20Abuse?uri=%2FAMADoc%2FHOD.xml-0-4664.xml
This policy document discusses the role of practitioners in lessening the prevalence, scope, and severity of child maltreatment, intimate partner violence, and elder abuse.

ADDRESSING ONESELF AND PATIENT

7

Medicine is not merely a science but an art. The practitioner's character may act more powerfully upon the patient than the drugs employed.
—Paracelsus

* * *

MYSTERY STORY

> **DECEPTIVE PRESCRIPTIONS: THE IMPORTANCE OF PROPER INTRODUCTIONS AND TITLES IN THE HEALTHCARE PROFESSION**
>
> In the tranquil cocoon of First Impression Harmony Hospital, an inexplicable tragedy pierced the routine humdrum. A patient, previously on a promising path to recovery, was discovered lifeless in his room, with no immediate clues pointing to his sudden demise. The hospital corridors reverberated with stunned whispers as the administration dialed the local police. Thus, the wheels of an unnerving investigation were set in motion.
>
> Detective Oliver Braxton, a seasoned sleuth with a keen eye for detail, was handed the reins of this baffling case. She embarked on her inquiry by interrogating the hospital staff, focusing particularly on the healthcare team attending to the deceased patient. As the layers of the mystery began to unfurl, a disconcerting piece of information came to light: The patient had shared his discomfort with his nurse about a supposed doctor who, upon closer inspection, was no more than a student and had handed him a prescription.
>
> This revelation set off alarm bells in Detective Braxton's mind, prompting her to summon the implicated student, Mr. Daniel Allen, for a grilling session. Initially defensive, Mr. Allen crumbled under relentless probing, confessing to having misrepresented his credentials and prescribing the medication, driven by a perilous curiosity.
>
> Scrutiny into Mr. Allen's past painted an unsettling portrait of a history marred by deception, punctuated with past reprimands for similar transgressions. Detective Braxton concluded that the patient's untimely death was a direct consequence of Mr. Allen's unethical conduct, and the medication he had cavalierly prescribed was not suitable for the patient's condition.
>
> The grim episode culminated in a court trial, convicting Mr. Allen of lethal healthcare malpractice. The trial underscored the gravity of authentic introductions and the appropriate use of titles within healthcare settings, reflecting the fundamental principles of integrity and trust that define the profession.
>
> This chilling incident served as a stark reminder for aspiring and existing healthcare practitioners of the inviolable importance of honesty and transparency in how they address themselves to their patients. The sacred bond between a practitioner and a patient, founded on mutual respect and trust, is too vital to be compromised. This case was a somber testament to all those at First Impression Harmony Hospital that healthcare practice is as much a moral obligation and ethical commitment as it is a science. Any transgressions against these principles can wreak devastating consequences, forever altering the course of innocent lives.

* * *

THINK

The initial patient–practitioner encounter establishes the relationship based on the fundamental tenets of patient confidentiality and privacy. This relationship establishes the patient's right to decisional authority in authorizing the practitioner to provide specific healthcare interventions and general healthcare.

This initial encounter begins with the disclosure of the following:

1. Who the practitioner is and who is the supervising attending
2. Whether or not the practitioner is a physician or what type of practitioner the provider is
3. What specialty the practitioner has
4. If the person is a student, what program year they have completed

Second, the practitioner should address the patient using the patient's first and last name as documented in the healthcare record. This allows

- the patient to determine whether or not they wish to communicate with the practitioner;
- the practitioner to determine if they are seeing the correct patient; and
- the patient to determine how they would like to be addressed and establish their gender identity and title of respect.

Example:

Hello, my name is Cary Porter, and I'm a third-year student at the Chicago Medical School, and my supervisor and your attending physician is Dr. Cure. May I please ask your name? ... How would you like me to address you, and do you have a title of respect you would like me to use?

Titles of respect might include Mister, Miss, Mix, Officer, Captain, Doctor, Professor, and others.

Socially, among equals, there is an unspoken rule of reciprocity when addressing others. If an addresser addresses an addressee using their first name, then based on reciprocity, the addresser has given permission to the addressee to address the addresser using their first name. However, in authoritative relationships, or relationships in which an element of distance and respect must be kept, titles can be essential for establishing social boundaries. Social relationships such as parents, teachers, police, judges, and practitioners have the following respective titles of respect: Mom or Dad, Professor, Officer, Your Honor, and Doctor. These titles of respect are socially necessary for establishing authoritative positions and social boundaries.

Under no circumstances should a student refer to themself directly or with the vague implication that they are a doctor or a licensed practitioner. Potentiality is not the same as actuality, so it is never permissible for students to refer to themselves as John Doe, MD class of (some future date), or MD Candidate (some future date). Instead, the student needs to be as transparent and honest as possible, addressing themself with their current status, such as John Doe, First Year Medical Student at the Chicago Medical School. Society gives an enormous amount of honor and respect to students—for many, if not most, patients, the respect is even more than that given to licensed practitioners. Students must acknowledge and embrace their student title as long as possible, as they will soon be lifetime practitioners.

ASSESS

Patient: Autonomy

Differences in title in the patient–practitioner relationship indicate a social and authoritative difference between the practitioner and the patient. Titles communicate social status, which can be antithetical in relation to patient decisional authority. Therefore, the practitioner must make it clear at the outset that the patient has the autonomy to provide informed consent that authorizes, refuses, or withdraws the practitioner's treatment options.

The practitioner informs the patient of various accepted healthcare options, including no treatment, along with their associated risks and benefits; answers all the patient's questions; and then it is the patient who has the legal and moral authority to determine what treatment the practitioner will ultimately be authorized to provide. Of course, this is with the implicit understanding that the practitioner has the most knowledge and experience with the treatment options. Therefore, practically, the practitioner's counseling and guidance are an indispensable part of what is defined as a shared decision-making process.

It behooves or is incumbent upon practitioners to introduce themselves accurately and then determine how the patient wishes to be addressed and identify themself. As a general rule, it is best to address the patient using a title of respect as social recognition of the patient's incontestable right and authority for making autonomous informed consent decisions.

Care must be taken to minimize the patient's desire to show and express deference to the practitioner. These submissive tendencies are a natural result of the one-sidedness of confidential and private information that the practitioner has about the patient; the patient being in a vulnerable and dependent state; psychological transference; and the patient's need to have complete trust in the practitioner's knowledge, skills, and competency.

The term *doctor* comes from the Latin word *docere*, which means "to teach," and was historically reserved for the academic degree of PhD, so it is a bit of a misnomer when applied to a professional degree such as MD. However, the use of the address Doctor can be justified when considering the role that practitioners have in not only teaching the art of healing to the next generation of practitioners but also teaching patients about their healthcare treatment options and helping them make appropriate treatment decisions using the patient's reasonable goals, values, and priorities.

Practitioner: Beneficence and Nonmaleficence

Although care must be taken to minimize the patient's submissive tendencies, titles of respect in healthcare are essential for setting up the necessary and expected professional boundaries that are part of the patient–practitioner relationship. Strong legal, professional, and moral mandates forbid boundary violations, such as sexual or romantic relations between practitioners and their patients and the healthcare treatment of self, family members, and friends. The following are some reasons for these professional boundaries:

1. Historical disclosure of essential information, such as the past and present use of drugs and medications, along with other necessary information for accurate, safe, and effective treatment, may be compromised.
2. Physical examination of sensitive or private parts may be compromised and perceived as an intimate violation.
3. Social and family structures are such that differences in social status and decisional authority can disrupt or invalidate any notion of valid informed consent and authorization.

Titles of respect help create boundaries for both the practitioner and the patient that are an effective means for establishing social limits both for the protection of the practitioner and, most important, for the protection of the patient.

Public Policy: Justice

The patient–practitioner relationship is not just the relationship between a single practitioner and a single patient but also a social contract between a member of the healthcare profession and a citizen of society. Any violation of this relationship has a cascading effect of diminishing society's trust in the profession.

If, for example, a practitioner's title of Doctor is used not to establish authority but, rather, to communicate the professional patient–practitioner boundaries, then such use is recommended and laudable. A patient's title of Mister, Miss, Professor, etc. communicates both the patient–practitioner boundaries and the patient's authority to consent and give authorization to the practitioner for treatment and care, and that, too, is recommended and laudable.

The social principle of justice is often equilibrated with the notion of reciprocity because reciprocity is considered socially fair. Also, the fair exchange of titles of respect reciprocally empowers patients to make autonomous decisions and establishes professional and social boundaries in the patient–practitioner relationship.

CONCLUDE

Students, resident doctors, and practitioners must introduce themselves honestly, accurately, and with no implied deception caused by vagueness. At no time is it permissible for students to refer to themselves as MD or Doctoral Candidate. Students must embrace the honor of being a healthcare student and communicate that they are students and the year of the program they are in. Titles of respect should establish the patient's authority for making consent decisions and for establishing professional boundaries. When used properly, titles of honor can maximize the patient's best interests, as determined by the patient's reasonable goals, values, and priorities.

In summary, the proper introduction and use of titles of respect play a pivotal role in establishing a strong patient–practitioner relationship, enabling patients to make autonomous decisions and setting up essential professional boundaries. Students and practitioners must introduce themselves honestly and accurately, embracing their current status and ensuring the patient's best interests are met through a shared decision-making process that respects the patient's reasonable goals, values, and priorities. (For more information on professional boundaries, see Chapters 40, 50, and 51.)

* * *

REVIEW QUESTIONS

1. The student must disclose to the patient that they are a student, the year in the program, and the supervising attending practitioner.
    A. True
    B. False

2. Titles of respect are essential for establishing social boundaries.
    A. True
    B. False

3. Based on the principle of beneficence, being vague about one's student status is permissible if doing so increases patient compliance.
    A. True
    B. False

4. It is essential to address the patient using a title of respect because that is a social recognition of the patient's decision-making authority.
    A. True
    B. False

5. The patient's natural tendency to express deference to the practitioner is caused by
    A. The patient is in a vulnerable condition.
    B. The patient is dependent on the practitioner for healing.
    C. Psychological transference.
    D. The patient must trust the practitioner's knowledge, skills, and competency.
    E. All of the above.

6. The term doctor comes from the Latin word *docere*, which means
   A. Practitioner
   B. Healthcare professional
   C. To teach
   D. Healthcare provider

7. The principle of justice is often equilibrated with reciprocity. Therefore, the fair exchange of titles of respect is essential for empowering the patient to make treatment decisions.
   A. True
   B. False

Answers: 1A, 2A, 3B, 4A, 5E, 6C, 7A

* * *

CLINICAL VIGNETTES

1. Mr. Max Taylor, a 66-year-old farmer, has scheduled an initial appointment with a practitioner. During this encounter, the following details will be communicated:
   A. The identity of the practitioner, their supervising attending, and their specialty will be disclosed, but the practitioner's status as a physician or student will not be shared.
   B. The identity and status as a physician or non-physician practitioner will be disclosed, but the specialty, supervising attending, and program year will not be shared.
   C. The identity, status as a physician or non-physician practitioner, specialty, and supervising attending will be disclosed, but the program year will not be shared.
   D. The practitioner's identity, status as a physician or non-physician practitioner, specialty, supervising attending, and program year (if applicable) will be disclosed.

   Explanation: The correct answer is to provide the most complete and transparent disclosure of the practitioner's identity, status as a physician or non-physician practitioner, specialty, supervising attending, and program year (if applicable). This level of disclosure is necessary for the patient to make an informed decision about their care and to ensure that the patient receives care from a qualified and appropriately supervised practitioner. Each of the other options omits important information that could impact the patient's decision-making process and potentially compromise the quality of their care. *Answer*: D

2. Ms. Brooke Campbell, a 26-year-old journalist, walks into a healthcare clinic for the first time to meet with a practitioner. Which actions best exemplify the unspoken rule of reciprocity and the importance of titles in establishing social boundaries in the patient–practitioner relationship?
   A. The patient addresses the practitioner as "Doctor," and the practitioner addresses the patient by their first name.
   B. The patient addresses the practitioner as "Doctor," and the practitioner addresses the patient as "Ms. [last name]."
   C. The patient addresses the practitioner by their first name, and the practitioner addresses the patient by their first name.
   D. The patient addresses the practitioner by their first name, and the practitioner addresses the patient as "Ms. [last name]."

   Explanation: There is an unspoken rule of reciprocity and the importance of titles in establishing social boundaries in the patient–practitioner relationship. Ideally, the patient demonstrates respect for the practitioner's title by addressing them as "Doctor," and in turn, the practitioner shows respect for the patient's social status by addressing them as "Mr./

Ms. [last name]." This reflects the importance of establishing clear social boundaries in the patient–practitioner relationship to ensure a professional and respectful environment. *Answer*: B

3. Mr. Wade Perez, a 22-year-old college student who identifies as nonbinary, walks into your clinic for a routine checkup. As a practitioner, you are responsible for accurately introducing yourself and determining how the patient wishes to be addressed. How should the practitioner address the patient during the appointment?
   A. Using their first name
   B. Using the title "Mr." or "Ms."
   C. Asking the patient how they prefer to be addressed
   D. Assuming the patient's gender and addressing them accordingly

   Explanation: As a general rule, it is best to address the patient using a title of respect as social recognition of the patient's incontestable right and authority for making autonomous informed consent decisions. Practitioners should always ask the patient how they prefer to be addressed rather than making assumptions based on the patient's appearance or gender. This shows the patient that their wishes and comfort are respected and valued, leading to a better patient–practitioner relationship and improved health outcomes. *Answer*: C

4. What is the origin of the term "Doctor," and why is it sometimes considered a misnomer when used in reference to an MD?
   A. The term "Doctor" comes from the Latin word *docere*, which means "to heal," and was historically reserved for professional degrees such as MD.
   B. The term "Doctor" comes from the Latin word *docere*, which means "to teach," and was historically reserved for academic degrees such as PhD.
   C. The term "Doctor" comes from the Greek word *dokein*, which means "to think," and was historically reserved for academic degrees such as PhD.
   D. The term "Doctor" comes from the Greek word *iatros*, which means "healer," and was historically reserved for professional degrees such as MD.

   Explanation: The term "Doctor" comes from the Latin word *docere*, which means "to teach." Originally, the title of "Doctor" was reserved for individuals who had earned academic degrees, such as a PhD, in recognition of their advanced knowledge and ability to teach others. In contrast, the title of "Physician" was historically reserved for individuals licensed to practice medicine and care for patients. Today, however, the term "Doctor" is often used interchangeably with "Physician," even though it is technically a misnomer when used in reference to an MD, as the MD degree is primarily a professional degree rather than an academic degree. Nonetheless, "Doctor" as a physician title has become widely accepted in many countries, including the United States. *Answer*: B

5. Ms. Irene Brown, a 42-year-old counselor, visits your clinic for a follow-up appointment. They introduce themself and inform you that they prefer to be addressed by their first name. How should a practitioner respond when a patient requests to be addressed by their first name instead of a formal title such as "Mr." or "Ms."?
   A. Refuse the patient's request and insist on using formal titles of respect as a professional boundary
   B. Agree to the patient's request and address them by their first name only
   C. Agree to the patient's request but also acknowledge the use of formal titles of respect in a professional setting for establishing decision-making authority and professional boundaries
   D. Agree to the patient's request on the condition that they also refer to you by first name

Explanation: It is important to respect the patient's preference for how they wish to be addressed, but it is also important to acknowledge the use of formal titles of respect in a professional setting for establishing decision-making authority and professional boundaries. The practitioner respects the patient's autonomy and personal preferences by agreeing to the patient's request and using their first name. However, by also acknowledging the importance of professional boundaries, the practitioner can help ensure that the patient understands the practitioner's role as a healthcare professional and the patient's role as a patient in the relationship. Disregarding the patient's preference and autonomy is inappropriate, and being too informal does not show proper respect for professional boundaries and runs the risk of blurring the professional–patient relationship. *Answer*: C

REFLECTION VIGNETTES

1. A third-year healthcare student enters the patient's room and introduces themself:

   Hello, my name is Cary Porter, and I'm a third-year doctoral candidate at the Chicago Medical School. May I confirm your name? Is your name John Doe? Great. How would you like me to address you, and do you have a title of respect or preferred pronoun you would like me to use?

   What is the ethical issue(s)?

   How should the ethical issue(s) be addressed?

2. Being accurate and truthful in professional communications is important, especially in healthcare. What is wrong with the following: A third-year healthcare student has an email signature card that states "Cary Porter, MD Candidate, Class of [some future date]."

   What is the ethical issue(s)?

   How should the ethical issue(s) be addressed?

* * *

WEB LINKS

1. *Code of Medical Ethics*. American Medical Association. https://code-medical-ethics.ama-assn.org
   The AMA *Code of Medical Ethics*, established by the American Medical Association, is a comprehensive guide for healthcare practitioners. It addresses issues and challenges, promotes adherence to standards of care, and is continuously updated to reflect contemporary practices and challenges.
2. "How to Introduce Yourself to Patients: Guide to Building Nurse–Patient Rapport." NurseMoneyTalk. https://nursemoneytalk.com/blog/how-to-introduce-yourself-to-patients
   This source provides a guide for nurses on how to introduce themselves to patients, emphasizing the importance of preparation and clear communication.
3. "How to Introduce Yourself to Patients." Mags Guest. 2016. https://pubmed.ncbi.nlm.nih.gov/27286624
   This article explores the process of introducing oneself to patients, highlighting that this interaction forms the basis of the therapeutic nurse–patient relationship.
4. "Proper Introductions at Your Practice Are Critical to Patient Relations." Sue Jacques. 2012. https://www.physicianspractice.com/view/proper-introductions-your-practice-are-critical-patient-relations

This source discusses the importance of proper introductions in a healthcare practice, emphasizing that the patient is always the person of highest precedence.
5. "Communication Skills: A Guide to Practice for Healthcare Professionals." Ausmed. 2019. https://www.ausmed.com/cpd/guides/communication-skills
This guide provides insights into communication skills in healthcare, including how to build patient rapport and introduce oneself.
6. "Preparing for Your First Patient Interaction." Wolters Kluwer. https://www.wolterskluwer.com/en/expert-insights/preparing-for-your-first-patient-interaction
This source provides advice on preparing for the first patient interaction, including how to introduce oneself and establish the interview order.
7. "How to Make a Great First Impression at Your Medical Healthcare Practice." Lucien Roberts and Logan Lutton. 2021. https://www.physicianspractice.com/view/how-to-make-a-great-first-impression-at-your-medical-practice
This article discusses how to make a great first impression at a healthcare practice, including when to introduce oneself to the patient.
8. "Introducing Yourself to Patients Is a Patient Safety Issue." Peter Pronovost. 2014. https://www.kevinmd.com/2014/04/introducing-patients-patient-safety-issue.html
This source discusses the importance of introducing oneself to patients as a patient safety issue, emphasizing the role of effective communication in patient safety.

ADVANCE DIRECTIVES— PLANNING

8

The practitioner should not treat the disease but the patient who is suffering from it.
—Maimonides

* * *

MYSTERY STORY

> ### DOCUMENTED WISHES: THE IMPORTANCE OF ADVANCE DIRECTIVES AND DURABLE POWER OF ATTORNEY FOR HEALTHCARE
>
> On an ordinary Wednesday amidst the rhythmic hum at Navigators' Guiding Care Hospital, an anomaly was found in a patient's electronic medical record (EMR). Mrs. Susana Smith was recently admitted for pneumonia, and there was no advance directive or durable power of attorney in her healthcare records. This absence compelled the hospital to ensure that all patients were informed of their right to dictate their healthcare through advance directives and durable power of attorney.
>
> The vigilant nursing staff quickly alerted Practitioner Samuel Johnson, Mrs. Smith's assigned healthcare provider. Practitioner Johnson acknowledged the importance of informing patients about their rights, documenting their surrogate preferences with the durable power of attorney, and promoting patient autonomy (informed consent) with the advance directive. Determined to rectify this oversight, Practitioner Johnson resolved to initiate a candid conversation with Mrs. Smith regarding her healthcare desires, guiding her through the process of crafting her durable power of attorney and advance directive.
>
> With a reassuring demeanor, Practitioner Johnson entered Mrs. Smith's room, gently broaching the sensitive subject. His approach was met with initial apprehension from Mrs. Smith, but with skilled navigation and compassionate assurance, he conveyed that these documents were her security shields, designed to protect her autonomy (informed consent) and honor her surrogate preferences.
>
> Practitioner Johnson vividly outlined the importance of advance directives and the durable power of attorney for healthcare. He depicted how these tools served as a beacon, guiding her healthcare treatment if she ever became incapacitated and unable to express her desires, thereby ensuring that the right decisions were made in alignment with her goals, values, and priorities.
>
> In the wake of his explanation, Mrs. Smith gave her consent. Practitioner Johnson provided her with the requisite forms and meticulously walked her through the procedure. Mrs. Smith, in her newly drafted durable power of attorney, nominated her daughter, Sarah, as her surrogate healthcare decision-maker and outlined her healthcare wishes in her advance directive. Practitioner Johnson ensured that Sarah comprehended her role and the guiding principles of beneficence (do good) and nonmaleficence (do no harm) that she needed to specify and balance while making treatment decisions on her mother's behalf, using her mother's preferences and her mother's goals, values, and priorities.
>
> Unfortunately, in the subsequent days, Mrs. Smith's condition deteriorated, rendering her unable to participate in decision-making. Sarah was thrust into her role as the surrogate decision-maker. Armed with her mother's documented wishes and the ethical principles Practitioner Johnson had ingrained in her, she faced the labyrinth of healthcare decision-making with remarkable confidence.

> Mrs. Smith's tale served as a stark reminder to the Navigator's Guiding Care Hospital staff about the imperative of addressing and documenting patients' healthcare preferences in their EMRs. Practitioner Johnson's proactivity and Sarah's assertive advocacy ensured that Mrs. Smith's care plan aligned seamlessly with her wishes. The encounter underscored the need for transparency, respect for patient autonomy (informed consent), and adherence to ethical norms in the intricate world of healthcare.

* * *

THINK

Autonomy, or self-determination, is valued precisely because the patient is in the best position to determine their own subjective goals, values, and priorities, which are necessary for determining what objective treatments will maximize the patient's best interests.

This recognition resulted in Congress passing the Patient Self-Determination Act of 1990, which mandates by federal law that all healthcare institutions that receive any Medicaid or Medicare payments must

- inform patients of their legal rights to make decisions concerning their healthcare;
- periodically inquire as to whether the patient has executed an advanced directive;
- document in the healthcare records the patient's healthcare wishes; and
- provide educational programs for staff, patients, and the community on ethical issues concerning patient self-determination and advance directives.

ASSESS

Patient: Autonomy

An advance directive, or living will, is a preemptive act of a patient with the decisional capacity to autonomously self-determine what treatment preferences they authorize under various conditions. A durable power of attorney for healthcare (DPOA) establishes who will serve as their healthcare decision-making surrogate if they ever lack decisional capacity. A DPOA is durable because it only comes into effect once the patient loses decisional capacity and continues throughout that duration of time.

Although the living will is the patient's most direct communication regarding treatments the patient would or would not want to give informed consent to, the living will can also be the least helpful if the patient cannot predict future events and changing desires. This can result in vague terms such as "no heroic measures" or "no extraordinary care," which is virtually no better than saying nothing. This lack of contextual precision is where surrogate decision-making can make its helpful contribution.

Practitioner: Beneficence and Nonmaleficence

The DPOA is accompanied by professional responsibilities for both the practitioner and the surrogate. As a matter of beneficence (do good) and nonmaleficence (do no harm), the practitioner must ensure that the surrogate is educated about making treatment decisions for the patient. First, the surrogate decision-maker should not make treatment decisions for the patient using their own goals, values, and priorities. Second, the surrogate, or proxy, must make treatment decisions that maximize the patient's best interests—beneficence (do good) and nonmaleficence (do no harm)—using the patient's reasonable goals, values, and priorities. In other words, the surrogate decision-maker must make treatment decisions for the patient as if the surrogate were the patient making the decisions. This is why the practitioner must include the surrogate in the patient's healthcare discussions as much as possible before they lose their decisional capacity, especially when directly discussing the patient's reasonable goals, values, and priorities. If the practitioner fully informs the surrogate of the circumstances and healthcare condition of the patient, and if the surrogate understands how to make decisions that will maximize the patient's best interests, then this will typically avoid some of the inherent problems of vagueness and imprecision that frequently occur with living wills.

However, if, for some reason, the patient's reasonable goals, values, and priorities are not determinable, then the responsibility of the surrogate, in consultation with the practitioner, will be to maximize the patient's best interests, using the moral principles of beneficence (do good) and nonmaleficence (do no harm).

Public Policy: Justice
The federal government enacted the Patient Self-Determination Act, which essentially makes every healthcare institution in the country responsible for

- enforcing the patient's right to autonomously choose what healthcare they will receive (living will);
- enforcing the patient's right to autonomously choose who their surrogate will be (DPOA); and
- documenting the living will and DPOA in the EMR.

CONCLUDE
Autonomous decisional authority is a universal patient right. Therefore, if a living will provides clear and distinct instructions regarding a foreseen healthcare circumstance, then that living will should be respected. Suppose there is no living will or the living will is not clear for a particular circumstance. In that case, the decision-making authority will go to whoever has been appointed to be the surrogate decision-maker by the patient's DPOA. When making treatment decisions, the surrogate has the legal and moral obligation to use the patient's reasonable goals, values, and priorities. In other words, the surrogate must make treatment decisions as if they were the patient, which is not necessarily how the proxy might make decisions for themself. The practitioner has a vital role in educating the patient's surrogate as to how the surrogate is to make treatment decisions by using the patient's reasonable goals, values, and priorities or, if those are not knowable, by maximizing the patient's best interests using beneficence (do good) and nonmaleficence (do no harm).

In summary, advance directives are essential for respecting and upholding patient autonomy in healthcare. Living wills provide guidance on the patient's desired treatment, whereas a DPOA for healthcare appoints a surrogate to make decisions on their behalf when necessary. It is crucial for practitioners to educate patients and surrogates about their roles and obligations, ensuring that treatment decisions are made based on the patient's reasonable goals, values, and priorities or, when these are not known, in the patient's best interest. By doing so, healthcare professionals can effectively support patients' rights to self-determination and maintain trust in the healthcare profession.

* * *

REVIEW QUESTIONS
1. Congress passed the Self-Determination Act of 1990, which federally mandates that any institution that receives Medicaid or Medicare payments must
    A. Inform patients of their legal rights to make decisions concerning their healthcare
    B. Periodically inquire as to whether the patient has executed an advanced directive
    C. Document in the healthcare records the patient's healthcare wishes
    D. Provide educational programs for staff, patients, and the community on ethical issues concerning patient self-determination and advance directives
    E. All of the above

2. A durable power of attorney for healthcare is durable because a power of attorney
   A. Only comes into effect once the patient loses decisional capacity
   B. Stays in effect even after the patient regains decisional capacity
   C. Cannot be rescinded
   D. Includes not just treatment decisions but also the authority to make estate decisions

3. Living wills can be the most direct patient communication, and they can also be the least helpful.
   A. True
   B. False

4. It is essential to select a trusted surrogate because the surrogate will be making treatment decisions using their own goals, values, and priorities.
   A. True
   B. False

5. Which of the following is the responsibility of the surrogate decision-maker according to the Patient Self-Determination Act of 1990?
   A. Make treatment decisions based on their own goals, values, and priorities
   B. Make treatment decisions based on the patient's reasonable goals, values, and priorities
   C. Document the patient's living will and durable power of attorney in the electronic healthcare records
   D. Provide educational programs for staff, patients, and the community on ethical issues concerning patient self-determination and advance directives

Answers: 1E, 2A, 3A, 4B, 5B

* * *

CLINICAL VIGNETTES

1. Mr. Gabriel Ahmed, a 52-year-old network administrator, is admitted and informed of their rights to make decisions about their healthcare. The practitioner periodically asks patients if they have an advanced directive and documents their wishes in the healthcare records. The healthcare institution also regularly offers educational programs for staff, patients, and the community on ethical issues related to self-determination and advance directives. Which of the following best describes the legal requirement imposed on the healthcare institution?
   A. The healthcare institution is not required to inform patients of their rights to make decisions about their healthcare.
   B. The healthcare institution is required to inform patients of their rights to make decisions about their healthcare; periodically inquire about advanced directives; document the patient's wishes in the healthcare records; and regularly provide educational programs for staff, patients, and community on ethical issues related to self-determination and advance directives.
   C. The healthcare institution is only required to provide educational programs for staff, patients, and community on ethical issues related to self-determination and advance directives.
   D. The healthcare institution is only required to document the patient's wishes in their healthcare records.

   Explanation: There are legal requirements imposed on healthcare institutions regarding patients' rights to make decisions about their healthcare. Informed consent and the right to make decisions about one's own healthcare are fundamental principles in bioethics and are protected by law. The healthcare institution is legally obligated to inform patients of their rights; periodically inquire about advanced directives; and provide educational programs for staff, patients, and the community on ethical issues related to self-determination and advance directives. Documenting the patient's wishes in their healthcare records is important but is not the only legal requirement imposed on the healthcare institution in this scenario. *Answer*: B

2. Ms. Sofia Ali, a 62-year-old banker, has a DPOA for healthcare in case Ms. Ali ever loses decisional capacity. This document designates a surrogate to make treatment decisions on their behalf if they cannot do so themself. What makes the DPOA durable?
   A. The DPOA only lasts for a specific period of time.
   B. The DPOA is only effective when the patient is conscious.
   C. The DPOA comes into effect only after the patient loses decisional capacity, and it ends once the patient regains decisional capacity or dies.
   D. The DPOA document is automatically revoked once the patient regains decisional capacity.

   Explanation: A DPOA for healthcare is a legal document that designates a surrogate decision-maker to make treatment decisions on behalf of a patient if the patient loses decisional capacity. It is called "durable" because it comes into effect after the patient loses decisional capacity, and it stays in effect until the patient regains decisional capacity or until the patient dies. This differs from a regular POA, which is invalid once the person granting the power becomes incapacitated. A DPOA is essential in ensuring that a person's wishes are carried out if they cannot make their own treatment decisions. *Answer*: C

3. Mr. Caden Chen, a 68-year-old marketing professional, has created a living will expressing their treatment preferences, but the document contains vague terms such as "no heroic measures" or "no extraordinary care." This lack of specific instructions can make it difficult for healthcare providers to make decisions on behalf of the patient. How can a surrogate be helpful?
   A. A surrogate can provide known patient preferences.
   B. A surrogate is not helpful in this scenario.
   C. A surrogate is only helpful if the patient's wishes are clear in the living will.
   D. A surrogate is only helpful if the patient has not created a living will.

   Explanation: Surrogate decision-making can provide specific treatment preferences based on the patient's reasonable goals, values, and priorities, filling in the gaps left by vague terms in the living will. A living will is a legal document that outlines a person's healthcare wishes if they cannot make decisions for themself. However, if the living will contains ambiguous or unclear terms, healthcare providers may have difficulty interpreting the patient's wishes. In such cases, a surrogate decision-maker can help provide additional information about the patient's reasonable goals, values, and priorities. The surrogate can work with the healthcare team to help interpret the patient's wishes and make decisions that align with the patient's reasonable goals, values, and priorities. Although a surrogate may not be able to change the terms of the living will, it can provide valuable context and additional information to help ensure that the patient's wishes are being followed. *Answer*: A

4. Ms. Leslie Hamilton, a 72-year-old retiree, is diagnosed with a progressive, degenerative illness that will eventually result in a loss of decisional capacity. The patient has a DPOA and appoints their spouse as the surrogate decision-maker. The patient has made it clear that they do not want any healthcare interventions that will prolong their life if they are in a persistent vegetative state, and that wish is documented in the healthcare records. What is the primary responsibility of the practitioner when working with the surrogate decision-maker in this situation?
   A. To ensure that the surrogate makes treatment decisions based on their goals, values, and priorities

B. To involve the surrogate in discussions about the patient's reasonable goals, values, and priorities and educate the surrogate on how to make treatment decisions that maximize the patient's best interests
C. To ensure that the surrogate makes treatment decisions that prolong the patient's life, regardless of the patient's wishes
D. To only discuss the patient's healthcare condition with the surrogate after the patient has lost their decisional capacity

Explanation: The surrogate decision-maker is responsible for making healthcare decisions on behalf of the patient if the patient becomes unable to make decisions for themself. The practitioner is responsible for involving the surrogate in discussions about the patient's reasonable goals, values, and priorities and educating the surrogate on making treatment decisions that maximize the patient's best interests. In this case, the patient has made it clear that they do not want any healthcare interventions that will prolong their life if they are in a persistent vegetative state, and this preference should be respected. The practitioner should ensure that the surrogate is aware of the patient's wishes and can make decisions that align with those wishes. The practitioner should also support and guide the surrogate to help them make informed decisions that promote the patient's well-being. *Answer*: B

5. Mr. Griffin Singh, an 82-year-old retiree, has been admitted with a serious illness and cannot make decisions regarding their healthcare. The healthcare team has discovered that the patient has a living will. The patient also has a durable power of attorney that appoints a surrogate decision-maker. Which of the following statements best describes the decision-making process in this situation?
   A. The healthcare team should rely solely on the patient's surrogate decision-maker.
   B. The healthcare team should rely on the patient's living will if it provides clear instructions.
   C. The healthcare team should consult both the patient's surrogate decision-maker and the patient's living will but ultimately make the decision based on the patient's best interests.
   D. The healthcare team should rely on the patient's living will unless it contradicts the patient's surrogate decision-maker.

Explanation: A living will is a legal document that outlines a person's healthcare wishes if they cannot decide for themself. If the living will provides clear instructions, the healthcare team should rely on them to guide their decision-making. However, if the living will is unclear or does not provide specific guidance, then the healthcare team should consult with the patient's surrogate decision-maker to determine the best course of action based on the patient's best interests. The healthcare team should make every effort to ensure that the patient's reasonable goals, values, and priorities are respected to maximize the patient's best interest. *Answer*: B

REFLECTION VIGNETTES
1. Mr. John Jameson, a 65-year-old retired construction worker, is admitted to the intensive care unit in the end stages of small cell lung cancer (SCLC), a particularly aggressive form of lung cancer. Mr. Jameson cannot communicate for himself, and his spouse and family want the practitioner to do everything possible to extend his life. However, a friend of Mr. Jameson, who is estranged from the family, presents a recently signed and witnessed DPOA authorizing them to be the surrogate decision-maker. The practitioner carefully reviews Mr. Jameson's healthcare records and notes his current symptoms, which include shortness of breath, coughing, chest pain, and significant weight loss. The differential diagnosis includes SCLC with metastases to other body parts, pneumonia, bronchitis, and chronic obstructive pulmonary disease. The practitioner discusses the situation with Mr. Jameson's spouse, family, and friend holding the DPOA. The family desires aggressive healthcare intervention to

extend Mr. Jameson's life. At the same time, the friend reiterates Mr. Jameson's wish to stop all healthcare interventions other than hydration, nutrition, and pain management.

What is the ethical issue(s)?

How should the ethical issue(s) be addressed?

2. Ms. Isabel Rodriguez, a 52-year-old schoolteacher, fills out a living will and states explicitly that she never wants to be on a heart–lung machine under any condition. The living will is included in Ms. Rodriguez's healthcare records. However, soon after completing the living will, a healthcare emergency occurs, and Ms. Rodriguez needs to be temporarily placed on a heart–lung machine to get through the emergency. Ms. Rodriguez is expected to recover and live everyday life. The practitioner faces a difficult decision as they review Ms. Rodriguez's living will and the immediate need for life-saving measures. They meet to discuss Ms. Rodriguez's wishes and the best course of action. Ms. Rodriguez's family is also consulted, and they express their desire to do everything possible to save her life. The practitioner carefully reviews the situation and determines that Ms. Rodriguez's condition is severe and requires immediate intervention.

What is the ethical issue(s)?

How should the ethical issue(s) be addressed?

3. The practitioner received a consult for a 65-year-old female patient, Ms. Sarah Johnson, who was admitted to the hospital with shortness of breath. Ms. Johnson has a history of metastatic pancreatic cancer, which has been treated with chemotherapy. On examination, the practitioner noted that Ms. Johnson appeared to be in significant distress, with a respiratory rate of 24 breaths per minute, oxygen saturation of 87% on room air, and use of accessory respiratory muscles. A chest X-ray was ordered, which revealed diffuse pulmonary metastases. The practitioner initiated oxygen therapy and started Ms. Johnson on a low-dose morphine infusion for pain control. The practitioner then contacted Ms. Johnson's family to discuss her care. The practitioner learned that Ms. Johnson did not have decisional capacity and that her youngest offspring held the healthcare power of attorney. The youngest offspring made the decision to make Ms. Johnson a do-not-resuscitate patient, given her poor prognosis and desire for comfort-focused care. However, Ms. Johnson's eldest offspring disagreed with this decision and wanted all possible measures taken to prolong her life.

What is the ethical issue(s)?

How should the ethical issue(s) be addressed?

* * *

WEB LINKS

1. *Code of Medical Ethics*. American Medical Association. https://code-medical-ethics.ama-assn.org
   The AMA *Code of Medical Ethics*, established by the American Medical Association, is a comprehensive guide for healthcare practitioners. It addresses issues and challenges, promotes adherence to standards of care, and is continuously updated to reflect contemporary practices and challenges.
2. "An Ethical Framework for Incorporating Digital Technology into Advance Directives: Promoting Informed Advance Decision Making in Healthcare." Sophie Gloeckler, Andrea Ferrario, and Nikola Biller-Andorno. 2022. https://www.ncbi.nlm.nih.gov/pmc/articles/PMC9511942

This article discusses the value of advance directives and the importance of informed decision-making in promoting patient autonomy.
3. "Advance Directives." American Medical Association. https://code-medical-ethics.ama-assn.org/ethics-opinions/advance-directives
This source provides an overview of advance directives as tools that allow patients to express their values, goals for care, and treatment preferences.
4. "Advance Directives as a Tool to Respect Patients' Values and Preferences: Discussion on the Case of Alzheimer's Disease." Corinna Porteri. 2018. https://bmcmedethics.biomedcentral.com/articles/10.1186/s12910-018-0249-6
This article discusses the value of advance directives, particularly in the context of dementia, and the ethical considerations surrounding personal identity and changing interests.
5. "Advance Directive: Does the GP Know and Address What the Patient Wants? Advance Directive in Primary Care." Guda Scholten, Sofie Bourguignon, Anthony Delonote, Bieke Vermeulen, Geert Van Boxem, and Birgitte Schoenmakers. 2018. https://bmcmedethics.biomedcentral.com/articles/10.1186/s12910-018-0305-2
This article explores the role of the general practitioner in discussing advance directives with patients and the need for advanced care planning.
6. "Patient Rights." American Medical Association. https://code-medical-ethics.ama-assn.org/ethics-opinions/patient-rights
This source discusses patient rights, including the use of advance directives to express treatment preferences and identify decision-makers.
7. "Overriding Advance Directives: A 20-Year Legal and Ethical Overview." John Banja and Michele Sumler. 2019. https://pubmed.ncbi.nlm.nih.gov/31433120
This article discusses the ethical and legal considerations when health professionals override patients' advance directives, particularly in the context of life-prolonging treatment.
8. "Advance Directives." Steven A. House, Caroline Schoo, and Wes A. Ogilvie. https://www.ncbi.nlm.nih.gov/books/NBK459133
This source provides an overview of advance directives as legal documents that specify healthcare options and designate individuals to speak on the patient's behalf.

ASSISTED SUICIDE

9

Act in such a way that you treat humanity, whether in your own person or in the person of any other, never merely as a means to an end, but always at the same time as an end.
—Immanuel Kant

* * *

MYSTERY STORY

A DANCE WITH MORALITY: THE CONUNDRUM OF ASSISTED SUICIDE IN HEALTHCARE

A sense of solemn duty hung heavily in the halls of Life's Embrace Hospital, where Practitioner Vincent Euthymos, revered for his compassionate demeanor, was a beacon of solace for those clinging to the precipice of life. Known for his unwavering dedication to his patients, Practitioner Euthymos found himself amidst a storm of ethical dilemmas with the admission of Mr. Thomas Walker. The frail silhouette of Mr. Walker, tormented by terminal illness, bore the burden of a profound wish—to meet death on his own terms, with dignity and grace.

Practitioner Euthymos became a confidante for the ailing Mr. Walker, engaging in somber dialogues about the options that lay before him. Initially, hesitation gnawed at Practitioner Euthymos. Yet, eventually, he found himself acquiescing to the man's dying wish, providing him with the grim tools to sculpt his own end.

Days later, the eerie silence that enveloped Mr. Walker's room, now a shrine of sorrow, confirmed the inevitable—he had departed from life. His devastated family sought answers, and the hospital administration, in response, unfurled an internal investigation into the circumstances of Mr. Walker's death. Meanwhile, Practitioner Euthymos, the architect of mercy in Mr. Walker's narrative, was suspended, his reputation hanging in the balance.

The relentless wheels of the investigation unearthed the truth: Practitioner Euthymos had indeed facilitated Mr. Walker's exit from life. Despite the practitioner's knowledge of the hospital's stringent prohibition of assisted suicide, he had chosen to walk the murky path, violating institutional policy and potentially overstepping legal boundaries.

The situation took a dire turn as the case was escalated to the state healthcare board. Charges of violating standards of care loomed over Practitioner Euthymos, causing a ripple in his career that saw him barred from practicing medicine pending a disciplinary hearing.

At the hearing, Practitioner Euthymos stood in the eye of the storm, fervently arguing his stance. He claimed he had only acted as a vessel of compassion for Mr. Walker, honoring his autonomy and personal wishes above all else. He asserted that, in his view, he had provided the best care for his patient, even if it meant straying from hospital policy.

Despite his impassioned defense, the state healthcare board found Practitioner Euthymos guilty, contending that the healthcare profession prides itself as a conduit of healing and restoration. The axioms of beneficence (do good) and nonmaleficence (do no harm) stand as the twin pillars of this noble field. Assisted suicide, regardless of its legal status, had no place within the healthcare's sacred bounds.

Practitioner Euthymos, they decreed, had not only breached public trust and hospital regulations but also defied the foundational principles that upheld the healthcare profession. The gavel came down hard—Practitioner Euthymos was stripped of his healthcare license, and the specter of criminal charges loomed ominously.

> The tragic tapestry of events at Life's Embrace Hospital served as a stark reminder of the complexities shrouding the issue of assisted suicide. Although patient autonomy is paramount, the healthcare profession is obligated to adhere to the principles of beneficence and nonmaleficence, staying true to its mission as a healing profession. The narrative underscored the critical need for robust education among healthcare providers regarding the legal and ethical implications of practitioner-assisted suicide to ensure an unwavering commitment to patient-centered care.

* * *

THINK

Practitioner-assisted death or "medical aid in dying" is legal in 11 jurisdictions: California, Colorado, the District of Columbia, Hawaii, Montana, Maine, New Jersey, New Mexico, Oregon, Vermont, and Washington. State law states, "Actions taken in accordance with [the Act] shall not, for any purpose, constitute suicide, assisted suicide, mercy killing, or homicide under the law." https://en.wikipedia.org/wiki/Assisted_suicide_in_the_United_States

Practitioner-assisted death is defined differently from "assisted suicide," which is prohibited by state law in 48 states and the District of Columbia. However, it appears that practitioner-assisted suicide is the same as practitioner-assisted death if they are both defined as the practitioner providing the patient with the knowledge and means of ending their own life voluntarily.

Although state law appears to contradict itself concerning whether practitioner-assisted death is the same as practitioner-assisted suicide and whether or not it should be allowed, the healthcare profession is consistent. The American Medical Association (AMA) states, "Physician [practitioner]-assisted suicide is fundamentally incompatible with the physician's [practitioner's] role as healer, would be difficult or impossible to control, and would pose serious societal risks." The American College of Physicians (ACP) states, "The College does not support the legalization of physician [practitioner]-assisted suicide or euthanasia. After much consideration, the College concluded that making physician [practitioner]-assisted suicide legal raised serious ethical, clinical, and social concerns."

ASSESS

Patient: Autonomy

If an adult patient, with decisional capacity, considers the risks of harm and the benefits of assisted suicide, along with all the other available alternatives using their reasonable goals, values, and priorities, and the patient decides that their best interests would be maximized if they were to die, then it is reasonable and rational for the patient to ask their practitioner for assistance. It is also clear that civil justice has supported such patient efforts for end-of-life assistance through the legislation of death with dignity laws in several states.

However, just because patient autonomy authorizes a practitioner to do something or to provide a particular treatment, that does not mean that the healthcare profession as a whole needs to comply with the patient's autonomous choice if the choice is not in accordance with professional standards of care.

Practitioner: Beneficence and Nonmaleficence

It is rationally consistent for the healthcare profession to define itself as a healing profession, thereby prohibiting any healthcare options that violate that definition of coherency. Professionally, there has been near-universal agreement that practitioner participation in assisted suicide would go against healthcare tradition, values of beneficence (do good) and nonmaleficence (do no harm). The healthcare professional's moral principles outweigh the autonomous patient requests and the civil justice permissibility.

This is also the case for the professional prohibition of any healthcare activity that supports or even gives an appearance of support concerning capital punishment, interrogation, torture, or any other activities that directly or indirectly harm others.

Public Policy: Justice

Federal laws are fairly consistent because there is less chance of independent and conflicting legislation occurring simultaneously. The United States is a federal republic consisting of 50 states, one federal district (Washington, DC), five major territories (American Samoa, Guam, the Northern Mariana Islands, Puerto Rico, and the U.S. Virgin Islands), and various minor islands (loosely clustered around Hawaii).

Each state, federal district, and major territory has its own court system. These 56 independent court systems generate numerous inconsistencies and incoherence regarding what healthcare practices are permissible or impermissible.

Regardless of what can sometimes appear to be an inconsistent legislative mess, the healthcare profession and society at large understand that the healthcare profession is a healing art and needs to protect its social image from being maligned. Public trust is essential for a patient-centered nonmaleficence (do no harm) profession, and this means that just because an action is legally permissible, that does not determine how the healthcare profession defines itself by professional standards of practice.

Assisted suicide is not part of the healthcare profession's standards of care, regardless of whether or not it is legally permissible, because the art of healing is one of beneficence (do good) and nonmaleficence (do no harm): *Primum non nocere* (Latin for "First do no harm").

CONCLUDE

The healthcare profession prohibits practitioner-assisted suicide because

- the public's perception of the profession as a whole could be maligned;
- the patient's trust in the practitioner could be diminished; and
- it would violate the healthcare profession's foundational principles of beneficence (do good) and nonmaleficence (do no harm).

In summary, the prohibition of practitioner-assisted suicide within the healthcare profession stems from a need to preserve public trust, uphold the profession's image, and adhere to the foundational principles of beneficence (do good) and nonmaleficence (do no harm). Despite the legal permissibility of the practice in some areas, the healthcare profession stands firm in its commitment to healing and avoiding harm. This stance ensures that the healthcare profession remains patient-centered and maintains society's trust as it continues to prioritize the well-being and healing of patients. (See also Chapters 10, 15, 21, 31, and 62.)

* * *

REVIEW QUESTIONS

1. Which of the following reasons is *not* cited as a rationale for the healthcare profession's prohibition of practitioner-assisted suicide?
    A. The public's perception of the profession as a whole could be maligned.
    B. Patient trust in the practitioner could be diminished.
    C. It would be a violation of the healthcare profession's foundational principles of beneficence and nonmaleficence.
    D. It is legally permissible in some jurisdictions.

2. The AMA and the ACP both agree that practitioner-assisted suicide is compatible with the healthcare profession's duty of nonmaleficence (do no harm) and beneficence (do good).
   A. True
   B. False

3. If society authorizes the permissibility of a healthcare procedure or treatment, then the practitioner has a responsibility, as a result of the social contract, to provide that procedure or treatment to the patient.
   A. True
   B. False

4. It is rationally consistent for the healthcare profession to define itself as a healing profession and, therefore, prohibit any healthcare options that violate that definition for the sake of coherency.
   A. True
   B. False

5. Assisted suicide is not a standard of care, regardless of whether or not it is legally permissible, because the art of healing is one of beneficence (do good) and nonmaleficence (do no harm).
   A. True
   B. False

Answers: 1D, 2B, 3B, 4A, 5A

* * *

CLINICAL VIGNETTES

1. What is the position of the AMA and the ACP regarding practitioner-assisted suicide or euthanasia?
   A. They support the legalization of practitioner-assisted suicide or euthanasia.
   B. They have no official stance on practitioner-assisted suicide or euthanasia.
   C. They are neutral toward practitioner-assisted suicide or euthanasia.
   D. They are against the legalization of practitioner-assisted suicide and euthanasia.

   Explanation: Both the AMA and the ACP have taken a firm stance against the legalization of practitioner-assisted suicide or euthanasia, viewing them as incompatible with the practitioner's role as a healer and with the ethical principles of medicine. They argue that such practices can erode the public's perception of medicine being a healing profession and undermine the trust between the practitioner and the patient. Instead, the AMA and the ACP support providing high-quality end-of-life care that seeks to alleviate pain and suffering, respects the patient's dignity, and supports the patient and their loved ones. *Answer*: D

2. Mr. Philip Hall, a 62-year-old graphic designer, requests a specific treatment that is within their legal right to receive, but it goes against the healthcare profession's standards of healthcare. What should the practitioner do?
   A. Refuse to provide the requested treatment due to the practitioner's personal beliefs
   B. Provide the requested treatment regardless of the professional standards of healthcare
   C. Provide the requested treatment while documenting their objections
   D. Explain to the patient the healthcare profession's objections and then offer alternative treatments that align with standards of care

   Explanation: As a healthcare professional, the practitioner has a duty to provide care that is consistent with accepted standards of care. However, the patient has the right to make informed decisions about their healthcare, including the right to refuse or request specific treatments, as long as they are within their legal right to receive the treatments and the treatments are in line with standards of care. In this situation, the practitioner should explain to the patient the healthcare profession's objections to the requested treatment and provide information on alternative treatments that align with standards of care. The practitioner should also ensure that the patient understands the potential risks and benefits of the requested treatment and of the alternatives so that they can make an informed decision about their care. The practitioner should document the discussion in the patient's healthcare record and work collaboratively to develop a care plan consistent with the patient's reasonable goals, values, and priorities while meeting accepted healthcare standards. *Answer*: D

3. What is the stance of the healthcare profession regarding practitioner-assisted suicide?
   A. It is acceptable as long as it aligns with the patient's autonomous request.
   B. It is acceptable as long as it is permissible by civil law.
   C. It is morally unacceptable due to its contradiction with the healthcare profession being a healing profession.
   D. It is acceptable because the autonomous patient requests outweigh the professional moral principles.

   Explanation: The healthcare profession views itself as a healing profession, and its primary goal is to relieve suffering and promote the health and well-being of patients. As such, the healthcare profession considers practitioner-assisted suicide morally unacceptable because it goes against the practitioner's role as a healer and the ethical principles of medicine. The healthcare profession believes that the appropriate response to a patient's suffering is to provide high-quality end-of-life care that seeks to alleviate pain and suffering while respecting the patient's dignity. The healthcare profession's stance on practitioner-assisted suicide is based on the belief that preserving life and promoting well-being are central to the goals and values of medicine and that practitioner-assisted suicide undermines these fundamental principles. *Answer*: C

4. Mr. Seth Johnson, an 82-year-old retiree with no health issues, requests practitioner-assisted suicide, which is legally permissible in their state. What should the practitioner do?
   A. Provide the requested assistance because it is legally permissible
   B. Refuse to provide the requested assistance because it is antithetical to medicine being defined as a healing art.
   C. Provide the requested assistance because the patient has the autonomous right to choose their treatment option
   D. Provide the treatment because it is legal and the patient has provided an informed consent decision

   Explanation: The healthcare profession views practitioner-assisted suicide as antithetical to healthcare, defined as a healing art. Therefore, a practitioner asked to assist in a patient's suicide should refuse to provide assistance because it goes against healthcare's fundamental values and goals. *Answer*: B

5. Ms. Faith Park, a 92-year-old retiree, comes to your clinic seeking advice on end-of-life options. Ms. Park has been diagnosed with a terminal illness and is in severe pain. Ms. Park expresses interest in assisted suicide as a means to end their suffering. What is the appropriate response for the practitioner?
   A. Provide information on assisted suicide and help the patient access resources to facilitate the process
   B. Refuse to discuss the topic and refer the patient to another practitioner who may be more willing to assist
   C. Explain that assisted suicide is not a professional standard of care, would diminish public trust, and offer alternative options for managing pain and addressing end-of-life concerns
   D. Agree to assist the patient in ending their life as a means of relieving their suffering

Explanation: The practitioner should explain to the patient that assisted suicide is not a professional standard of care and goes against healthcare's fundamental values and goals. The practitioner should also discuss alternative options for managing pain and addressing end-of-life concerns, such as palliative care, hospice care, and other comfort-focused treatments. The practitioner should offer support and guidance to the patient as they navigate these options and work collaboratively to develop a care plan consistent with the patient's reasonable goals, values, and priorities. It is important for the practitioner to provide accurate information, support, and compassion to the patient while also ensuring that their ethical obligations are met. Refusing to discuss the topic or agreeing to assist the patient in ending their life are inappropriate responses because they do not align with the fundamental values and goals of healthcare and may compromise the practitioner's ethical obligations. *Answer*: C

REFLECTION VIGNETTES

1. Ms. Elizabeth Rodriguez, a 72-year-old retired schoolteacher, visited her practitioner with a unique request. She expressed her desire to end her life voluntarily and requested her practitioner's assistance. Ms. Rodriguez stated that she had a fulfilling life and was not suffering from mental illness or depression. She had a long-standing patient–practitioner relationship with her doctor, which she described as excellent. Ms. Rodriguez has decisional capacity and is fully aware of the implications of her request and states that she wants to end her life in a dignified manner.

   What is the ethical issue(s)?

   How should the ethical issue(s) be addressed?

2. Mr. John Smith, a 60-year-old retired accountant, visited his practitioner and expressed his desire for assisted suicide and argued that it would not be a violation of nonmaleficence (do no harm) because death is not harm but, rather, a transition to either a state of nonexistence or a state of transcendent existence, and either state fulfills the moral principle of beneficence (do good). Mr. Smith communicates that he has been diagnosed with terminal cancer and is experiencing intolerable pain and suffering. Mr. Smith states that his quality of life has significantly deteriorated and that he no longer has any hope of recovery and wishes for assisted suicide.

   What is the ethical issue(s)?

   How should the ethical issue(s) be addressed?

* * *

WEB LINKS

1. *Code of Medical Ethics.* American Medical Association. https://code-medical-ethics.ama-assn.org
   The AMA *Code of Medical Ethics*, established by the American Medical Association, is a comprehensive guide for healthcare practitioners. It addresses issues and challenges, promotes adherence to standards of care, and is continuously updated to reflect contemporary practices and challenges.
2. "Choosing Death Over Suffering." Amber Comer. 2019. https://bioethics.hms.harvard.edu/journal/choosing-death
   This source discusses the ethical standard of discussing physician aid in dying during healthcare decision-making, emphasizing that physicians are not ethically obligated to assist a patient in ending their life.

3. *Code of Medical Ethics.* American Medical Association. https://code-medical-ethics.ama-assn.org/sites/default/files/2022-08/5.7%20Physician-assisted%20suicide%20--%20background%20reports.pdf
   This source provides an overview of physician-assisted suicide, including the process and ethical considerations.
4. "Physician-Assisted Suicide." American Medical Association. https://code-medical-ethics.ama-assn.org/ethics-opinions/physician-assisted-suicide
   This source provides an overview of physician-assisted suicide, including the process and ethical considerations.
5. "Euthanasia and Assisted Suicide: An In-Depth Review of Relevant Historical Aspects." Yelson Alejandro Picón-Jaimes, Ivan David Lozada-Martinez, Javier Esteban Orozco-Chinome, et al. 2022. https://www.ncbi.nlm.nih.gov/pmc/articles/PMC8857436
   This article discusses euthanasia and assisted suicide in the context of life as a human right, highlighting the social, moral, and ethical conflicts.
6. "Mortal Responsibilities: Bioethics and Medical-Assisted Dying." Courtney S. Campbell. 2019. https://www.ncbi.nlm.nih.gov/pmc/articles/PMC6913808
   This article discusses the role of patient agency in physician-assisted death, distinguishing this practice from medical-assisted dying provided by a healthcare professional.
7. "Ethical Considerations at the End-of-Life Care." Melahat Akdeniz, Bülent Yardımcı, and Ethem Kavukcu. 2021. https://pubmed.ncbi.nlm.nih.gov/33786182
   This article discusses the ethical difficulties for healthcare professionals in end-of-life care decisions, including euthanasia and physician-assisted suicide.
8. "Autonomy, Voluntariness and Assisted Dying." Ben Colburn. 2020. https://jme.bmj.com/content/46/5/316
   This article discusses the role of patient autonomy in the context of assisted dying, presenting an argument for legalizing assisted dying.

CAPITAL PUNISHMENT— EXECUTIONS

10

In nothing do practitioners more nearly approach the gods than in giving health.
—Marcus Tullius Cicero

* * *

MYSTERY STORY

LIFE AND DEATH IN THE GRAY ZONE: PROBING THE HEALTHCARE PROFESSION'S INVOLVEMENT IN CAPITAL PUNISHMENT

An eerie stillness gripped the autopsy room at Justice Scales Medical Center. Detective Andrew Garcia penetrated its sterile silence, his scrutiny falling on Practitioner Adrienne Mortis. Her gloved hands were engrossed in a somber dance, charting the chilling path of life's departure in the body of an executed prisoner. A life extinguished by the relentless flow of lethal injection. As Detective Garcia sought insights into the complexities of death, the room's chilling tableau unfolded its grim narrative.

Practitioner Mortis, an orchestra of precision and clinical elucidation, guided Detective Garcia through the twisted intricacies of the execution. Amidst the stark, detached specifics, Detective Garcia's thoughts spiraled toward a contentious paradox—the healthcare profession's standpoint on the volatile issue of capital punishment.

Capital punishment, the menacing specter of the death penalty, casts its ominous shadow over the legal terrains of 24 states, American Samoa, the federal government, and the military. However, it is an unsettling anomaly within the healthcare ethos, a practice deemed unethical when healthcare practitioners are called to facilitate its execution.

Authorities within the healthcare domain, notably the American Medical Association (AMA) and the American College of Physicians (ACP), staunchly decry practitioner involvement in executions. The roles they might be drawn into—determining competence, delivering the lethal amalgam of drugs, and monitoring the fading lifelines—ignite a fierce conflict with the profession's guiding lights of beneficence (do good) and nonmaleficence (do no harm). These actions starkly contrast with the profession's emphasis on patient-centered care and the amplification of patients' best interests within the healing sphere.

Beyond the guiding beacons of beneficence and nonmaleficence, the issue of patient autonomy (informed consent), a fundamental pillar of the healthcare profession, is unequivocally breached within the confines of capital punishment. The absence of a prisoner's choice in this irrevocable verdict veers sharply away from the professional mandate of prioritizing patient health and well-being over political ideologies of retribution.

As Detective Garcia absorbed Practitioner Mortis' revelations, he wrestled with a burgeoning epiphany. The potential engagement of healthcare practitioners in executions was not merely an ethical labyrinth—it was a precarious balancing act, teetering on the edge of their sacred duty to prioritize patient-centered care and uphold the "do no harm" mantra. The risks, he realized, were ominously palpable despite the legal permission surrounding capital punishment.

Exiting the Justice Scales Medical Center's autopsy room, the echoes of Practitioner Mortis' discourse reverberated within Detective Garcia. His resolve was fortified—he knew the healthcare profession's foundational values demanded staunch guardianship against

> the tides of legal and political pressures. He pledged to spotlight the issue of practitioner involvement in capital punishment to the appropriate authorities. After all, the principles of the healthcare profession must persist in their resonant echoes, uncompromised and relentless, amidst the cacophony of legal controversy and societal law.

* * *

THINK

Capital punishment or the death penalty is a government-sanctioned punishment for committing a capital offense. Etymologically, the word "capital" comes from the Latin word *caput*, which means head, and was used to describe an execution by beheading. Historically, executions have been carried out by hanging, shooting, lethal injection, stoning, electrocution, and gassing.

Twenty-three states and Washington, DC, have abolished the death penalty, and three states have a governor-imposed moratorium. Virginia was the 23rd state to abolish the death penalty as of March 2, 2022. As a result, there are currently only 24 states in which the death penalty is still legal.

Capital punishment is not coherent with the patient–practitioner relationship. The patient-centered purpose and function of the social contract are to maximize the patient's best interests as determined by the patient's reasonable goals, values, and priorities. For a practitioner to participate in capital punishment would contradict the social contract, potentially damaging the social trust in the healthcare profession as a healing art.

ASSESS

Patient: Autonomy

Patient healthcare decision-making focuses on the principle of autonomy and acquiring informed consent. Capital punishment has no process or interest in getting the prisoner's informed consent authorization to be executed. Because there is no prisoner choice, there is no disclosure to the prisoner of risks and benefits, and there is no rational assessment for determining what will maximize the patient's or, in this case, the prisoner's best interests; therefore, it logically follows that capital punishment has no regard for the principle of patient autonomy.

Practitioner: Beneficence and Nonmaleficence

Patient-centered evidence-based care for health and wellness is how the art of healing is defined and understood by the healthcare profession and the community. Capital punishment and executions, even if legal, are an incontrovertible contradiction with

- the art of healing's goals and objectives of health and wellness; and
- the professional principles of beneficence (do good) and nonmaleficence (do no harm).

The ACP prohibits any practitioner engagement in cruel or unusual punishment, such as capital punishment and other disciplinary activities beyond those permitted by the United Nations Standard Minimum Rules for the Treatment of Prisoners.

The AMA states, "A physician [practitioner] must not participate in a legally authorized execution." And the ACP states, "Participation by physicians [practitioners] in the execution of prisoners except to certify death is unethical."

But even in the certification of death, the AMA clarifies this statement by stating that the certification of death can only occur if the condemned prisoner has already been declared dead by another person.

The AMA defines practitioner participation in executions to include the following prohibitory categories:

1. Causing death
2. Assist, supervise, or contribute in any way toward the ability of another to cause death
3. Automatic causation of death

The AMA's list of forbidden practices includes, but is not limited to, the following:

1. Determining the competence of a person for execution
2. Treating an incompetent person to make the person competent enough to be executed
3. Administering any drugs for the use and purpose of execution
4. Monitoring vital signs
5. Attending or even observing an execution
6. Providing technical advice
7. Selecting injection sites or starting an intravenous line
8. Prescribing, preparing, administering, or supervising injection of drugs
9. Inspecting or maintaining lethal injection devices
10. Consulting or supervising personnel who perform executions

Although the condemned are under a legal court order to die, it is still unethical and unprofessional for any practitioner to participate or even be associated with an execution. This is because the healthcare profession's foundational principles of beneficence (do good) and nonmaleficence (do no harm) would be violated or contradicted by any form of participation in an execution.

Public Policy: Justice

It is essential to understand that just because "something" is illegal, that does not necessarily make that action immoral or unprofessional. Similarly, just because "something" is legal, that does not necessarily make that action moral or professional. Capital punishment is in this latter category. Although it is a legal penalty in 24 states, American Samoa, the federal government, and the military, that legality in and of itself is not a mandate for practitioners or the healthcare profession to be federal or state henchmen.

Healthcare's professional ideology as a healing art is not redefined by a political ideology of retribution. It is imperative that everyone in society can always trust that their practitioners will only provide appropriate healthcare for the benefit of the sick and that the practitioner's focus will be patient-centered for the maximization of the patient's best interests as determined by the patient's reasonable goals, values, and priorities for healing.

CONCLUDE

Participation in capital punishment or executions is a fundamental contradiction with the healthcare profession being a healing art and a violation of the professional principle of nonmaleficence (do no harm). Under no circumstances is a practitioner to participate in or even be associated with an execution. Social trust, reputation, and professional obligations are paramount for promoting the art of healing.

In summary, the healthcare profession's participation in capital punishment or executions is a fundamental contradiction to its role as a healing art and a violation of its principles of beneficence (do good) and nonmaleficence (do no harm). Practitioners are strictly prohibited from participating in or even being associated with executions. Upholding social trust, reputation, and professional obligations is crucial in promoting the healthcare profession's primary objective—the healing and well-being of patients. (See also Chapters 9, 15, 21, 31, and 62.)

* * *

REVIEW QUESTIONS

1. If a practitioner is practicing medicine in a state in which capital punishment is mandated and if capital punishment does not violate the practitioner's conscience, then it is permissible for the practitioner to participate in legal and humane executions.
    A. True
    B. False

2. Which of the following activities is *not* mentioned as a forbidden practice for practitioners in relation to capital punishment?
   A. Administering drugs for the purpose of execution
   B. Monitoring vital signs during an execution
   C. Observing an execution
   D. Providing healthcare to a prisoner on death row

3. Although the condemned are under a legal court order to die, it is still unethical and unprofessional for any practitioner to participate or even be associated with an execution.
   A. True
   B. False

4. It is imperative that everyone in society can always trust that their practitioners will only provide appropriate healthcare for the benefit of the sick and that the practitioner's focus will be patient-centered for the maximization of the patient's best interests as determined by the patient's reasonable goals, values, and priorities.
   A. True
   B. False

5. Under no circumstances is a practitioner to participate in or even be associated with an execution. Social trust, reputation, and professional obligations are paramount for promoting the art of healing.
   A. True
   B. False

Answers: 1B, 2D, 3A, 4A, 5A

* * *

CLINICAL VIGNETTES

1. A practitioner is practicing medicine in a state in which capital punishment is mandated. Although capital punishment does not violate the practitioner's conscience, it is still impermissible for the practitioner to participate in legal executions because (a) capital punishment is not compatible with the patient–practitioner relationship; (b) capital punishment is not compatible with the patient's best interests as determined by the patient's reasonable goals, values, and priorities; and (c) capital punishment is not compatible with the social contract of the healthcare profession being a healing art. Which of the following is true:
   A. Capital punishment is not antithetical to the patient–practitioner relationship.
   B. Capital punishment is not antithetical to the patient's best interests.
   C. Capital punishment is not antithetical to the social contract of medicine being a healing art.
   D. Capital punishment is incompatible with all four principles of bioethics.

   Explanation: The four principles of bioethics are autonomy (informed consent), beneficence (do good), nonmaleficence (do no harm), and justice (be fair); capital punishment is incompatible with all four principles. Capital punishment is incompatible with the principle of autonomy because it denies the condemned individual the right to decide about their own life and death. It is not compatible with the principle of beneficence because it does not promote the condemned individual's best interests. It is not compatible with the principle of nonmaleficence because it intentionally causes harm, suffering, and death. It is incompatible with the principle of justice because it disproportionately affects marginalized and vulnerable populations and does not contribute to the overall well-being of society. Therefore, a practitioner practicing medicine in a state in which capital punishment is mandated is faced with an ethical dilemma because they must balance their legal and ethical obligations as a practitioner. Although capital punishment may not violate the practitioner's conscience, it is still incompatible with the fundamental values and goals of the healthcare profession. *Answer*: D

2. Mr. Blake Shah, a 38-year-old prisoner, is about to undergo a state-sanctioned execution. The practitioner is asked to be involved in some capacity. Which of the following actions is *not* prohibited by the AMA?
   A. Determining competence of the person to be executed
   B. Prescribing or administering drugs for execution
   C. Monitoring vital signs or observing an execution
   D. Providing technical advice or supervising personnel for executions
   E. Promotion of legislation that advances medicine as a healing art

   Explanation: Any participation or appearance of approval of capital punishment is prohibited by the AMA and the ACP, as well as the healthcare profession as a whole. Competency is a legal definition, not a healthcare determination. Practitioners determine decisional capacity, not competency. The AMA prohibits prescribing or administering a lethal drug. Assisting with an execution is prohibited because it violates the principle of nonmaleficence and goes against the fundamental goals of medicine. Monitoring vital signs or observing an execution is prohibited because it involves a practitioner participating in a nonmedical procedure incompatible with the patient–practitioner relationship. Providing technical advice or supervising is also prohibited because it involves a practitioner assisting in a procedure incompatible with the professional obligations of medicine. Therefore, the only correct answer is one not related to the ethical dilemma of practitioners' involvement in state-sanctioned executions, which in this case is the promotion of medicine as a healing art. *Answer*: E

3. Mr. Nathan Perez, a 52-year-old prisoner, is a death row inmate who has exhausted all legal appeals and is scheduled to be executed in 2 weeks. The prison warden approached a practitioner to administer the lethal injection for the execution. What should the practitioner do?
   A. Agree to administer the lethal injection because they are obligated to follow the law
   B. Refuse to participate in the execution because it goes against their professional and ethical responsibilities as a practitioner
   C. Seek advice from a bioethics committee to determine the best course of action
   D. Find a different practitioner who is willing to administer the lethal injection

   Explanation: Administering the lethal injection goes against the fundamental principles of medicine and the ethical responsibilities of practitioners. Practitioners have a duty to prioritize their patients' well-being, do no harm, and adhere to professional standards of care. Administering the lethal injection for an execution violates these principles because it intentionally causes harm and suffering to a patient and is not a medically necessary or beneficial procedure. The AMA and other professional healthcare organizations have specifically prohibited practitioners from participating in executions, recognizing that it violates the patient–practitioner relationship and undermines public trust in the healthcare profession. Therefore, in this scenario, the practitioner should refuse to participate in the execution and inform the prison warden of their ethical responsibilities and professional obligations as a practitioner. *Answer*: B

4. Mr. Marcus Evans, a 28-year-old prisoner, presents to the clinic with chronic pain and discomfort history. The patient's healthcare history reveals that the patient has been convicted of a crime and is facing the possibility of capital punishment. Despite their current situation, the patient's main concern is finding relief from pain and discomfort. Which of the following best reflects the practitioner's role?
   A. The practitioner should prioritize the patient's punishment and ignore their healthcare needs.

B. The practitioner should provide standards of care based on the patient's best interests as determined by the patient's reasonable goals, values, and priorities, regardless of their criminal history.
C. The practitioner should prioritize their own political beliefs over the patient's healthcare needs.
D. The practitioner should deny healthcare treatment to the patient due to their criminal history of crimes against society.

Explanation: Practitioners have a fundamental duty to provide healthcare based on the patient's best interests as determined by the patient's reasonable goals, values, and priorities. The fact that the patient is a prisoner facing the possibility of capital punishment does not change this fundamental obligation. Practitioners are bound by professional and ethical standards that prioritize the well-being of the patient, and the AMA specifically states that practitioners must provide healthcare without discrimination or prejudice based on factors such as race, gender, or criminal history. Therefore, the practitioner's role in this scenario is to provide the patient with the standard of healthcare they would provide to any other patient experiencing chronic pain and discomfort. The practitioner should not prioritize the patient's punishment, deny treatment, or allow their political beliefs to influence their healthcare decision-making. Instead, the practitioner should prioritize the patient's health and well-being and work with them to determine the best course of treatment based on their healthcare needs and values. *Answer*: B

5. Ms. Olivia Turner, a 54-year-old prisoner, is scheduled for lethal injection as capital punishment. The patient's practitioner is asked to assist in the execution. The practitioner recognizes that participating in the execution would go against the healthcare profession's definition of a healing art, contradict the professional principles of beneficence and nonmaleficence, and socially weaken the community's trust in the healthcare profession. What should the practitioner do in this situation?
   A. Participate in the execution as requested
   B. Refuse to participate in the execution but remain associated with it
   C. Refuse to participate in the execution and disassociate themself from it
   D. Agree to participate in the execution, but only if it is in the best interest of the patient

Explanation: Participating in a lethal injection as capital punishment would violate the healthcare profession's core principles and ethical responsibilities. As a healthcare professional, the practitioner's primary duty is to promote the patient's health and well-being, and participating in the execution goes against this obligation. The AMA has clarified that practitioners should not participate in executions because it undermines the patient–practitioner relationship and weakens public trust in the healthcare profession. Therefore, the practitioner should refuse to participate in the execution and disassociate themself from it entirely. Refusing to participate in the execution sends a message that the healthcare profession does not condone such practices, and it reinforces the fundamental principles of medicine. *Answer*: C

REFLECTIONS VIGNETTES

1. Ms. Sarah Thompson, a 40-year-old practitioner, received a request from the state to provide instructions to correctional personnel on how to start an intravenous drip for legally authorized executions. Ms. Thompson has a critical care background and has provided intravenous therapy instruction to healthcare professionals. Ms. Thompson is politically in favor of the death penalty and would like to assist in the instruction.

   What is the ethical issue(s)?

   How should the ethical issue(s) be addressed?

2. Ms. Rachel Lee, a 35-year-old registered practitioner, received a request from the state to monitor prisoners' vital signs during executions, provide comfort, and ensure that the prisoners experience minimal suffering while dying. The request has put Ms. Lee in a difficult position because she has ethical concerns about participating in executions. Ms. Lee knows that executions are controversial, with different opinions on their use. She understands that her role as a healthcare professional is to prioritize the patient's well-being and adhere to ethical principles.

   What is the ethical issue(s)?

   How should the ethical issue(s) be addressed?

* * *

WEB LINKS

1. *Code of Medical Ethics.* American Medical Association. https://code-medical-ethics.ama-assn.org
   The AMA *Code of Medical Ethics*, established by the American Medical Association, is a comprehensive guide for healthcare practitioners. It addresses issues and challenges, promotes adherence to standards of care, and is continuously updated to reflect contemporary practices and challenges.
2. "The Prisoner's Dilemma: The Role of Medical Professionals in Executions." L. Elisabeth Armstrong. 2021. https://blogs.bmj.com/medical-ethics/2021/09/07/the-prisoners-dilemma-the-role-of-medical-professionals-in-executions
   This source discusses the role of practitioners in executions and the ethical implications of leveraging healthcare education and credentials to punish or harm.
3. "The Death Penalty: A Breach of Human Rights and Ethics of Care." *The Lancet.* 2023. https://www.thelancet.com/journals/lancet/article/PIIS0140-6736(23)01004-8/fulltext
   This article discusses the death penalty as a relic from the past that should be shed in the 21st century, with its history seeped in politics and discrimination.
4. "Capital Punishment." American Medical Association. https://code-medical-ethics.ama-assn.org/ethics-opinions/capital-punishment
   This source overviews the AMA's stance on capital punishment, emphasizing that a physician must not participate in a legally authorized execution.
5. "AMA to Supreme Court: Doctor Participation in Executions Unethical." American Medical Association. 2018. https://www.ama-assn.org/delivering-care/ethics/ama-supreme-court-doctor-participation-executions-unethical
   This source discusses the AMA's stance that physicians should not participate in state-ordered executions, emphasizing their role as healers rather than executioners.
6. "AMA *Code of Medical Ethics*' Opinions Related to Health Care for Incarcerated People." Annalise Norling. 2017. https://journalofethics.ama-assn.org/article/ama-code-medical-ethics-opinions-related-health-care-incarcerated-people/2017-09
   This source discusses the AMA's ethical opinions related to healthcare for incarcerated people, including the prohibition of physician participation in executions.
7. "Should Physicians Participate in State-Ordered Executions?" James K. Boehnlein. 2013. https://journalofethics.ama-assn.org/article/should-physicians-participate-state-ordered-executions/2013-03
   This source discusses the ethical dilemma of physician involvement in state-ordered executions.

CHAPERONES AND PERSONAL PRIVACY

11

A practitioner who is a lover of wisdom is equal to a god.
—Hippocrates

* * *

MYSTERY STORY

PRIVACY BETRAYAL: UNRAVELING THE INTERSECTION OF CHAPERONES AND PERSONAL PRIVACY IN THE HEALTHCARE SYSTEM

In the hushed solace of her office, Detective Katherine Hawthorne found herself riveted to a chilling file—the narrative of a murder case whose grisly details resonated with an unsettling echo. A life ruthlessly strangled within the fortress of their home, the victim was found to be connected to Practitioner Benjamin Foster, a stalwart in the local healthcare community. Resolving to shed light on this gloomy mystery, she decided to pay the practitioner a visit.

The sterile ambiance of Guardian Angel Health Facility greeted Detective Hawthorne. She introduced herself to Practitioner Foster, unraveled the grim reason for her visit, and plunged into her inquiries. Practitioner Foster, visage ashen at the mention of his patient's demise, offered his cooperation. He confirmed having seen the victim a day prior to the death for a routine physical examination.

Steering the conversation toward the protocol of employing a chaperone during examinations, Practitioner Foster detailed his meticulous practice of using a chaperone during sensitive procedures, such as pelvic exams, as a means to uphold patient comfort and dignity. During the victim's last visit, Practitioner Foster had appointed Sarah O'Neil, a nurse's aide with years of collaborative experience with him, as the chaperone.

As Detective Hawthorne probed into the privacy protocols during the examination, Practitioner Foster's demeanor wavered. Pressed by her relentless questioning, he conceded to having faltered—failing to provide the victim with private space to undress and neglecting to maintain appropriate coverage during the exam.

Detective Hawthorne, perturbed by his confessions, questioned Practitioner Foster about handling confidential patient information during the exam. He admitted to discussing the victim's healthcare history with Ms. O'Neil, the chaperone, in the open and to not offering the patient a private discussion for sensitive information.

Spurred by this revelation, Detective Hawthorne decided to engage Ms. O'Neil next. Her observations of Ms. O'Neil's anxious demeanor and evasive gaze were soon confirmed: Ms. O'Neil confessed to harboring a grudge against the victim. She admitted to abusing her position, leveraging overheard confidential information to confront and subsequently murder the victim.

As Detective Hawthorne exited the clinic sterility, she carried a crystallized understanding of the sequence leading to the victim's demise. Ms. O'Neil was subsequently arrested and charged with murder. Practitioner Foster was held accountable for violating HIPAA and patient privacy laws and his healthcare license was revoked.

Reflecting on the events as she departed from the Guardian Angel Health Facility, Detective Hawthorne was starkly reminded of the indispensable nature of chaperones and personal privacy in healthcare delivery. The haunting question persisted: If only Practitioner Foster had strictly adhered to these paramount principles of the healthcare profession, could a life have been spared from such a gruesome fate?

* * *

THINK

Chaperones can significantly aid in the comfort of the patient and the practitioner under various circumstances and conditions. This respect for the patient's dignity, through the use of a chaperone, is in line with the professional moral duty of beneficence (do good).

ASSESS

Patient: Autonomy

Clear communication should be made to the patient that they may choose to have a family member chaperone or a non-family member chaperone present during their examination. Any institutional or legal requirements for using a chaperone should also be disclosed.

Practitioner: Beneficence and Nonmaleficence

Based on the professional principle of beneficence (do good), private facilities should be provided to the patient for undressing and putting on a healthcare gown. Appropriate covering and drapery should also be used during the healthcare examination. At no time should private areas of the patient's body ever be exposed to family members or the chaperone because that could be a violation of the patient's dignity, could seriously disrupt the family social structure, and, as such, could be a violation of the professional principle of nonmaleficence (do no harm).

The American College of Physicians states, "The location and degree of privacy should be appropriate for the examination being performed, with chaperone services as an option."

Public Policy: Justice

All discussions of confidential patient information, such as inquiries into the patient's protected health information (PHI), healthcare history, and other sensitive information, must be kept to a minimum while the chaperone or family member(s) is present. Failure to do so could result in Health Insurance Portability and Accountability Act (HIPAA) confidentiality violations and a violation of the principle of nonmaleficence (do no harm). HIPAA is a federal statute with fines of up to $250,000 and a jail term of up to 5 years.

Communication between the patient and the practitioner is highly privileged, a central component of the patient–practitioner relationship, and is legally protected. Therefore, the practitioner must always provide the patient with the opportunity for a private conversation in which all confidential PHI can be freely discussed. Such practices align with all four bioethical principles: patient autonomy (informed consent), practitioner beneficence (do good), practitioner nonmaleficence (do no harm), and public justice (be fair).

CONCLUDE

Chaperones and personal privacy are essential to the healthcare culture and healthcare delivery. Not violating any of the HIPAA confidentiality requirements and recognizing the importance of the opportunity for the patient to be able to have a private conversation with the practitioner, without any family members or chaperone being present, are imperative so that the patient can have a free and open discussion about their personal and healthcare history with the practitioner.

In summary, chaperones and personal privacy are integral components of healthcare delivery. Adhering to HIPAA confidentiality requirements and providing opportunities for private conversations between patients and practitioners without the presence of chaperones or family members are critical for fostering open and honest discussions about personal lifestyles and healthcare history. Respecting patient autonomy and upholding the principles of beneficence, nonmaleficence, and justice are essential for maintaining trust and ensuring the best possible patient care. (See also Chapters 12 and 35.)

* * *

REVIEW QUESTIONS

1. Chaperones can significantly aid in the comfort of both the patient and the practitioner under various circumstances and conditions. This respect for the patient's dignity, through the use of a chaperone, is in line with the professional moral duty of beneficence (do good).
   A. True
   B. False

2. Because a chaperone is a family or non-family member, it is permissible for the chaperone to see the patient's exposed private areas, ensuring no impropriety occurs.
   A. True
   B. False

3. Which of the following is *not* an appropriate practice during a healthcare examination with a chaperone present?
   A. Providing private facilities for the patient to undress
   B. Discussing confidential patient information in front of the chaperone
   C. Allowing the patient to choose a family or non-family member chaperone
   D. Ensuring appropriate covering and drapery during the examination

4. A chaperone is always legally authorized to hear discussions regarding the patient's PHI.
   A. True
   B. False

5. The practitioner is protected from HIPAA violations when using chaperones.
   A. True
   B. False

Answers: 1A, 2B, 3B, 4B, 5B

* * *

CLINICAL VIGNETTES

1. Ms. Kaitlyn Clark, a 23-year-old salesperson who identifies as nonbinary, is coming in for a routine physical examination. They have expressed concerns about privacy and modesty during the examination. As a practitioner, it is important to ensure the comfort and dignity of all patients during healthcare procedures. Which of the following is the most appropriate action for the practitioner to take in this situation?
   A. Proceed with the examination without a chaperone
   B. Offer the patient the option of having a chaperone present during the examination
   C. Require the patient to have a chaperone present during the examination

   Explanation: Offering the patient the option of having a chaperone present during the examination is the most appropriate action for the practitioner to take in this situation. It is important for practitioners to respect the patient's privacy and modesty concerns and to provide options that allow the patient to feel comfortable and respected during the examination. Requiring the patient to have a chaperone present during the examination may not be necessary or further contribute to the patient's discomfort and may not respect the patient's concerns for privacy and modesty. Using a chaperone is in line with the professional moral duty of beneficence (do good) and can significantly aid in the comfort of the patient and the practitioner. *Answer*: B

2. Ms. Kimberly Chen, a 21-year-old college student, has come in for a healthcare examination and requested that a chaperone be present during the procedure. The practitioner follows the professional principle of beneficence (do good) by providing private facilities for the patient to undress and put on a healthcare gown and by using appropriate covering and drapery during

the examination. However, during the examination, the practitioner realizes that the chaperone has scooted their seat to a vantage point that allows them to see the patient's private areas. What is the most appropriate course of action for the practitioner to take to maintain the patient's dignity and abide by the professional principle of nonmaleficence (do no harm)?
A. Continue the examination
B. Stop the examination and provide additional privacy measures
C. Explain the situation to the patient and proceed with the examination

Explanation: The most appropriate course of action for the practitioner to take to maintain the patient's dignity and abide by the professional principle of nonmaleficence is to stop the examination and provide additional privacy measures. The practitioner must protect the patient's privacy and dignity, and any perceived violation must be addressed immediately. Creating a safe and secure patient environment and respecting their autonomy and preferences are essential. Therefore, the practitioner should explain the situation to the patient and offer them the option to continue the examination with a different chaperone or without a chaperone, as per their preference. The importance of using chaperones in healthcare examinations to aid comfort and respect for the patient's dignity should not be underestimated. The practitioner should also take necessary steps to address the issue with the chaperone to prevent similar incidents in the future. *Answer*: B

3. Ms. Carmen Hernandez, a 32-year-old human resources professional, comes to the clinic for a routine checkup in an open examination room with only thin curtains separating patients. During the examination, the practitioner notices that the patient seems uncomfortable discussing certain healthcare information. The practitioner recognizes the importance of protecting patient privacy and the privileged communication between the patient and practitioner. What is the best course of action for the practitioner to take?
   A. Insist on discussing the healthcare information in the open examination room
   B. Offer to discuss the healthcare information in a private room
   C. Refuse to discuss the healthcare information privately, as everyone else is in the same situation
   D. Offer to discuss the healthcare information over the phone

   Explanation: The professional principles of beneficence and nonmaleficence prioritize the patient's privacy and comfort during the examination. Insisting on discussing healthcare information in the open examination room could be a violation of the patient's privacy and could make them feel uncomfortable, which would be against the professional principle of nonmaleficence. Refusing to discuss the healthcare information privately or offering to discuss it over the phone could also be problematic because they would not provide the patient with adequate privacy, which is necessary for the patient–practitioner relationship. Thus, offering to discuss the healthcare information in a private room is the most appropriate action for the practitioner to take in this situation. *Answer*: B

4. Ms. Zoe Lee, a 33-year-old retail worker, comes to the clinic for a routine checkup accompanied by their spouse. During the examination, the practitioner must discuss confidential healthcare information with the patient, including sensitive information about their healthcare history and current health status. What is the best course for the practitioner to avoid HIPAA confidentiality violations and protect patient privacy?
   A. Discontinue the examination and reschedule for a private appointment
   B. Discuss the confidential information in front of the patient's spouse
   C. Ask the patient's spouse to leave the examination room during the discussion
   D. Refuse to discuss the confidential information

   Explanation: To maintain patient privacy and confidentiality, practitioners should ensure that only necessary people can access the patient's confidential healthcare information. In this case, the patient's spouse is not required to be present during the discussion of sensitive healthcare information, and the practitioner should ask them to leave the examination room. By doing so, the practitioner would protect the patient's privacy, maintain confidentiality, and abide by the professional principle of nonmaleficence. Discontinuing the examination or refusing to discuss the information altogether would not necessarily address the issue, whereas discussing the information in front of the spouse would violate the patient's privacy and confidentiality. *Answer*: C

5. Mr. Jeremiah Adams, a 28-year-old architect, is undergoing a healthcare examination with a practitioner. The patient's chaperone or family member(s) is in the room during the examination. Which of the following is the most appropriate way for the practitioner to handle confidential patient information during the examination?
    A. Share all details of the patient's PHI, healthcare history, and other sensitive information with the chaperone or family member(s) without any restrictions
    B. Confidential patient information (PHI) may be necessary to share with the chaperone when conducting the examination
    C. Avoid discussing PHI in the presence of the chaperone or family member(s) when conducting the examination
    D. Share all confidential patient information with the chaperone or family member(s) with no restrictions, as they are present during the examination

Explanation: The practitioner should respect the patient's privacy and maintain the confidentiality of PHI during the examination. Although the chaperone or family member(s) may be present in the examination room, the practitioner should still maintain confidentiality of the patient's PHI during the examination. By doing so, the practitioner is upholding the principle of beneficence by respecting the patient's dignity, promoting trust in the patient–practitioner relationship, and following professional and legal standards for protecting patient privacy. *Answer*: C

REFLECTION VIGNETTES

1. Mr. Michael Johnson, a 27-year-old accountant, agreed to have a chaperone present during his routine well-patient visit. During the checkup, the practitioner noticed some concerning symptoms that may indicate that Mr. Johnson has contracted a sexually transmitted disease. The practitioner knows that discussing personal questions about sexual activity can be uncomfortable for some patients, but diagnosing and treating any potential health issues are essential. The differential diagnosis for Mr. Johnson's symptoms includes various sexually transmitted diseases, such as chlamydia, gonorrhea, and syphilis. The practitioner must obtain a thorough healthcare history from Mr. Johnson and ask sensitive questions about his sexual history to determine the appropriate course of treatment.

    What is the ethical issue(s)?

    How should the ethical issue(s) be addressed?

2. Ms. Jennifer Lee, a 42-year-old elementary schoolteacher, agreed to have her spouse act as a chaperone during her routine well-patient visit. During the checkup, the practitioner noticed old and new bruises on Ms. Lee's arms and legs. When the practitioner inquired about the bruises, Ms. Lee's spouse interjected and explained that Ms. Lee is a bit clumsy and prone to falls. The practitioner observed that Ms. Lee did not respond or add anything to the explanation. The differential diagnosis for Ms. Lee's symptoms includes various possible causes for the bruises, such as accidental falls, physical abuse, or self-harm. The practitioner must obtain a thorough healthcare history from Ms. Lee and ask sensitive questions about her home life to determine the cause of the bruises.

    What is the ethical issue(s)?

    How should the ethical issue(s) be addressed?

* * *

WEB LINKS

1. *Code of Medical Ethics.* American Medical Association. https://code-medical-ethics.ama-assn.org

   The AMA *Code of Medical Ethics*, established by the American Medical Association, is a comprehensive guide for healthcare practitioners. It addresses issues and challenges, promotes adherence to standards of care, and is continuously updated to reflect contemporary practices and challenges.

2. "1.2.4 Use of Chaperones." American Medical Association. https://code-medical-ethics.ama-assn.org/sites/default/files/2022-08/1.2.4%20Use%20of%20chaperones%20--%20background%20reports_0.pdf

   This source discusses the benefits of using a chaperone during the physical examination of a patient, including providing reassurance of the professional character of the exam.

3. "Use of Chaperones." American Medical Association. https://code-medical-ethics.ama-assn.org/ethics-opinions/use-chaperones

   This source provides an overview of chaperones, emphasizing the importance of creating a comfortable and considerate atmosphere for the patient and the physician.

4. "The Case for Medical Chaperones." Allen L. Pimienta and Rachel E. Giblon. 2018. https://www.aafp.org/pubs/fpm/issues/2018/0900/p6.html

   This source discusses the American Medical Association Code of Ethics recommendations on notifying patients about chaperone guidelines and always honoring a patient's request to have a chaperone.

5. "Federal Privacy Protections: Ethical Foundations, Sources of Confusion in Clinical Medicine, and Controversies in Biomedical Research." Mary Anderlik Majumder and Christi J. Guerrini. 2016. https://journalofethics.ama-assn.org/article/federal-privacy-protections-ethical-foundations-sources-confusion-clinical-medicine-and/2016-03

   This source discusses the harms that can result from the disclosure of patients' private information, including economic harm and social harm.

6. "The Use of Chaperones During Sensitive Examinations and Procedures." https://www.uofmhealth.org/patient-visitor-guide/patients/use-chaperones-during-sensitive-examinations-and-procedures

   This source discusses the use of chaperones during sensitive examinations or procedures to help protect and enhance the patient's comfort, safety, privacy, security, and/or dignity.

7. "Use of Chaperones: Code of Medical Ethics Opinion 1.2.4." https://www.ama-assn.org/print/pdf/node/5986

   This source discusses the importance of establishing clear expectations that chaperones will uphold professional standards of privacy and confidentiality.

CONFIDENTIALITY

12

As a member of the healthcare profession, I will respect the secrets that are confided in me, even after the patient has died.
—Declaration of Geneva

* * *

MYSTERY STORY

WHISPERED BETRAYALS: NAVIGATING THE TREACHEROUS WATERS OF CONFIDENTIALITY BREACHES

The rhythmic, staccato symphony of a bustling hospital surrounded Practitioner Alexander Swift as he navigated the chaotic labyrinth that was the emergency department at Whispering Pines Privacy Hospital. Suddenly, the cacophony escalated as the shrill wail of an ambulance siren heralded the arrival of a casualty—a man crumpled by the ferocious assault of a car accident, clinging to life's fragile thread.

As Practitioner Swift, with the laser focus of his experience and training, launched into action to map the injuries and orchestrate a lifesaving treatment, a clandestine conversation prickled at his periphery. Two nurses, caught in a hushed discussion, speculated about the identity behind the patient's masked, pain-wrought face and dissected the vivid tableau of injuries marking his body.

The echo of their whispered speculations triggered Practitioner Swift's alarm. He intervened, his words slicing through their speculation like a scalpel, reminding them of the sacrosanct covenant of confidentiality guarding the patient's information. Unauthorized sharing could fracture the bedrock of trust underpinning their healthcare role, he cautioned, triggering a landslide of dire consequences.

However, the seeds of indiscretion had been sown, and they bore a bitter fruit days later. A cold, impersonal letter from the hospital's legal department pierced Practitioner Swift's peace. Allegations of confidentiality breach, lobbed by the patient's family, cast a long, damning shadow over the sanctity of his professional integrity.

Gutted and beleaguered by the accusation, Practitioner Swift grappled with the potential fallout. His career, meticulously sculpted with steadfast dedication to patient confidentiality, now teetered on the precipice. The threat of a tarnished reputation and a revoked healthcare license hung over him like a guillotine.

As the hospital unveiled a rigorous inquisition, probing the timeline of the patient's treatment and interrogating the emergency department's staff members, the villainous architect of the breach emerged. One of the nurses, a cog in the lifesaving machine, had siphoned off the patient's information to her friend, an outsider to the healthcare battlefield.

The hospital enacted swift justice, severing ties with the guilty nurse and extending a heartfelt apology to the patient's family. Chastened by the incident's severity, Practitioner Swift hardened his resolve, forging a new commitment to the fortress of patient information protection. He echoed the gravity of maintaining confidentiality with his team, imprinting the destructive potential of a single breach.

This harrowing ordeal at Whispering Pines Privacy Hospital underscored that patient confidentiality, often perceived as an ethical principle, is a fierce legal guardian of patient dignity and trust. In the complex theater of healthcare, practitioners at Whispering Pines

> and beyond must vigilantly shield patient information, for a breach can spark a devastating domino effect, impacting the patient, the healthcare provider, and the sanctity of the healthcare relationship.

* * *

THINK

Confidentiality is a core precept of the healthcare profession and is manifested in the practical application of the bioethical principles of autonomy (informed consent), beneficence (do good), nonmaleficence (do no harm), and justice (be fair).

The Hippocratic Oath states,

> Whatever I see or hear in the lives of my patients, whether in connection with my professional practice or not, which ought not to be spoken of outside, I will keep secret, as considering all such things to be private.

ASSESS

Patient: Autonomy

Patients must trust that the practitioner and other care professionals will keep all healthcare information confidential. These confidential assurances are essential so the patient will be willing to seek healthcare and candidly communicate with their practitioner, both being the necessary conditions for attaining proper healthcare. Patient healthcare information, also known as privileged communications, and patient's protected health information (PHI) should, under normal circumstances, never be disclosed to any other person outside of the patient's treatment team without the patient's or proxy's explicit consent.

This means that all patient healthcare information must be kept in confidence, even from family and friends, unless the patient or proxy has first given their autonomous permission as to what healthcare information is permissible to share with others or forced by law.

Practitioner: Beneficence and Nonmaleficence

The principle of beneficence (do good) is exemplified by the precept of confidentiality because it results in the patient's willingness to trust the practitioner enough to disclose privileged communication so that the practitioner can be empowered to provide the appropriate kind and amount of care necessary for the professional goal of maximizing the patient's best interests.

The principle of nonmaleficence (do no harm) is exemplified with the precept of confidentiality because failing to keep confidentiality could result in

- the irreparable harm to the reputation of the healthcare profession;
- less patient willingness to seek care;
- less willingness to provide truthful and honest information to the practitioner;
- an increase in ineffective and harmful treatments; and
- an increase in personal and interpersonal social harms to the patient.

Public Policy: Justice

The principle of justice (be fair) is also exemplified by the precept of confidentiality in that patients have the legal right to choose who should have access to their healthcare information. The Health Insurance Portability and Accountability Act of 1996 (HIPAA) provides rules for protecting

- privacy;
- healthcare record confidentiality; and
- patient PHI.

The Privacy Rule and the Health Information Technology for Economic and Clinical Health Act of 2009 extends HIPAA's privacy and security provisions. Society has prohibited the violation of healthcare confidentiality based on the public policy principle of justice (be fair).

HIPAA is a federal statute with fines of up to $250,000 and a jail term of up to 5 years. Exceptions to the patient's right to be provided with their healthcare information occur

- when such disclosure would induce the patient to harm themself—violating the principle of nonmaleficence (do no harm); or
- if the patient rationally chooses not to be informed—the principle of autonomy (informed consent).

Exception to the right of the patient to have full confidentiality and privacy of their healthcare information would be justified

- if the revealing of the patient's information is required by law, such as in response to a court order that subpoenas the healthcare information—the principle of justice (be fair); or
- because the patient would inflict serious physical harm to an identifiable individual or group—the principles of beneficence (do good), by preventing harm to others, and justice (be fair), regarding the distribution of benefits and burdens.

CONCLUDE

It is never permissible to talk about a patient in public areas in which the confidentiality of the patient could be compromised, and it is never permissible to reveal information about a particular patient to those not directly involved in the patient's care. If the practitioner presents the patient's case for educational or advisory purposes, then due diligence must be exercised to ensure that all patient identifiers have been removed and not revealed.

In summary, maintaining confidentiality is essential to healthcare. Patient information should never be publicly discussed or revealed to those not involved in the patient's care. Patient identifiers must be removed to protect their privacy when a patient's case is presented for educational or advisory purposes. Upholding confidentiality strengthens patient trust, ensures compliance with ethical principles, and contributes to the overall integrity of the healthcare profession. (See also Chapter 35.)

* * *

REVIEW QUESTIONS

1. Like the principle of autonomy, the healthcare profession's concept of confidentiality is also relatively new.
    A. True
    B. False

2. A patient's PHI should, under normal circumstances, never be disclosed to anyone outside the patient's treatment team without the patient's or proxy's explicit consent.
    A. True
    B. False

3. Because the family is considered a familial relationship, the practitioner is legally, professionally, and ethically permitted to share patient healthcare information without the patient's or proxy's explicit consent.
    A. True
    B. False

4. Which of the following is *not* a consequence of failing to maintain confidentiality in the healthcare profession?
   A. Irreparable harm to the reputation of the healthcare profession
   B. Less patient willingness to seek care
   C. Improved patient trust in their practitioner
   D. An increase in personal and interpersonal social harms to the patient

5. Under which conditions is it permissible to deny the patient's right to be provided their healthcare information or share it with others without the patient's informed consent?
   A. If doing so would induce the patient to harm themself
   B. If the patient rationally chooses not to be informed
   C. If required by law
   D. If the patient would, as a result, inflict serious physical harm to an identifiable individual or group
   E. All of the above

Answers: 1B, 2A, 3B, 4C, 5E

* * *

CLINICAL VIGNETTES

1. Ms. Eleanor Zhang, a 44-year-old teacher, visits a clinic for healthcare attention and shares personal and sensitive information with their practitioner. The practitioner understands the importance of trust in the patient–practitioner relationship that dates back to the Hippocratic Oath. The Oath states that the practitioner should keep secret and not disclose any information they hear or see in their practice, considering it private. What principle is being referred to?
   A. Responsibility
   B. Transparency
   C. Confidentiality
   D. Immutability

   Explanation: Confidentiality is the principle of keeping personal and sensitive information about a patient private and not disclosing it to anyone without the patient's explicit consent. In the given scenario, the patient shares personal and sensitive information with their practitioner, who understands the importance of trust in the patient–practitioner relationship, which dates back to the Hippocratic Oath. The Oath states that the practitioner should keep secret and not disclose any information they hear or see in their practice, considering it private. Therefore, the principle being referred to in this scenario is confidentiality. *Answer*: C

2. Ms. Caroline Phillips, a 26-year-old speech therapist, is seeking healthcare and has shared sensitive information with their practitioner. It is important for the patient to trust that their healthcare information will be kept confidential. What is the appropriate action for the practitioner to take with the patient's sensitive healthcare information?
   A. Share the information with friends and family without the patient's consent
   B. Disclose the information to other healthcare professionals without the patient's consent
   C. Keep the information confidential and only share it with others involved in the patient's treatment unless the patient has given explicit consent to do otherwise

   Explanation: The appropriate action for the practitioner to take with the patient's sensitive healthcare information is to keep it confidential and only share it with others involved in the patient's treatment unless the patient has given explicit consent to do otherwise.

Confidentiality is an essential principle in the patient–practitioner relationship, and it is crucial for patients to trust that their healthcare information will be kept private. Sharing the information with friends and family without the patient's consent, as described, would breach confidentiality and violate the patient's trust. Disclosing the information to other healthcare professionals without the patient's consent may also violate confidentiality unless it is necessary for the patient's treatment and the patient has not explicitly objected to the disclosure. Therefore, the appropriate action is to keep the information confidential and only share it with others involved in the patient's treatment unless the patient has given explicit consent to do otherwise. *Answer*: C

3. Mr. Grant Johnson, a 43-year-old police officer, has shared sensitive healthcare information with their practitioner. According to the patient's rights, what should be done with this information?
   A. The right to privacy is nullified by the need to share the information with the patient's family and friends without consent.
   B. The right to confidentiality is not public policy, allowing disclosure of the information to others without the patient's consent.
   C. The patient's right to privacy is upheld, and the information is safeguarded as confidential, only allowing sharing with the treatment team unless explicit consent has been given to share with specific individuals or groups or if legally compelled to share.
   D. The public's right to be informed of public servants' healthcare information holds more weight than the officer's personal right to confidentiality.

   Explanation: In this case, the patient has shared sensitive healthcare information with their practitioner, and it is essential to respect their privacy rights. Strict confidentiality laws and ethical standards bind practitioners, so the patient's right to privacy is of utmost importance. The practitioner must keep the patient's healthcare information confidential and only share it with other treatment team members when necessary or if the patient has given explicit consent for specific individuals or groups to receive the information. In some cases, practitioners may also be legally required to disclose the information to authorized personnel, but such disclosures should be limited to what is necessary and legal. The other options are incorrect because they violate the patient's right to privacy or are incorrect by misrepresenting the right to confidentiality for public servants. *Answer*: C

4. Mr. Curtis Roberts, a 67-year-old retiree, presents to the clinic with symptoms of severe abdominal pain. The patient is clearly in distress, but they inform the practitioner that they do not wish to be informed of their diagnosis or any further details about their healthcare condition. They want to undergo the necessary treatments but do not want to be informed of any information that could cause undue stress or harm. What is the most appropriate action for the practitioner to take in this situation?
   A. Provide the patient with all healthcare information, regardless of their wishes
   B. Provide the patient with the necessary information for their treatment, but withhold the diagnosis
   C. Respect the patient's wishes and do not provide them with any information about their healthcare condition, in accordance with the principle of autonomy (informed consent)

   Explanation: The most appropriate action for the practitioner to take in this situation is to respect the patient's wishes and not provide them with any information about their healthcare condition in accordance with the principle of autonomy (informed consent). Autonomy is a fundamental principle of bioethics, and it states that patients have the right to make their own decisions about their healthcare, including the right to refuse information or treatments. In this scenario, the patient has informed the practitioner that the patient does not want to be informed of their diagnosis or any further details about their healthcare condition. It is essential for the practitioner to respect the patient's autonomy and not provide them with any information they do not wish to receive. *Answer*: C

5. Ms. Vanessa Wilson, a 34-year-old chemical engineer, is seen at a primary care clinic for the first time. During the initial assessment, the patient reports a history of domestic violence, substance abuse, and a recent suicide attempt. The primary care practitioner explains the importance of privacy and confidentiality but also informs the patient that there may be circumstances in which the practitioner is legally required to disclose information about the patient. In which circumstances can the primary care practitioner break patient confidentiality and disclose their healthcare information?
    A. The patient's family wishes to be fully informed of the patient's healthcare condition
    B. The patient's employer requests a copy of the patient's healthcare records
    C. It is required by law, or the patient would inflict serious physical harm to an identifiable individual or group
    D. The patient is a politician, and the media wishes to be informed about the patient's healthcare condition

    Explanation: The primary care practitioner can break patient confidentiality and disclose the patient's healthcare information if it is required by law or the patient inflicts serious physical harm to an identifiable individual or group. Confidentiality is an essential principle in the patient–practitioner relationship, and patients have the right to expect their personal and sensitive information to be kept private. However, there are specific circumstances in which the practitioner may be legally required to disclose patient information. For example, if the patient poses an imminent threat to themselves or others, the practitioner may be obligated to warn or protect the potential victim. Similarly, if the law requires mandatory reporting of certain diseases, such as sexually transmitted infections, the practitioner must disclose the information. The other options are inappropriate circumstances for practitioners to break patient confidentiality and disclose healthcare information. The patient's family does not have an automatic right to be fully informed of the patient's healthcare condition, and the patient's employer does not have the right to access their healthcare records without the patient's explicit consent. In addition, disclosing the patient's healthcare information to the media for political reasons would violate the patient's privacy and confidentiality. *Answer*: C

REFLECTION VIGNETTES

1. Ms. Emily Nguyen, a 58-year-old retired librarian, is currently under the care of a practitioner for an undisclosed healthcare condition. Ms. Nguyen's family members are very concerned about her condition and would like to provide as much assistance as possible to help her recover. However, Ms. Nguyen has not given any explicit permission to disclose her healthcare information to her family members.

    What is the ethical issue(s)?

    How should the ethical issue(s) be addressed?

2. Mr. David Kim, a 45-year-old software engineer, has a unique healthcare condition that would make for an interesting grand rounds presentation. The practitioner is excited about the potential to share this case with colleagues to further the healthcare knowledge and understanding of the condition. However, Mr. Kim has yet to give explicit permission for the practitioner to make such a presentation.

    What is the ethical issue(s)?

    How should the ethical issue(s) be addressed?

3. Ms. Mei Ling, a 67-year-old retired seamstress who only speaks Chinese, is admitted to the hospital with nonproductive coughing. Healthcare communication has been conducted through the use of a healthcare interpreter. Magnetic resonance imaging is conducted, and the results show a peripheral lung tumor with a probable diagnosis of adenocarcinoma. However, before the test results came back, family members requested that any negative information be withheld from the patient because informing the patient of negative test results would not help the patient but would only emotionally harm them. The interpreter communicates this conversation with the patient, and the patient agrees not to be informed.

What is the ethical issue(s)?

How should the ethical issue(s) be addressed?

* * *

WEB LINKS

1. *Code of Medical Ethics*. American Medical Association. https://code-medical-ethics.ama-assn.org
   The AMA *Code of Medical Ethics*, established by the American Medical Association, is a comprehensive guide for healthcare practitioners. It addresses issues and challenges, promotes adherence to standards of care, and is continuously updated to reflect contemporary practices and challenges.
2. "Principles of Clinical Ethics and Their Application to Practice." Basil Varkey. 2021. https://www.ncbi.nlm.nih.gov/pmc/articles/PMC7923912
   This source discusses sthe inherent and inseparable part of clinical medicine as the physician has an ethical obligation to benefit the patient, avoid or minimize harm, and respect the values and preferences of the patient.
3. "Patient Rights and Ethics." Jacob P. Olejarczyk and Michael Young. https://www.ncbi.nlm.nih.gov/books/NBK538279
   This source discusses the right to informed consent in the United States, protected to some degree by state and federal legislation.
4. "Confidentiality and the Ethics of Bioethics." W. A. Rogers and H. Draper. 2003. https://jme.bmj.com/content/29/4/220
   This source discusses the ethical implications of confidentiality in bioethics.
5. "Ethics of Privacy, Confidentiality and Medical Records." American Medical Association. https://www.ama-assn.org/topics/ethics-privacy-confidentiality-medical-records
   This source provides guidance to help physicians strike the balance with patient's rights and privacy when dealing with contact tracing, isolation, or quarantine.
6. "Confidentiality." American Medical Association. https://code-medical-ethics.ama-assn.org/ethics-opinions/confidentiality
   This source discusses the importance of confidentiality in the care of a patient, emphasizing that disclosing information to third parties for commercial purposes without consent undermines trust and violates principles of informed consent and confidentiality.
7. "Patient Rights." American Medical Association. https://code-medical-ethics.ama-assn.org/ethics-opinions/patient-rights
   This source discusses the rights of a patient to make decisions about the care the physician recommends and to have those decisions respected, including the right to privacy and confidentiality.

CONFLICT OF INTEREST

13

But tell me, is this practitioner a moneymaker, an earner of fees, or a healer of the sick?
—Plato

* * *

MYSTERY STORY

VEILED TIES: A DANCE WITH TRANSPARENCY IN THE LABYRINTH OF CONFLICTS OF INTEREST

The sun had dipped below the horizon, and Practitioner Julian Crossroads was ready to wrap herself in the comfort of home after a grueling day in the heart of the bustling Harmony Bridge Hospital. Just as she was about to surrender herself to the lure of relaxation, a call punctured the tranquility. The voice on the other end belonged to the hospital's stern administrator, Ms. Maria Ramirez.

"Practitioner Crossroads, we've found ourselves in a quagmire," Ms. Ramirez's voice echoed with apprehension. "One of our patients lodged a complaint against you, alluding to a conflict of interest."

The accusation left Practitioner Crossroads breathless. Her practice was an altar to ethical conduct, and the insinuation that she might have betrayed a patient's trust felt like a gut punch.

"What manner of conflict?" Practitioner Crossroads inquired, her mind churning.

"The patient insinuates that you forwarded him to a specialist who is not just a professional associate but a close friend," Ms. Ramirez clarified. "The grievance alleges that you clouded your ties with the specialist, and the referral was devoid of his best interests."

Stunned, Practitioner Crossroads grappled with the allegations. Indeed, she had referred the patient to her friend, Practitioner Suresh Patel, a luminary in his field. But her referrals were built on a foundation of trust and full disclosure. She staunchly recommended Patel only when her patient's well-being and best interests demanded his expertise.

"My relationship with Practitioner Patel is never veiled," Practitioner Crossroads retorted, her voice imbued with conviction. "My patients' care is my compass, and I would never allow personal gains to lead me astray."

"Indeed, Practitioner Crossroads," Ms. Ramirez responded. "Nonetheless, we must cast a thorough gaze over the situation to dispel any shadows of conflict."

Practitioner Crossroads agreed to participate in the investigation, although a gnawing unease suggested a sinister attempt to besmirch her reputation.

The subsequent weeks saw Practitioner Crossroads navigating the investigation's choppy waters. She furnished proof of her association with Practitioner Patel and her disclosure practices while recounting her dealings with the accusing patient in meticulous detail.

The dust eventually settled, and the investigation exonerated Practitioner Crossroads, revealing the patient's misunderstanding about her ties with Practitioner Patel and his misguided interpretation of her intentions. Yet, the episode left an indelible mark.

The experience at Harmony Bridge Hospital had cost Practitioner Crossroads more than just stress: It was a stark reminder of the labyrinthine nature of conflicts of interest. She vowed henceforth to double down on her transparency, ensuring crystal-clear communication with her patients at Harmony Bridge and reiterating her oath to put their care on a pedestal. The harrowing ordeal had underlined the necessity of navigating potential conflicts of interest with the unwavering torch of transparency.

* * *

THINK

A practitioner's *conflict of interest* is any actual or perceived influence that diminishes the practitioner's judgment and patient-centered focus away from maximizing the patient's best interests as determined by the patient's reasonable goals, values, and priorities.

ASSESS

Patient: Autonomy

Patients have the right to make autonomous informed consent decisions about their healthcare treatment options. Practitioners must reveal any real or apparent conflict of interest to the patient regarding treatment options so that the patient can provide legitimate informed consent in accordance with the principle of autonomy. Without a conflict of interest disclosure to the patient, the practitioner would be less prone to mitigate and eliminate the conflict, and it would be difficult, if not impossible, for the patient to determine if the practitioner is consciously or unconsciously biased in the treatment options being offered and biased in the assessment of benefits and risks of the various treatment options being made available to the patient.

Practitioner: Beneficence and Nonmaleficence

The professional principle of beneficence (do good) requires that the practitioner always put the patient's welfare before all other personal interests. This does not mean that the practitioner cannot have personal interests; rather, it means that the practitioner needs to be self-reflecting enough to recognize if and when those personal interests start to negatively affect the practitioner's judgment, which may negatively affect patient care, violating the professional principle of nonmaleficence (do no harm).

Public Policy: Justice

The social principle of justice (be fair) argues that conflict of interest unfairly burdens the patient and unfairly benefits the practitioner. Undisclosed practitioner conflict of interest will augment the already unequal distribution of benefits and burdens.

Conflict of interest may arise for numerous reasons. Institutional structures such as fee-for-service, where the practitioner gets X amount of dollars per service delivered, tend to cause practitioners to overutilize services. Institutional structures such as a capitation, where the practitioner or practice gets X amount of dollars per year per patient regardless of how many services are provided, tend to cause the practitioner to underutilize services.

Other examples of conflict of interest arise when companies provide

- gifts;
- hospitality trips; and
- subsidies for services.

Another temptation for the practitioner is fee-splitting, in which a professional colleague or institution shares the patient's compensation for healthcare services in exchange for the referral.

Personal conflict of interest may also occur when the practitioner is engaged in

- the sale of nonessential products to patients;
- self-referring a patient to another site in which the practitioner has a financial relationship; or, even more personally,
- the providing of healthcare to oneself, family, or friends.

CONCLUDE

The practitioner with a conflict of interest must recuse themself from the activity and disclose the conflict of interest to those affected.

In summary, practitioners must recognize and address conflicts of interest to uphold ethical principles in healthcare. They should recuse themselves from activities where a conflict exists and disclose it to those affected. By doing so, healthcare professionals maintain the trust of their patients and ensure that the quality of care provided remains the highest priority. (See also Chapters 22, 25, 50, and 51.)

* * *

REVIEW QUESTIONS

1. A practitioner's *conflict of interest* is any actual or perceived influence that diminishes the practitioner's judgment and patient-centered focus away from maximizing the patient's best interests as determined by the patient's reasonable goals, values, and priorities.
   A. True
   B. False

2. The professional principle of beneficence (do good) requires that the practitioner always put the patient's welfare before all other personal interests.
   A. True
   B. False

3. Institutional structures such as fee-for-service, where the practitioner gets X amount of dollars per service delivered, tend to cause practitioners to overutilize services. Institutional structures such as a capitation, where the practitioner or practice gets X amount of dollars per year per patient regardless of how many services are provided, tend to cause the practitioner to underutilize services.
   A. True
   B. False

4. If a practitioner exercises due diligence and stays patient-focused, then gifts, hospitality trips, and subsidies for services will not be considered a conflict of interest.
   A. True
   B. False

5. Which of the following is *not* an example of a potential conflict of interest for a practitioner?
   A. Fee-splitting with a professional colleague
   B. Receiving gifts from pharmaceutical companies
   C. Providing care to a friend or family member
   D. Regularly attending educational conferences

Answers: 1A, 2A, 3A, 4B, 5D

* * *

CLINICAL VIGNETTES

1. Mr. Jake Jones, a 42-year-old technical writer, presents to a practitioner complaining of persistent headaches and fatigue. After a thorough examination, the practitioner recommends using a certain medication brand for treatment. However, the practitioner fails to disclose that they receive monetary compensation from the manufacturer for promoting their products. What is the practitioner's conflict of interest?
   A. The practitioner has a financial relationship with the manufacturer of the medication they recommend.
   B. The practitioner is using a medication that is not commonly prescribed.

C. The practitioner is not considering other treatment options.
D. The practitioner does not believe the patient's symptoms are genuine.

Explanation: The practitioner's conflict of interest is their financial relationship with the medication manufacturer they recommend. A conflict of interest occurs when a practitioner's professional judgment or actions may be influenced by their personal interests or relationships. In this scenario, the practitioner recommends a certain brand of medication for the patient's treatment without disclosing that they receive monetary compensation from the medication manufacturer for promoting its products. This failure to disclose creates a potential conflict of interest because the practitioner's financial relationship with the manufacturer could influence their treatment recommendation. The practitioner's duty is to act in the best interest of their patients, and their financial relationship with the medication manufacturer could compromise this duty. *Answer*: A

2. Ms. Melody Taylor, a 52-year-old data analyst, presents to the clinic with persistent headaches. During the consultation, the patient said they had been stressed and overwhelmed with work. The practitioner recognizes that the patient needs stress management techniques and refers them to a therapist. However, the therapist happens to be the practitioner's spouse. The practitioner realizes that this referral may be perceived as a conflict of interest and is faced with a dilemma. What should the practitioner do in this situation?
    A. Proceed with the referral because it is in the best interest of the patient
    B. Refer the patient to a different therapist to avoid any perceived conflicts of interest
    C. Disclose the relationship to the patient and proceed with the referral if the patient agrees
    D. Do not refer the patient to a therapist at all

Explanation: The most appropriate action for the practitioner in this situation is to refer the patient to a different therapist to avoid any perceived conflicts of interest. Referring the patient to the practitioner's spouse as a therapist creates a potential conflict of interest because the practitioner's personal relationship with the therapist could influence the patient's treatment. To avoid any perceived conflicts of interest, the practitioner should refer the patient to another therapist who is not associated with them in any way. This approach ensures that the patient receives unbiased and objective care and prevents potential harm from a conflict of interest. Proceeding with the referral, as it is in the patient's best interest, is not the best course of action in this situation because it creates a potential conflict of interest. Disclosing the relationship with the patient and proceeding with the referral if the patient agrees may not be sufficient to address the potential conflict of interest because the patient may still perceive bias or prefer not to continue with the referral. Not referring the patient to a therapist at all would not be appropriate because it would deny them access to the care they need. *Answer*: B

3. A practitioner has been assigned to care for a 42-year-old female under a capitation payment model. The practitioner knows that under this model, they will receive a set amount of money per year for the patient, regardless of the number of services provided. Which of the following is a likely result of the practitioner working under a capitation payment model?
    A. Overutilization of services
    B. Appropriate utilization of services
    C. Underutilization of services

Explanation: A likely result of the practitioner working under a capitation payment model is the underutilization of services. Capitation is a payment model in which the practitioner or healthcare organization receives a set amount per year for each patient, regardless of the number of services provided. In this model, the practitioner assumes the financial risk for the patient because the practitioner will not receive any additional payment if the patient requires more services than anticipated. To manage this risk, practitioners may limit the number of services provided to the patient, leading to underutilization of services. They may avoid necessary tests, procedures, or referrals to specialists because these would increase their costs and reduce their profits. This can result in patients not receiving appropriate care, leading to potential negative health outcomes. *Answer*: C

4. Mr. Kaleb Evans, a 42-year-old hair stylist, complains of fatigue and difficulty sleeping. The practitioner performs a thorough evaluation and determines that the patient would benefit from a sleep study. The practitioner owns a sleep study facility and would receive a fee for each study performed. What potential conflict of interest may arise?
   A. The practitioner may be more likely to overutilize services for the patient.
   B. The practitioner may be more likely to underutilize services for the patient.
   C. The patient may be more likely to receive inadequate care.
   D. The practitioner may not have the patient's best interests in mind.

   Explanation: The potential conflict of interest in this situation is that the practitioner may be more likely to overutilize services for the patient. The practitioner owns a sleep study facility and would receive a fee for each study performed, which creates a financial incentive for the practitioner to recommend the sleep study to the patient. This could result in overutilization of services because the practitioner may recommend the sleep study even if it is not necessary or appropriate for the patient's condition. This could lead to additional costs, inconvenience for the patient, and potential risks associated with unnecessary tests or treatments. It is essential for the practitioner to consider the patient's best interests and provide care that is appropriate and necessary for their condition rather than being influenced by financial incentives. *Answer*: A

5. A practitioner is faced with a situation in which a pharmaceutical company has offered them gifts, hospitality trips, and subsidies for services. What would best describe the ethical dilemma the practitioner faces in this situation?
   A. The practitioner should accept all offers because it is a common practice in the industry.
   B. The practitioner should decline all offers because it may be perceived as a conflict of interest.
   C. The practitioner should accept gifts and hospitality trips but decline subsidies for services.
   D. The practitioner should accept subsidies for services but decline gifts and hospitality trips.

   Explanation: The best option that describes the ethical dilemma faced by the practitioner is to decline all offers because it may be perceived as a conflict of interest. Accepting gifts, hospitality trips, and subsidies for services from pharmaceutical companies creates a potential conflict of interest because it may influence the practitioner's judgment or actions regarding patient care. It is essential to ensure that any decision regarding patient care is based solely on the patient's best interest and not influenced by external incentives. Therefore, declining all offers is the best course to avoid potential conflicts of interest, maintain professional integrity, and ensure patients' highest quality of care. Accepting all offers as a common practice in the industry is inappropriate because it could compromise the practitioner's professional judgment and ethics. Accepting some offers and declining others may not be sufficient to avoid potential conflicts of interest and may create confusion or doubts in patients' minds. *Answer*: B

REFLECTION VIGNETTES

1. Ms. Sarah Jones, a 35-year-old graphic designer, has successfully gone into remission after a particularly difficult chemotherapy regimen. The practitioner is thrilled to share the good news with Ms. Jones and her family. The family members are so grateful for the practitioner's help that they give her a $200 gift card to a local restaurant they own.

   What is the ethical issue(s)?

   How should the ethical issue(s) be addressed?

2. Mr. John Smith, a 47-year-old construction worker, comes to the practitioner's local practice with symptoms that suggest he may benefit from a magnetic resonance imaging (MRI) scan. The practitioner recently purchased an MRI machine for the practice because there are no other machines in town. The differential diagnosis for Mr. Smith's symptoms includes various healthcare conditions, such as a herniated disc or a spinal cord injury. The practitioner believes an MRI scan would provide valuable information for a more accurate diagnosis and prognosis.

   What is the ethical issue(s)?

   How should the ethical issue(s) be addressed?

* * *

WEB LINKS

1. *Code of Medical Ethics.* American Medical Association. https://code-medical-ethics.ama-assn.org
   The AMA *Code of Medical Ethics*, established by the American Medical Association, is a comprehensive guide for healthcare practitioners. It addresses issues and challenges, promotes adherence to standards of care, and is continuously updated to reflect contemporary practices and challenges.
2. "Conflicts of Interest in Patient Care." American Medical Association. https://code-medical-ethics.ama-assn.org/ethics-opinions/conflicts-interest-patient-care
   This source discusses the primary objective of the healthcare profession to render service to humanity and how physicians may not place their own financial interests above the welfare of their patients.
3. "Principles of Clinical Ethics and Their Application to Practice." Basil Varkey. 2021. https://www.ncbi.nlm.nih.gov/pmc/articles/PMC7923912
   This source discusses conflicts between ethical principles in patient care situations, especially between beneficence and autonomy.
4. "What Should Physician–Researchers Tell Patient–Subjects About Their Relationships with Industry?" Jeffrey R. Botkin. 2021. https://journalofethics.ama-assn.org/article/what-should-physician-researchers-tell-patient-subjects-about-their-relationships-industry/2021-09
   This source discusses the management and reporting of financial conflicts of interest and the importance of not letting these conflicts interfere with patient care.
5. "11.2.2 Conflicts of Interest in Patient Care." American Medical Association. https://code-medical-ethics.ama-assn.org/sites/default/files/2022-09/11.2.2%20Conflicts%20of%20interest%20in%20patient%20care%20--%20background%20reports.pdf
   This source discusses the importance of prioritizing patient welfare over the economic interests of the hospital, healthcare organization, or other entity.
6. "Ethical Conflicts in Patient-Centered Care." Sven Ove Hansson and Barbro Froding. 2021. https://journals.sagepub.com/doi/full/10.1177/1477750920962356
   This source discusses the lack of a single, generally accepted definition or description of what patient-centered care entails and how this can lead to ethical conflicts.
7. "Cases." American Medical Association. https://journalofethics.ama-assn.org/cases
   This source discusses how conflicts of interest must be acknowledged with sincerity and earnestness and managed such that the conflicts are eliminated or, at least, credibly mitigated.

CONTRACEPTION

14

The best practitioner is the one who is the most ingenious inspirer of hope.
—Samuel Taylor Coleridge

* * *

MYSTERY STORY

> **VEILED AGENDAS: THE ETHICAL DIVIDE IN THE CONTRACEPTION CONUNDRUM**
>
> Detective Laura Sinclair, a seasoned sleuth with a sharp eye for unraveling enigmatic puzzles, found herself summoned to a distressing tableau of chaos and despair at Tapestry Family Planning Clinic. Her trained gaze scanned the upheaval, finally settling on the lifeless form of Practitioner Isabel Chambers, an esteemed OB/GYN and a tireless champion for women's health rights. The exam room was in turmoil, strewn with scattered papers and abandoned healthcare instruments—a chaotic still life capturing the frenzied moment of an untimely demise.
>
> Diving headlong into the investigation, Detective Sinclair found herself swept into the tumultuous currents of a contentious debate that had shadowed Practitioner Chambers. The eye of the storm was a legal yet controversial case—the provision of contraception to a minor without parental consent. Unflinching in the face of criticism, Practitioner Chambers had upheld patient autonomy as her guiding beacon, ensuring minors had access to contraceptives upon legal request.
>
> The investigation's path led Detective Sinclair into the labyrinth of discord among the clinic's healthcare fraternity. Certain colleagues, advocates of a parental consent requisite for minors accessing contraceptives, viewed Practitioner Chambers's approach as a violation of their professional ethos. One such dissenting voice was that of Practitioner Ethan Strangeway, who had brandished his opposition with fervor against the wind of Practitioner Chambers' convictions.
>
> As Detective Sinclair delved deeper, the murkiness of Practitioner Strangeway's involvement thickened. Witness accounts painted a volatile picture of escalating confrontations between him and Practitioner Chambers. Adding a further layer of suspicion was the revelation of Practitioner Strangeway's financial stake in a company producing a rival contraceptive product—a clandestine conflict of interest woven into the narrative.
>
> Although Practitioner Strangeway initially cloaked his actions behind a façade of denial, the relentless tide of incriminating evidence wore him down. He eventually yielded, confessing his heinous act. Driven by frustration over Practitioner Chambers's approach to minor contraception and perceiving her actions as a transgression of the principle of nonmaleficence (do no harm), he believed her stance posed a perilous threat to minor safety.
>
> Detective Sinclair, aghast at the twisted logic used to justify the act, underscored the pillars of healthcare ethics: patient autonomy (informed consent), beneficence (do good), nonmaleficence (do no harm), and justice (be fair). She implored the healthcare community, and specifically those at Tapestry Family Planning Clinic, to heed Practitioner Chambers's tragic fate as a stark reminder of the sanctity of patients' rights and the responsibility to dispense care prioritizing their best interests, rather than personal ideologies or hidden agendas. The ethereal echo of Practitioner Chambers' advocacy for patient rights, silenced prematurely, would continue to reverberate within Tapestry Family Planning Clinic's corridors, guiding its path toward a more ethically harmonious future.

* * *

THINK

In 1977, the U.S. Supreme Court affirmed in *Carey v. Population Services International* the constitutional right for all citizens, including minors, to use contraceptives in all states. The due process clause of the 14th Amendment was used to establish the fundamental liberty of the right to use contraception:

> Fourteenth Amendment Section 1:
> No state shall deprive any person of life, liberty, or property without due process of law.

Minors are considered partially emancipated regarding contraception, meaning that minors do not need parental consent (Table 14.1).

Table 14.1 Circumstances in Which Minors Do Not Require Parental Consent

Healthcare	Emergency care
	Sexually transmitted infections
	Contraception
	Substance abuse (most states)
	Prenatal care
Emancipated minor	Homeless
	Parent
	Married
	Military
	Financially independent

ASSESS

Patient: Autonomy

Patient autonomy is a fundamental right, and whenever there is a right, others are obligated toward those who have the right. There are two categories of rights: negative rights and positive rights. A *negative right* is when others have an obligation to "not interfere." This obligation of non-interference is why this right is considered a negative right. In the context of autonomy, this means that others have an obligation not to interfere in the minor's right to choose to have contraception. A *positive right* is when others have an obligation to "provide" something. This obligation to provide is why this right is considered a positive right. In the context of autonomy, others have an obligation to provide the minor with contraception (Table 14.2).

Table 14.2 Civil Rights

Negative right	Obligation of others to not interfere
Positive right	Obligation of others to provide something

Practitioner: Beneficence and Nonmaleficence

Professionally, the practitioner must respect the minor's choice for contraception and provide the minor with the contraception in the same manner as a practitioner would provide an adult with the same request. This respect and provision are based on the professional principles of beneficence (do good) and nonmaleficence (do no harm).

Not allowing minors the choice and access to contraception would increase unwanted pregnancies, which has historically increased high school dropout rates, single-mother families, unemployment, and social welfare spending. Because all of these consequences are considered harmful to the patient, it therefore follows that not providing access to contraception would clearly violate the principle of nonmaleficence (do no harm).

However, whenever a practitioner treats a minor patient, it is generally considered healthcare best practice to communicate to the minor the benefits of having an open and candid discussion about contraceptives with their parent(s). Communication, providing access to contraceptives, and helping prevent unwanted pregnancy are all part of the professional principle of beneficence (do good).

Public Policy: Justice
The principle of justice requires the fair distribution of benefits and burdens. To not allow minors access to contraception would be to burden minors unfairly. This is why the Supreme Court affirmed a minor's constitutional right to privacy to obtain contraceptives in all states. Therefore, as a matter of justice (be fair), practitioners must allow minors the right to choose contraceptives and help provide contraceptive access.

CONCLUDE
The correct response regarding contraception or prenatal care of a minor will always be for the practitioner to "provide the access or the care" and "encourage the minor to discuss contraception or prenatal care with the minor's parent(s)."

In summary, with regard to contraception or prenatal care for a minor, the appropriate response for a practitioner is to provide access to care and encourage the minor to discuss the matter with their parent(s). By doing so, practitioners uphold the principles of patient autonomy (informed consent), beneficence (do good), nonmaleficence (do no harm), and justice (be fair), ensuring a fair distribution of benefits and burdens while promoting open communication between minors and their parents.

* * *

REVIEW QUESTIONS
1. Minors are considered to be partially emancipated when making treatment decisions for
   A. Emergency care
   B. Sexually transmitted infections
   C. Substance abuse
   D. Prenatal care
   E. All of the above

2. A negative right is a right that is wrong.
   A. True
   B. False

3. Whenever someone has a right, it always means that others have an obligation.
   A. True
   B. False

4. A minor's right to contraception is only a negative right.
   A. True
   B. False

5. Whenever treating a minor, it is always important to encourage the minor to discuss their decisions with their parent(s).
   A. True
   B. False

Answers: 1E, 2B, 3A, 4B, 5A

* * *

CLINICAL VIGNETTES

1. Ms. Angelica Jackson, a 17-year-old landscape architect, presents to the emergency department with severe abdominal pain. The patient reports having unprotected sexual intercourse a week prior. The patient also reports a history of substance abuse and is currently pregnant. What type of healthcare treatment is appropriate for this patient?
   A. Parental consent is required for all healthcare treatment.
   B. Minors and practitioners do not require parental consent for emergencies, treatment for sexually transmitted infections, treatment for substance abuse, and prenatal care.
   C. Parental consent is required for all healthcare treatments except for substance abuse treatment.
   D. Parental consent is required for all healthcare treatments except for emergencies.

   Explanation: The type of healthcare treatment appropriate for this patient is that minors and practitioners do not require parental consent for emergencies, treatment for sexually transmitted infections, treatment for substance abuse, and prenatal care. In most states, minors have the right to consent to their own healthcare treatment for certain conditions, such as emergencies, sexually transmitted infections, treatment for substance abuse, and prenatal care. These conditions are often considered confidential and protected by law to encourage minors to seek appropriate healthcare without fear of judgment or retribution. In this case, the patient presents to the emergency department with severe abdominal pain and reports a history of substance abuse and an unplanned pregnancy. The patient also reports having unprotected sexual intercourse a week prior, indicating a risk of sexually transmitted infections. Therefore, the patient would require immediate healthcare attention and treatment for sexually transmitted infections, substance abuse, and prenatal care, if necessary. Parental consent is not required for these treatments, and practitioners can provide these treatments to minors without fear of legal action. *Answer*: B

2. Ms. Noelle Harris, a 16-year-old high school student, presents to the clinic seeking contraception. The practitioner must consider the patient's negative right to obtain contraception. Which of the following best describes the practitioner's obligation toward the patient?
   A. To provide the minor with contraception
   B. To not interfere with the minor's choice to have contraception
   C. To provide the minor with both contraception and information on the potential consequences of using contraception
   D. To provide information on the potential consequences of using contraception

   Explanation: The practitioner's obligation toward the patient in this situation is to not interfere (negative right) with the minor's choice of contraception. In most states, minors have the right to access contraception without parental consent. The practitioner must respect the patient's negative right to obtain contraception, which means that the practitioner should not interfere with the patient's ability to obtain contraception. The practitioner should provide the patient with the necessary information about contraception, its use, and its potential benefits and risks, but ultimately it is up to the patient to make the decision regarding their own healthcare. Therefore, the correct option is to not interfere with the minor's choice to have contraception. *Answer*: B

3. Ms. Eileen Clark, a 26-year-old aerospace engineer, visits the clinic for a routine checkup. The patient is in a sexually active relationship and wants to discuss options for contraception. The practitioner discusses the various methods of contraception available, including hormonal methods, barrier methods, and intrauterine devices (IUDs). The patient expresses interest in an IUD, and the practitioner explains the benefits and risks of this method. The practitioner also emphasizes the importance of contraception to prevent unintended pregnancy and protect against sexually transmitted infections. If the patient has a right to contraception, what does that socially mean?

A. It means that others are not obligated to do anything.
B. It means the patient is obligated to respect others and their social standing.
C. It means that others have an obligation either to not interfere or to provide something.
D. It means that the patient is obligated to get contraception.

Explanation: If the patient has a right to contraception, it means that others have an obligation either to not interfere (negative right) or to provide something (positive right). Socially, the right to contraception implies that individuals have the right to access and use contraception without fear of judgment, discrimination, or interference from others. This right is supported by the healthcare community and many legal frameworks that recognize the importance of sexual and reproductive health for individuals and societies. By recognizing this right, others, including healthcare providers and policymakers, have an obligation to support and promote access to contraception, whether by providing information and resources or by refraining from interference. Therefore, the correct option is that others have an obligation either to not interfere or to provide something. *Answer*: C

4. Ms. Veronica Davis, a 16-year-old high school student, presents to the clinic seeking contraception. The patient expresses concern about their parents finding out and potentially interfering with their ability to access contraception. What is the patient's right to contraception?
    A. The patient only has a negative right to contraception, meaning that others are obligated not to interfere with their access to contraception.
    B. The patient only has a positive right to contraception, meaning that others are obligated to provide contraception.
    C. The patient has both a negative and a positive right to contraception, meaning that others are obligated not to interfere and to provide contraception.
    D. The patient has no right to contraception in this scenario.

Explanation: In this scenario, the patient has both a negative and a positive right to contraception, meaning that others are obligated not to interfere with the patient's access to contraception and are also obligated to provide contraception, if necessary. In most states, minors have the right to access contraception without parental consent, and healthcare providers are obligated to respect this right and maintain confidentiality as long as the patient is deemed mature enough to make informed decisions about their own healthcare. The patient's concern about their parents finding out and interfering with their access to contraception highlights the importance of protecting the patient's confidentiality and ensuring they have access to contraception without fear of judgment or interference. Therefore, the correct option is that the patient has both a negative and a positive right to contraception, meaning that others are obligated not to interfere and to provide contraception. *Answer*: C

5. Ms. Judith Miller, a 17-year-old high school student, visits the clinic seeking information about contraception options. The practitioner has the following options to discuss with the patient:
    A. Provide a prescription for the birth control pill and advise the patient to start taking it immediately
    B. Provide a prescription for a long-acting reversible contraceptive (LARC) method, such as an IUD, and advise the patient to have it inserted at their next visit
    C. Provide information on both the birth control pill and LARC methods and advise the patient to discuss the options with their parent(s) before making a decision
    D. Advise the patient to consider using a barrier method, such as a condom, and schedule a follow-up appointment in a few weeks

Explanation: The most appropriate option for the practitioner to discuss with the patient is to provide information on the birth control pill and LARC methods and advise them to discuss the options with their parent(s) before deciding. This option allows the patient to be fully informed about the available contraception options, including short-term and long-term methods. The practitioner also recognizes that the patient is a minor and may benefit from discussing their decision with their parent(s) before making a final choice. By providing the patient with information and encouraging the patient to discuss their options with their parent(s), the practitioner can support the patient's right to access contraception while respecting their family's role in decision-making. Therefore, the correct option is to provide information on the birth control pill and LARC methods and advise the patient to discuss the options with their parent(s) before deciding. *Answer*: C

REFLECTION VIGNETTES

1. Ms. Ashley Lee is a 16-year-old high school student who has come in for her annual well-patient care with her mother. Practitioner Maria Rodriguez, a primary care practitioner specializing in adolescent health, sees Ms. Lee. During the private portion of the visit, Ms. Lee expressed to Practitioner Rodriguez that she is sexually active and would like to start using contraception. She also requests that Practitioner Rodriguez inform her mother that she needs to take the pill to control menstrual flow rather than revealing the true reason for wanting contraception.

   What is the ethical issue(s)?

   How should the ethical issue(s) be addressed?

2. Mr. James Brown is a school superintendent who has come in to see Practitioner Elizabeth Harper, a primary care practitioner specializing in adolescent health. Mr. Brown is interested in discussing the benefits and risks of making condoms available to junior high and high school students through vending machines on school grounds.

   What is the ethical issue(s)?

   How should the ethical issue(s) be addressed?

* * *

WEB LINKS

1. *Code of Medical Ethics*. American Medical Association. https://code-medical-ethics.ama-assn.org
   The AMA *Code of Medical Ethics*, established by the American Medical Association, is a comprehensive guide for healthcare practitioners. It addresses issues and challenges, promotes adherence to standards of care, and is continuously updated to reflect contemporary practices and challenges.
2. "Access to Contraception." American College of Obstetricians and Gynecologists. 2015. https://www.acog.org/clinical/clinical-guidance/committee-opinion/articles/2015/01/access-to-contraception
   This Committee Opinion states that although most U.S. women have used contraception, various barriers hinder consistent and effective use, advocating for unimpeded, affordable access to all U.S. Food and Drug Administration–approved contraceptives. It also provides strategies to improve this access.
3. "New ACOG Guidance on Contraceptive Counseling Emphasizes a Patient-Centered Framework." American College of Obstetricians and Gynecologists. https://www.acog.org/news/news-releases/2022/01/new-acog-guidance-contraceptive-counseling-emphasizes-patient-centered-framework
   This source discusses how practitioners who apply a patient-centered, reproductive justice framework to contraceptive counseling can help patients achieve their reproductive goals.

4. "Principles of Clinical Ethics and Their Application to Practice." Basil Varkey. 2021. https://www.ncbi.nlm.nih.gov/pmc/articles/PMC7923912
   This source discusses the ethical obligation of the physician to benefit the patient, avoid or minimize harm, and respect the values and preferences of the patient.
5. "Reproductive Rights and Access to Reproductive Services for Women with Disabilities." Anita Silvers, Leslie Francis, and Brittany Badesch. 2016. https://journalofethics.ama-assn.org/article/reproductive-rights-and-access-reproductive-services-women-disabilities/2016-04
   This source discusses whether women with disabilities are owed equitable access to reproductive health services, including family planning, contraception, screening for sexually transmitted infections, maternal health services, and fertility services.
6. "Adolescents' Right to Consent to Reproductive Medical Care: Balancing Respect for Families with Public Health Goals." Margaret Moon. 2012. https://journalofethics.ama-assn.org/article/adolescents-right-consent-reproductive-medical-care-balancing-respect-families-public-health-goals/2012-10
   This source discusses the legal framework that supports a limited right for adolescents to consent to care, including contraception.
7. "Patient Rights and Ethics." Jacob P. Olejarczyk and Michael Young. https://www.ncbi.nlm.nih.gov/books/NBK538279
   This source discusses the right to informed consent in the United States, which is somewhat protected by state and federal legislation.
8. "Women's Views About Contraception Requirements for Biomedical Research Participation." Kristen A. Sullivan, Margaret Olivia Little, Nora E Rosenberg, et al. 2019. https://pubmed.ncbi.nlm.nih.gov/31067273
   This source discusses the scientific and ethical importance of including women of reproductive age in biomedical research and the concerns about preventing fetal exposure to research interventions.

DEATH WITH DIGNITY AND ASSISTED DYING

15

A practitioner is obligated to consider more than a diseased organ, more even than the whole patient—the practitioner must view the patient within the patient's world.
—Harvey Cushing

* * *

MYSTERY STORY

THE VEIL OF SERENITY: A DANCE WITH ETHICAL SHADOWS OF ASSISTED DYING

Detective Olivia Sterling was a silhouette against the backdrop of the bustling Sanctuary of Hope Hospital, her thoughts spiraling in the vortex of a troubling investigation. She had been summoned to unravel a complex puzzle of assisted dying, more eloquently known as death with dignity, at this location. The victim, Practitioner Benjamin Gallagher, was an icon in the field of healthcare, his practice spanning three decades. In the austere silence of his office at Sanctuary of Hope, his life had found closure, leaving behind a note that breathed a cryptic testament: "I have taken the final step toward my death with dignity."

The case teetered on the razor's edge of paradox. Practitioner Gallagher had been a vocal adversary of the death with dignity laws, a stance that appeared in stark contrast to the nurturing aura he emanated as a practitioner, always prioritizing his patients' needs. The solution to this enigmatic scenario, Detective Sterling deduced, lay in peeling back the layers of ethical nuance underpinning death with dignity.

Peering through the lens of Practitioner Gallagher's professional life, Detective Sterling discovered a focal point: Ms. Sarah Bennett, a patient caught in the relentless clasp of terminal illness. Ms. Bennett had implored Practitioner Gallagher for assistance in orchestrating her end, a plea he had denied, citing his ethical obligations. Yet, fate had granted Ms. Bennett a grim alternative, securing the lethal potion elsewhere and executing her own finale. Detective Sterling wondered whether this harrowing episode could have ignited a change in Practitioner Gallagher's convictions.

Her investigation navigated the labyrinth of local laws surrounding Practitioner Gallagher's practice. Death with dignity was deemed legal, but participation remained optional for practitioners. For those choosing to step into this ethically charged arena, rigorous compliance to guidelines was mandated, ensuring the sanctity of patient autonomy and adherence to the primal healthcare oath: Do no harm.

The narrative of Practitioner Gallagher, woven by his colleagues and patients, painted the portrait of a man brimming with compassion, putting his patients at the heart of his practice. Yet, within the echo of their praises, a subtle note of discord arose. It was evident that the resonance of Ms. Bennett's ordeal had stirred turbulent questions about death with dignity in Practitioner Gallagher's conscience.

The final act of her inquiry led Detective Sterling to the heartbroken family of Practitioner Gallagher. His demise had blindsided them; there were no forewarnings that their beloved patriarch was dancing on the precipice of assisted dying. They remembered him as a loving family man, as dedicated to his hearth as he was to his patients.

> After meticulously weaving together the strands of her investigation, Detective Sterling discerned the transformation in Practitioner Gallagher's views. Ms. Bennett's demise had cast a profound ripple effect, pushing him to navigate the tumultuous waters of assisted dying himself. His parting note was an echo of his journey, a plea to his fellow practitioners to deliberate the ethical maze of death with dignity, with vigilance and discernment.
>
> As Detective Sterling drew the curtains on this case at Sanctuary of Hope Hospital, she mulled over the words of Harvey Cushing: "A practitioner must view the patient within the patient's world." She acknowledged that death with dignity extended beyond the realm of healthcare, dipping into the profound pools of a patient's reasonable goals, values, and priorities. It was a poignant reminder that healthcare providers at Sanctuary of Hope and elsewhere should focus not only on the disease or condition but also on the entirety of the patient, appreciating the holistic human canvas painted by life's myriad hues.

* * *

THINK

Death with dignity laws, also known as "assisted dying" or "aid-in-dying," are based on the individual principle of autonomy (informed consent), in which it is believed that the patient should make end-of-life decisions. Death with dignity is legal in 11 jurisdictions: California, Colorado, the District of Columbia, Hawaii, Maine, Montana, New Jersey, New Mexico, Oregon, Vermont, and Washington. However, just because some action is legally permissible for a practitioner to participate in does not necessarily mean that the practitioner must or is obligated to participate in that action.

ASSESS

Patient: Autonomy

There is no contesting that there are many competent patients or persons, not necessarily patients at the time of request with full decisional capacity, who are autonomously asking for healthcare assistance in the planning and control of their death, and no one contests the legality of states to make such actions permissible as a social attempt to help actualize the fair distribution of benefits and burdens for the patient as a matter of justice (be fair).

Practitioner: Beneficence and Nonmaleficence

The central issue concerning the death with dignity laws or practitioner-assisted dying topic is whether or not such participation by the practitioner is compatible with

- the healthcare profession's code of ethics;
- the core tenets of the healthcare profession; and
- the promotion of patient trust in the healthcare profession.

Starting with the question of compatibility with the healthcare profession's code of ethics, no entries in the American Medical Association (AMA) *Code of Medical Ethics* directly address death with dignity or practitioner-assisted suicide, and the American College of Physicians (ACP) also does not address these issues. The closest reference to the ethical issues related to this topic is euthanasia and capital punishment.

The AMA *Code of Medical Ethics* 5.8 defines euthanasia as "the administration of a lethal agent by another person to a patient for the purpose of relieving the patient's intolerable and incurable suffering." Euthanasia refers to the deliberate and direct causation of death by a practitioner, and the AMA *Code of Medical Ethics* 5.8 states, "Permitting physicians [practitioners] to engage in euthanasia would ultimately cause more harm than good. Euthanasia is fundamentally incompatible with the physician's [practitioner's] role as healer."

However, death with dignity or practitioner-assisted suicide is argued to be significantly different from euthanasia. Death with dignity only refers to giving a patient a prescription of lethal medication to be voluntarily self-administered by the patient, not to be administered by the practitioner.

The ACP prohibits practitioners from engaging in cruel or unusual punishment, such as capital punishment or other disciplinary activities beyond those permitted by the United Nations Standard Minimum Rules for the Treatment of Prisoners.

What is difficult in regard to death with dignity is if (a) the patient is autonomously choosing to have practitioner assistance in death, in accordance with their reasonable goals, values, and priorities—autonomy (informed consent); (b) under the principles of beneficence (do good) and nonmaleficence (do no harm), the practitioner determines that assistance in death would be in accordance with the core professional tenets of the patient's best interests as determined by the patient's reasonable goals, values, and priorities; and (c) the patient and practitioner are in a state of jurisdiction in which death with dignity is legally permissible (justice—be fair). In this case, all four principles of autonomy (informed consent), beneficence (do good), nonmaleficence (do no harm), and justice (be fair) would have been satisfied, making it morally permissible for practitioner participation in the giving to a patient a lethal prescription to be self-administered by the patient.

Currently, 60% of all Americans will die in acute care hospitals, and 20% will die in nursing homes, all under the care of a practitioner. The question arises: Is death with dignity "fundamentally incompatible with the physician's [practitioner's] role as a healer"? (AMA *Code of Medical Ethics* 5.8).

Is medicine only a healing art, or should the art of dying also be part of healthcare practice? Some will argue that the problem is that of the healthcare provider's deadly intent and that the ensuing negative professional reputation could result in irreparable harm to the social perception of the professional function and role of the healthcare profession as a healing art.

Public Policy: Justice

With public policy, the issue is whether or not practitioner-assisted death would harm the patient–practitioner relationship and decrease the social trust in the healthcare profession. Logically, if (a) no violations are being made concerning the patient's autonomous choices (informed consent), (b) the practitioner is only engaged in patient-centered practices that maximize the patient's best interests in accordance with beneficence (do good) and nonmaleficence (do no harm), and (c) these practices are consistent with state laws that have been enacted through the process of democratic representation of the people—justice (be fair)—then it reasonably follows that there would be no adverse effect on the patient–practitioner relationship or social trust of the healthcare profession for the practitioner to participate in death with dignity where it is legal to do so.

CONCLUDE

Death with dignity or practitioner-assisted suicide is state-dependent. If the participation by the healthcare professional is not compatible with (a) the healthcare profession's code of ethics of nonmaleficence (do no harm), (b) the core tenets of the healthcare profession being a healing profession, and (c) the promotion of patient trust that the healthcare profession will only pursue life, then there would be a sufficient argument to oppose being complicit in activities that do not promote the healthcare profession's goals, values, and priorities.

In summary, the topic of death with dignity and practitioner-assisted suicide is complex and state-dependent. If practitioner participation in assisted dying does not align with the healthcare profession's code of ethics, core tenets, and the promotion of patient trust, then there is a sufficient argument to oppose such activities. However, if these practices are consistent with state laws and do not adversely affect the patient–practitioner relationship or social trust in the healthcare profession, it may be considered morally permissible for practitioners to participate in death with dignity where it is legal to do so. (See also Chapters 9, 10, 21, 31, and 62.)

* * *

REVIEW QUESTIONS

1. Death with dignity laws are based on the practitioner's professional obligation to help the patient make end-of-life decisions.
   A. True
   B. False

2. Which of the following are the central issues concerning the death with dignity laws:
   A. The healthcare profession's code of ethics
   B. The core tenets of the healthcare profession
   C. The promotion of patient trust in the healthcare profession
   D. All of the above

3. There are no entries in the codes of ethics of the AMA and the ACP on the topic of death with dignity.
   A. True
   B. False

4. Which of the following statements about euthanasia is true according to the AMA *Code of Medical Ethics*?
   A. Euthanasia is the same as death with dignity.
   B. Euthanasia is compatible with the practitioner's role as a healer.
   C. Euthanasia is administering a lethal agent by another person to a patient to relieve the patient's intolerable and incurable suffering.
   D. Euthanasia involves the patient self-administering lethal medication with a prescription.

5. Death with dignity only refers to giving a patient a prescription of lethal medication to be voluntarily self-administered by the patient, not administered by the practitioner:
   A. True
   B. False

Answers: 1B, 2D, 3A, 4C, 5A

* * *

CLINICAL VIGNETTES

1. Mr. Emery Wright, an 82-year-old retiree, is diagnosed with a terminal illness and wants to end their life peacefully. Mr. Wright has read about death with dignity laws and has asked their practitioner about the possibility of this option. In states that have death with dignity laws, what is the fundamental premise of these laws?
   A. That it is the practitioner who should be making end-of-life decisions
   B. That it is the patient who should be making end-of-life decisions
   C. That the practitioner should refuse to provide the lethal prescription
   D. That medicine is defined as a healing art

   Explanation: The fundamental belief of death with dignity laws is that the patient should be making end-of-life decisions. Death with dignity laws, also known as practitioner-assisted dying or aid-in-dying laws, allow terminally ill patients to request a prescription for medication that they can self-administer to end their lives peacefully. These laws are based on the principle of patient autonomy and the belief that individuals can make decisions about their healthcare and end-of-life care. Practitioners participating in death with dignity laws must inform patients about their diagnosis, prognosis, treatment options, and the potential risks and benefits of the lethal medication. They must also ensure that patients are mentally competent and have made an informed decision about their end-of-life care. *Answer*: B

2. Ms. Athena Adams, a 79-year-old retiree, is a terminally ill patient and has requested assistance in ending their life. The patient's practitioner is faced with the decision of whether to comply with the request or not. The practitioner is concerned that complying with the request would be incompatible with the healthcare profession's code of ethics, the core tenets of the healthcare profession, and the promotion of patient trust.
    A. The central issue is that death with dignity laws and practitioner-assisted dying are illegal in many states.
    B. The central issue is that death with dignity laws and practitioner-assisted dying can lead to increased feelings of guilt and moral distress among practitioners.
    C. The central issue is that death with dignity laws and practitioner-assisted dying are incompatible with the healthcare profession's code of ethics, the core tenets of the healthcare profession, and the promotion of patient trust in the healthcare profession.
    D. The central issue is that death with dignity laws and practitioner-assisted dying are supported by evidence-based research.

    Explanation: The central issue in this scenario is that death with dignity laws and practitioner-assisted dying are incompatible with the healthcare profession's code of ethics, the core tenets of the healthcare profession, and the promotion of patient trust in the healthcare profession. Healthcare has traditionally been focused on saving lives and preserving health, and practitioners are typically held to high ethical and moral standards in their practice. Practitioner-assisted dying, which involves intentionally helping a patient end their own life, goes against these traditional goals and can raise ethical and moral concerns for practitioners. Moreover, this can potentially erode patients' trust in the healthcare profession, which depends on the belief that practitioners prioritize the patient's best interests and work to promote their well-being. Although the legality of death with dignity laws and practitioner-assisted dying can vary by state, the central issue in this scenario is not whether it is legal or not. Similarly, although complying with a patient's request for assistance in ending their life can lead to feelings of guilt and moral distress for practitioners, this is not the central issue in this scenario. Finally, although there may be evidence-based research to support or oppose practitioner-assisted dying, this is not the central issue in this scenario because the main concern is the ethical and moral implications of the practice for practitioners and the healthcare profession as a whole. *Answer*: C

3. Mr. Abel Nelson, an 81-year-old retiree, has requested assistance ending their life due to a terminal illness. The patient's practitioner is unsure about the ethical guidance provided by the AMA *Code of Medical Ethics* and the ACP regarding this matter.
    A. The AMA and ACP have clear guidance on the ethical considerations of death with dignity and practitioner-assisted suicide.
    B. The AMA and ACP do not directly address death with dignity or practitioner-assisted suicide, but they provide guidance on related issues, such as euthanasia and capital punishment.
    C. The AMA and ACP prohibit death with dignity and practitioner-assisted suicide under all circumstances.
    D. The AMA and ACP support death with dignity and practitioner-assisted suicide legislation.

    Explanation: The scenario describes a practitioner who is unsure about the ethical guidance provided by the AMA *Code of Medical Ethics* and the ACP regarding the request of an 81-year-old retiree for assistance in ending their life due to a terminal illness. The correct option is that the AMA and ACP do not directly address death with dignity or practitioner-assisted suicide. Still, they provide guidance on related issues, such as euthanasia and capital punishment. This means that the practitioner will need to carefully consider the ethical implications of the situation and consult with other sources to make an informed decision. *Answer*: B

4. Ms. Jessica Hassan, an 86-year-old retiree, has been diagnosed with a terminal illness and has a prognosis of only a few months to live. Ms. Hassan expresses to their practitioner that they would like to have control over the manner and timing of their death. The patient asks about the option of "death with dignity." What is the correct definition of "death with dignity"?
   A. Administering lethal medication to the patient by the practitioner
   B. A process in which the practitioner assists the patient in ending their life through active means
   C. Giving a patient a prescription for a lethal medication to be voluntarily self-administered by the patient
   D. Withholding or withdrawing life-sustaining treatment at the request of the patient or surrogate decision-maker

5. Mr. Nathanael King, a 42-year-old sound engineer, comes to a practitioner to request treatment. However, the treatment being asked for by the patient goes against the healthcare profession's code of ethics of nonmaleficence (do no harm) and the core tenets of being a healing profession. The treatment would also negatively impact society's trust in the healthcare profession. Given this information, which of the following would be the most appropriate course of action for the practitioner to take?
   A. Go ahead with the recommended treatment regardless of the ethical implications
   B. Refuse to participate in the treatment because it goes against the healthcare profession's goals, values, and priorities
   C. Discuss alternative treatment options with the patient that align with the healthcare profession's code of ethics and core tenets before providing the requested treatment
   D. Seek guidance from a bioethics committee to determine the best course of action

Explanation: The scenario describes a situation in which a patient requests a treatment that conflicts with the healthcare profession's code of ethics and the core tenets of being a healing profession. The most appropriate course of action for the practitioner is to refuse to participate in the treatment because it goes against the healthcare profession's goals, values, and priorities. This is because the healthcare profession is responsible for prioritizing patient well-being and maintaining society's trust, and participating in a treatment that goes against these principles would violate professional ethics. Going ahead with the recommended treatment regardless of the ethical implications is inappropriate because it would violate the core principles of bioethics. Discussing alternative treatment options with the patient may be appropriate in some cases but not in cases in which the requested treatment is unethical. Last, seeking guidance from a bioethics committee may sometimes be appropriate. Still, the practitioner's first responsibility is to uphold bioethics principles and refuse to participate in unethical practices. *Answer*: B

REFLECTION VIGNETTES
1. Mr. Michael Johnson is a 70-year-old retired accountant who has been struggling with terminal cancer for 3 years. He has undergone multiple rounds of chemotherapy and radiation therapy, but the cancer has spread to various organs, and his condition has deteriorated rapidly. He is under the care of Practitioner Emily White, an attending practitioner who specializes in palliative care. During a routine visit, Mr. Johnson expresses his desire to end his life on his own terms. He states that the pain and discomfort are unbearable, and he does

not want to suffer anymore. He requests a prescription for medication to allow him to end his life peacefully and with dignity.

What is the ethical issue(s)?

How should the ethical issue(s) be addressed?

2. Ms. Patricia Brown is a 76-year-old retired librarian under the care of Practitioner Rachel Johnson, an attending practitioner specializing in hospice and palliative care. Ms. Brown has been battling a terminal illness for several years, and her condition has recently worsened, with increasing pain and discomfort. Despite the best efforts of Practitioner Johnson and her team, Ms. Brown's symptoms are becoming increasingly difficult to manage. During a visit with Practitioner Johnson, Ms. Brown expresses her desire to die with dignity in the presence of her family and friends. She requests a prescription for medication that would allow her to do so.

What is the ethical issue(s)?

How should the ethical issue(s) be addressed?

* * *

WEB LINKS

1. *Code of Medical Ethics*. American Medical Association. https://code-medical-ethics.ama-assn.org
The AMA *Code of Medical Ethics*, established by the American Medical Association, is a comprehensive guide for healthcare practitioners. It addresses issues and challenges, promotes adherence to standards of care, and is continuously updated to reflect contemporary practices and challenges.
2. "Ethical Considerations at the End-of-Life Care." Melahat Akdeniz, Bülent Yardımcı, and Ethem Kavukcu. 2021. https://www.ncbi.nlm.nih.gov/pmc/articles/PMC7958189
This source discusses the process of end-of-life care, starting with the diagnosis of a fatal disease, and includes the dignity death that the patient desires and the post-death mourning period.
3. "Patient Rights at the End of Life: The Ethics of Aid-in-Dying." Mary Atkinson Smith, Lisa Torres, and Terry C. Burton. 2020. https://pubmed.ncbi.nlm.nih.gov/32000206
This source discusses the same standard of practice as for clinical care providers regarding promoting the biomedical ethical points of autonomy, beneficence, nonmaleficence, justice, and fidelity.
4. "Choosing Death Over Suffering." Amber Comer. 2019. https://bioethics.hms.harvard.edu/journal/choosing-death
This source discusses the ethical standard of discussing physician aid in dying during healthcare decision-making.
5. "Mortal Responsibilities: Bioethics and Medical-Assisted Dying." Courtney S. Campbell. 2018. https://www.ncbi.nlm.nih.gov/pmc/articles/PMC6913808
This source discusses the agency of the terminally ill patient in physician-assisted death.
6. "Patient Rights at the End of Life: The Ethics of Aid-in-Dying." Mary Atkinson Smith, Lisa Torres, and Terry C. Burton. 2020. https://journals.lww.com/professionalcasemanagementjournal/Abstract/2020/03000/Patient_Rights_at_the_End_of_Life__The_Ethics_of.4.aspx
This source discusses the importance of following ethical principles for case managers when supporting the desires and preferences of terminally ill patients.
7. "Death with Dignity." Peter Allmark. 2002. https://jme.bmj.com/content/28/4/255
This source develops a conception of death with dignity and examines whether it is vulnerable to the sort of criticisms that have been made of other conceptions.

8. "Ethical Considerations at the End-of-Life Care." Melahat Akdeniz, Bülent Yardimci, and Ethem Kavukcu. 2021. https://pubmed.ncbi.nlm.nih.gov/33786182
This source discusses the main situations that create ethical difficulties for healthcare professionals at the end of life.

DISAGREEMENTS

16

ATTENDING VERSUS RESIDENT

Practitioners are, in general, the most amiable companions and the best friends, as well as the most learned individuals I know.
—Alexander Pope

* * *

MYSTERY STORY

AN UNRULY ACCORD: THE CRUCIAL DANCE OF COLLABORATION AND RESPECT IN QUELLING TUMULTS BETWEEN ATTENDING PHYSICIANS AND RESIDENTS

Nestled in the heart of the city, United Resolve Healthcare Center was a crucible where residents learned the essence of medicine under the tutelage of accomplished attending physicians. Among the mentors, Practitioner Ethan Shephard stood out, a master of his craft, a sentinel of knowledge and professionalism, guiding the fledgling residents with a firm yet nurturing hand. However, a simmering discord threatened to rupture this revered dynamic, coalescing around a complex patient case that split opinions like a surgeon's scalpel.

The heart of the contention was a young woman, her body invaded by an elusive heart malady, a marauder that held her life captive. For weeks, she had been carefully treated under the watchful guidance of Practitioner Shephard, with the assisting residents following his lead as if they were musicians in an orchestra, and he their diligent conductor. Amid this symphony of care, a dissonant note reverberated, resonating from Resident Ethan Crossfield, a skilled player who found himself questioning the conductor's chosen score.

Resident Crossfield, bold in his beliefs, held a conviction that the melody of treatment needed a heightened tempo and a more assertive rhythm to overpower the patient's malady. Yet, Practitioner Shephard, a seasoned maestro, advocated for a gentler harmony, a slower crescendo to stave off potential complications. The clashing symphony of views crescendoed into a standoff. Resident Crossfield, certain of his solo performance's merit, took the baton and altered the patient's medication regimen, a rogue composition.

As the patient's health deteriorated, an ominous dissonance filled the ward. Practitioner Shephard, now the conductor of a faltering symphony, discovered the rogue crescendo that had usurped his composition. Seething with indignation, he rallied the administrative powers of the hospital, sparking an investigation that swept through the hospital like a whirlwind.

The investigative storm soon broke over Resident Crossfield, revealing his unauthorized overture. In the face of this transgression, the hospital exacted its justice, severing Resident Crossfield from his duties with a 2-week suspension and placing him under probation's shadow for his remaining residency.

This incident rippled through the resident ensemble at United Resolve Healthcare Center, an unexpected encore to their learning. They confronted the reality that attending-resident conflicts were not a discordant note but, rather, a part of the larger symphony of medicine. The incident underscored the weight of the attending's authority, the importance of adhering to the written score, and the necessity for due process when altering it.

> The young woman, the unwitting centerpiece to this tumultuous symphony, eventually found her health restored. However, the echoes of the incident lingered as a powerful coda to the residents' journey, reminding them that medicine is a delicate dance of collaboration, respect, and humility. In the end, this discordant note within United Resolve Healthcare Center's symphony became a valuable lesson in harmony for all those who played there.

* * *

THINK

Healthcare is an interprofessional activity requiring legal, institutional, professional, and evidence-based healthcare decision-making. As with all multi-individual decision-making, disagreements will arise for many reasons, including disagreements on what the objective facts are; disagreements on what the patient's subjective goals, values, and priorities are; disagreements on what the proper treatment or amount of treatment should be; disagreements on what will maximize the patient's best interests; and disagreements on what actions are morally permissible for a practitioner.

This is a partial list of possible categories of disagreements. Still, it shows that one must accept that various types of disagreements can and will occur due to the human condition. The question that needs to be addressed is, What professional and institutional procedures need to be followed when a disagreement about patient management does occur?

ASSESS

Patient: Autonomy

The patient–practitioner relationship is between the licensed practitioner (i.e., the attending) and the patient. Confidentiality and privacy must be respected. If the patient with decisional capacity has autonomously provided informed consent, giving the attending the authorization to provide treatment, then it follows that the resident, or others attending to the patient, in disagreeing with the attending, is in effect disagreeing with the patient's decision. The resident has no authority to usurp the patient–practitioner relationship by discussing the disagreement with the patient independently of the attending. Rather, the resident should discuss the disagreement with the attending practitioner.

Practitioner: Beneficence and Nonmaleficence

The best patient care usually happens when an interprofessional team of healthcare providers work together and communicate effectively with each other toward the goal of maximizing the patient's best interests, as determined by the patient's reasonable goals, values, and priorities, and through the effective implementation of the professional principles of beneficence (do good) and nonmaleficence (do no harm). Although patient care and management is a team activity, the ultimate responsibility and accountability lie squarely on the shoulders of the licensed attending practitioner. Therefore, if an attending practitioner disagrees with how the resident manages the attending practitioner's patient, the attending should respectfully discuss the matter with the resident. Then the resident must legally, institutionally, and morally comply with the attending practitioner's corrections. If a resident disagrees with how an attending practitioner manages their patient, then the resident should respectfully discuss the matter with the attending. However, in the end, the resident must comply with the attending's decision because the attending practitioner is ultimately responsible and accountable for patient care.

Certainly, there are times when a resident may be correct in their assessment of mismanagement by an attending practitioner. If after the resident has had a respectful and thorough discussion with the attending, and if the disagreement may have a serious adverse effect on the patient's best interests, as determined by the patient's reasonable goals, values, and priorities, then it is professionally incumbent on the resident to take the disagreement to the next level for further discussion. However, in the meantime, the resident must not change the patient management without the attending practitioner's approval because the resident has no authority to make such decisions; has no authority to practice unsupervised healthcare; and will be held legally, institutionally, and morally accountable for any harms that might occur to the patient due to the unauthorized changes in patient management.

Public Policy: Justice

If a serious adverse effect occurs to the patient without some intervention, even after having a respectful and thorough discussion with the attending, the institutional policy will direct the resident to whom they should report the dispute. Generally, the resident would be obliged to bring the dispute to the attention of the next institutional level above the attending, such as the division head or department chair. It would be professionally inappropriate to undermine the attending practitioner's reputation by discussing the disagreement with other non-practitioner members of the healthcare team. It would be inappropriate to discuss this discord with the patient independent of the attending because that could diminish the patient's trust in the patient–practitioner relationship and in the interprofessional healthcare team.

CONCLUDE

It is not legal for a student to practice medicine without a license by making a healthcare judgment independent of the attending practitioner. If a resident disagrees with the attending practitioner's patient management, then the resident must discuss this and not change the patient management without the attending's authorization. At no time should the student or resident inform the patient about the disagreement; instead, the student or resident should have a respectful and thorough discussion with the attending practitioner. The same applies to other members of the care team.

In summary, residents are not legally allowed to practice medicine independently and must discuss any disagreements with the attending practitioner. They should not change patient management without the attending's authorization or discuss the disagreement with the patient, as this could damage the patient's trust in the healthcare team. In cases in which the disagreement may seriously affect the patient, institutional policy should guide the resident in reporting the dispute. Effective communication and collaboration among the healthcare team are essential for providing the best patient care.

* * *

REVIEW QUESTIONS

1. Disagreements can and will occur as a matter of the human condition. For the practitioner, these disagreements consist of
   A. Objective facts
   B. Subjective goals, values, and priorities
   C. The maximization of the patient's best interests
   D. Legal interpretations
   E. All of the above

2. The resident has no authority to usurp the patient–practitioner relationship by discussing their disagreement with the attending practitioner with the patient.
   A. True
   B. False

3. Although patient care and management is a team activity, the ultimate responsibility and accountability lie squarely on the shoulders of the licensed attending practitioner.
   A. True
   B. False

4. The resident must not change the patient management without the attending practitioner's approval because the resident has no authority to make such decisions; has no authority to practice unsupervised healthcare; and will be held legally, institutionally, and morally accountable for any harms that might occur to the patient due to the unauthorized changes in patient management.
   A. True
   B. False

5. If a serious adverse effect occurs to the patient without some intervention, even after having a respectful and thorough discussion with the attending, then the resident would be obliged to bring the dispute to the attention of the next institutional level above the attending, such as the division head or department chair.
   A. True
   B. False

Answers: 1E, 2A, 3A, 4A, 5A

* * *

CLINICAL VIGNETTES
1. Mr. Neil Mohamed, a 68-year-old animator, is admitted to the hospital with a complex healthcare condition that requires a multidisciplinary approach. Practitioners have different opinions on managing patients during treatment. Which is the most appropriate way to resolve these disagreements between practitioners?
   A. The licensed practitioner should discuss the disagreements with the patient to get their input.
   B. The disagreements should be resolved through respectful communication between the licensed practitioner and the other healthcare providers.
   C. The licensed practitioner should consult with a supervisor to make the final decision.
   D. The licensed practitioner should avoid discussing disagreements with the other healthcare providers and make the final decision independently.

   Explanation: The scenario describes a situation in which different practitioners have different opinions on managing a complex healthcare condition of a patient, which requires a multidisciplinary approach. The most appropriate way to resolve these disagreements between practitioners is through respectful communication between the licensed practitioner and the other healthcare providers. Effective communication is essential to ensure that all practitioners involved in the patient's care know the treatment plan and that it is agreed upon by all. This can be achieved through collaboration and respectful communication, where all the practitioners involved in the patient's care can share their opinions and expertise. Discussing disagreements with the patient to get their input is inappropriate in this situation because the patient may not have the healthcare knowledge to make an informed decision. The option of consulting with a supervisor to make the final decision may be appropriate in some situations, but it is not always feasible or necessary. The option of avoiding discussing disagreements with other healthcare providers and making the final decision independently can lead to suboptimal care and should be avoided.
   Answer: B

2. A resident works with a licensed attending practitioner at a hospital. One day, a 52-year-old female comes to the hospital with a complaint. The resident examines the patient and makes a diagnosis, but the attending practitioner disagrees with the diagnosis and orders different tests to be done. What should the resident do in this situation?
   A. Refuse to comply with the attending practitioner's orders and continue to diagnose the patient based on their assessment
   B. Explain their reasoning to the attending practitioner and try to convince them to change their mind
   C. Comply with the attending practitioner's orders and work together to make a final diagnosis for the patient
   D. Contact the hospital administration to discuss the disagreement between the resident and the attending practitioner

   Explanation: The scenario describes a situation in which a resident and a licensed attending practitioner have different opinions on the diagnosis and management of a patient. In this situation, the resident should comply with the attending practitioner's orders and they should work together to make a final diagnosis for the patient. This is because the attending practitioner has more experience and expertise, and their input is valuable in ensuring the best possible care for the patient. The option of refusing to comply with the attending practitioner's orders is inappropriate because it can lead to suboptimal care for the patient and can also be considered insubordination. The option of explaining their reasoning to the attending practitioner and trying to convince them to change their mind is a good approach, but ultimately, the attending practitioner has the final say. The option of contacting the hospital administration to discuss the disagreement between the resident and the attending practitioner is unnecessary unless the disagreement cannot be resolved through communication and collaboration. *Answer*: C

3. Mr. Joseph James, a 68-year-old economist, presents to the emergency department with complaints of chest pain and shortness of breath, requiring a multidisciplinary approach. The patient has a history of asthma but has not had an attack in several months. The patient's vitals are stable, but their oxygen saturation is low at 92% on room air. The licensed attending practitioner orders a 12-lead electrocardiogram and a chest X-ray, which reveal no acute abnormalities. Who is ultimately responsible and accountable for the patient's care and management?
   A. The resident
   B. The respiratory therapist
   C. The registered nurse
   D. The licensed attending practitioner

   Explanation: The scenario describes a patient with chest pain and shortness of breath, requiring a multidisciplinary approach, and the licensed attending practitioner orders tests to help diagnose the condition. In this situation, the licensed attending practitioner is ultimately responsible and accountable for the patient's care and management. This is because the attending practitioner is the leader of the healthcare team and is responsible for coordinating the patient's care and making the final decisions regarding their management. The other healthcare providers—such as the resident, respiratory therapist, and registered nurse—are responsible for carrying out the orders of the attending practitioner and providing their professional expertise and knowledge to ensure optimal care for the patient. However, the attending practitioner is ultimately responsible and accountable for the overall care and management of the patient. *Answer*: D

4. A healthcare resident works in a busy hospital, rotating through various departments and gaining practical experience. One day, while on duty, the resident notices that a patient's condition has significantly worsened and decides to change the treatment plan without

consulting the attending practitioner. The resident believes that a change in treatment is necessary to save the patient's life. What is the most important action the resident should have taken?

A. Notify the attending practitioner immediately and follow their orders
B. Proceed with the change in treatment plan without informing the attending practitioner
C. Contact the patient's family members to seek their consent for the change in the treatment plan
D. Consult with a senior resident before making any changes to the patient's treatment plan

Explanation: The scenario describes a healthcare resident who notices that a patient's condition has significantly worsened and decides to change the treatment plan without consulting the attending practitioner. The most important action the resident should have taken is to notify the attending practitioner immediately and follow their orders. This is because the attending practitioner is responsible for the overall care and management of the patient, and their experience and expertise can help ensure that the best course of treatment is followed. The resident should have discussed their concerns and suggestions with the attending practitioner, and together, they could have come up with the best course of action for the patient. The option of proceeding with the change in treatment plan without informing the attending practitioner is inappropriate because it can lead to suboptimal care for the patient and can also be considered insubordination. The option of contacting the patient's family members to seek their consent for the change in the treatment plan is not necessary in this situation because the attending practitioner should be the first point of contact for any changes to the patient's treatment plan. The option of consulting with a senior resident before making any changes to the patient's treatment plan can be helpful in some situations, but the attending practitioner should always be informed and involved in the decision-making process.
Answer: A

5. Ms. Carl Lewis, a 72-year-old retiree, experiences a serious adverse effect. Despite the resident having a respectful and thorough discussion with the attending, the resident remains concerned about the patient's well-being and believes that some intervention is necessary. What should the resident do in this situation?

A. Ignore the adverse effect and continue to follow the attending's orders
B. Document the adverse effect and move on to the next patient
C. Report the adverse effect to the nursing staff and hope they will address it
D. Bring the dispute to the attention of the next institutional level above the attending, such as the division head or department chair

Explanation: The scenario in question involves a serious adverse effect on a patient, which leads the resident to disagree with the attending practitioner. The resident is responsible for advocating for the patient's well-being and ensuring appropriate care is provided. In this situation, the resident should bring the dispute to the attention of the next institutional level above the attending, such as the division head or department chair. If the resident remains concerned about the patient's well-being, they should escalate the situation to the next level of institutional authority to ensure that the best possible care is provided. Ignoring the adverse effect and continuing to follow the attending's orders is inappropriate because it can lead to suboptimal care for the patient and can also be considered a violation of professional ethics. The option of documenting the adverse effect and moving on to the next patient is inappropriate because the resident has a responsibility to ensure that appropriate care is provided to the patient. Reporting the adverse effect to the nursing staff and hoping they will address it is inappropriate because the resident is part of the healthcare team and is responsible for ensuring that appropriate care is provided to the patient.
Answer: D

REFLECTION VIGNETTES

1. Resident Olivia Carter is a 28-year-old resident in training who has been working in the hospital for the past year. She works with Practitioner Andrew Jameson, an attending practitioner, on the healthcare–surgical floor. One of their patients is a 60-year-old man admitted with a bowel obstruction. The patient has a history of hypertension and has been taking medication to control his blood pressure for the past 5 years. Upon admission, the patient's blood pressure was high, and Practitioner Jameson initiated treatment with intravenous antihypertensive medication. However, Resident Carter disagreed with Practitioner Jameson's choice of medication and suggested an alternative medication she believed would be more appropriate for the patient's condition.

   What is the ethical issue(s)?

   How should the ethical issue(s) be addressed?

2. Practitioner David Delano is a 42-year-old attending practitioner who has worked at the hospital for 10 years. On the healthcare–surgical floor, he works with Resident Samantha Parker, a resident in training. One of their patients is a 70-year-old woman who was admitted with pneumonia. The patient has a history of asthma and chronic obstructive pulmonary disease and has been taking medications to manage these conditions. Upon admission, the patient's respiratory rate was elevated, and Resident Parker initiated treatment with bronchodilators and corticosteroids. However, Practitioner Delano disagreed with Resident Parker's choice of medication and suggested an alternative treatment plan that he believed would be more appropriate for the patient's condition. Resident Parker disagreed with Practitioner Delano's explanation of her reasoning for selecting the current medication.

   What is the ethical issue(s)?

   How should the ethical issue(s) be addressed?

* * *

WEB LINKS

1. *Code of Medical Ethics*. American Medical Association. https://code-medical-ethics.ama-assn.org
   The AMA *Code of Medical Ethics*, established by the American Medical Association, is a comprehensive guide for healthcare practitioners. It addresses issues and challenges, promotes adherence to standards of care, and is continuously updated to reflect contemporary practices and challenges.

2. "What Should an Intern Do When She Disagrees with the Attending?" Timothy Crisci, Zeynep N. Inanc Salih, Ndidi Unaka, Jehanna Peerzada, and Armand H. Matheny Antommaria. 2021. https://publications.aap.org/pediatrics/article/147/3/e2020049646/33373/What-Should-an-Intern-Do-When-She-Disagrees-With
   This source discusses a case in which an intern and attending disagree about discharging the patient; the attending recommends that the patient be hospitalized longer without providing evidence to support his recommendation.

3. "What Should Physicians Do When They Disagree, Clinically and Ethically, with a Surrogate's Wishes?" Terri Traudt and Joan Liaschenko. 2017. https://journalofethics.ama-assn.org/article/what-should-physicians-do-when-they-disagree-clinically-and-ethically-surrogates-wishes/2017-06
   This source discusses situations in which patients' surrogates and physicians disagree about the appropriateness of aggressive treatment in intensive care units. Physicians can experience surrogates' demands as sources of moral distress.

4. "9.2.2 Resident & Fellow Physicians' Involvement in Patient Care." American Medical Association. https://code-medical-ethics.ama-assn.org/sites/default/files/2022-08/9.2.2%20Resident%20physicians%27%20involvement%20in%20patient%20care%20--%20background%20reports.pdf

   This source discusses how to address the patient refusal of care from a resident or fellow. If a patient does not want to participate in training after discussion, the physician may exclude residents or fellows from the patient's care.

5. "Residents' Ethical Disagreements with Attending Physicians: An Unrecognized Problem." J. G. Shreves and A. H. Moss. 1996. https://pubmed.ncbi.nlm.nih.gov/9177647

   This source evaluates the frequency and nature of ethical disagreements over patient care between house staff and attending physicians.

6. "Principles of Clinical Ethics and Their Application to Practice." Basil Varkey. 2021. https://www.ncbi.nlm.nih.gov/pmc/articles/PMC7923912

   This source discusses a model for patient care, with caring as its central element, that integrates ethical aspects (intertwined with professionalism) with clinical and technical expertise desired of a physician.

DO NOT RESUSCITATE ORDER, DO NOT ATTEMPT RESUSCITATION ORDER, AND ALLOW NATURAL DEATH ORDER

Nearly all men die of their remedies, and not of their illnesses.
—Molière

* * *

MYSTERY STORY

VANISHED DIRECTIVE: NAVIGATING THE DIGITAL LABYRINTH AND UPHOLDING DIGNITY AT ECHOLIFE DECISION HOSPITAL

On an otherwise uneventful Tuesday, an intricate enigma was unraveled in the high-stakes world of EchoLife Decision Hospital. The emergency room, a swirling vortex of healthcare crises, was under the vigilant supervision of Practitioner Jessica Wu. Amidst the cacophony of healthcare alarms, a code blue echoed from room 302, immediately propelling Practitioner Wu into action.

The distressed patient, Mrs. Donna Wilson, was an octogenarian valiantly wrestling with pneumonia. As the specter of cardiac arrest reared its head, Practitioner Wu, guided by years of healthcare practice, instinctively commenced the lifesaving dance of cardiopulmonary resuscitation CPR. But as her eyes darted over the digital landscape of Mrs. Wilson's Electronic Medical Records (EMR), a disconcerting absence screamed at her from the screen—where was the do not resuscitate (DNR) order?

As the drama unfolded, a vital piece of information charged into the scene with Mrs. Wilson's daughter, Emily. "My mother signed a living will," she exclaimed, casting the mystery of the missing DNR order under a more intense spotlight.

Alarmed but composed, Practitioner Wu suspended her resuscitation efforts, plunging once more into the sprawling network of electronic data. Hidden in an unexpected alcove of the digital records, the critical DNR directive lay silent, its crucial message almost muted amidst the sea of clinical notes and healthcare histories.

The key to the riddle lay in the inexperienced hands of a novice resident, who, still fumbling with the labyrinthine EMR system, had misfiled the critical document. This error had birthed a whirlwind of confusion, nearly compromising the fine balance of life, dignity, and patient autonomy.

An immediate course correction was imperative. Seizing the reins, Practitioner Wu promptly initiated a robust protocol that demanded the meticulous inspection of crucial documents in the EMR, reinforcing the vital importance of accurately capturing a patient's treatment preferences.

The incident served as a striking reminder that a DNR directive does not equate to a withdrawal of care; rather, it is a shift in the healthcare approach, focusing on comfort, dignity, and respect for the patient's autonomy. Mrs. Wilson was no less a priority; her care

> was uncompromised, maintaining her comfort, managing her symptoms, and ensuring her dignity, all while upholding her wish not to undergo CPR.
>
> The unraveling of the mystery of Mrs. Wilson's elusive DNR order at EchoLife Decision Hospital reiterated a poignant truth. Healthcare transcends clinical acumen and treatment procedures; it is an embodiment of respect for the patient's autonomy and preferences. The story at EchoLife underscored the need for meticulous record-keeping and patient-centered communication, and it emphasized the reality that in the dynamic theater of healthcare, every detail, every directive, and every patient wish commands the utmost significance. In the wide expanse of the digital EMR, nothing, especially not a DNR order, should ever dissolve into obscurity.

* * *

THINK

Do not resuscitate order, do not attempt resuscitation order (DNAR), and allow natural death order (AND) are all patient or surrogate decisions not to be subjected to resuscitative procedures if that patient were to have a cardiopulmonary arrest. CPR procedures typically include chest compressions, artificial ventilation, electrical cardioversion, and anti-arrhythmic medications.

ASSESS

Patient: Autonomy

Because the patient will not have the decisional capacity to provide or refuse consent for CPR while having a cardiopulmonary arrest, it is essential for the practitioner to have a candid discussion about CPR procedures before such an event occurs. Having a prior discussion recognizes the patient's autonomous right to self-determination. As for any other healthcare treatment, the patient has the legal and moral right to consent or decline the CPR treatment. If the CPR treatment is declined, a DNR order must be entered into the healthcare records.

Practitioner: Beneficence and Nonmaleficence

Normally, there is no time to determine the patient's decisional capacity in a healthcare emergency or get a patient's informed consent without increasing a serious risk of harm or death for the patient. Therefore, the moral principle of beneficence (do good) has the most weight, justifying the performance of the standards of care. Cardiopulmonary arrest perfectly fits this healthcare emergency criterion, and the standard of care is CPR. However, if, before the CPR procedure, a DNR order is revealed, then CPR should not commence. If the DNR order is revealed while CPR is in progress, then CPR should be stopped. The basis of this judgment is that the patient has the legal and moral right to autonomous self-regulation regarding what others are allowed to do or are forbidden to do to the patient's body. Reasons for a DNR may be because CPR was determined to be futile; more harmful and burdensome than beneficial; or not in line with the patient's reasonable goals, values, and priorities. Under those conditions, a surrogate who authorizes a CPR order or a healthcare provider who provides CPR against a patient's wishes would be in violation of nonmaleficence (do no harm).

A common misperception regarding DNR is the false belief that it will result in inferior or less quantitative and qualitative healthcare treatment or care than if the patient were not DNR. All healthcare treatments and patient management are identical for both the non-DNR patient and the DNR patient, except CPR.

Public Policy: Justice

If the patient does not have decisional capacity, there is no living will, no durable power of attorney for healthcare (DPOA), no available surrogate, and no way of determining the patient's preferences, then the practitioner should provide the standards of care.

CONCLUDE

A common misperception regarding DNR is that the DNR patient will receive less care or diminished patient management. The DNR patient should receive exactly the same care and management as the non-DNR patient, other than no CPR. DNR decisions are generally made by the patient as determined by the patient's DPOA or by the patient's surrogate as determined by state law, such as the hierarchical list of the patient's spouse, adult offspring, either parent, or adult sibling.

In summary, DNR patients should receive the same care and management as non-DNR patients, except for CPR. The patient generally makes decisions regarding DNR orders as expressed with a living will or by their designated surrogate as determined by a DPOA or state law. It is essential for practitioners to engage in open communication with patients about their preferences for CPR and DNR orders to respect their autonomy and provide appropriate care.

* * *

REVIEW QUESTIONS

1. Because a patient will not have decisional capacity when having a cardiopulmonary arrest, it is essential for the practitioner to have a candid discussion about CPR before such an event occurs.
   A. True
   B. False

2. Once CPR has commenced, it is against the standards of care to stop, even if it is revealed that there is a DNR order.
   A. True
   B. False

3. It is imperative that the practitioner gets the patient to provide informed consent for a DNR order because surrogates do not have the authority to authorize a DNR order.
   A. True
   B. False

4. Although having a DNR order will diminish the standard healthcare provided to the patient, at least the patient will not be subjected to futile treatment.
   A. True
   B. False

5. If the patient does not have decisional capacity and there is no living will, no DPOA, no available surrogate, and no way of determining the patient's preferences, then the practitioner should provide the standards of care.
   A. True
   B. False

Answers: 1A, 2B, 3B, 4B, 5A

* * *

CLINICAL VIGNETTES

1. Ms. Unity Jones, an 82-year-old retiree, presents to the clinic for a routine checkup. During the appointment, the practitioner realizes that the patient does not have an advanced directive in place, which includes their wishes regarding CPR in the event of a cardiac or respiratory arrest. What is the best approach for the practitioner to take in this situation?
   A. Document that the patient declined to discuss CPR and move on with the appointment
   B. Wait until the patient experiences a cardiac or respiratory arrest to make decisions about CPR

C. Inform the patient about the importance of discussing CPR and encourage them to make an informed decision
D. Assume that the patient would not want CPR and do not discuss it further

Explanation: The scenario describes a patient who does not have an advanced directive in place, which includes their wishes regarding CPR in the event of a cardiac or respiratory arrest. In this situation, the best approach for the practitioner is to inform the patient about the importance of discussing CPR and encourage them to make an informed decision. This is because the patient has the right to decide about their own healthcare, and it is important for the practitioner to provide the patient with all the necessary information to make an informed decision about their wishes for CPR. The option of documenting that the patient declined to discuss CPR and moving on with the appointment is inappropriate because it does not address the patient's lack of advanced directive and leaves the patient at risk of receiving unwanted or unnecessary healthcare interventions. The option of waiting until the patient experiences a cardiac or respiratory arrest to make decisions about CPR is inappropriate because it can lead to suboptimal care for the patient and can also be considered a violation of professional ethics. The option of assuming that the patient would not want CPR and not discussing it further is inappropriate because it is impossible to know what the patient's wishes are without discussing the matter with them. *Answer*: C

2. Mr. Tanner Taylor, an 83-year-old retiree, presents to the emergency department with sudden cardiac arrest. The healthcare team quickly assesses the patient and initiates CPR. During resuscitation, the healthcare team discovers a DNR order in the patient's healthcare chart. What should the healthcare team do in this situation?
   A. Continue the CPR despite the DNR order
   B. Stop the CPR immediately and honor the patient's DNR request
   C. Consult with the patient's family before making a decision
   D. Obtain a second opinion from another healthcare team before making a decision

   Explanation: The scenario describes a patient who presents to the emergency department with sudden cardiac arrest. The healthcare team quickly assesses the patient and initiates CPR, but during resuscitation, the team discovers a DNR order in the patient's healthcare chart. In this situation, the healthcare team should stop the CPR immediately and honor the patient's DNR request. This is because a DNR order is a legal document that expresses a patient's wishes for end-of-life care, and it is the healthcare team's responsibility to honor the patient's wishes. Continuing CPR despite a DNR order violates the patient's autonomy and constitutes healthcare malpractice. The option of consulting with the patient's family before making a decision is inappropriate in this situation because the DNR order is a legal document that expresses the patient's wishes and does not require input from the patient's family. The option of obtaining a second opinion from another healthcare team before making a decision is inappropriate in this situation because the DNR order is a legal document and does not require input from another healthcare team. *Answer*: B

3. Ms. Iris Thompson, a 68-year-old fashion designer, has been admitted to the hospital with a serious healthcare condition, does not have decisional capacity, and is not expected to recover. The patient has previously designated a surrogate decision-maker through a POA. The healthcare team is considering a DNR order. What is the proper process for determining the patient's DNR status?
   A. The surrogate decision-maker should make the decision based solely on their own beliefs and values.
   B. The healthcare team should make the decision based on the team's clinical judgment and expertise.
   C. The surrogate decision-maker should make the decision based on the patient's reasonable goals, values, and priorities. If those are not known, then the surrogate decision-maker should make the decision based on the patient's best interests.

D. The patient should decide if they can communicate and participate in the decision-making process.

Explanation: The scenario describes a patient who has designated a surrogate decision-maker through a POA, does not have decisional capacity, and is not expected to recover. The healthcare team is considering a DNR order. In this situation, the surrogate decision-maker should make the decision based on the patient's reasonable goals, values, and priorities. If those are not known, the surrogate decision-maker should make the decision based on the patient's best interests. This is because the surrogate decision-maker is responsible for making decisions that are consistent with the patient's values and preferences, and if those are not known, then decisions should be based on the patient's best interests. The option of the surrogate decision-maker making the decision based solely on their own goals, values, and priorities is inappropriate because it does not consider the patient's wishes and preferences. The option of the healthcare team making the decision based on the team's clinical judgment and expertise is inappropriate because it does not consider the patient's wishes and preferences or the role of the surrogate decision-maker. The option of the patient deciding if they can communicate and participate in the decision-making process is inappropriate because the scenario describes a patient who does not have decisional capacity. *Answer*: C

4. Ms. Amber Ali, an 85-year-old retiree, is admitted to the hospital with a serious healthcare condition, and the healthcare team is considering a DNR order. The patient's surrogate decision-maker is concerned about the quality of care the patient will receive if a DNR is in place. What is the relationship between a DNR order and a patient's healthcare treatment and care quality?
   A. DNR patients receive lower quality healthcare treatment and care compared to non-DNR patients.
   B. DNR patients receive identical healthcare treatment and care as non-DNR patients, except for CPR.
   C. DNR patients receive higher quality healthcare treatment and care compared to non-DNR patients.
   D. The quality of healthcare treatment and care a patient receives is related to their DNR status.

Explanation: A DNR order is a healthcare order which indicates that CPR should not be performed in the event of a cardiac or respiratory arrest. However, a DNR order does not limit or affect the quality of healthcare treatment and care a patient receives, except for withholding CPR. The patient should continue to receive all other necessary healthcare treatments and care indicated for their condition, and their care plan should be based on their individual needs and goals of care. The option stating that DNR patients receive lower quality healthcare treatment and care than non-DNR patients needs to be more is inaccurate because DNR status does not affect the quality of healthcare treatment and care a patient receives. The option stating that DNR patients receive higher quality healthcare treatment and care than non-DNR patients is also inaccurate because DNR status does not affect the quality of healthcare treatment and care a patient receives. The option stating that the quality of healthcare treatment and care a patient receives is related to their DNR status is also inaccurate because the quality of care a patient receives should be based on their needs and care goals, regardless of their DNR status. *Answer*: B

5. Mr. Winston Ahmed, an 88-year-old retiree, has been admitted to the hospital and cannot decide about their healthcare due to a serious healthcare condition. The patient does not have a living will, DPOA, or any available surrogate to make decisions on their behalf. In addition, there is no way to determine the patient's preferences. What should the practitioner do regarding the patient's healthcare?
   A. Withhold all healthcare treatment until a surrogate or living will is located
   B. Provide the standards of care because they generally are in the patient's best interests

C. Provide only comfort measures and withhold all curative treatments
D. Provide experimental or non-standard treatments because there is no way to determine the patient's preferences

Explanation: When a patient is unable to make decisions about their healthcare and there is no living will, DPOA, or any available surrogate to make decisions on their behalf, the practitioner should provide the standards of care because they generally are in the patient's best interests. This means that the practitioner should provide all necessary healthcare treatment and care indicated for the patient's condition and consistent with the healthcare profession's code of ethics. The option stating that all healthcare treatment should be withheld until a surrogate or living will is located is not accurate because it would be unethical to withhold necessary healthcare treatment and care from a patient, even if no surrogate or living will is available. The option stating that only comfort measures should be provided is also inaccurate because the patient may require curative treatments, depending on their condition. The option stating that experimental or non-standard treatments should be provided is also inaccurate because the practitioner should follow the standards of care indicated for the patient's condition. *Answer*: B

REFLECTION VIGNETTES

1. The attending practitioner, Practitioner Benjamin Hawthorne, cared for Mr. Robert Williams, a 65-year-old man admitted to the hospital with severe pneumonia. Mr. Williams had a history of heart disease, and his condition had worsened, with respiratory distress and sepsis. Practitioner Hawthorne reviewed Mr. Williams's healthcare records and noted that he had a DNR order in his chart. During his rounds, Practitioner Hawthorne explained the DNR order to Mr. Williams, informing him that they would not perform CPR to revive him if his heart stopped. Mr. Williams listened carefully and nodded in agreement, indicating he understood what the DNR order meant. However, soon after Practitioner Hawthorne left, Mr. Williams' wife, Mrs. Elizabeth Williams, arrived at the hospital. Mrs. Williams informed the nursing staff that Mr. Williams had changed his mind about the DNR order and wanted everything done if his heart stopped. She mentioned that Mr. Williams had been adamant about having a DNR in his healthcare records to avoid becoming a vegetable. However, after further discussion with Mrs. Williams, it became clear that Mr. Williams had not understood the implications of the DNR order and had agreed to it without fully comprehending its meaning. Later that day, Mr. Williams' condition deteriorated rapidly, and he went into cardiac arrest. The healthcare team was notified, and they rushed to his bedside. The nursing staff informed Practitioner Hawthorne that Mr. Williams's wife had requested that everything be done to save her husband's life. Practitioner Hawthorne spoke with Mrs. Williams and explained that the DNR order was in Mr. Williams's healthcare records and that they could not perform CPR without violating the order. Mrs. Williams became emotional, stating that Mr. Williams had not fully understood the implications of the DNR order and had agreed to it without knowing what he was agreeing to.

   What is the ethical issue(s)?

   How should the ethical issue(s) be addressed?

2. The patient in question is a 68-year-old man with a history of coronary artery disease, hypertension, and type 2 diabetes. He has been admitted to the hospital to treat an exacerbation of his heart failure. His healthcare record includes a DNR order, which was discussed and signed by the patient and his family members during a previous hospitalization. During this hospitalization, the patient's care team managed his heart failure with diuretics, angiotensin-converting enzyme inhibitors, and beta-blockers. However, due to a medication error, the patient received an incorrect dose of his beta-blocker medication, which

caused his heart rate to drop rapidly. This led to the patient going into cardiac arrest. The attending, Practitioner Austin Meyers, was called to the patient's bedside when the code blue was announced. Upon arrival, he reviewed the patient's healthcare record and noted that the patient had a DNR order.

What is the ethical issue(s)?

How should the ethical issue(s) be addressed?

3. Practitioner Natalie Grant, an emergency medicine practitioner, was working in the emergency department when a patient was brought in by ambulance in cardiac arrest. The patient, an elderly man, was unconscious and had no pulse. Practitioner Grant and her team immediately began resuscitation efforts, performing chest compressions and administering medications. As they worked to stabilize the patient, they noticed a tattoo across the patient's chest that read "Do Not Resuscitate."

What is the ethical issue(s)?

How should the ethical issue(s) be addressed?

4. Practitioner Christopher Morgan, an emergency medicine practitioner, was working in the emergency department when an unconscious patient was brought in by ambulance in cardiac arrest. The young adult patient had attempted suicide, and it was unclear what had caused the cardiac arrest. As the healthcare team worked to stabilize the patient, the team noticed that the patient's body was covered in writing, including the phrase "Do Not Resuscitate" written in Sharpie pen all over the patient's body. Practitioner Morgan immediately assessed the situation and consulted with his colleagues to understand the legal and ethical implications of the patient's situation. They discovered no living will or DPOA on record, and no surrogates were available to make treatment decisions for the patient.

What is the ethical issue(s)?

How should the ethical issue(s) be addressed?

* * *

WEB LINKS

1. *Code of Medical Ethics*. American Medical Association. https://code-medical-ethics.ama-assn.org
   The AMA *Code of Medical Ethics*, established by the American Medical Association, is a comprehensive guide for healthcare practitioners. It addresses issues and challenges, promotes adherence to standards of care, and is continuously updated to reflect contemporary practices and challenges.
2. "Ethical Competence in DNR Decisions: A Qualitative Study of Swedish Physicians and Nurses Working in Hematology and Oncology Care." Mona Pettersson, Mariann Hedström, and Anna T. Höglund. 2018. https://bmcmedethics.biomedcentral.com/articles/10.1186/s12910-018-0300-7
   This source discusses ethical competence in DNR decisions and how it can be improved through ethical guidelines and education.
3. "Ethical Considerations at the End-of-Life Care." Melahat Akdeniz, Bülent Yardimci, and Ethem Kavukcu. 2021. https://www.ncbi.nlm.nih.gov/pmc/articles/PMC7958189
   This source discusses the ethical considerations at end-of-life care, including decisions regarding resuscitation.
4. "Bioethics in Practice: Unilateral Do-Not-Resuscitate Orders." Meredith Miceli. 2016. https://www.ncbi.nlm.nih.gov/pmc/articles/PMC4896650

This source discusses the ethical considerations when physicians make unilateral DNR orders.

5. "DNR and COVID-19: The Ethical Dilemma and Suggested Solutions." Hala Sultan, Razan Mansour, Omar Shamieh, Amal Al-Tabba', and Maysa Al-Hussaini. 2021. https://pubmed.ncbi.nlm.nih.gov/34055703

   This source discusses the ethical dilemmas posed by DNR decisions in the context of the COVID-19 pandemic.

6. "Influence of Institutional Culture and Policies on Do-Not-Resuscitate Decision Making at the End of Life." Elizabeth Dzeng, Alessandra Colaianni, Martin Roland, et al. 2015. https://jamanetwork.com/journals/jamainternalmedicine/fullarticle/2212265

   This source discusses the ethical considerations in DNR decision-making, including the balance between autonomy and professional guidance.

7. "Do-Not-Resuscitate (DNAR) Orders." Clarence H. Braddock, III, and Jonna Derbenwick Clark. https://depts.washington.edu/bhdept/ethics-medicine/bioethics-topics/articles/do-not-resuscitate-dnar-orders

   This source provides an overview of DNAR orders, including the role of the patient or surrogate decision-maker in the process.

8. "ANA Position Statement: Nursing Care and Do-Not-Resuscitate (DNR) Decisions." Liz Stokes. 2020. https://ojin.nursingworld.org/MainMenuCategories/ANAMarketplace/ANAPeriodicals/OJIN/TableofContents/Vol-26-2021/No1-Jan-2021/Nursing-Care-and-DNR-Decisions.html

   The ANA revised position statement from March 12, 2012, by the ANA Center for Ethics and Human Rights, advocates for nurses' crucial role in initiating discussions on Do Not Resuscitate (DNR) and Allow Natural Death (AND) decisions, emphasizing education and policy reform to support informed end-of-life choices and respect for patients' wishes.

THE DOCTRINE OF DOUBLE EFFECT

18

The best practitioner is the one who can distinguish the possible from the impossible.
—Herophilos

* * *

MYSTERY STORY

> ### DILEMMA OF DUAL OUTCOMES: NAVIGATING THE RAZOR-EDGE PATH OF ETHICAL CONSEQUENCES IN ONCOLOGY
>
> In the intricate labyrinth of healthcare at Quadrant Cancer Institute, Practitioner Evelyn Pierce stood as a beacon of unwavering dedication and expertise. An esteemed oncologist at Dual Paths, she devoted her life to equipping her patients with the most potent defenses in their battles against the unrelenting adversary of cancer. Yet, one daunting day, she found herself on the precipice of an ethical paradox at the institute that would test her cognitive prowess and moral integrity like never before.
>
> Her patient, the resolute expectant mother, Ms. Megan Taylor, was caught in a devastating struggle with a formidable foe—a deadly uterine cancer. The standard healthcare protocol necessitated a drastic solution: the removal of the uterus and, tragically, the fetus within, to ensure Ms. Taylor's survival. Burdened with the enormity of the situation, Practitioner Pierce unveiled the grim diagnosis, prognosis, and the severe potential repercussions of the procedure. Although devastated, Ms. Taylor, fortified by determination, granted her informed consent to the life-preserving, albeit heart-wrenching, procedure.
>
> In the solemn quietude preceding the operation, Practitioner Pierce reflected on the age-old doctrine of "double effect." This philosophical concept, deeply embedded in Catholic moral tradition, posited the moral acceptability of an action that generated both beneficial and detrimental effects, provided the latter was unintended. Yet, Practitioner Pierce understood that the complex realm of modern bioethics demanded a more nuanced interpretation.
>
> Guided by the four cardinal pillars of bioethics—autonomy, beneficence, nonmaleficence, and justice—Practitioner Pierce meticulously navigated her ethical dilemma. She validated Ms. Taylor's autonomy via her informed consent, pondered the delicate balance between benefiting the patient and minimizing harm, and acknowledged Ms. Taylor's unequivocal right to lifesaving treatment—a salute to the principle of justice.
>
> Empowered by these principles, Practitioner Pierce conducted the formidable operation, sparing Ms. Taylor's life but resulting in the tragic, albeit anticipated, loss of the fetus. It was a heartbreaking balancing act but one that aligned with healthcare standards and the ethical compass of bioethics.
>
> The enigmatic narrative of this case highlights the inherent limitations of the doctrine of double effect in contemporary healthcare practice. Although the doctrine emphasizes the moral weight of a practitioner's paternalistic intentions, enacted independently of any patient autonomy, it often neglects certain consequential outcomes. In contrast, the four pillars of bioethics provide a comprehensive guideline that holds practitioners accountable on legal, professional, and ethical fronts. This framework accentuates the consequential outcomes of actions that are undertaken with the patient's informed consent, irrespective of the practitioner's intentions.

> This ethically charged scenario at Quadrant Cancer Institute serves as a compelling learning backdrop for students and providers. It highlights the need to transition beyond the doctrine of double effect and embrace the complexities of the four principles of bioethics. By navigating the processes of specification, weighing, and balancing, practitioners can steer through the turbulent waters of modern healthcare, ensuring they deliver the best possible care to their patients.

* * *

THINK

The doctrine of double effect has historically been defined by the medieval Natural Law tradition of Thomas Aquinas and by contemporary Catholic moral theologians as being the following:

> If, when doing something morally good, it is also accompanied by a foreseen but unintended harmful effect, then the action is morally permissible if the intention was only for good.

What is essential to notice is that the moral permissibility of action is determined solely by the agent's intention rather than by the consequences of the actions. Using this type of moral analysis, if a person intended to make a significant economic gain (the good), with the foreseen but unintended consequence of polluting the environment (harmful effect), then using the doctrine of double effect would make such actions morally permissible if the intention was only for good.

The cynic's snark reply is the old proverb, "The road to hell is paved with good intentions." However, the Catholic magisterium has found it necessary to have this doctrine so that the Church can justify moral actions that are impermissible within Catholic theology. One example that most rational agents would consider morally permissible and in line with the standards of care would be the intentional removal of a life-threatening cancerous uterus, even if its intentional removal would also result in a fetus' foreseen but unintended death. The doctrine of double effect was put in place precisely so Catholics could justify such a procedure.

However, it is essential to notice that the doctrine of double effect precedes any moral recognition of patient autonomy (informed consent). Rather, the doctrine of double effect reflects the paternalistic medieval age, in which patient choice was not recognized. Because of this paternalism, the doctrine of double effect only focuses on the practitioner's intent and does not recognize that it is the patient who has the authority to authorize the practitioner to perform the healthcare procedure.

The field of bioethics uses a different evaluative process of specifying and balancing the four principles of bioethics to determine accepted moral actions. Specifying analytically establishes how a general moral principle relates to a particular circumstance or situation. Balancing analyzes how two or more specified principles should be weighed and balanced for a particular circumstance or situation.

Using the previous example of the life-threatening cancerous uterus, the principlist approach would be the following.

ASSESS

Patient: Autonomy

Using the principle of patient autonomy (informed consent), the practitioner informs the patient with decisional capacity of the following:

1. The diagnosis of having a cancerous uterus.
2. The prognosis of the condition is life-threatening.
3. The standard of care is the removal of the uterus and fetus.
4. Benefit: Saving the patient's life.
5. Burdens: Reasonable risks associated with having surgery and the de facto abortion of the fetus.

The practitioner then needs to answer all the patient's questions and get informed consent authorizing the practitioner to perform the procedure.

Practitioner: Beneficence and Nonmaleficence

Professional beneficence (do good) requires the practitioner to empathetically assess what standards of care options would maximize the patient's best interests as determined by the patient's reasonable goals, values, and priorities.

Nonmaleficence (do no harm) is the obligation to mitigate harm to the patient. Because the fetus is part of the patient's body, the patient can authorize healthcare treatment with or without regard to the fetus. In this case, the fetus might not come to term even if the procedure was not performed.

Public Policy: Justice

With the principle of justice (be fair), it is necessary to determine what social resources and treatment options are legally permissible and available. If the procedure is legal, which it is, then the patient has the right to decide whether or not to have the cancerous uterus and fetus removed; the practitioner is obligated to maximize the patient's best interests as determined by the patient's reasonable goals, values, and priorities; and public policy, as a matter of social justice, is obliged to provide the patient with healthcare access.

Weighing and Balancing

Weighing and balancing determine the rank and order of specified moral principles. For this particular case, and because of regulatory state laws, the order of priority would be as follows:

1. Public policy (justice)
2. Patient autonomy (informed consent)
3. Practitioner professionalism (beneficence and nonmaleficence)

Using the four principles of bioethics and specifying, weighing, and balancing circumvents any need for the medieval natural law tradition of the doctrine of double effect. In modern society, regardless of intentions, foreseen but unintended consequences will hold the practitioner professionally, legally, and morally accountable. Legally, healthcare malpractice is based on a deviation from standards of care due to the practitioner being negligent in an action or inaction that, as a consequence, harms the patient. This legal, professional, and moral accountability is independent of the practitioner's intentions.

CONCLUDE

The doctrine of double effect always focuses on the importance of the healthcare provider's beneficent (good) intentions. Under no circumstance is it ever permissible for a healthcare provider to have the intention of being maleficent (do harm) to the patient. Beneficent intent (do good), with an unfortunate double effect of harm, is accepted if it is in accordance with standards of care. However, if an action or inaction that deviates from standards of care is defined as negligent, and if, as a result, the action or inaction is accompanied with patient harm, then that action, regardless of intent, would be grounds for healthcare malpractice.

In summary, although the doctrine of double effect emphasizes the importance of a practitioner's good intentions, it does not account for the consequences of their actions. Modern bioethics, using the four principles of bioethics and the processes of specifying, weighing, and balancing, holds practitioners legally, professionally, and morally accountable for the consequences of their actions, regardless of their intentions. This approach is more comprehensive and better suited to address the complexities of modern healthcare practice. (See also Chapters 9, 10, 15, and 21.)

* * *

REVIEW QUESTIONS

1. The doctrine of double effect establishes that the permissibility of an action is dependent on the consequences of the action, not the intention.
   A. True
   B. False

2. The doctrine of double effect is thoroughly paternalistic and does not recognize patient autonomy (informed consent).
   A. True
   B. False

3. Specifying and balancing the four principles of bioethics have superannuated or surpassed any need for the doctrine of double effect.
   A. True
   B. False

4. If a practitioner can establish that their intention was solely for something morally good, in the patient's best interests, then that will legally absolve the practitioner from any foreseen but unintended consequences.
   A. True
   B. False

5. Although the intent of the doctrine of double effect was for good, that does not absolve the practitioner from legal jeopardy for foreseen unintended effects.
   A. True
   B. False

Answers: 1B, 2A, 3A, 4B, 5A

* * *

CLINICAL VIGNETTES

1. Ms. Serenity Smith, a 72-year-old urban planner, presents with a painful terminal illness and seeks relief from their suffering. The practitioner faces a difficult decision: They can prescribe a medication that will alleviate the patient's pain but may also hasten their death. According to the doctrine of double effect, what is the practitioner's ethical responsibility in this situation?
   A. To prescribe the medication, as the intention is to relieve the patient's pain, and any consequences are incidental
   B. To refuse to prescribe the medication, as the consequences of hastening the patient's death are not morally acceptable
   C. To seek a second opinion, as the practitioner is unsure of the appropriate course of action
   D. To consult with a bioethicist, as the practitioner is still determining the ethical implications of their actions

   Explanation: According to the doctrine of double effect, an action that has two effects—one good and one bad—may be permissible if four conditions are met: The action is intrinsically good or indifferent, the good effect is intended, the bad effect is not intended but merely foreseen, and the good effect outweighs the bad effect. In this case, the practitioner's intention is to alleviate the patient's pain, which is an intrinsically good act. The hastening of the patient's death is an unintended but foreseen effect of the medication, and the good effect of relieving the patient's pain outweighs the bad effect of hastening their death. Therefore, medication prescribing is ethically permissible according to the doctrine of double effect. The option of refusing to prescribe the medication is not ethically justifiable because it would cause the patient unnecessary suffering. The options of seeking a second opinion or consulting with a bioethicist may be appropriate, but they do not address the ethical responsibility of the practitioner in this situation. *Answer*: A

2. Mr. Felix Evans, a 62-year-old stockbroker, comes to you with a complex healthcare condition and asks for your opinion on a treatment the patient has researched. The patient knows the treatment may have unintended consequences, but they believe the benefits outweigh the risks. You are familiar with the doctrine of double effect, which states that an action with both good and bad consequences can be morally justifiable if the practitioner intends for the good. Which of the following statements about the doctrine of double effect is most accurate?
   A. The doctrine of double effect recognizes the importance of patient autonomy and informed consent.
   B. The doctrine of double effect prioritizes the practitioner's intention and does not consider patient autonomy and informed consent.
   C. The doctrine of double effect prioritizes the consequence of the action, not the intention of the action.
   D. The doctrine of double effect is consistent with healthcare malpractice law.

   Explanation: The doctrine of double effect prioritizes the practitioner's intention and does not consider patient autonomy and informed consent. The doctrine of double effect holds that an action is morally justifiable if the practitioner's intention is good, with no mention of the patient's intention, even if the action has negative consequences. In this case, the patient researched a treatment with potential negative consequences but believes that the benefits outweigh the risks. The doctrine of double effect does not consider the patient's beliefs or values; therefore, patient autonomy and informed consent are not part of the evaluation. The practitioner's intention is the most important consideration, and if the intention of the practitioner is to do good, then the action can be considered morally justifiable. *Answer*: B

3. Ms. Fiona Woods, a 48-year-old investment banker, recognizes that the proposed treatment option may have unintended consequences, but the patient believes the benefits outweigh the burdens. As a practitioner, you are familiar with the four principles of biomedical ethics: autonomy (informed consent), beneficence (do good), nonmaleficence (do no harm), and justice (be fair). Which of the following statements about the four principles is most accurate?
   A. The four principles are outdated and do not consider the importance of the doctrine of double effect.
   B. The four principles have superannuated or surpassed any need for the doctrine of double effect.
   C. The four principles and the doctrine of double effect are equally important in guiding ethical decision-making in healthcare.

   Explanation: The correct answer is that the four principles have superannuated or surpassed any need for the doctrine of double effect. The four principles of biomedical ethics—autonomy (informed consent), beneficence (do good), nonmaleficence (do no harm), and justice (be fair)—are widely used to guide ethical decision-making in healthcare. The doctrine of double effect is an ethical principle that pertains to actions that have both good and bad consequences and is a specific application of the principle of nonmaleficence solely from the practitioner's perspective. Although the doctrine of double effect may be relevant in certain situations, the four principles are generally considered more comprehensive and applicable to a wider range of ethical issues in healthcare. Therefore, the four principles have superseded any need for the practitioner-centered doctrine of double effect. *Answer*: B

4. Mr. Solomon Hassan, a 59-year-old financial analyst, presents to their primary care practitioner complaining of chest pain and shortness of breath. The practitioner orders a battery of tests, including a stress test and an electrocardiogram, but fails to diagnose the patient with acute myocardial infarction (heart attack). The patient is sent home with a diagnosis of indigestion and instructed to follow up in a week. Two days later, the patient returns to the emergency room with severe chest pain and is diagnosed with a heart attack. The patient undergoes emergency coronary angioplasty and is recovering for several days in the hospital. What is the basis for a healthcare malpractice claim?
   A. Deviation from standards of care
   B. Practitioner intention
   C. Doctrine of double effect
   D. Patient noncompliance

   Explanation: The basis for a healthcare malpractice claim in this situation is a deviation from standards of care. Healthcare malpractice occurs when a practitioner fails to meet the accepted standard of care, resulting in harm or injury to the patient. In this case, the primary care practitioner may have deviated from the accepted standard of care by failing to diagnose the patient with a heart attack despite their symptoms and the test results, resulting in a delay in treatment and harm to the patient. The other options are not relevant in this situation. The practitioner's intention and the doctrine of double effect are related to ethical decision-making, and patient noncompliance would only be relevant if the patient had failed to follow the practitioner's instructions or treatment plan. *Answer*: A

5. Ms. Katherine Brooks, a 63-year-old sales manager, comes to the emergency room, and the practitioner faces a difficult ethical dilemma. The practitioner must choose between two options with potential benefits and harm:
   Option 1: Standards of care: The practitioner provides a treatment that has been proven effective according to standards of care, but it may also have negative side effects for the patient.
   Option 2: Doctrine of double effect: The practitioner with good intentions deviates from standards of care, resulting in foreseen but unintended harm to the patient.
   Which of the following options would the practitioner be considered to have committed healthcare malpractice if harm resulted from their decision?
   A. Option 1: Standards of care.
   B. Option 2: Doctrine of double effect.
   C. Neither option would be considered healthcare malpractice.
   D. Both options could be considered healthcare malpractice.

   Explanation: The correct answer is "Option 2, the doctrine of double effect. Although the practitioner's intentions may be good, deviating from established standards of care is generally not ethically justifiable. The doctrine of double effect, an archaic ethical theory, fails to consider the importance of patient autonomy and informed consent, highly valued principles in contemporary bioethics. The focus should be on providing the most effective treatment for the patient while minimizing harm and upholding their autonomy. Therefore, it is essential for practitioners to follow established standards of care that are in line with the patient's reasonable goals, values, and priorities to ensure they are providing the most effective treatment for the patient while minimizing harm. Deviating from these standards can result in healthcare malpractice because of negligence and harm to the patient. *Answer*: B

REFLECTION VIGNETTES

1. Mr. Walter Jenkins, a 65-year-old retired carpenter, was hospitalized with severe chest pain and shortness of breath. Mr. Jenkins has smoked for more than 30 years and has been experiencing a persistent cough and occasional hemoptysis for the past 6 months. His healthcare history is otherwise unremarkable, and he has no known medication allergies. On

examination, the practitioner noticed that Mr. Jenkins' respiratory rate was 30 breaths per minute, and his oxygen saturation was 88% on room air. The chest X-ray revealed a mass in the left lung, measuring approximately 8 cm in diameter, and the computed tomography scan confirmed that the mass was malignant. The practitioner initiated treatment with supplemental oxygen, nebulizers, and intravenous pain medications, including opioids, to manage Mr. Jenkins' symptoms. However, despite aggressive management, Mr. Jenkins continued to experience significant pain and discomfort, affecting his quality of life. During the rounds, Mr. Jenkins expressed his desire to increase the pain medication dose, understanding that it could hasten his death.

What is the ethical issue(s)?

How should the ethical issue(s) be addressed?

2. Mrs. Alice Jones is an 85-year-old woman with a history of advanced dementia who was admitted to the hospital with aspiration pneumonia. She has been bedridden for the past year, and her oral intake has declined significantly over the past few months. She is now dependent on tube feeding for her nutrition and hydration. During the course of her hospitalization, Mrs. Jones developed severe pain and discomfort despite receiving high doses of pain medication. The practitioner discussed the situation with Mrs. Jones' family and explained that increasing the pain medication could potentially hasten her death, given her advanced age and underlying healthcare condition.

What is the ethical issue(s)?

How should the ethical issue(s) be addressed?

* * *

WEB LINKS

1. *Code of Medical Ethics*. American Medical Association. https://code-medical-ethics.ama-assn.org
   The AMA *Code of Medical Ethics*, established by the American Medical Association, is a comprehensive guide for healthcare practitioners. It addresses issues and challenges, promotes adherence to standards of care, and is continuously updated to reflect contemporary practices and challenges.
2. *A Hand-Book of Proverbs*. Hery Bohn. 1855. https://archive.org/details/ahandbookprover01raygoog/page/n525/mode/2up
   Proverbs as of sayings, sentences, maxims, and phrases.
3. "Principles of Clinical Ethics and Their Application to Practice." Basil Varkey. 2021. https://www.ncbi.nlm.nih.gov/pmc/articles/PMC7923912
   This source discusses integrating ethical aspects with clinical and technical expertise in patient care.
4. "Goods, Causes and Intentions: Problems with Applying the Doctrine of Double Effect to Palliative Sedation." Hannah Faris, Brian Dewar, Claire Dyason, et al. 2021. https://bmcmedethics.biomedcentral.com/articles/10.1186/s12910-021-00709-0
   This source discusses the challenges of applying the doctrine of double effect in palliative sedation.
5. "Doctrine of Double Effect." *Stanford Encyclopedia of Philosophy*. https://plato.stanford.edu/entries/double-effect
   This source provides a comprehensive overview of the doctrine of double effect, including its historical origins and applications.
6. "The Rule of Double Effect." Mary Katherine Brueck and Daniel P. Sulmasy. 2020. https://bioethics.hms.harvard.edu/journal/rule-double-effect
   This source discusses the formulation and application of the rule of double effect in bioethics.

7. "A Double Dose of Double Effect." C. E. Kendall. 2000. https://jme.bmj.com/content/26/3/204
   This source discusses the doctrine of double effect in the context of treatment outcomes, including the balance between beneficial and harmful effects.
8. "The Doctrine of Double Effect." Suzanne Uniacke. In Richard E. Ashcroft, Angus Dawson, Heather Draper, and John R. McMillan (Eds.), *Principles of Health Care Ethics*, 2nd ed. https://onlinelibrary.wiley.com/doi/pdf/10.1002/9780470510544.ch34
   This source provides an in-depth discussion of the doctrine of double effect in the context of healthcare ethics.
9. "The Double Effect Effect." Charles Foster, Jonathan Herring, Karen Melham, and Tony Hope. 2011. https://www.cambridge.org/core/journals/cambridge-quarterly-of-healthcare-ethics/article/abs/double-effect-effect/E22C1E9B0E34AC0D9EBC60D0E6073D1C
   This source discusses the controversy surrounding the doctrine of double effect in end-of-life issues.

EMAIL AND ELECTRONIC COMMUNICATION

19

The best practitioner is also a philosopher.
—Galen

* * *

MYSTERY STORY

BREACH OF TRUST: THE HIGH-STAKES FALLOUT OF A DIGITAL MISSTEP IN HEALTHCARE

On a monotonous Monday morning, Practitioner Thomas Jameson found himself locked in a weary duel with a backlog of paperwork. A grueling weekend shift at Digital Harmony Hospital had left him drained, his energy ebbing with each passing minute. Amidst the numb silence, the piercing beep of a new email notification punctuated the air, jolting him from his languid state.

With the reluctant curiosity of a man bracing for bad news, Practitioner Jameson unfurled the digital letter. He half-expected another routine plea for pain medication refills, but the actual contents made his heart skip a beat. The patient, Mr. Christopher Thompson, was grappling with an alarming triad of symptoms—chest pain, shortness of breath, and a creeping dread. His written plea for guidance echoed ominously in Practitioner Jameson's mind.

Wasting no time, Practitioner Jameson volleyed back an email, probing for more details. In light of the gathered information, he diagnosed a potential emergency, urging Mr. Thompson to seek immediate assistance either through an ambulance or a dash to the nearest hospital. As a footnote to the unfolding drama, he inscribed a reminder in Mr. Thompson's electronic health record to circle back for a follow-up.

Exhaling a sigh of relief, Practitioner Jameson paused to review the lightning-fast exchange. His relief curdled into a wave of dread as he realized his unintentional blunder—he had unwittingly breached patient confidentiality. The absence of written consent for email communication and lack of encryption in the correspondence constituted a stark violation of Health Insurance Portability and Accountability Act (HIPAA) regulations.

The magnitude of the mistake hung heavy on his conscience. Taking immediate action, he reported the incident to the hospital's compliance officer and mustered the courage to confront Mr. Thompson, confessing the mishap and expressing heartfelt apologies.

The ensuing investigation painted a grim picture: Practitioner Jameson's well-intentioned email had been snatched by an unauthorized entity, leaving Mr. Thompson's personal and healthcare data exposed to the unforgiving winds of the internet. The extent of the resulting damage remained a dreadful unknown, but the guilt and responsibility bore down on Practitioner Jameson.

In the wake of this digital debacle, Practitioner Jameson weathered a reprimand from the hospital, a hefty fine, a grueling refresher on HIPAA training, and a yearlong probation. The bitter fallout, however, sparked a profound realization among his colleagues about the inherent risks of electronic communication.

Emerging from the ashes of the incident, Practitioner Jameson and his team adopted a more cautious approach toward digital communication. Implementing encrypted

> protocols, obtaining written consent, and adhering to stringent guidelines to safeguard patient confidentiality and privacy became the team's newfound ethos.
>
> In the grand scheme at Digital Harmony Hospital, the episode, as unfortunate as it was, resulted in a crucial paradigm shift. It underscored the importance of rigorous privacy protocols in an increasingly digitized healthcare world, transforming a harsh lesson into a beacon for improved patient safety and privacy.

* * *

THINK

Communication is central to all patient–practitioner relationships and establishing an effective interprofessional treatment team. Just as the healthcare structure and implementation are rapidly changing, so is the evolution of communication technology and social expectations. Email, text messaging, and other electronic communications fully integrate into our social structures and personal communicative relationships.

ASSESS

Patient: Autonomy

With the increase in diverse forms of communication comes the necessity of increased responsibility to ensure that patient–practitioner communications are confidentially safe, secure, and consensual. Practitioners and patients must be aware of electronic communications' limitations and risks.

Face-to-face communications have multiple advantages over electronic communications such as email and texts. First, human presence and touch help establish and solidify the patient–practitioner relationship. Communication is interactive, where vocal inflections, gestures, and body movements work together synergistically to create shared intersubjective experiences among people. Vocal communication is unrecorded and private, creating an environment of trust and security. As a result, patients are more willing to share their deepest intimate secrets, and this protected health information (PHI) is necessary to establish an accurate diagnosis and develop an agreed-upon effective treatment plan. The history and physical will then be summarized in the healthcare record, which can be considered a safe, secure, and private location where only those involved in the patient's healthcare treatment will have legal access as needed.

Electronic communications have multiple advantages over face-to-face communications. Emails and texts fit well into the busy and chaotic schedules of the practitioner and patient and can span the reach of the globe. It is fast, short, and to the point and can occur at any hour of the day. A permanent record of all communications is there to be retrieved and reviewed by the patient or the practitioner. This allows patients to reflect on what has been communicated and search online for more information about their condition and treatment, providing even more educated, informed consent.

Practitioner: Beneficence and Nonmaleficence

Unlike face-to-face communications, however, before having any electronic communications with patients, the healthcare provider must get written consent to use and clarify what type of information is permissible for the practitioner to discuss with the patient. Patients must be fully aware of the risks of breaches in confidentiality, the possibility of someone reading their text messages on the patient's or practitioner's phone, or the possibility of someone getting access to their email on their computer. The patient must also know that confidentiality could be compromised if there is a breach. In addition, there is the problem of ensuring the person being communicated with is the intended person. Someone could theoretically communicate with the practitioner without the practitioner being aware that they are mistakenly communicating with someone other than the intended patient.

Electronic communications, such as email and texts, can also result in a much higher risk of miscommunication of factual information and misunderstanding of personal intentions. Professional standards and boundaries must be vigilantly kept, ensuring that the patient–practitioner relationship never crosses over into anything other than a professional relationship.

Public Policy: Justice

Electronic health records with PHI are no longer the central focus for institutional privacy risk management because electronic communication is now also part of that focus. Electronic communications that contain PHI have become a significant privacy risk. If an employer owns a patient's email and if an email with PHI is sent to a patient, then the employer would have access to that information and could, as a consequence, use the patient's PHI for hiring, firing, and promotion decisions. This would be a HIPAA confidentiality and privacy violation punishable by fines of up to $250,000 and a jail term of up to 5 years. However, emails with PHI can be safely sent with encryption.

It is permissible for a patient to send PHI by email to a practitioner without encryption. However, if the practitioner responds to the unencrypted email, then the patient's PHI will have been sent out again as part of the original email thread, resulting in an inadvertent but negligent HIPAA privacy violation. All electronic messaging with PHI must be securely encrypted.

Short message service texting is never HIPAA compliant because any text can be read and forwarded to anyone. Before using email and electronic communication, it is safest to get HIPAA authorization from the patient in writing and document the authorization in the healthcare record. In addition, encrypted communication protocols should always be used to ensure security, confidentiality, and privacy.

CONCLUDE

With electronic communications, the practitioner must always focus on the importance of security, confidentiality, privacy, and patient–practitioner relationship boundaries.

In summary, as electronic communication becomes more prevalent in healthcare, practitioners must maintain security, confidentiality, and privacy while using these methods. Obtaining written consent from patients, informing them of the risks, and always using encrypted communication protocols are essential steps to ensure the safe and responsible use of electronic communication in healthcare. In addition, maintaining professional boundaries and prioritizing the patient–practitioner relationship are vital in this evolving landscape of healthcare communication. (See also Chapters 7, 12, 51, 53, and 59.)

* * *

REVIEW QUESTIONS

1. Which of the following are advantages of face-to-face communications over electronic communications such as email and texts?
   A. Human presence and touch help establish and solidify the patient–practitioner relationship.
   B. Communication is interactive, where vocal inflections, gestures, and body movements work together synergistically to create shared intersubjective experiences among people.
   C. Vocal communication is unrecorded and private, creating an environment of trust and security.
   D. Patients are more willing to share their deepest intimate secrets.
   E. All of the above.

2. What are the advantages of emails and electronic communication over face-to-face communication?
   A. Emails and texts fit well into busy and chaotic schedules.
   B. Communication can span the reaches of the globe.
   C. Communication is fast, short, and to the point and can occur at any hour of the day.
   D. A permanent record of all the communications is there to be retrieved or reviewed by the patient or the practitioner.
   E. All of the above.

3. Which of the following is a negative aspect of emails and electronic communications?
   A. Someone else could get access to text messages or emails sent by the patient or practitioner.
   B. It can be difficult to establish that the person the patient or practitioner is communicating with is the intended person.
   C. There is a higher risk of miscommunication.
   D. There is a higher risk of professional boundary violation.
   E. All of the above.

4. Protected health information could be accessed by an employer if the employer owns the patient's email account.
   A. True
   B. False

5. If a patient sends a practitioner an unencrypted email with PHI and the practitioner replies with a response that does not have PHI, then that is not a HIPAA violation.
   A. True
   B. False

Answers: 1E, 2E, 3E, 4A, 5B

* * *

CLINICAL VIGNETTES
1. Ms. Kendra Jenkins, a 26-year-old advertising executive, presents to the clinic with symptoms of depression and anxiety. The practitioner considers the best way to communicate with the patient to establish a strong therapeutic relationship and gather information about their symptoms. Which of the following forms of communication is most likely to provide the greatest advantage in this situation?
   A. Electronic communication through a secure messaging platform
   B. Face-to-face communication in the clinic
   C. Telephone communication
   D. Video conferencing

   Explanation: Face-to-face communication in the clinic is likely to provide the greatest advantage because it provides the most comprehensive form of communication regarding gathering information about the patient's symptoms and establishing a strong therapeutic relationship. In a face-to-face interaction, the practitioner can use nonverbal cues to communicate empathy and understanding. Observing the patient's body language and other nonverbal cues may provide insight into their emotional state. In addition, in-person communication allows for the opportunity to ask follow-up questions and actively listen, facilitating a deeper understanding of the patient's concerns. Although electronic communication and other forms of telecommunication may be convenient in certain situations, they lack the richness and nuance of in-person interaction and may be less effective for establishing a strong therapeutic relationship and gathering comprehensive information about the patient's symptoms. *Answer*: B

2. Ms. Emma Coleman, a 32-year-old petroleum engineer, presents concerns about communication with their practitioner. The patient is often busy, and their chaotic schedule makes it difficult to schedule appointments or follow-up visits. The patient mentions that they prefer email and text messaging to communicate with their practitioner. Which of the following is the most likely advantage of the patient's preferred method of communication with their practitioner?
   A. Electronic communication is only possible during office hours.
   B. Electronic communication is limited to face-to-face interactions.
   C. Electronic communication is limited to the patient's local area.
   D. Electronic communication can occur at any hour of the day and can span the reaches of the globe.
   E. Electronic communication provides no permanent record for review.

   Explanation: Electronic communication can occur at any hour of the day and span the globe's reaches. This is because electronic communication, such as email and text messaging, allows asynchronous communication outside of traditional office hours and does not require the patient to be physically present in the same location as their practitioner. This can be a significant advantage for patients with busy schedules or living far from their practitioner. Electronic communication can also facilitate the exchange of information and updates, enabling the patient to ask questions and receive prompt responses. In addition, electronic communication allows for a record of the communication to be kept, which can be helpful for both the patient and practitioner to refer back to at a later time. *Answer*: D

3. Mr. Enoch Ortiz, a 52-year-old help desk analyst, presents concerns about electronic communication in healthcare. The patient has heard negative aspects of using emails and text messages for communication between healthcare providers and patients. Which of the following is *not* an aspect of using emails and text messages for communication?
   A. The possibility of someone else accessing the messages
   B. The difficulty in verifying the identity of the recipient
   C. The decreased risk of miscommunication
   D. The increased risk of professional boundary violations

   Explanation: The decreased risk of miscommunication is not an aspect of using emails and text messages for communication in healthcare. Emails and text messages can increase the risk of miscommunication due to the lack of nonverbal cues and the potential for messages to be misunderstood or misinterpreted. Some negative aspects of using emails and text messages for communication in healthcare include the possibility of someone else accessing the messages, difficulty verifying the recipient's identity, and increased risk of professional boundary violations. Electronic messages can be intercepted or accessed by unauthorized individuals, potentially compromising patient privacy and confidentiality. In addition, it may be difficult to confirm the identity of the recipient of the message, which can lead to errors or misunderstandings. Finally, electronic communication can increase the risk of professional boundary violations, such as using informal or inappropriate language, which can compromise the professional relationship between the patient and the practitioner. *Answer*: C

4. Ms. Nina Rivera, a 45-year-old film director, visits their primary care practitioner for a routine checkup. During the visit, the practitioner orders several laboratory tests, the results of which are later emailed to the patient. The practitioner notices that the email account used by the patient is a business account. Which of the following is a unique risk when using a business email account without encryption to communicate PHI?
   A. The email may be lost or delayed in transit.
   B. An unauthorized third party may intercept the email.

C. The employer may access the patient's email account and view the PHI.
D. The email may be stored on an insecure server.

Explanation: When a patient uses a business email account to receive PHI, their employer may access it and view the information, compromising the patient's privacy and confidentiality. This is especially true if the employer owns the email account and has policies allowing them to monitor employee communication. Other risks that are not unique when using an unencrypted email account to communicate PHI include the possibility of the email being intercepted by an unauthorized third party, the email being stored on an insecure server, and the email being lost or delayed in transit. It is important for healthcare providers to use secure methods of communication when transmitting PHI to protect the privacy and confidentiality of their patients, and this is now part of the standards of care. Failure to use encryption with PHI is a HIPPA violation. *Answer*: C

5. Mr. Oscar Powel, a 62-year-old nuclear engineer, emails their practitioner PHI in an unencrypted format. The practitioner replies to the patient's email with a response that contains no PHI but is also sent in an unencrypted format. Which of the following statements best describes this situation regarding HIPAA regulations?
   A. The practitioner's actions do not violate HIPAA regulations because the practitioner did not include any PHI in their response.
   B. The practitioner's actions are not considered a HIPAA violation because the patient initiated the email exchange by sending unencrypted PHI.
   C. The practitioner's actions are a HIPAA violation because the entire email thread, including the patient's PHI, gets resent when the practitioner replies in an unencrypted format.
   D. The practitioner's actions are not a HIPAA violation because the email exchange was between the practitioner and the patient and did not involve any third parties.

Explanation: The practitioner's actions are a HIPAA violation because the entire email thread, including the patient's PHI, gets resent when the practitioner replies in an unencrypted format. HIPAA regulations require that PHI be protected and secured when transmitted; using unencrypted email to transmit PHI violates this requirement. Even if the practitioner's response did not contain PHI, the fact that the entire email thread was resent in an unencrypted format means that the patient's PHI was also transmitted in an unsecured manner. Both the patient and the practitioner are responsible for protecting PHI's privacy and confidentiality, and using unencrypted email to transmit PHI violates HIPAA regulations. It is important for healthcare providers to use secure methods of communication when transmitting PHI to protect the privacy and confidentiality of their patients. *Answer*: C

REFLECTION VIGNETTES

1. Practitioner Jennifer Smith had cared for Ms. Lisa Johnson, a 65-year-old woman with a history of hypertension and diabetes, for several years. Ms. Johnson deeply trusted Practitioner Smith, who had developed a strong patient–practitioner relationship. Recently, Ms. Johnson has been experiencing several concerning symptoms and has undergone various tests, imaging studies, and other diagnostic procedures to determine the underlying cause. In the past, Practitioner Smith discussed results with Ms. Johnson during office visits. However, Ms. Johnson's busy schedule and mobility limitations made it increasingly difficult to visit the office for follow-up appointments. Ms. Johnson requested that Practitioner Smith communicate with her through email to discuss the results of her tests and treatment options.

   What is the ethical issue(s)?

   How should the ethical issue(s) be addressed?

2. Practitioner Emily Davis participated in the Doctors Without Borders program. Doctors Without Borders is a humanitarian organization that provides healthcare assistance to people in distress, conflict, and post-conflict environments. During her time with the organization, Practitioner Davis witnessed firsthand the devastating impact of war and conflict on communities and the importance of providing healthcare to those in need. As a way of sharing her experience with others and raising awareness about the work of Doctors Without Borders, Practitioner Davis began posting images and stories of the patients she treated on her social media accounts.

What is the ethical issue(s)?

How should the ethical issue(s) be addressed?

* * *

WEB LINKS

1. *Code of Medical Ethics*. American Medical Association. https://code-medical-ethics.ama-assn.org
   The AMA *Code of Medical Ethics*, established by the American Medical Association, is a comprehensive guide for healthcare practitioners. It addresses issues and challenges, promotes adherence to standards of care, and is continuously updated to reflect contemporary practices and challenges.
2. "American College of Physicians Ethical Guidance for Electronic Patient–Physician Communication: Aligning Expectations." Wei Wei Lee and Lois Snyder Sulmasy, for the American College of Physicians Ethics, Professionalism and Human Rights Committee. 2020. https://www.ncbi.nlm.nih.gov/pmc/articles/PMC7459080
   This position paper from the American College of Physicians offers recommendations for navigating e-communication, focusing on ethics and professionalism in patient care, privacy, and confidentiality; practice considerations; and alignment of patient and physician expectations.
3. "Electronic Communication with Patients." American Medical Association. https://code-medical-ethics.ama-assn.org/ethics-opinions/electronic-communication-patients
   This source discusses the importance of informed consent and confidentiality in electronic patient communication.
4. "2.3.1 Electronic Communication with Patients." American Medical Association. https://code-medical-ethics.ama-assn.org/sites/default/files/2022-08/2.3.1%20Electronic%20communication%20with%20patients%20--%20background%20reports.pdf
   This source discusses the use of electronic communication, such as email or text messaging, as a useful tool in the practice of healthcare and in facilitating communication within a patient–physician relationship.
5. "Examination and Diagnosis of Electronic Patient Records and Their Associated Ethics: A Scoping Literature Review." Tim Jacquemard, Colin P. Doherty, and Mary B. Fitzsimons. 2020. https://bmcmedethics.biomedcentral.com/articles/10.1186/s12910-020-00514-1
   This source discusses the ethical implications of electronic patient record technology as a key enabler for improvements to healthcare service and management.
6. "American College of Physicians Ethical Guidance for Electronic Patient–Physician Communication: Aligning Expectations." Wei Wei Lee and Lois Snyder Sulmasy, for the American College of Physicians Ethics, Professionalism and Human Rights Committee. https://www.acponline.org/sites/default/files/documents/clinical_information/ethics-professionalism/ethical-guidance-for-electronic_-patient-physician-communications-jgim2020.pdf
   This source discusses the benefits of electronic communication between patients and their physicians, including improved patient care, patient satisfaction, and clinical outcomes.

ERRORS

20

A careful practitioner... before attempting to administer a remedy to your patient, you must investigate the patient's malady you wish to cure, the patient's habits when in health, and the patient's physical constitution.
—Marcus Tullius Cicero

* * *

MYSTERY STORY

A WINTER NIGHT'S WAKE-UP CALL: TRANSPARENCY, ACCOUNTABILITY, AND THE COLD TRUTH IN SURGICAL ERRORS

The bitter chill of winter made its presence known, setting the backdrop for an event that would forever alter the professional path of Practitioner Nicholas Baxter at TrustShield Hospital. The quiet of the night was shattered by a call from the emergency department, urgently beckoning him to attend to a patient in distress. His patient, Mr. Jonathan Kennedy, had arrived in the grip of severe abdominal pain, the cause of which was quickly identified—an inflamed appendix in desperate need of extraction.

As Practitioner Baxter began the operation, the calm, routine rhythm of the procedure was disrupted. A chilling realization took hold: His scalpel had inadvertently grazed Mr. Kennedy's colon, causing a minor but potentially harmful tear. Swiftly, his skilled surgical team repaired the damage and completed the appendectomy, but the haunting echo of the mistake remained, a stark reminder of the gravity of Practitioner Baxter's error.

The following days saw Practitioner Baxter grappling with the weight of his misstep. He found himself suspended in a precarious balance between his professional obligation to disclose the error and his fear of the potential consequences—losing his healthcare license, facing litigation, or confronting the erosion of his reputation. Seeking guidance, he confided in a trusted colleague, Practitioner Harrison, who reinforced their collective duty as healthcare practitioners—always to prioritize the welfare of the patient.

Embracing transparency over evasion, Practitioner Baxter finally disclosed the error to Mr. Kennedy and his family. His sincere apology and acceptance of accountability were met with understandable initial shock and disappointment from Mr. Kennedy. Still, in the honesty and integrity shown by Practitioner Baxter, Mr. Kennedy found a measure of reassurance and trust.

The commitment to patient safety drove the hospital to conduct a root cause analysis. The inquiry revealed a systemic breakdown in communication within the surgical team, which had ultimately led to the unfortunate event. This revelation resulted in the hospital implementing rigorous protocols for communication and double-checking surgical procedures, acting as a bulwark to prevent a recurrence of such errors.

Despite facing repercussions such as additional training and increased supervision at TrustShield Hospital, Practitioner Baxter viewed these measures not as a punishment but, rather, as an opportunity for growth. The experience had imparted a powerful lesson: Transparency and accountability in healthcare are not just expectations but necessities. The fear of consequences should never overshadow the obligation to provide the highest standards of patient care and safety.

Through his journey at TrustShield Hospital, Practitioner Baxter became a living testament to the ethos of honesty and accountability, fostering a culture that prioritizes patient safety and trust in the complicated labyrinth of healthcare.

* * *

THINK

The patient–practitioner relationship is a social contract that establishes a relationship grounded on protected health information, confidentiality, and mutual trust. Honesty and transparency reflect this trust, and lying and not revealing information relevant to the patient's healthcare, such as a healthcare error, violate this patient–practitioner relationship, which is the very foundation of the art of healing. Therefore, regardless of the practitioner's sentiments, the practitioner must inform the patient about all errors that might have a direct healthcare effect on the patient and communicate all information needed for the patient to make an autonomous informed consent decision.

ASSESS

Patient: Autonomy

The patient will only be substantially autonomous if they know healthcare errors. The provision of manipulative information, meaning the willful disclosure of only selective information to determine a patient's decision, is a direct violation of "freedom," a necessary condition of the principle of autonomy or informed consent. The patient's consent is not substantially informed under manipulation conditions, such as concealed errors. Not revealing healthcare errors violates patient autonomy's legal, professional, and moral rights.

A patient has the negative right of autonomy, meaning that others are obligated not to interfere with the patient's informed consent decision-making process. A practitioner not revealing healthcare errors violates the patient's negative rights.

A patient has the positive right of autonomy, meaning that others have an obligation to provide the patient with information relevant to making informed consent decisions. A practitioner not revealing healthcare errors violates the patient's positive rights. The practitioner not revealing healthcare errors violates the patient's negative and positive rights to make an autonomous, informed consent decision.

Because the patient did not provide informed consent authorizing the practitioner to commit the treatment error, it follows that all healthcare errors violate informed consent.

Practitioner: Beneficence and Nonmaleficence

The professional principles of beneficence (do good) and nonmaleficence (do no harm) require that the practitioner maximizes the patient's best interests using the patient's reasonable goals, values, and priorities. Except for very unusual circumstances, such as placebos for clinical trials or therapeutic privilege for situations in which the information would cause serious harm to the patient or others, it is professionally expected that all errors which relate to the patient's healthcare be disclosed to the patient in a timely fashion. Being honest and informative about errors is considered professionally necessary to maximize the patient's best interests.

There are therapeutic benefits when the practitioner apologizes for errors made by one or more members of the practitioner's healthcare team and when the practitioner takes full responsibility for errors. The art of healing as a practice of evidence-based care, with its probabilistic uncertainties and complexities, makes it inevitable that errors will occasionally happen. What the practitioner does when such errors occur will reflect the practitioner's character and ability to behave professionally and not violate the patient–practitioner relationship. Not revealing the error to the patient would violate beneficence (do good) and nonmaleficence (do no harm).

A healthy patient–practitioner relationship psychologically results in more open and honest discussions by the patient and physiologically results in better patient healthcare outcomes, maximizing the patient's best interests.

Public Policy: Justice

As a matter of justice (be fair), the patient has the right to know and be substantially informed of all healthcare errors related to the patient's healthcare treatment. As a negative right, nobody should interfere with the patient gaining truthful information about their healthcare. As a positive right, the practitioner and healthcare team must provide all relevant information to the patient's care. Anything less than communicating the error to the patient would violate justice.

The practitioner has a duty of nonmaleficence (do no harm) and beneficence (do good) to report any errors to the institution so that a root cause analysis can be conducted and mitigation policies implemented. *Root cause analysis* is a quality improvement measure that identifies what, how, and why a preventable error occurred so that the institution can create policies and procedures to prevent similar errors from happening again. Examples of error mitigation procedures are the implementation of checklists, double-checking doses, and armband patient identification.

Failure to report a healthcare error to the patient and the institution would be a legal, professional, moral, and institutional policy violation. Honest communication and personal accountability for healthcare errors are necessary for the social contract between the institution, practitioner, and patient.

CONCLUDE

There is no equivocation regarding the practitioner's professional and moral obligation to disclose a healthcare error, apologize for the error, and take full responsibility for the error, regardless of whether or not the error had a negative impact on the patient's healthcare.

In summary, disclosing healthcare errors and addressing them transparently are crucial to preserving the patient–practitioner relationship and ensuring patients can make informed decisions about their care. Practitioners must prioritize honesty, accountability, and institutional reporting to facilitate quality improvement measures and maintain a high standard of care. Upholding these values demonstrates a practitioner's commitment to professional and moral obligations and promotes a fair and just healthcare environment.

* * *

REVIEW QUESTIONS

1. Not reporting an error to a patient violates the patient–practitioner relationship, which is the foundation of the art of healing.
   A. True
   B. False

2. A patient has a positive right of others providing information necessary for informed consent and a negative right of others not interfering with the patient's informed consent decision-making process. A practitioner not revealing healthcare errors violates the patient's negative and positive rights.
   A. True
   B. False

3. An error is not a violation of informed consent and should be considered part of the probabilistic uncertainties associated with the practice of healthcare.
   A. True
   B. False

4. Which of the following is applicable to root cause analysis?
   A. It is a quality improvement measure.
   B. It identifies what, how, and why an error occurred.
   C. It is mandated by institutional policy.
   D. The purpose is to implement mitigation procedures.
   E. All of the above.

5. Which of the following best describes a practitioner's responsibility when a healthcare error occurs?
   A. The practitioner should only disclose the error if it significantly impacts the patient's health.
   B. The practitioner should keep the error confidential to maintain the patient's trust.
   C. The practitioner should disclose the error, apologize, and take responsibility for it, regardless of its impact on the patient's healthcare.
   D. The practitioner should report the error to the institution rather than disclose it to the patient.

Answers: 1A, 2A, 3B, 4E, 5C

* * *

CLINICAL VIGNETTES

1. Ms. Stella Reyes, a 48-year-old network security specialist, visits their practitioner for a routine checkup. During the examination, the practitioner realizes that they made a mistake in the patient's previous test results, and the patient has not been informed about it. What is the practitioner's most appropriate course of action?
   A. Keep the error a secret and not inform the patient because it might cause unnecessary anxiety
   B. Inform the patient about the error, but only if they ask directly
   C. Inform a supervisor about the error and let them handle the situation
   D. Inform the patient about the error and communicate all necessary information for the patient to make an informed decision

   Explanation: The most appropriate course of action is to inform the patient about the error and communicate all necessary information for the patient to make an informed decision. It is essential to inform the patient of the mistake, which could impact their health and well-being. The practitioner should explain the error and the potential consequences to the patient and provide them with all the necessary information to make an informed decision about their care. The patient has a right to know about any errors that may affect their health, and it is the practitioner's ethical and legal responsibility to inform the patient of the mistake. Keeping the error a secret would violate the patient's rights and could lead to further harm. Similarly, only informing the patient if they ask is insufficient because the patient may not be aware of the mistake and may not think to ask about it. It is also inappropriate to shift the responsibility to someone else. Therefore, the most appropriate course of action is to inform the patient directly and provide them with all the necessary information to make an informed decision. *Answer*: D

2. Ms. Lana Ahmed, a 66-year-old lighting designer, presents to the clinic for a routine checkup. During the visit, the practitioner realizes that a mistake was made during a previous procedure that could impact the patient's health. The patient has a right to be informed of any healthcare errors and make informed decisions about their care, but the practitioner faces a dilemma. If the error is disclosed, it could violate the patient's right to make informed decisions because of the lack of trust in the practitioner that might ensue from the disclosure. However, if the error is not disclosed, it would interfere with the patient's positive right to be informed. Which of the following actions is the most appropriate for the practitioner in this situation?
   A. Do not disclose the error to the patient
   B. Disclose the error to the patient but emphasize that it is not the practitioner's fault
   C. Disclose the error to the patient, apologize for the mistake, and provide all necessary information for the patient to make an informed decision
   D. Consult with a colleague before making a decision to disclose

Explanation: The practitioner needs to disclose the error to the patient, apologize for the mistake, and provide all necessary information for the patient to make an informed decision. Although it is understandable that the practitioner may be hesitant to disclose the error due to concerns about damaging the patient's trust, it is essential that the patient be informed of any errors that may impact their health. This allows the patient to make informed decisions about their care and take necessary steps to address the error. In addition, withholding information about the error can violate the patient's right to be informed. Disclosing the error but emphasizing that it is not the practitioner's fault is inappropriate because it can be seen as an attempt to shift blame or avoid taking responsibility for the error. Similarly, not disclosing the error is inappropriate because it violates the patient's right to be informed. Consulting with a colleague before making a decision to disclose may be appropriate in some situations. Still, ultimately it is the practitioner's responsibility to inform the patient of any errors that may impact their health. *Answer*: C

3. Ms. Adriana Temple, a 72-year-old marine engineer, presents to the emergency department with severe abdominal pain. The treating practitioner orders a computed tomography (CT) scan to evaluate the source of the pain. During the scan, the technologist accidentally injects the patient with a contrast agent intended for another patient. The patient experiences an allergic reaction and requires healthcare treatment. Which of the following statements best describes the situation?
    A. The healthcare error did not violate informed consent because the patient consented to the CT scan.
    B. The healthcare error was not a violation of informed consent because the patient did not experience any permanent harm from the reaction.
    C. The healthcare error violated informed consent because all healthcare errors violate informed consent.
    D. The healthcare error was not a violation of informed consent because errors are just part of the probabilistic uncertainties associated with medicine.

Explanation: The healthcare error violated informed consent because all healthcare errors violate informed consent. Informed consent is a fundamental ethical and legal principle that requires healthcare providers to inform patients of the risks and benefits of healthcare interventions and obtain the patient's agreement before proceeding with any treatment or diagnostic procedure. In this case, the patient consented to the CT scan but did not consent to being injected with a contrast agent intended for another patient. The healthcare error resulted in an allergic reaction and required healthcare treatment, an adverse outcome that the patient did not consent to. Therefore, the healthcare error violated informed consent because the patient was unaware of the risks of injecting a contrast agent intended for another patient. The fact that the patient consented to the CT scan does not justify the violation of informed consent related to injecting the contrast agent intended for another patient. The fact that the patient did not experience any permanent harm from the reaction does not justify the violation of informed consent. Errors are not an inevitable part of medicine and do not excuse the violation of informed consent. Healthcare providers are responsible for taking all necessary steps to prevent healthcare errors and inform patients of any risks associated with healthcare interventions. *Answer*: C

4. Mr. Isaiah Lee, a 61-year-old surveyor with a history of hypertension and diabetes, presents to the hospital complaining of chest pain and shortness of breath. The patient is diagnosed with acute myocardial infarction. Despite appropriate management, the patient deteriorates and passes away. As per institutional policy, an institutional root cause analysis (RCA) is conducted to identify the causes of the adverse event and implement quality improvement measures. The RCA team found that the delay in the administration of thrombolytic therapy was a significant factor in the unfavorable outcome of the patient. Which of the following best describes the purpose of conducting an institutional RCA?

A. To assign blame to individual practitioners
B. To identify areas for improvement in patient care so that such events do not happen again
C. To comply with regulatory requirements
D. To assess the cost-effectiveness of care delivery

Explanation: The RCA team needs to identify areas for improvement in patient care so that such events do not happen again. An institutional RCA aims to identify the underlying causes of adverse events, such as delays in treatment, errors in communication, or system failures, to implement quality improvement measures to prevent such events from happening again. The RCA team investigates the event by examining the actions leading up to the adverse event, identifying contributing factors, and developing recommendations for improvements. The ultimate goal of an RCA is to identify areas for improvement in patient care and to implement changes to prevent future adverse events. To assign blame to individual practitioners is not a correct answer because the RCA process is not intended to be a punitive exercise. The focus is on improving the system, not on blaming individuals. To comply with regulatory requirements is not a complete answer because although regulatory requirements may mandate conducting an RCA, the primary purpose is to identify areas for improvement in patient care, not simply to comply with regulations. Assessing the cost-effectiveness of care delivery is incorrect because although cost-effectiveness is an important consideration in healthcare, there are other purposes of an RCA. *Answer*: B

5. Ms. Kelly Baker, a 29-year-old mobile app developer, presents to their primary care provider complaining of persistent headaches and fatigue. During the examination, the practitioner realizes they had made a mistake in the patient's previous diagnosis and treatment plan. Which actions should the practitioner take in response to this realization?
    A. Deny responsibility for the error and continue with the previously prescribed treatment plan
    B. Apologize to the patient for the error, take full responsibility for it, and take steps to prevent similar errors from occurring in the future
    C. Wait for the patient to bring up the error before acknowledging it
    D. Blame a colleague for the mistake and offer no apology or corrective action

Explanation: The most appropriate course of action is apologizing to the patient, taking responsibility for the error, and taking steps to prevent similar errors. In the case of a healthcare error, it is important for healthcare providers to acknowledge it, take responsibility for it, and apologize to the patient. This helps build trust between the patient and provider and demonstrates a commitment to transparency and honesty. The practitioner should explain the error to the patient, apologize for any harm caused by the mistake, and discuss any necessary changes to the treatment plan. The practitioner should also take steps to prevent similar errors from occurring in the future. Denying responsibility for the error and continuing with the previously prescribed treatment plan are inappropriate actions because they ignore the impact of the error on the patient's health and well-being. Waiting for the patient to bring up the error before acknowledging it is inappropriate because it does not demonstrate a commitment to transparency and honesty. Blaming a colleague for the mistake and offering no apology or corrective action are inappropriate because they shift the responsibility for the error to someone else and do not address the impact of the error on the patient's health and well-being. *Answer*: B

REFLECTION VIGNETTES

1. Ms. Anna Davis, a 45-year-old kindergarten teacher, was admitted to the hospital with symptoms of dehydration, including fatigue, dizziness, and dry mouth. The Attending, Benjamin Wallace, conducted a physical examination and ordered an intravenous drip to

rehydrate Ms. Davis. However, when filling out the prescription, Attending Wallace made a typo and inadvertently underdosed the prescription. Nurse Laura Franklin, reviewing the orders, noticed the error and immediately informed Attending Wallace. Attending Wallace acknowledged the mistake and thanked Nurse Franklin for her vigilance. Nurse Franklin corrected the dosage error, and the medication was administered to Ms. Davis without harm.

What is the ethical issue(s)?

How should the ethical issue(s) be addressed?

2. Ms. Sarah Palmer, a 75-year-old retired nurse, was under hospice care due to a terminal illness. She had a do not attempt resuscitation order in her healthcare records. The attending practitioner was administering an intravenous drip to Ms. Palmer when they inadvertently made an error, resulting in Ms. Palmer going into cardiac arrest.

What is the ethical issue(s)?

How should the ethical issue(s) be addressed?

3. Mr. Charles Cody, a 70-year-old retired mechanic with multiple comorbidities, had been admitted to the hospital four times last month due to his worsening overall condition. Hospice had been recommended for Mr. Cody, but his offspring wanted aggressive management. During the current admission, the attending practitioner, in an attempt to be aggressive, gave Mr. Cody a dose of medication that was too high, resulting in cardiac arrest and his subsequent death. The differential diagnosis for Mr. Cody's multiple healthcare problems could include heart failure, chronic obstructive pulmonary disease, and renal failure. However, the medication error ultimately led to his death. The attending practitioner refuses to divulge the mistake to Mr. Cody's offspring.

What is the ethical issue(s)?

How should the ethical issue(s) be addressed?

* * *

WEB LINKS

1. *Code of Medical Ethics*. American Medical Association. https://code-medical-ethics.ama-assn.org
   The AMA *Code of Medical Ethics*, established by the American Medical Association, is a comprehensive guide for healthcare practitioners. It addresses issues and challenges, promotes adherence to standards of care, and is continuously updated to reflect contemporary practices and challenges.
2. "The Best Response to Medical Errors? Transparency." Dinah Wisenberg Brin. 2018. https://www.aamc.org/news-insights/best-response-medical-errors-transparency
   This article highlights the University of Michigan Health System's effective approach to healthcare errors, focusing on transparency, communication, and patient safety, which has influenced a wider adoption of communication and resolution programs in healthcare.
3. "Principles of Clinical Ethics and Their Application to Practice." Basil Varkey. 2021. https://www.ncbi.nlm.nih.gov/pmc/articles/PMC7923912
   This source discusses integrating ethical aspects with clinical and technical expertise in patient care.
4. "The Ethics of Error in Medicine." Shamai A. Grossman, Kiersten L. Gurley, and Richard E. Wolfe. 2020. https://www.ncbi.nlm.nih.gov/pmc/articles/PMC7571429
   This source discusses the definition of "error" in the context of healthcare practice and the ethical implications of such errors.

5. "Ethics: Ethical Issues with Medical Errors: Shaping a Culture of Safety." Jeanne Merkle Sorrell. 2017. https://ojin.nursingworld.org/MainMenuCategories/ANAMarketplace/ANAPeriodicals/OJIN/TableofContents/Vol-22-2017/No2-May-2017/Ethical-Issues-with-Medical-Errors.html
   This source discusses the ethical issues associated with healthcare errors and the importance of creating a culture of safety in healthcare.
6. "Patient Rights." American Medical Association. https://code-medical-ethics.ama-assn.org/ethics-opinions/patient-rights
   This source discusses patients' rights in bioethics, including the right to informed consent and privacy.
7. "Mistakes." Douglas S. Diekema. https://depts.washington.edu/bhdept/ethics-medicine/bioethics-topics/detail/70
   This source discusses the occurrence of mistakes in healthcare practice and the ethical duty of physicians to disclose information about healthcare mistakes to their patients.
8. "Biomedical Ethics." Princeton University. https://hpa.princeton.edu/sites/g/files/toruqf2006/files/biomedical_ethics.pdf
   This source discusses the ethics of the physician–patient relationship, including the duty to help the patient and avoid harming them, as well as duties of confidentiality.

EUTHANASIA

21

Cure sometimes, treat often, and comfort always.
—Hippocrates

* * *

MYSTERY STORY

EDGE OF EXISTENCE: AN UNVEILING OF THE ETHICAL QUANDARY IN MERCY'S CLOSURE

As dawn's first light splashed across LifeGuard Sanctity Hospital, the serene silence was disrupted by the relentless pulse of life and death. Amidst this unyielding cycle, Practitioner Jane Sanders stood as a bulwark of empathetic resolve. An esteemed oncologist, her reputation as a harbinger of hope in the shadow of terminal afflictions was unparalleled. In the grueling battle against stage 4 lung cancer, Mr. John Parker, a 65-year-old gentleman, was her latest patient. Despite Practitioner Sanders' relentless efforts, the cruel scourge showed no mercy, dragging Mr. Parker into a vortex of persistent torment and agony.

One day, the distressed Mr. Parker requested a private dialogue with Practitioner Sanders. The conversation that followed left Practitioner Sanders grappling with an ethical paradox that would test the boundaries of her professional convictions. Mr. Parker, despair etched across his face, divulged his research into euthanasia—an act of intentionally ending a life to relieve unbearable suffering—and beseeched her to liberate him from his agony. The plea caught Practitioner Sanders off guard; her instincts swiftly guided her to explain the illegality and unethical nature of the practice.

As days turned into nights, Practitioner Sanders bore witness to Mr. Parker's dwindling spirit and the exacerbation of his pain. Her desperate consultation with a palliative care team and mental health professional failed to turn the tide in favor of Mr. Parker's peace. One fateful morning, Practitioner Sanders arrived to find the life extinguished from Mr. Parker's eyes. The cause of death was identified as a lethal dose of medication, not on any prescribed regimen.

The hospital's corridors echoed with the shock wave of the revelation. An intensive investigation culminated in the unraveling of a chilling truth: The lethal dosage was administered by a nurse, known for her controversial stance on euthanasia. The ensuing fallout saw the nurse dismissed and handed over to the law enforcement agencies. The investigation further unraveled the probability of coercion by Mr. Parker, who had been an active part of a pro-euthanasia group.

Shaken to her core, Practitioner Sanders was left teetering on the precipice of self-doubt and regret. Could she have been a better shield against Mr. Parker's suffering, thereby averting this tragic outcome? The entire incident served as a catalyst for robust policy alterations in the hospital to preclude similar occurrences in the future.

Haunted yet motivated by the incident, Practitioner Sanders transformed into an ardent champion for palliative care and the dignity of patients in their twilight journey. She spearheaded initiatives to educate her peers about euthanasia's ethical implications and the criticality of providing compassionate care.

The chilling narrative of Mr. Parker, treated at LifeGuard Sanctity Hospital, serves as an urgent clarion call in the realm of healthcare, highlighting the imperative of upholding the pillars of beneficence (do good) and nonmaleficence (do no harm). It also accentuates the necessity of open, honest, and empathetic communication between patients and practitioners, especially when the patient is imprisoned in the clutches of unbearable pain.

* * *

THINK

Etymologically, the term *euthanasia* comes from the Greek *Eu*, meaning "good," and *Thanatos*, meaning "death." Euthanasia means "good death."

The American Medical Association (AMA) *Code of Medical Ethics* 5.8 defines euthanasia as "the administration of a lethal agent by another person to a patient for the purpose of relieving the patient's intolerable and incurable suffering."

However, there is an incontrovertible contradiction between the role and function of euthanasia and the role and function of the healthcare profession. Euthanasia is the "art of dying," whereas medicine is the "art of healing." Rationally, the art of healing and the art of dying are a logical contradiction. Empirically, it is argued that the healthcare practice of euthanasia would negatively affect the patient–practitioner relationship and the public perception of the healthcare profession.

The AMA *Code of Medical Ethics* 5.8 concludes that "permitting physicians [practitioners] to engage in euthanasia would ultimately cause more harm than good. Euthanasia is fundamentally incompatible with the physician's [practitioner's] role as healer, would be difficult or impossible to control, and would pose serious societal risks."

The American College of Physicians (ACP) *Ethics Manual* states, "The College does not support the legalization of physician [practitioner]-assisted suicide or euthanasia. After much consideration, the College concluded that making physician [practitioner]-assisted suicide legal raised serious ethical, clinical, and social concerns."

Because of rational incoherence and the possible harmful consequences, the healthcare practice of euthanasia has been deemed professionally unacceptable.

ASSESS

Patient: Autonomy

If a patient has autonomously chosen to be euthanized, then it is understandable that the patient would want their practitioner to do the euthanization. From the patient's perspective, the practitioner would know the best method for euthanasia and have the best skills for delivering the method to the patient without error.

The autonomous informed consent process is a collaborative or joint decision-making process in that the diagnosis, prognosis, treatment options, information of risks and benefits, and the answering of patient's questions are provided by the practitioner to aid the patient in the making of an autonomous informed consent decision for authorizing the practitioner to provide treatment. In other words, the treatment options offered by the practitioner to the patient will only be those options that are standards of care. Futile treatments and treatments not professionally accepted are not options for authorization. Euthanasia is not within the standards of care; therefore, there can be no patient authorization of a treatment option that will not be provided.

Practitioner: Beneficence and Nonmaleficence

Regardless of whether or not the patient autonomously chooses to be euthanized or whether or not euthanasia is legal, those facts do not determine if practitioners should be permitted by the healthcare profession to be engaged in euthanasia. For example, capital punishment is legal and enforced in 27 states. Yet, the healthcare profession still forbids practitioners from participating in the killing of a convicted prisoner because of the logical contradictions and harmful consequences. Similarly, even if euthanasia became legal in some states, it would not become professionally permissible for practitioners to participate or provide assistance.

If it is rationally true that euthanasia and the art of healing together result in a logical contradiction, such as being both a healing art and dying art, and if euthanasia would result in unacceptable harmful consequences, such as violating the patient–practitioner relationship or diminishing the public's trust of the healthcare profession, then it would be reasonable for the healthcare profession to forbid practitioners from such activities, even if practitioners were to personally or individually believe in its legal and moral permissibility.

Public Policy: Justice

Practitioner-assisted death or *death with dignity* is defined as the practitioner providing the patient with the knowledge or means of ending their own life voluntarily.

Euthanasia differs from practitioner-assisted death. With assisted death, the patient is the one who takes the medicine or administers the method that results in death, whereas with euthanasia, the individual does not directly end their own life. Instead, another person acts to cause the individual's death. Because of this distinction, euthanasia is illegal throughout the United States. However, euthanasia is legal in Belgium, Canada, Luxembourg, the Netherlands, New Zealand, Spain, and several states of Australia.

Death with dignity, also known as assisted death, is legal in 11 U.S. jurisdictions: California, Colorado, the District of Columbia, Hawaii, Maine, Montana, New Jersey, New Mexico, Oregon, Vermont, and Washington.

CONCLUDE

In the United States, the healthcare profession has made it clear that euthanasia is never permissible for a practitioner to engage in, even if it were to become legal. If the practitioner is having difficulty in controlling a patient's pain and suffering, then they need to seek out a palliative care consultation while providing a supportive and caring patient environment.

In summary, the healthcare profession in the United States strongly opposes euthanasia, even if it becomes legal. The focus should be on providing palliative care and a compassionate patient environment to manage pain and suffering. Upholding the role of a practitioner as a healer is essential in maintaining the trust and integrity of the patient–practitioner relationship and preserving the healthcare profession's commitment to the principles of beneficence and nonmaleficence. (See also Chapters 9, 10, 15, 31, and 62.)

* * *

REVIEW QUESTIONS

1. The term *euthanasia* comes from Greek, meaning
   A. Children in Asia
   B. Good death
   C. Bad death
   D. Compassion

2. Both the AMA and the ACP will allow euthanasia but only under some very restricted conditions.
   A. True
   B. False

3. In the states where euthanasia is legal, it is professionally permissible for practitioners to participate under strict limitations.
   A. True
   B. False

4. Treatment options offered by the practitioner to the patient are those that meet standards of care. Futile treatments and treatments not professionally accepted are not options for patient authorization.
   A. True
   B. False

5. With practitioner-assisted death or death with dignity, the practitioner is the one who administers the method that results in death.
   A. True
   B. False

Answers: 1B, 2B, 3B, 4A, 5B

* * *

CLINICAL VIGNETTES

1. Ms. Matilda Patel, an 82-year-old retiree, has been diagnosed with a debilitating illness and is experiencing incurable and intolerable suffering. The patient is requesting euthanasia as a means to end their suffering. Which actions would be considered the most ethical for a practitioner to take?
   A. Administering a lethal agent to end the patient's suffering
   B. Referring the patient to a practitioner who will assist with euthanasia
   C. Explaining to the patient that euthanasia is not a permissible option and offering alternative forms of pain management and end-of-life care
   D. Ignoring the patient's request and not discussing the option of euthanasia

   Explanation: The most ethical action for a practitioner in this situation is explaining to the patient that euthanasia is not a permissible option and offering alternative forms of pain management and end-of-life care. Euthanasia, or intentionally ending a patient's life, is illegal in most jurisdictions and is considered ethically and morally controversial. Healthcare providers are responsible for providing the best possible care to their patients, including comfort care and pain management, but not intentionally ending a patient's life. Administering a lethal agent to end the patient's suffering is unethical because it violates the principles of bioethics, which prohibit intentional killing or harm to patients. Referring the patient to a practitioner who will assist with euthanasia is also not ethical because it can be seen as an endorsement of euthanasia and can put the practitioner at risk of legal and ethical violations. Ignoring the patient's request and not discussing the option of euthanasia is unethical because this approach fails to address the patient's concerns and needs and can lead to further suffering. *Answer*: C

2. Mr. Nathaniel Thomas, a 78-year-old retiree with end-stage cancer, presents with unrelenting pain and suffering to the practitioner. The patient requests the practitioner to assist in ending their life peacefully. Which actions are considered most ethical according to the AMA and ACP guidelines?
   A. The practitioner should comply with the patient's request and assist in ending their life.
   B. The practitioner should refer the patient to another practitioner who may assist in ending their life.
   C. The practitioner should alleviate the patient's physical symptoms while addressing their psychological and personal needs through palliative care.
   D. The practitioner should offer the patient a sedative.

   Explanation: The practitioner should alleviate the patient's physical symptoms while addressing their psychological and personal needs through palliative care. This involves providing comfort care and pain management while addressing the patient's emotional, spiritual, and social needs. Palliative care aims to improve the quality of life for patients with serious illnesses and does not intentionally end the patient's life. Complying with the patient's request and assisting in ending their life is unethical because it violates the principles of bioethics, which prohibit intentional killing or harm to patients. Referring the patient to another practitioner who may assist in ending their life is also not ethical because it can be seen as an endorsement of euthanasia and can put the practitioner at risk of legal and ethical violations. Offering the patient a sedative may be appropriate in some situations, but it is not the primary approach recommended by the AMA and ACP guidelines. The most ethical action for the practitioner to take in this situation is to provide palliative care, which involves alleviating the patient's physical symptoms while addressing their psychological and personal needs. This approach is consistent with bioethics principles and provides the best possible care for the patient while respecting their dignity and autonomy. *Answer*: C

3. Ms. Claire Allen, a 92-year-old retiree, suffers from a terminal illness with no chance of recovery and is experiencing significant pain and suffering. The patient's family seeks assistance ending their loved one's life. The practitioner is conflicted about what to do and turns to the AMA and the ACP for guidance. This scenario occurs in a state where euthanasia is legal and under strict limitations. Which of the following actions aligns with the ethical principles established by the AMA and ACP in the context of euthanasia or assisted suicide?
   A. The practitioner agrees to administer a lethal dose of medication to the patient at the request of the patient and the family.
   B. The practitioner refers the patient to another practitioner who is known to participate in euthanasia or assisted suicide.
   C. The practitioner provides palliative care of comfort and support to the patient and the family but declines to participate in any form of euthanasia or assisted suicide.
   D. The practitioner provides the patient with information about how to obtain a lethal dose of medication but declines to participate in the administration of the medication.

Explanation: The practitioner should provide palliative care of comfort and support to the patient and the family but decline to participate in euthanasia or assisted suicide. The AMA and the ACP prohibit intentional killing or harm to patients, and the practitioner's primary responsibility is to provide compassionate care and alleviate the patient's pain and suffering. In a state where euthanasia is legal, the AMA and ACP oppose these practices, even under strict limitations. The practitioner should discuss the patient's prognosis and treatment options, including palliative care and hospice care, with the patient and the family and provide emotional support and counseling to the patient and the family during this difficult time. Referring the patient to another practitioner who is known to participate in euthanasia or assisted suicide or providing the patient with information about how to obtain a lethal dose of medication but declining to participate in the administration of the medication can be viewed as a tacit endorsement of euthanasia and can put the practitioner at risk of legal and ethical violations. *Answer*: C

4. Ms. Mariah Phillips, a 54-year-old meteorologist, presents to the practitioner with a terminal illness and seeks treatment options. The practitioner explains to the patient that they can only offer treatments that meet standards of care and that futile treatments and treatments not professionally accepted are not options for patient authorization. Which treatments would the practitioner be ethically and professionally able to offer the patient?
   A. A treatment that is ineffective in treating the patient's condition and has significant adverse effects
   B. A treatment that has not been scientifically tested and lacks evidence of efficacy
   C. A treatment that is in line with established standards of care and is effective for the patient's condition
   D. A treatment that is considered experimental and has not been widely adopted by the healthcare community

Explanation: The practitioner can ethically and professionally offer the patient treatments that align with established standards of care and are effective for the patient's condition. The ethical principle of beneficence requires that the practitioner act in the patient's best interest and provide treatments that will benefit the patient. In addition, nonmaleficence requires that the practitioner avoid harm to the patient and not provide treatments that have significant adverse effects or are futile. Treatments that have not been scientifically tested and lack evidence of efficacy, treatments that are ineffective in treating the patient's condition, and treatments that are considered experimental and have not been widely adopted by the healthcare community are not options that meet standards of care and are not ethically or professionally appropriate for the practitioner to offer to the patient. *Answer*: C

5. Mr. Leonardo Wright, a 78-year-old retiree with a terminal illness, seeks a peaceful and dignified death. The patient asks the practitioner for assistance in ending their life. The practitioner explains that with practitioner-assisted death, or death with dignity, they do not administer the method that results in death. Still, they will only provide the patient with a lethal prescription, which is to be taken by the patient. Although it is professionally questionable for practitioners to engage in practitioner-assisted death or death with dignity, it is an option that some patients may choose in the face of terminal illness and unbearable suffering. Which of the following actions would align with the concept of practitioner-assisted death or death with dignity?
    A. The practitioner agrees to administer a lethal dose of medication to the patient.
    B. The practitioner provides the patient with a prescription for a lethal dose of medication and instructs the patient on how to self-administer the medication.
    C. The practitioner engages in end-of-life care, such as providing comfort measures and symptom management, but declines to participate in practitioner-assisted death or death with dignity.
    D. The practitioner refers the patient to a support organization that provides information and resources on end-of-life options.

Explanation: The practitioner who participates in practitioner-assisted death or death with dignity would provide the patient with a prescription for a lethal dose of medication and instruct the patient on how to self-administer the medication. In this scenario, the practitioner does not administer the method that results in death but only provides the patient with the means to end their life. Although it is professionally questionable for practitioners to engage in practitioner-assisted death or death with dignity, it is an option that some patients may choose in the face of terminal illness and unbearable suffering. Providing the patient with a prescription for a lethal dose of medication and instructing them on how to self-administer align with practitioner-assisted death or death with dignity. The practitioner agrees to provide the patient with a means to end their life, but the patient takes the final action. The other options—administering the medication, engaging in end-of-life care but declining to participate in practitioner-assisted death, or referring the patient to a support organization—are not considered practitioner-assisted death or death with dignity. *Answer*: B

REFLECTION VIGNETTES

1. Mr. David Brown, a 60-year-old retired teacher, is currently in the intensive care unit (ICU) with symptoms of a terminal illness, including severe pain and suffering. Despite the best efforts of the attending practitioner to manage his symptoms, Mr. Brown's condition has continued to deteriorate. He has decisional capacity and requests that the practitioner perform euthanasia, citing his unacceptable pain and suffering and a lack of joyful life prospects.

    What is the ethical issue(s)?

    How should the ethical issue(s) be addressed?

2. Ms. Rachel Lee, a 55-year-old lawyer, is currently in the ICU with a terminal illness, and her condition has continued to worsen despite aggressive healthcare intervention. Ms. Lee has decisional capacity and requests that the practitioner perform euthanasia, arguing that there is no moral difference between having the right to end her life by having someone

withdraw life-sustaining treatment and having someone administer a lethal dose. Ms. Lee argues that both actions are equal in that they each require someone else to do something, resulting in the patient's death.

What is the ethical issue(s)?

How should the ethical issue(s) be addressed?

* * *

WEB LINKS

1. *Code of Medical Ethics.* American Medical Association. https://code-medical-ethics.ama-assn.org
   The AMA *Code of Medical Ethics*, established by the American Medical Association, is a comprehensive guide for healthcare practitioners. It addresses issues and challenges, promotes adherence to standards of care, and is continuously updated to reflect contemporary practices and challenges.
2. "Ethical Considerations at the End-of-Life Care." Melahat Akdeniz, Bülent Yardimci, and Ethem Kavukcu. 2021. https://www.ncbi.nlm.nih.gov/pmc/articles/PMC7958189
   This source discusses the ethical difficulties for healthcare professionals regarding decisions on euthanasia and physician-assisted suicide.
3. "Relational Autonomy in End-of-Life Care Ethics: A Contextualized Approach to Real-Life Complexities." Carlos Gómez-Vírseda, Yves de Maeseneer, and Chris Gastmans. 2020. https://bmcmedethics.biomedcentral.com/articles/10.1186/s12910-020-00495-1
   This source discusses the principle of autonomy in contemporary bioethics and its influence on end-of-life care decisions.
4. "Nursing and Euthanasia: A Narrative Review of the Nursing Ethics Literature." Barbara Pesut, Madeleine Greig, Sally Thorne, et al. 2020. https://www.ncbi.nlm.nih.gov/pmc/articles/PMC7323743
   This source discusses the ethical considerations in nursing regarding euthanasia, highlighting the principles of nonmaleficence and beneficence.
5. "Euthanasia." American Medical Association. https://code-medical-ethics.ama-assn.org/ethics-opinions/euthanasia
   This source discusses the principle of informed consent in healthcare treatment and its relevance to decisions on euthanasia.
6. "Ethical Justifications for Voluntary Active Euthanasia." University of Richmond. https://scholarship.richmond.edu/jolpi/vol3/iss1/6/
   This source discusses the ethical principle of nonmaleficence and its application to justify a doctor's role in euthanasia.
7. "Nursing and Euthanasia: A Narrative Review of the Nursing Ethics Literature." Barbara Pesut, Madeleine Greig, Sally Thorne, et al. 2020. https://journals.sagepub.com/doi/10.1177/0969733019845127
   This source discusses key considerations in nursing ethics regarding euthanasia, including the nurse–patient relationship and the potential impact on the profession.
8. "Euthanasia and Assisted Suicide: An In-Depth Review of Relevant Historical Aspects." Yelson Alejandro Picón-Jaimes, Ivan David Lozada-Martinez, Javier Esteban Orozco-Chinome, et al. 2022. https://pubmed.ncbi.nlm.nih.gov/35242326
   This source discusses the relevance of euthanasia and assisted suicide in end-of-life care, highlighting advances in biomedical research and bioethics.

FINANCIAL DISCLOSURES

22

Is it not also true that no practitioner considers or enjoins what is for the practitioner's good, but that all seek the best interests of their patients? The practitioner is a ruler of bodies, not moneymaker.
—Plato

* * *

MYSTERY STORY

THE TEMPEST OF HIDDEN AGENDAS: NAVIGATING THE MURKY WATERS OF FINANCIAL DISCLOSURES IN HEALTHCARE ETHICS

As the storm's rage painted the glass windows of Crystal Clear Transparency Hospital with rivulets of rain, Practitioner Sebastian Montgomery, a renowned endocrinologist, stood before an eager crowd of colleagues. They had gathered for the grand rounds, a pivotal academic event, their interest piqued by Practitioner Montgomery's recently published, ground-breaking paper on an innovative diabetes treatment.

The atmosphere within the conference room was electric with anticipation. Practitioner Montgomery, a stalwart in his field, delivered a riveting presentation, illuminating the promising vistas of diabetes treatment. His professional acumen was admired, his words hung onto by those who sought to learn and incorporate this new knowledge into their practice.

Just as the echoes of applause started to recede and Practitioner Montgomery prepared to leave the platform, an unexpected voice broke the satisfaction-laden silence. It was Practitioner Fairchild, the stern, principled head of the ethics committee.

"Practitioner Montgomery," he began, his voice carrying an edge of seriousness, "Could you enlighten us on any financial conflicts of interest associated with your research?"

Caught off guard, Practitioner Montgomery stuttered a quick "No," claiming no financial conflicts existed. However, Practitioner Fairchild countered, revealing a recent report alleging Montgomery's substantial grant from a pharmaceutical company producing diabetes medication.

Practitioner Montgomery's charismatic facade slipped, his face a kaleidoscope of shock and guilt. A hushed pall descended on the conference room. The revelation was like a thunderclap, casting a gloomy shadow over Montgomery's glittering presentation. His peers' trust was shaken, their respect tainted with disappointment.

Weeks later, a patient prescribed the praised medication suffered severe undisclosed side effects, leading to a complaint against Practitioner Montgomery. A hospital inquiry unveiled the depth of his deceit—monetary grants, speaking fees, consulting payments—from the same pharmaceutical company. His silence on these financial ties had clouded his professional judgment and integrity.

The once-respected Practitioner Montgomery, affiliated with Crystal Clear Transparency Hospital, was stripped of his healthcare license, facing the sharp teeth of legal reprisals. His career, once soaring, plummeted into the abyss of professional disgrace. The incident served as a grim reminder to all healthcare professionals of the paramount importance of transparent financial disclosures and the undying commitment to patient-centric care. It starkly underscored that trust, once shattered, is challenging to restore, and the line between ethical obligations and personal gain should never blur.

* * *

THINK

Financial disclosures are essential because they are a significant source of real and perceived conflicts of interest. In the clinical context, a conflict of interest affects the practitioner's clinical judgments and redirects the practitioner's focus from patient-centered to practitioner-centered. The law, institutional policies, the healthcare profession, and moral mandates emphasize the importance of honesty and transparency about any financial interests that might be perceived as a conflict of interest. Financial interests include monetary values or interests that could directly or indirectly affect practitioners' judgment regarding patient management and professional presentations.

ASSESS

Patient: Autonomy

Patient autonomy is only possible within a patient–practitioner relationship.

Informed consent depends on the patient trusting the practitioner with their protected health information (PHI) necessary for the practitioner to provide accurate patient diagnosis, prognosis, and treatment options. Failure of the practitioner to provide financial disclosures of conflict of interest would compromise the patient's trust in the practitioner's healthcare judgments. Without patient trust, the patient will not freely disclose necessary information.

Informed consent also depends on the practitioner's use of PHI to make unbiased judgments on the patient's diagnosis, prognosis, treatment options, and risks and benefits. Failure to provide financial disclosures of conflict of interest compromises the practitioner's healthcare judgments, thus compromising the patient's informed consent.

Failure to provide financial disclosures violates the patient–practitioner relationship for both the patient and the practitioner, and the patient–practitioner relationship is a necessary condition for patient autonomy (informed consent).

Practitioner: Beneficence and Nonmaleficence

Regarding beneficence (do good) and nonmaleficence (do no harm), both principles are violated by the practitioner who does not provide financial disclosures of conflict of interest. A conflict of interest, by definition, can affect the practitioner's judgment of what is beneficial or risky for the patient, violating the professional end of maximizing the patient's best interests. The patient's best interests are determined by the patient's reasonable goals, values, and priorities, not those of the practitioner. Practitioners have a professional obligation to be patient-centered, not practitioner-centered. Conflicts of interest are influences that shift the practitioner's focus from being patient-centered to practitioner-centered.

Public Policy: Justice

Justice (fair social distribution of benefits and burdens) has resulted in legal and institutional mandates that practitioners provide both a conflict of interest and financial disclosure. All publications in healthcare journals and all presenters at healthcare grand rounds must disclose no conflicts of interest or financial interests related to the publication or presentation.

With regard to any lifelong learning or continuous professional development, it is mandatory to clearly and explicitly communicate any potential conflicts of interest as a prerequisite for qualifying for any lifelong learning credits or points associated with continuous professional development, be it continuing healthcare education credits, continuing nursing education credits, continuing pharmacy education credits, continuing dental education credits, continuing education units, maintenance of certification points, performance improvement continuing healthcare education, continuing professional development points, or quality improvement points. The commitment to transparency and the absence of conflicts of interest underpins the trust our healthcare community places in our educational activities, thereby ensuring that we consistently provide the highest standards of patient care.

CONCLUDE

It is always required that a practitioner disclose any financial conflict of interest. For grand rounds, there must always be a conflict of interest and financial disclosure, either affirming or denying, for any presentation intended for healthcare lifelong learning or continuing professional development credit.

In summary, practitioners and healthcare providers must always disclose any financial conflict of interest to uphold bioethics principles and maintain trust within the patient–practitioner relationship. This transparency is required for publications in healthcare journals, lifelong learning, and continuing professional development. By disclosing financial conflicts of interest, healthcare professionals can ensure that their focus remains patient-centered and that the best interests of their patients are prioritized. (See also Chapters 13, 25, 32, 50, and 51.)

* * *

REVIEW QUESTIONS

1. Conflict of interest
    A. Redirects focus to practitioner-centered
    B. Affects the selection and interpretation of data
    C. Affects clinical judgment
    D. All of the above

2. Failure of the practitioner to provide financial disclosures of conflict of interest compromises the patient's trust in the practitioner's healthcare judgments. Without patient trust, the patient will not freely disclose necessary information.
    A. True
    B. False

3. Failure of the practitioner to provide financial disclosures of conflict of interest may compromise the practitioner's healthcare judgments, compromising the patient's informed consent.
    A. True
    B. False

4. A conflict of interest violates the professional obligation of beneficence (do good) and nonmaleficence (do no harm) by changing the practitioner's focus from being patient-centered to practitioner-centered.
    A. True
    B. False

5. There is no correlation between conscious and unconscious biases related to the selection and interpretation of data.
    A. True
    B. False

6. Every grand grounds presentation that provides CME credit must have a conflict of interest notice.
    A. True
    B. False

Answers: 1D, 2A, 3A, 4A, 5B, 6A

* * *

CLINICAL VIGNETTES

1. Mr. Alijah Johnson, a 62-year-old male help desk analyst, presents to your clinic complaining of chest pain and shortness of breath. The patient has a history of coronary artery disease and has previously undergone percutaneous coronary intervention. The practitioner is accompanied by a pharmaceutical company representative promoting a new drug for treating coronary artery disease. Which of the following best describes the situation regarding the presence of the pharmaceutical representative during the patient's visit?
   A. There is no conflict of interest because the representative is simply providing information about a new treatment option.
   B. The presence of the representative may redirect the focus of the patient–practitioner interaction to practitioner-centered and negatively affect the practitioner's clinical judgment.
   C. The presence of the representative may affect the practitioner's selection and interpretation of data but will not negatively affect the practitioner's clinical judgment.
   D. The presence of the representative will not have any impact on the patient–practitioner interaction or the practitioner's clinical judgment.

   Explanation: The presence of a pharmaceutical representative during a patient's visit may create a conflict of interest for the practitioner. The representative may influence the practitioner's clinical judgment and divert the focus of the patient–practitioner interaction from patient-centered to practitioner-centered. The representative may present data selectively, leading to biased interpretation by the practitioner, which could impact the quality of care delivered to the patient. It is crucial for practitioners to recognize and manage potential conflicts of interest to ensure the delivery of high-quality, patient-centered care. *Answer*: B

2. Ms. Zoey Jenkins, a 48-year-old cloud computing specialist, is seen in a clinic for a routine follow-up visit. During the visit, the patient expresses concerns about the financial relationships the practitioner may have with pharmaceutical companies. What is the most appropriate response by the practitioner?
   A. Reassure the patient that these relationships do not influence healthcare decision-making
   B. Acknowledge the patient's concerns and provide transparency by disclosing all financial relationships
   C. Ignore the patient's concerns and proceed with the scheduled appointment
   D. Defend the practitioner's financial relationships and explain why they are necessary

   Explanation: The most appropriate response by the practitioner in this scenario is to acknowledge the patient's concerns and provide transparency by disclosing all financial relationships. Practitioners are ethically obligated to disclose any financial relationships they may have with pharmaceutical companies to their patients to promote transparency and maintain their patients' trust. The practitioner can reassure the patient that these relationships do not influence their healthcare decision-making but should also be transparent about any financial relationships. Ignoring the patient's concerns or defending the practitioner's financial relationships would not address the patient's concerns or promote transparency, potentially leading to a loss of trust and negatively impacting the patient–practitioner relationship. *Answer*: B

3. Ms. McKenzie Parker, a 52-year-old civil engineer, visits a practitioner for a diagnosis and treatment plan. The practitioner must access the patient's PHI to provide the best care. Still, the patient is concerned about the financial disclosures of the practitioner and wants to know if there are any potential conflicts of interest. What is the most important factor for ensuring the patient is willing to disclose PHI?
   A. The practitioner's ability to diagnose the patient accurately
   B. The patient's knowledge of their right to provide informed consent
   C. The practitioner's ability to ensure that there are no financial conflicts of interests
   D. The practitioner's ability to provide treatment options

Explanation: The most important factor for ensuring that a patient is willing to disclose PHI is the practitioner's ability to ensure that there are no financial conflicts of interest. Patients have a right to know about any financial relationships the practitioner may have with pharmaceutical companies or healthcare device manufacturers that could affect the treatment decisions made by the practitioner. This transparency can help build trust between the patient and practitioner and enhance the patient's willingness to provide the necessary PHI. In addition, it is important for the practitioner to disclose any financial relationships with the patient because this may impact the patient's decision to accept or refuse certain treatments. *Answer*: C

4. A pharmaceutical company approaches a practitioner to participate in a clinical trial for a new drug. The practitioner is offered a large sum of money to enroll patients in the trial and report the results. How does the financial conflict of interest violate the practitioner's professional obligations?
   A. By making the practitioner patient-centered
   B. By making the practitioner practitioner-centered
   C. By promoting beneficence (do good)
   D. By promoting nonmaleficence (do no harm)

Explanation: A conflict of interest changes the practitioner's focus from patient-centered to practitioner-centered. The practitioner must prioritize the patient's needs and well-being above the practitioner's financial interests. This violates the professional obligation of beneficence (do good) and nonmaleficence (do no harm) because the practitioner's decisions and actions are now motivated by personal gain rather than the patient's best interests. This can harm the patient and undermine the trust between the practitioner and the patient. *Answer*: B

5. A practitioner gives a grand rounds presentation on a new healthcare device. The institution requires all grand rounds speakers to disclose any financial conflicts of interest before the presentation. What is the correct method for fulfilling the financial conflict of interest disclosure requirement?
   A. The practitioner must not write a statement declaring that there is no financial conflict of interest and include it in the presentation slides.
   B. The person introducing the practitioner must not verbally communicate that there is no financial conflict of interest before the presentation begins.
   C. The practitioner must personally declare that there is no financial conflict of interest during the presentation.
   D. The after-presentation survey which confirms that there was a disclosure of any financial conflict of interest is not required for CME credit.

Explanation: The correct method for fulfilling the financial conflict of interest disclosure requirement is for the practitioner to personally declare that there is no financial conflict of interest during the presentation. This is important because it allows the audience to make an informed judgment about the potential influence of any financial ties on the presentation's content. Writing a statement declaring that there is no financial conflict of interest and including it in the presentation slides may not be sufficient because this method does not allow for any questioning or follow-up from the audience. The person introducing the practitioner should verbally communicate the financial conflict of interest before the presentation begins to add to the transparency of the presentation. In addition, the after-presentation survey which confirms that there was a disclosure of any financial conflict of interest is usually required for CME credit. *Answer*: C

REFLECTION VIGNETTES

1. Ms. Sarah Johnson, a 65-year-old retired schoolteacher, comes to the clinic complaining of dizziness, fatigue, and shortness of breath. Upon examination, the practitioner suspects she may have a slow heart rate and requires a pacemaker. The practitioner recommends the company that they believe has the best pacemaker on the market, which coincidentally has been donating funds for educational purposes for healthcare professionals at the institution where the practitioner works. The differential diagnosis includes aortic stenosis, heart block, and sick sinus syndrome.
   What is the ethical issue(s)?
   How should the ethical issue(s) be addressed?

2. Mr. John Smith, a 45-year-old construction worker, is referred to the practitioner to evaluate chronic lower back pain. The practitioner orders magnetic resonance imaging, which shows a herniated disc at L5–S1. The differential diagnosis includes discogenic pain, spinal stenosis, and facet joint syndrome. During a grand rounds presentation, the practitioner presents a study on a new pain medication without disclosing that the pharmaceutical company provided the research subjects and published the data in their in-house journal.
   What is the ethical issue(s)?
   How should the ethical issue(s) be addressed?

* * *

WEB LINKS

1. *Code of Medical Ethics.* American Medical Association. https://code-medical-ethics.ama-assn.org
   The AMA *Code of Medical Ethics*, established by the American Medical Association, is a comprehensive guide for healthcare practitioners. It addresses issues and challenges, promotes adherence to standards of care, and is continuously updated to reflect contemporary practices and challenges.
2. "Federal Privacy Protections: Ethical Foundations, Sources of Confusion in Clinical Medicine, and Controversies in Biomedical Research." Mary Anderlik Majumder and Christi J. Guerrini. 2016. https://journalofethics.ama-assn.org/article/federal-privacy-protections-ethical-foundations-sources-confusion-clinical-medicine-and/2016-03
   This source discusses the potential harms of disclosing patients' private information, including economic harm and social harm.
3. "What Should Physician–Researchers Tell Patient–Subjects About Their Relationships with Industry?" Jeffrey R. Botkin. 2021. https://journalofethics.ama-assn.org/article/what-should-physician-researchers-tell-patient-subjects-about-their-relationships-industry/2021-09
   This source discusses the legal requirements for manufacturers to report all payments to physicians and for physicians to confirm or dispute the data.
4. "Harmonization of Financial Disclosure Reporting in Biomedical Journals: A Shared Responsibility." Association of American Medical Colleges. 2021. https://www.aamc.org/media/46506/download?attachment
   This source discusses the requirement for manuscript authors to disclose all financial interests and professional relationships related to healthcare, research, and life sciences.
5. "Patient Rights and Ethics." Jacob P. Olejarczyk and Michael Young. https://www.ncbi.nlm.nih.gov/books/NBK538279
   This source discusses the right to informed consent in the United States, which is protected by state and federal legislation.

6. "Principles of Clinical Ethics and Their Application to Practice." Basil Varkey. 2021. https://www.ncbi.nlm.nih.gov/pmc/articles/PMC7923912
   This source discusses the ethical obligation of physicians to benefit the patient, avoid or minimize harm, and respect the values and preferences of the patient.
7. "Considerations for Applying Bioethics Norms to a Biopharmaceutical Industry Setting." Luann E. Van Campen, Tatjana Poplazarova, Donald G. Therasse, and Michael Turik on behalf of the Biopharmaceutical Bioethics Working Group. 2021. https://bmcmedethics.biomedcentral.com/articles/10.1186/s12910-021-00600-y
   This source discusses the ethical challenges in the biopharmaceutical industry, which operates at the intersection of life sciences, clinical research, clinical care, public health, and business.

FUTILE TREATMENT

23

What sense would it make, or what would it benefit a practitioner if they discover the origin of diseases but cannot cure or alleviate them?
—Paracelsus

* * *

MYSTERY STORY

ON THE BRINK OF DESPERATION: NAVIGATING THE ETHICAL LABYRINTH OF FUTILE TREATMENTS IN MEDICINE

In the tranquil haven of Clearwater, a town renowned for its placid landscapes, resided Practitioner Caroline Harris. An oncologist of impeccable repute and two decades of unwavering service at Evidence Base Care Center, she was a beacon of hope and solace for her patients. Little did she know, a storm was brewing that would upend her harmonious world.

The epicenter of the imminent tempest was Mr. Robert Mitchell, a patient wrestling with the cruel grasp of advanced-stage lung cancer. Despite multiple rounds of radiation and chemotherapy, his disease remained a formidable foe. His spirit, however, remained unvanquished, leading him to plead for an experimental treatment, a beacon he saw in his dark times.

The challenge was that this treatment, despite tales of miraculous recovery, held no scientific merit. It was a wild goose chase, one that Practitioner Harris knew could only exacerbate Mr. Mitchell's suffering. However, he clung onto this phantom of a cure with desperation, blind to reason.

Practitioner Harris found herself caught in an ethical quagmire. On the one hand, her commitment as a practitioner compelled her to ensure the best possible care for Mr. Mitchell; on the other hand, her ethical duty forbade causing him harm. To endorse the futile treatment was to walk a tightrope between beneficence (do good) and nonmaleficence (do no harm).

Despite her attempts to impart her wisdom, the pleas fell on deaf ears. The involvement of the family yielded no different result; their faith was shackled to this fallacious hope. With a heavy heart, Practitioner Harris realized she could no longer shoulder Mr. Mitchell's care and transferred him to another practitioner, standing firmly by her ethical principles.

However, the world she had always known seemed to crumble when she received a call from the state healthcare board. Mr. Mitchell had filed a complaint against her for refusing his preferred treatment. As a whirlwind of accusations, investigations, and a suspension from her life's calling engulfed her, Practitioner Harris was left to defend her integrity.

For the first time in her illustrious career, she was forced to wear the uncomfortable mantle of a malpractice suspect. Yet, she knew that in the face of unethical practice, she had stood tall. She had prioritized her patient's well-being, even if it had cost her a formidable storm.

The healthcare board's investigation led to the reaffirmation of her actions as aligned with evidence-based care and the upholding of bioethical principles. With all charges dropped and her professional license reinstated, Practitioner Harris emerged from the storm stronger.

> This incident left an indelible mark on her and the Evidence Base Care Center. She realized the criticality of understanding patients' motivations toward futile treatments and her role in correcting misconceptions. Committed to the principles of informed consent and patient autonomy, she realized that they were the fulcrum of the patient–practitioner relationship and equitable distribution of benefits and burdens. In the end, this storm had only steeled her resolve to serve her patients better, no matter the odds.

* * *

THINK

Futile treatment is any treatment determined by evidence-based care not to benefit the patient or achieve the intended healthcare outcomes. Standards of care are peer-reviewed, evidence-based care recognized and practiced by the healthcare community and established by the healthcare profession.

ASSESS

Patient: Autonomy

The patient–practitioner relationship is central to patient autonomy, as expressed with informed consent decision-making. Informed consent is a shared decision-making process because of the patient's positive right to be informed by the practitioner, in a manner that is understandable to the patient, of the diagnosis, prognosis, treatment options, benefits, and risks of each treatment option, including no treatment, and to have their questions answered. With this information, the patient can provide informed consent authorizing the practitioner to provide healthcare treatment in accordance with standards of care.

It is not the patient who presents treatment options to the practitioner; rather, the practitioner presents them to the patient. All treatment options must comply with standards of care. Futile treatments violate such parameters and should not be part of the patient's decision-making options.

If a patient chooses a futile treatment independently from the practitioner, then the practitioner should discuss why the patient wants the futile treatment. This will allow the practitioner to

- correct any misinformation regarding the practitioner's proposed standards of care treatment options; and
- provide the opportunity to correct any false beliefs and misinformation the patient may have regarding the futile, ineffective treatment.

This is a necessary corrective exchange because with correct healthcare knowledge, the patient is informed, making it possible to provide informed consent. Without having correct healthcare knowledge, the patient, by definition, lacks decisional capacity. It is not that the patient only has decisional capacity if they choose the right course of treatment; instead, it is the requirement to be logically consistent and coherent with evidence-based medicine and the patient's reasonable goals, values, and priorities. A patient's informed consent is the authorization for the practitioner to provide one of the standards of care options that the practitioner has proposed for maximizing the patient's best interests as determined by the patient's reasonable goals, values, and priorities.

Practitioner: Beneficence and Nonmaleficence

The practitioner has a professional obligation of beneficence (do good) only to provide evidence-based, peer-reviewed treatment options to benefit the patient. The practitioner also has the obligation of nonmaleficence (do no harm) to minimize the risk of harm to the patient. For the practitioner to provide futile treatment would violate beneficence (do good) and nonmaleficence (do no harm).

Sometimes, the patient's reasonable goals, values, and priorities may make the patient want to try one or more futile treatments independent of practitioner care. Not all futile treatments outside of evidence-based care are objectionable if the ineffective, futile treatments have minimal risks of harm, even if they have no hope of benefits.

As part of a complete healthcare history and physical, it is incumbent upon the practitioner to inquire into non-Western medicines, cultural treatments, and complementary approaches the patient has been engaged in. Some of these ineffective treatments may have a physiological and healthcare impact on evidence-based care, and it is essential to know this information to avoid the risk of adverse reactions. The practitioner is the one who offers standards of care treatment options for the patient to choose from, based on what will maximize the patient's best interests using the patient's reasonable goals, values, and priorities.

However, as a matter of healthcare professionalism, it is never permissible for a practitioner to offer or agree to provide futile treatment that would result in an increased risk of harm to the patient, as that would be a violation of the principle of nonmaleficence (do no harm) and a violation of standards of care. Offering a patient a knowingly ineffective or futile treatment violates healthcare professionalism.

Public Policy: Justice

Justice is defined as the fair distribution of societal benefits and burdens. Society has given the healthcare profession enormous resources and trust to help society address healthcare disparities. In response, society expects that when futile treatments harm or prevent or delay the patient from getting the proper effective healthcare treatment required, the practitioner has a social obligation to help the patient make rational and informed treatment decisions. Futile treatments should never delay proper patient treatment or drain society's limited healthcare resources.

CONCLUDE

It is important that the practitioner discusses with the patient why they want an ineffective and futile treatment. This will allow the practitioner to correct any misinformation the patient may have regarding the standards of care and also any misinformation the patient may have regarding ineffective and futile treatments.

In summary, practitioners play a vital role in helping patients make informed and rational treatment decisions. It is essential for practitioners to discuss with patients their reasons for wanting a futile treatment, correct any misinformation, and ensure that informed consent is obtained. By doing so, they uphold the principles of healthcare professionalism, ensure patient autonomy, and contribute to the fair distribution of benefits and burdens within society.

* * *

REVIEW QUESTIONS

1. A practitioner is responsible for respecting a patient's authority to choose a futile healthcare treatment and to provide it even if the practitioner disagrees.
    A. True
    B. False

2. A practitioner must respect the patient's right to choose a healthcare treatment that they believe will be best. It would be disrespectful to be so bold as to correct misinformation and false beliefs.
    A. True
    B. False

3. If a patient has misinformation and false beliefs related to treatment options, then the patient, by definition, lacks decisional capacity.
   A. True
   B. False

4. The practitioner has the professional obligation of nonmaleficence (do no harm) not to administer futile treatments that only increase the risk of harm.
   A. True
   B. False

5. Not all futile treatments outside of evidence-based care are objectionable if the ineffective treatments have minimal risks of harm, even if they have no hope of benefits.
   A. True
   B. False

6. Some ineffective treatments may have a physiological and healthcare impact on evidence-based care, and it is essential to know this information to avoid the risk of adverse reactions.
   A. True
   B. False

Answers: 1B, 2B, 3A, 4A, 5A, 6A

* * *

CLINICAL VIGNETTES

1. Ms. Natalia Miller, a 68-year-old astronomer, presents to their practitioner with a chronic healthcare condition that has not responded to previous treatments. The patient expresses interest in a new, experimental treatment they have read about online. Which of the following is the most appropriate response for the practitioner?
   A. Provide the patient with information about the experimental treatment and let them decide if they would like to proceed with it
   B. Explain to the patient that the experimental treatment is not an evidence-based standard of care and that it is unlikely to be effective, but offer it as an option for their consent
   C. Help the patient understand why the experimental treatment is not a viable option and discuss alternative treatments that are in line with evidence-based standards of care
   D. Offer to refer the patient to another practitioner who is experienced in the experimental treatment

   Explanation: The most appropriate response for the practitioner is to help the patient understand why the experimental treatment is not a viable option and discuss alternative treatments that align with evidence-based standards of care. Although patients may express interest in new, experimental treatments they have read about online, the practitioner must ensure that the treatment offered is safe, effective, and in line with evidence-based standards of care. The practitioner should explain to the patient why the experimental treatment may not be a viable option and discuss alternative treatments that are effective. Providing the patient with information about the experimental treatment and leaving the decision to them may not be the best course of action because the patient may need more healthcare knowledge or expertise to assess the risks and benefits of the treatment. *Answer*: C

2. Mr. Lazarus Wilson, a 73-year-old retiree, presents to the clinic with a complex healthcare condition, and the practitioner presents several standards of care treatment options. The patient is still trying to determine which option to choose and expresses confusion

about each option's potential risks and benefits. Which of the following best represents the requirement for informed consent?
   A. The practitioner provides the patient with limited information about one treatment option and encourages the patient to choose that option.
   B. The practitioner presents the information about each treatment option in a manner that is understandable to the patient and helps the patient make an informed decision.
   C. The practitioner decides on the treatment option and informs the patient what option to choose.
   D. The practitioner pressures the patient into choosing a particular treatment option without fully explaining the risks and benefits.

Explanation: Informed consent requires the practitioner to present the information about each treatment option in a manner that is understandable to the patient and helps the patient make an informed decision. Informed consent is a critical component of healthcare practice and involves providing understandable information about the nature, risks, benefits, and alternatives of a proposed treatment. This information allows the patient to make an informed decision about their healthcare. Providing understandable information empowers patients to decide what is best for them based on their goals, values, and priorities. It is incorrect to provide the patient with limited information on only one treatment option because the result is no choice. It is incorrect not to involve the patient in decision-making, which is essential for informed consent. Pressuring the patient is usually not in the patient's best interest and violates the requirement for informed consent. *Answer*: B

3. Ms. Yoko O'Rylee, a 59-year-old environmental engineer, has a serious healthcare condition. The practitioner's professional obligations include both beneficence (do good) and nonmaleficence (do no harm). Which of the following best describes those ethical obligations?
   A. To provide any treatment option requested by the patient, regardless of its potential benefits or risks
   B. To provide only treatments that are proven to be effective, even if they carry a high risk of harm
   C. To provide only treatments that offer a reasonable chance of benefiting the patient while minimizing the risk of harm
   D. To provide only futile treatments because they are the only options available

Explanation: The ethical obligation is to provide treatments that offer a reasonable chance of benefiting the patient while minimizing the risk of harm. This obligation is rooted in beneficence and nonmaleficence, which require practitioners to act in the patient's best interest and avoid harm. Providing any treatment option requested by the patient, regardless of its potential benefits or risks, would not necessarily be in the patient's best interest and could potentially cause harm. Similarly, providing only treatments that are proven to be effective, even if they carry a high risk of harm, would not necessarily minimize the risk of harm. Finally, providing only futile treatments is inconsistent with the principles of beneficence and nonmaleficence because it would not offer a reasonable chance of benefiting the patient. *Answer*: C

4. Mr. Kurt Chen, a 51-year-old costume designer, has been diagnosed with a terminal illness, and the healthcare team has exhausted all evidence-based treatment options. The patient's family wishes to try alternative treatments unsupported by scientific evidence and that carry minimal risk of harm. Which of the following statements best describes the ethical considerations in this scenario?
   A. It is unethical to provide futile treatments outside of evidence-based care, even if they carry minimal risks of harm.
   B. It is ethical to provide treatments that have little hope of benefit as long as they also carry minimal risk of harm.
   C. It is ethical to provide treatments that carry a risk of harm and no hope of benefit.
   D. It is unethical to provide treatments that have low hope of benefit and low risk of harm.

Explanation: In this scenario, the patient has been diagnosed with a terminal illness, and all evidence-based treatments have been exhausted. The patient's family expresses their wish to try alternative treatments that carry minimal risk of harm but are not supported by scientific evidence. The best statement that describes the ethical considerations in this scenario is that it is ethical to provide treatments with little hope of benefit as long as they also carry minimal risk of harm. This approach aligns with the ethical principles of beneficence, which requires practitioners to take action that promotes the patient's well-being, and nonmaleficence, which requires practitioners to avoid causing harm to the patient. In this situation, the patient and their family have the right to make an informed decision about their treatment options, even if they are not based on evidence-based care. It is important for the practitioner to provide the family with clear and accurate information about the potential risks and benefits of the alternative treatments while also making clear that they are not evidence-based. The patient and their family should make the ultimate decision in collaboration with the practitioner. The other options are incorrect because they do not align with the ethical principles of beneficence and nonmaleficence. *Answer*: B

5. Ms. Norah Henderson, a 32-year-old advertising executive, presents to the clinic for a routine checkup. The practitioner is conducting a complete healthcare history and physical examination. Although all of the following are important, which is the most critical factor for the practitioner to consider when determining prescriptions and dosages?
   A. The patient's occupation
   B. The patient's use of non-Western medicines, cultural treatments, and complementary approaches
   C. The patient's recent travel history
   D. The patient's social habits

Explanation: When taking a patient's healthcare history, it is critical for the practitioner to consider the patient's use of non-Western medicines, cultural treatments, and complementary approaches. This information is essential to assess the patient's overall health, including any potential interactions with prescribed treatments or drugs, which may affect the efficacy and outcome of the treatment. Non-Western medicines, cultural treatments, and complementary approaches can include herbal remedies, acupuncture, and meditation, affecting the patient's well-being and treatment outcomes. In addition, this information can provide insights into the patient's cultural background and beliefs, allowing the practitioner to provide culturally sensitive and appropriate care. Although the other options can be important, they are not as critical as the patient's use of non-Western medicines, cultural treatments, and complementary approaches. *Answer*: B

REFLECTION VIGNETTES
1. Mr. Michael Brown, a 60-year-old retired accountant, has been hospitalized for an extended period due to complications from cancer treatment. He expresses to the practitioner that he feels frustrated with the slow recovery and wonders if there are any alternative treatments that could help. He mentions that he has been reading about herbal remedies and vitamin supplements online and asks if it would be safe to start a regimen. The practitioner explains that although some supplements may have potential benefits, they may also interact with other medications or exacerbate certain conditions. The differential diagnosis includes drug–herb interactions, malnutrition, and the placebo effect.
   What is the ethical issue(s)?
   How should the ethical issue(s) be addressed?

2. Ms. Maria Rodriguez, a 30-year-old stay-at-home mother, is brought to the emergency room by her family due to severe abdominal pain. The practitioner diagnoses her with acute appendicitis and recommends immediate surgery. However, Ms. Rodriguez says she would like to pursue "prayer" and the "laying on of hands" instead of the accepted healthcare treatment. The practitioner explains that although the patient has the right to make their own decisions, informed consent requires a clear understanding of the risks and benefits of the chosen treatment. The differential diagnosis includes cultural beliefs, fear of surgery, and lack of knowledge.
What is the ethical issue(s)?
How should the ethical issue(s) be addressed?

* * *

WEB LINKS

1. *Code of Medical Ethics.* American Medical Association. https://code-medical-ethics.ama-assn.org
The AMA *Code of Medical Ethics*, established by the American Medical Association, is a comprehensive guide for healthcare practitioners. It addresses issues and challenges, promotes adherence to standards of care, and is continuously updated to reflect contemporary practices and challenges.
2. "Medical Futility." Nancy S. Jecker. https://depts.washington.edu/bhdept/ethics-medicine/bioethics-topics/articles/medical-futility
This source discusses the ethical authority of physicians to withhold or withdraw medically futile interventions and the importance of communication with professional colleagues, patients, and family.
3. "Ethical Considerations at the End-of-Life Care." Melahat Akdeniz, Bülent Yardimci, and Ethem Kavukcu. 2021. https://journals.sagepub.com/doi/full/10.1177/20503121211000918
This source discusses the ethical difficulties for healthcare professionals regarding decisions on resuscitation, mechanical ventilation, artificial nutrition and hydration, terminal sedation, withholding and withdrawing treatments, euthanasia, and physician-assisted suicide.
4. "Futility." Nancy S. Jecker. https://depts.washington.edu/bhdept/ethics-medicine/bioethics-topics/detail/65
This source discusses the term medical futility and how it applies when a treating healthcare provider determines that an intervention is no longer beneficial.
5. "Medical Futility: Legal and Ethical Analysis." Peter A. Clark. 2007. https://journalofethics.ama-assn.org/article/medical-futility-legal-and-ethical-analysis/2007-05
This source discusses the ethical argument that patients and surrogates have the right to request certain healthcare treatments based on their best interest.
6. "AMA *Code of Medical Ethics*' Opinions Related to Ethics of Life-Sustaining Technologies.." Rachel F. Harbut. 2019. https://journalofethics.ama-assn.org/article/ama-code-medical-ethics-opinions-related-ethics-life-sustaining-technologies/2019-05
This source discusses the ethical responsibility of a physician to use sound healthcare judgment on patients' behalf and to advocate for their patients' welfare.
7. "Meaningful Futility: Requests for Resuscitation Against Medical Recommendation." Lucas Vivas and Travis Carpenter. 2021. https://jme.bmj.com/content/47/10/654
This source discusses the term futility and how requests for futile care often cover a range of personal, emotional, cultural, and spiritual needs.

GENETIC INFORMATION NONDISCRIMINATION ACT

24

Wrong does not cease to be wrong because the majority share in it.
—Leo Tolstoy

* * *

MYSTERY STORY

GENETIC SECRETS: GENETIC DISCRIMINATION AND THE PURSUIT OF JUSTICE

In the heart of the thriving metropolis, where mysteries are layered as deep as the city's foundations, a relentless hunt was underway. Detective Olivia Bennett, a human riddle-solver with a reputation that resonated in every shadowy corner of the city, found herself in a vortex of perplexing circumstances, circling the untimely demise of the famed geneticist, Practitioner Eleanor Reynolds, at GeneGuard Integrity Hospital. The lost luminary had been on the precipice of revealing a pioneering discovery in the field of genetic markers for a particular rare disease.

As the veil of tranquility was lifted, chaos descended. A cascade of threatening messages had flooded Practitioner Reynolds' inbox, compelling her to abandon her momentous research. The safe haven of her laboratory fell prey to an unsolicited intrusion, and significant research samples evaporated. The climax of this horrifying series was the discovery of Practitioner Reynolds, cold and lifeless in her office.

Every person connected to the dead scientist found themselves face-to-face with Detective Bennett's probing inquiries, but the murky mystery remained untouched. That was until an anonymous call snaked its way to Bennett, claiming to hold the keys to the enigma. A rendezvous was set in the city's secluded edges, revealing the anonymous informant to be a nurse who had once worked closely with Practitioner Reynolds at GeneGuard.

The nurse wove a tale of Practitioner Reynolds steering a controversial project, a mission that flirted with genetic testing in the realms of insurance and employment. Drawing from her patients' DNA samples, Reynolds had conjured a storm she could not quell. The nurse hinted at a possible motive: The practitioner's groundbreaking work could have precipitated her death.

Emboldened, Detective Bennett plunged into the depths of Reynolds' past, unearthing a series of lawsuits related to genetic testing. Tales of a man deprived of insurance coverage and a woman dismissed from her job due to genetic predispositions emerged, painting a portrait of Reynolds as a relentless fighter against genetic discrimination.

Connecting the scattered dots, Bennett uncovered a chilling truth. Practitioner Reynolds had become a thorn in the side of those benefiting from genetic discrimination, eventually becoming a target to silence. The mastermind behind the tragic act was none other than the chief executive officer of a colossal insurance corporation, whose discriminatory practices were threatened by Reynolds' cutting-edge research.

> Yet, the game was up. Detective Bennett's dogged pursuit brought the orchestrator to justice, and the ripples of Practitioner Reynolds' undying legacy swept across the healthcare community, including GeneGuard Integrity Hospital, where she had worked. Her enduring struggle culminated in her influence in the passing of the Genetic Information Nondiscrimination Act, an edifice of protection against genetic discrimination. It was an echo of Practitioner Reynolds' spirit, a battle cry in the fight for justice and equality.

* * *

THINK

In the 1970s, some states started requiring African Americans to undergo genetic testing for sickle cell anemia to deny health insurance and employment. Congress, in response, enacted the 1972 National Sickle Cell Anemia Control Act, which withheld federal funding from any state that required sickle cell testing. Congress then passed Title II of the 2008 Genetic Information Nondiscrimination Act (GINA), which focused on preventing discrimination in the contexts of health insurance and employment:

1. Illegal to require individuals to purchase genetic tests
2. Illegal for insurance companies to use genetic information to
    - adjust premiums;
    - deny coverage; or
    - impose restrictions that relate to preexisting conditions
3. Illegal for companies with 15 employees or more to require or use genetic information, including healthcare history, for hiring, firing, job placement, or promotion decisions

ASSESS

Patient: Autonomy

Requiring genetic testing for insurance and employment is a type of coercion (a credible threat) that violates the principle of autonomy (informed consent). In this context, the patient–practitioner relationship is nonexistent. The practitioner does not disclose any information necessary for the patient to make informed consent, nor does the practitioner request any authorization from the patient. Mandatory genetic testing violates the patient–practitioner relationship and violates patient autonomy.

Practitioner: Beneficence and Nonmaleficence

Requiring genetic testing for insurance and employment violated the professional principles of beneficence (do good) and nonmaleficence (do no harm). The practitioner was not ordering the healthcare tests to maximize the patient's best interests as determined by the patient's reasonable goals, values, and priorities, violating beneficence (do good). Rather, the practitioner ordered the healthcare tests for the company's best interests, which employed the practitioner even if that meant harming the patient, violating nonmaleficence (do no harm). The practitioners who complied with these unethical practices were not acting according to the principles of beneficence (do good) and nonmaleficence (do no harm), did not maximize the patient's best interests, and did not act according to the standards of care.

Public Policy: Justice

Requiring genetic testing for insurance and employment violated the principle of justice (be fair). Genetic testing did not promote eliminating or reducing healthcare disparities but instead promoted and increased healthcare disparities.

CONCLUDE

It is never acceptable for workplace practitioners to require genetic tests or healthcare history for employment decisions.

In summary, mandatory genetic testing for insurance and employment purposes is an unacceptable practice that undermines patient autonomy, practitioner ethics, and social justice. The implementation of GINA ensures that individuals are protected from discrimination based on their genetic information, upholding the principles of autonomy (informed consent), beneficence (do good), nonmaleficence (do no harm), and justice (be fair) in the healthcare field.

* * *

REVIEW QUESTIONS

1. Congress enacted the 1972 National Sickle Cell Anemia Control Act, which withheld federal funding from any state that required sickle cell testing.
    A. True
    B. False

2. Congress passed Title II of the 2008 Genetic Information Nondiscrimination Act (GINA), which focused on preventing discrimination in health insurance and employment contexts.
    A. True
    B. False

3. Requiring genetic testing for insurance and employment is a type of coercion (a credible threat) that violates the principle of autonomy (informed consent).
    A. True
    B. False

4. Mandatory genetic testing violates the patient–practitioner relationship and is a violation of patient autonomy.
    A. True
    B. False

5. Requiring genetic testing for insurance and employment violates the professional principles of beneficence (do good) and nonmaleficence (do no harm).
    A. True
    B. False

Answers: 1A, 2A, 3A, 4A, 5A

* * *

CLINICAL VIGNETTES

1. Mr. Ramsey King, a 28-year-old African American mechanical engineer, presents to their practitioner for a routine health examination. The patient is asymptomatic and has no healthcare complaints. The patient has heard that in the past, some states required African Americans to undergo genetic testing for sickle cell anemia. Which of the following statements best describes why some states started requiring African Americans to undergo genetic testing for sickle cell anemia in the 1970s?
    A. To provide early diagnosis and treatment for individuals with sickle cell anemia
    B. To determine the prevalence of sickle cell anemia in the African American population
    C. To identify carriers of sickle cell anemia for the purpose of genetic counseling
    D. To deny health insurance and employment to African Americans with sickle cell anemia

Explanation: In the 1970s, some states required African Americans to undergo genetic testing for sickle cell anemia as a pretext for denying them access to health insurance and employment. This practice was part of a broader pattern of racism and discrimination in the United States that disadvantaged African Americans in many areas of life. The requirement for genetic testing was not intended to provide early diagnosis and treatment for sickle cell anemia, was not intended to determine the prevalence of the disease in the African American population, nor was the purpose of genetic testing to identify carriers of sickle cell anemia for the purpose of genetic counseling. Rather, the testing was used for discriminatory purposes to deny employment and insurance. *Answer*: D

2. Ms. Raquel Baker, a 42-year-old African American technical writer, visits their practitioner for a routine health examination. The patient has recently undergone genetic testing and is concerned about potential health insurance and employment discrimination based on their genetic information. Which of the following statements best describes the legislation to prevent health insurance and employment discrimination based on genetic information?
    A. The Americans with Disabilities Act (ADA)
    B. The Family and Medical Leave Act (FMLA)
    C. The Health Insurance Portability and Accountability Act (HIPAA)
    D. Title II of the 2008 Genetic Information Nondiscrimination Act (GINA)

Explanation: Title II of GINA is legislation that prevents discrimination in health insurance and employment based on genetic information. GINA prohibits health insurance companies and employers from requesting, requiring, or using genetic information to make decisions related to coverage, rates, or employment. It also prohibits employers from retaliating against individuals who file a complaint of discrimination under GINA or who participate in GINA proceedings. ADA and FMLA provide protections for individuals with disabilities and for employees taking leave for healthcare or family reasons, respectively. HIPAA provides privacy protections for healthcare information but does not specifically address genetic information discrimination. *Answer*: D

3. Mr. Augustus Carter, a 39-year-old unemployed carpenter, has been informed that they must undergo genetic testing to secure insurance coverage and employment. This requirement has raised ethical concerns for the patient because it involves coercion, violating multiple bioethics principles. Which of the following bioethics principles are violated by the requirement of genetic testing as a condition for insurance and employment?
    A. Autonomy and justice
    B. Beneficence and nonmaleficence
    C. Autonomy, beneficence, nonmaleficence, and justice
    D. Nonmaleficence and justice

Explanation: The requirement of genetic testing as a condition for insurance and employment violates multiple principles of bioethics, including autonomy (informed consent), beneficence (do good), nonmaleficence (do no harm), and justice (be fair). Autonomy refers to the patient's right to make decisions about their health without coercion. Coercion requiring genetic testing for health insurance or employment can limit the patient's autonomy. Beneficence and nonmaleficence are related to the obligation to do good and avoid harm. Genetic testing can result in psychological harm and discrimination and may not be in the patient's best interest. Justice refers to fair and equal treatment of all individuals. Requiring genetic testing for insurance or employment can create unequal opportunities and unfair treatment, particularly for marginalized groups at a higher risk for certain genetic conditions. Therefore, the requirement of genetic testing for insurance and employment violates multiple principles of bioethics. *Answer*: C

4. The 1970s requirement for African Americans to get genetic testing as a condition for insurance coverage and employment had what effect on healthcare disparities?
   A. It reduced disparities.
   B. It had no impact on disparities.
   C. It increased disparities.
   D. It created new disparities.

   Explanation: The requirement for African Americans to undergo genetic testing as a condition for insurance coverage and employment in the 1970s had the effect of increasing healthcare disparities. This discriminatory requirement contributed to mistrust of the healthcare system among African Americans, leading to a reluctance to seek healthcare and participate in clinical research. In addition, it created a disparity in access to employment and health insurance because individuals who tested positive for sickle cell anemia were often denied coverage and employment opportunities. Overall, the requirement for genetic testing had a negative impact on the health and well-being of African Americans and increased healthcare disparities. *Answer*: C

5. A workplace practitioner is considering requiring genetic testing or healthcare history as part of the employment decision process. This has raised ethical concerns regarding the acceptability of such practices. Is it acceptable for workplace practitioners to require genetic tests or healthcare history for employment decisions?
   A. Yes, it is acceptable in certain circumstances.
   B. No, it is never acceptable.
   C. It is acceptable, but only with the employee's consent.
   D. It depends on the specific situation.

   Explanation: Requiring genetic tests or healthcare history for employment violates several bioethics principles, including autonomy, beneficence, nonmaleficence, and justice. Using genetic information or healthcare history to discriminate against individuals in the workplace is unethical. Such requirements may also discourage individuals from seeking necessary healthcare or genetic testing, which can negatively affect their health outcomes. Furthermore, it is important to ensure that employment decisions are based on an individual's qualifications, skills, and experience rather than their genetic makeup or healthcare history. *Answer*: B

REFLECTION VIGNETTES

1. Mr. William Thompson, a 35-year-old software engineer, has received a job offer from a large multinational company with full benefits, including healthcare and retirement. However, to qualify for a special low-risk insurance premium, he must submit a complete healthcare history and a genetic screening test that clears him from a list of high-risk indicators and factors.

   What is the ethical issue(s)?

   How should the ethical issue(s) be addressed?

2. Ms. Emily Lee, a 40-year-old radiologist, was diagnosed with colon cancer and took healthcare leave for treatment and recovery. Before returning to work, her employer required an institutional healthcare checkup. During the checkup, the practitioner noticed that the employer was administering genetic tests on employees to determine genetic predisposition to cancers as a risk management strategy for workers' compensation claims.

   What is the ethical issue(s)?

   How should the ethical issue(s) be addressed?

* * *

WEB LINKS

1. *Code of Medical Ethics.* American Medical Association. https://code-medical-ethics.ama-assn.org

   The AMA *Code of Medical Ethics*, established by the American Medical Association, is a comprehensive guide for healthcare practitioners. It addresses issues and challenges, promotes adherence to standards of care, and is continuously updated to reflect contemporary practices and challenges.

2. "The Genetic Information Nondiscrimination Act of 2008." U.S. Equal Employment Opportunity Commission. https://www.eeoc.gov/statutes/genetic-information-nondiscrimination-act-2008

   GINA is a federal law that protects individuals from genetic discrimination in health insurance and employment. It prohibits employers, employment agencies, and labor organizations from requesting, requiring, or purchasing genetic information and bars them from using an individual's genetic information in employment decisions. The act also restricts health insurers from using genetic information to determine eligibility, premium calculations, or the extent of coverage and requesting or requiring genetic testing. GINA thus ensures genetic privacy and safeguards against discrimination based on genetic predisposition to diseases or disorders.

3. "Genetic Discrimination." National Human Genome Research Institute. https://www.genome.gov/about-genomics/policy-issues/Genetic-Discrimination

   This source discusses GINA's protections in health insurance and employment.

4. "Genetic Information Nondiscrimination Act Patient Resource." National Human Genome Research Institute. https://www.genome.gov/sites/default/files/media/files/2020-09/GINA_patient_resource.pdf

   This source provides a comprehensive overview of GINA and its protections against genetic discrimination in employment and health insurance.

5. "Genetic Discrimination and Misuse of Genetic Information: Areas of Possible Discrimination, Current Legislation, and Potential Limitations." Seon Gyu Lee, W. David Dotson, and Leonard Ortmann. 2022. https://blogs.cdc.gov/genomics/2022/10/03/genetic-discrimination

   This source discusses the protections provided by GINA and its focus on managing and using genetic information.

6. "Genetic Information." U.S. Department of Health and Human Services. https://www.hhs.gov/hipaa/for-professionals/special-topics/genetic-information/index.html

   This source provides an overview of GINA and its protections against discrimination based on genetic information in health coverage and in employment.

7. "Genetic Discrimination." American Medical Association. https://www.ama-assn.org/delivering-care/precision-medicine/genetic-discrimination

   This source discusses GINA's protections against genetic discrimination by health insurers and employers.

8. "Genetic Information Nondiscrimination Act (GINA): OHRP Guidance (2009)." U.S. Department of Health and Human Services. https://www.hhs.gov/ohrp/regulations-and-policy/guidance/guidance-on-genetic-information-nondiscrimination-act/index.html

   This source provides background information regarding GINA and its application to non-exempt human subjects research conducted or supported by the U.S Department of Health and Human Services.

9. "Fact Sheet: Genetic Information Nondiscrimination Act." U.S. Equal Employment Opportunity Commission. https://www.eeoc.gov/laws/guidance/fact-sheet-genetic-information-nondiscrimination-act

   This source discusses GINA's protections against employment discrimination based on genetic information.

GIFTS

25

Only those who regard healing as the ultimate goal of their efforts can, therefore, be designated as practitioners.
—Rudolf Virchow

* * *

MYSTERY STORY

ETHICAL DILEMMA: THE CASE OF THE MYSTERIOUS GIFTS AND THE IMPORTANCE OF HEALTHCARE ETHICS

An air of normalcy enveloped the buzzing corridors of Heartfelt Generosity Medical Center until an unusual incident disrupted the rhythm. Practitioner William Jameson, a pillar of the healthcare community, found himself caught at a complex ethical intersection. His patient, Mrs. Penelope Johnson, a radiant elderly woman with an abundance of gratitude, presented him with an offering—a luxury timepiece that sparkled with an undeniable extravagance.

Moved by her sincerity, Practitioner Jameson was nonetheless troubled. The seemingly benign act of kindness was laden with ethical land mines that threatened to destabilize their professional relationship. He gently declined the opulent token, invoking the sacred principles of bioethics that underscored the preservation of patient autonomy.

However, the ripple effect of this occurrence was far-reaching. His colleague, Practitioner Christopher Smith, was discovered to have been luxuriating in a sea of extravagant gifts from multiple patients, including Mrs. Johnson. With a sense of foreboding, Practitioner Jameson confronted Practitioner Smith about his questionable actions, only to be met with a stonewall of defiance and an air of hostility.

As Practitioner Jameson untangled the web, he stumbled upon a deeply troubling revelation. Practitioner Smith had been drawn into the shadowy embrace of pharmaceutical companies, receiving generous gifts in return for prescribing their medications. This dangerous quid pro quo was not only a gross violation of bioethics but also a significant threat to the well-being of their patients.

Refusing to stand by and let this ethical breach continue, Practitioner Jameson escalated the matter to the hospital's administrative hierarchy. A subsequent investigation unveiled a disconcerting reality: Practitioner Smith was not the lone wolf. Other practitioners had also been dancing to the tune of pharmaceutical companies, creating a stain on the fabric of the healthcare profession.

Driven by Practitioner Jameson's resolute adherence to ethical conduct, a wave of accountability swept through the hospital. Stricter policies regarding gift acceptance were put in place to prevent a recurrence of such unethical behavior.

The storm eventually passed, leaving in its wake renewed faith in the healthcare profession and a heightened awareness of the ethical boundaries. The practitioners were served a poignant reminder of their sacred duty: The health and safety of their patients must always eclipse personal gain or outside influence.

The mysterious case of extravagant gift-giving, particularly involving the Heartfelt Generosity Medical Center, underscored a profound lesson for all healthcare practitioners about the critical importance of maintaining the highest ethical standards. It was a stark reminder of their obligation to uphold the trust placed in them by patients and society, highlighting the thin line between human temptation and ethical responsibilities.

* * *

THINK

The giving of gifts is a long and deep tradition in human societies. Humans celebrate birthdays; anniversaries; Valentine's Day; religious holidays; tithes and sacrifices to the gods; tokens of thanks; and all manner of toasts with family, friends, and colleagues. These gift exchanges are tools and methods of developing social bonds, friendship, trust, solidarity, and reciprocal debt.

Recovering from an illness or an injury is, for many people, a qualitative gift that exceeds any quantitative value. Therefore, it should be of no surprise that in response to being cared for and healed, patients and their families may be culturally and even morally compelled to express some form or token of appreciation, often as a gift.

ASSESS

Patient: Autonomy

Certainly, out of professional courtesy and human decency, healthcare providers may judiciously accept from patients low-value token gifts and cards that do not imply reciprocity or exchange and that will have no implication of affecting the care of a patient.

However, all gifts that diminish a patient's autonomy (informed consent) by influencing treatment options and prescriptions, compromise the practitioner's and patient's decision-making judgment, or manipulate information and choices must not be accepted.

Practitioner: Beneficence and Nonmaleficence

Within the social traditions of gift-giving, there is also the notion of gift exchange or reciprocity. Professionalism must step in and set clear boundaries for all to know. For reciprocity or the exchange for "better" service, gifts violate healthcare professionalism, and practitioners must clarify that gift exchanges for such purposes are impermissible and unethical for the practitioner to accept. All patients must know and trust that their practitioner will always provide them with the best care possible, regardless of gifts and exchanges.

All gifts that diminish the professional principles of beneficence (do good) and nonmaleficence (do no harm) by influencing the practitioner's judgments, either consciously or unconsciously, must not be accepted.

Public Policy: Justice

Industry gifts have no purpose but to influence and alter practitioners' behavior and judgment when treating patients. Therefore, in this context, industry gifts should be considered a clear violation of the healthcare professional's social contract with the community and a direct attack on the patient–practitioner relationship. However, like most rules, there are exceptions to accepting industry gifts if they do not violate or appear to violate the patient–practitioner relationship and are carefully restricted to kind and amount. An example of such an exception is when the industry gift is of minimal value and directly benefits patients. Another example is when the industry gift specifically funds patient or healthcare education, the industry has no influence on the educational content, and there is no attribution to the sponsors. Practitioners are also generally allowed to accept meals associated with these educational lectures and conferences. All other nonmedical and noneducational gifts should be declined.

All gifts that violate social justice (fair distribution of benefits and burdens) and decrease society's trust in the healthcare profession must not be accepted.

CONCLUDE

The practitioner must be willing to gracefully and tactfully decline all corporate gifts that have no educational value and all patient gifts that are more than low value, such as token gifts and cards.

In summary, practitioners must navigate the delicate balance of gift-giving by accepting only low-value token gifts from patients and declining gifts that undermine professional ethics, patient autonomy, or social justice. By adhering to these guidelines, practitioners can maintain the trust and integrity of the healthcare profession while honoring the tradition of gift-giving as an expression of gratitude and social connection. (See also Chapter 13, 22, 50, and 51.)

* * *

REVIEW QUESTIONS

1. Gift exchanges are tools and methods of developing social bonds, friendship, trust, solidarity, and reciprocal debt.
   A. True
   B. False

2. Recovering from an illness or an injury is, for many people, a qualitative gift that exceeds any quantitative value.
   A. True
   B. False

3. Certainly, out of professional courtesy and human decency, healthcare providers may judiciously accept from patients low-value token gifts and cards that do not imply reciprocity or exchange and that will have no implication of affecting the care of a patient.
   A. True
   B. False

4. Within the social traditions of gift-giving, there is also the notion of gift exchange or reciprocity.
   A. True
   B. False

5. All gifts that diminish the professional principles of beneficence (do good) and nonmaleficence (do no harm) by influencing the practitioner's judgments, either consciously or unconsciously, must not be accepted.
   A. True
   B. False

6. All gifts that violate social justice (fair distribution of benefits and burdens) and decrease society's trust in the healthcare profession must not be accepted.
   A. True
   B. False

Answers: 1A, 2A, 3A, 4A, 5A, 6A

* * *

CLINICAL VIGNETTES

1. A practitioner is seeking to enhance their relationships with colleagues and patients. The belief is that the act of giving gifts is a means to foster social bonds, friendship, trust, and solidarity in the healthcare profession. The individual is considering giving gifts for various reasons, including birthdays, anniversaries, and tokens of appreciation with colleagues and patients. Which of the following best describes the primary motivation for the practitioner's desire to give gifts?
   A. To demonstrate adherence to bioethics
   B. To repay a debt owed to colleagues or patients
   C. To strengthen relationships and foster social bonds in the healthcare profession
   D. To display financial success

Explanation: The practitioner's primary motivation to give gifts is to foster social bonds and strengthen relationships with colleagues and patients. Although giving gifts can be a way to demonstrate adherence to bioethics, such as showing gratitude for patient trust and support, the primary reason for giving gifts should be to enhance social connections rather than an obligation. Repaying debts and displaying financial success are inappropriate reasons for giving gifts professionally because they can create conflicts of interest and undermine the practitioner's integrity. Ultimately, giving gifts should be guided by a commitment to professionalism and ethical conduct, aiming to build trust and mutual respect with colleagues and patients. *Answer*: C

2. A practitioner is approached by Ms. Daisy Penny, a 53-year-old hair stylist who offers a small token gift to show appreciation for their care. The practitioner values professional courtesy and human decency and wants to ensure that accepting the gift does not compromise the care they provide to their patients. Which of the following best describes the practitioner's approach to accepting the gift from the patient?
   A. Refuse the gift to maintain ethical integrity and avoid conflicts of interest
   B. Accept the gift without hesitation as a sign of gratitude from the patient
   C. Accept the gift judiciously, ensuring it is a low-value token that does not imply reciprocity or exchange for better treatment, care, or prescriptions
   D. Accept the gift as a way to improve the patient's satisfaction with their care

   Explanation: Accepting patient gifts can be a delicate issue that requires balancing professional boundaries and maintaining a positive relationship with patients. Whereas refusing the gift may be perceived as cold and uncaring, accepting it can be seen as a conflict of interest that compromises the practitioner's ethical integrity. The most appropriate approach is to accept the gift judiciously, ensuring it is a low-value token that does not imply any exchange for better care or prescriptions. This approach shows respect and appreciation for the patient's gesture while upholding the ethical standards of the profession. Accepting gifts without hesitation or hoping to improve patient satisfaction is unethical. Refusing all gifts might be overly rigid and a failure to consider nuances of the patient–practitioner relationship. *Answer*: C

3. A practitioner is offered a gift by Mr. Cedric Wright, a 69-year-old restaurant owner who has recently been under their care. The patient mentioned that they would like to provide the gift as a token of appreciation for the "better" care they have received. The practitioner values professionalism and wants to ensure that the patient understands that gift exchanges for the purpose of reciprocity or in exchange for "better" service are impermissible and unethical. Which of the following best describes the practitioner's approach to accepting the gift from the patient?
   A. Accept the gift to maintain ethical integrity and avoid conflicts of interest
   B. Accept the gift without hesitation as a sign of gratitude from the patient
   C. Refuse the gift, explaining to the patient that gift exchanges for the purpose of reciprocity or in exchange for "better" care, treatment, or prescriptions are impermissible and would be unprofessional for the practitioner to accept
   D. Accept the gift because it is a common practice in the healthcare industry

   Explanation: Although it is not uncommon for patients to offer gifts to their healthcare providers, accepting gifts from patients can create ethical dilemmas and the appearance of impropriety. In this scenario, the practitioner should refuse the gift and explain to the patient that gift exchanges for the purpose of reciprocity or in exchange for "better" care, treatment, or prescriptions are impermissible and would be unprofessional for the practitioner to accept. This will help maintain the practitioner's professional integrity and avoid conflicts of interest. Accepting gifts for better care, treatment, or prescriptions could create a power dynamic that may undermine the patient's trust and confidence in the provider. The practitioner should ensure that the patient understands that gift-giving should be solely based on the patient's gratitude and not as a way to gain favor or better treatment. *Answer*: C

4. A pharmaceutical company representative offers a gift to a practitioner. The gift is valued at $100 and is in the form of a pen set. The representative mentions that the gift shows appreciation for the practitioner's continued use of the company's products. Which of the following statements best reflects the ethical implications of the industry gift offered to the practitioner in this scenario?
   A. Industry gifts are a common and accepted practice in the healthcare field and should not be a cause for concern.
   B. Industry gifts are a clear violation of the healthcare professional's social contract with the community and a direct attack on the patient–practitioner relationship.
   C. Industry gifts can be beneficial for practitioners because they provide practitioners valuable resources and information that can improve patient care.
   D. The value of the gift does not affect the ethical implications of the industry gift offered to the practitioner.

   Explanation: The purpose of industry gifts is to buy influence and alter the practitioner's behavior and judgment when treating patients. This undermines the trust between the practitioner and the patient and the ethical principles of healthcare practice. In this context, industry gifts should be considered a clear violation of the healthcare professional's social contract with the community and a direct attack on the patient–practitioner relationship. *Answer*: B

5. Ms. Rosemary Jackson, a 63-year-old marketing manager, presents to the clinic for a routine checkup. The patient hands the practitioner a gift basket filled with expensive gourmet chocolates and a bottle of wine. The practitioner is still determining what to do, as accepting patient gifts can sometimes lead to ethical concerns. Which of the following is the most appropriate course of action for the practitioner to take in this situation?
   A. Accept the gift basket because it is a kind gesture from the patient
   B. Decline the gift basket because it is more than a low-value gift and could be perceived as a conflict of interest
   C. Accept the gift basket but donate it to a local charity
   D. Keep the gift basket but give the patient a written disclosure of the gift and its value

   Explanation: The best thing to do would be to decline the gift basket because it is more than a low-value gift and could be perceived as a conflict of interest. Healthcare professionals are responsible for maintaining objectivity and acting in their patients' best interests. Accepting expensive gifts from patients can potentially create a conflict of interest, undermining the trust and confidence of other patients in the profession. As such, many professional associations, including the American Medical Association, recommend that healthcare professionals decline gifts from patients that exceed a low monetary value or that create a conflict of interest. By declining the gift basket, the practitioner upholds professional standards and avoids ethical dilemmas that could arise from accepting patient gifts. *Answer*: B

REFLECTION VIGNETTES

1. Ms. Rachel Thompson, a 50-year-old business owner, was diagnosed with breast cancer and underwent a difficult radiation and chemotherapy regime. The practitioner monitored her progress and celebrated with her when she successfully went into remission. As a token of appreciation, Ms. Thompson brought the practitioner a platter of 36 homemade chocolate chip cookies, a thank-you card, and a $500 gift card to an exclusive local restaurant.

   What is the ethical issue(s)?

   How should the ethical issue(s) be addressed?

2. Mr. David Kim, a 55-year-old accountant, presents to the practitioner's clinic with symptoms of hypertension. The practitioner conducts a thorough physical examination and prescribes medication to manage his condition. During the visit, Mr. Kim expresses gratitude and hands the practitioner a small gift bag with a thank-you card and a box of chocolates.

   What is the ethical issue(s)?

   How should the ethical issue(s) be addressed?

* * *

WEB LINKS

1. *Code of Medical Ethics.* American Medical Association. https://code-medical-ethics.ama-assn.org
   The AMA *Code of Medical Ethics*, established by the American Medical Association, is a comprehensive guide for healthcare practitioners. It addresses issues and challenges, promotes adherence to standards of care, and is continuously updated to reflect contemporary practices and challenges.
2. "Humanizing the Physician–Patient Relationship: How Gift-Giving and –Receiving Can Be Ethical." Gyan Moorthy. 2022. https://journals.library.columbia.edu/index.php/bioethics/article/view/9958
   This source discusses the ethical considerations of accepting patient gifts and the potential influence on healthcare judgment.
3. "Is It Ethical to Accept Gifts from Patients?" Kehinde Eniola. 2018. https://www.aafp.org/pubs/fpm/issues/2018/0100/p40.html
   This source discusses the reasons patients offer gifts and the ethical considerations for physicians in accepting them.
4. "Should Physicians Accept Gifts from Patients?" Laurie J. Lyckholm. 1998. https://jamanetwork.com/journals/jama/fullarticle/188239
   This source discusses the tradition of patients' gifts to healthcare institutions and the importance of acknowledging gratitude.
5. "1.2.8 Gifts from Patients." American Medical Association. https://code-medical-ethics.ama-assn.org/sites/default/files/2022-08/1.2.8%20Gifts%20from%20patients%20-%20background%20reports.pdf
   This source discusses the reasons patients offer gifts and how accepting these gifts can enhance the patient–physician relationship.
6. "Gifts from Patients." American Medical Association. https://code-medical-ethics.ama-assn.org/ethics-opinions/gifts-patients
   This source provides the American Medical Association's opinion on accepting patient gifts.
7. "Healthcare Dilemma Towards Gift Giving by Patients." Yusrita Zolkefli. 2021. https://www.ncbi.nlm.nih.gov/pmc/articles/PMC8793965
   This source discusses the acceptance of gifts by healthcare professionals and the ethical scrutiny required to avoid unethical or unprofessional behavior.
8. "A Medical Resident's Guide to Gifts from Industry." Brendan Murphy. 2019. https://www.ama-assn.org/delivering-care/ethics/medical-resident-s-guide-gifts-industry
   This source guides healthcare residents on accepting gifts from the industry.

GUNSHOT WOUNDS

26

The practitioner who cures a disease may be the most skillful, but the practitioner that prevents disease is the safest.
—Thomas Fuller

* * *

MYSTERY STORY

> **CODE NAME CONFIDENTIAL GUNSHOTS: THE DELICATE BALANCING ACT OF MANDATORY REPORTING AND PATIENT CONFIDENTIALITY**
>
> On a day shadowed by the sinister weight of unsolved cases, Detective Ava Martinez found herself shackled to a sea of convoluted paperwork. Suddenly, the solemn silence of her office was shattered by the shrill ring of her desk phone, bearing the caller ID "Detective Robert Wilson."
>
> "Detective Martinez, a conundrum unfolds that might pique your interest," Detective Wilson's voice sliced through the line, stringing along a thread of intrigue.
>
> Detective Martinez probed, "What's the situation, Detective Wilson?"
>
> "A civilian, Mr. Michael Davis, was escorted to the ER at EthicalResponse Emergency Hospital with a gunshot wound marring his leg," Detective Wilson delineated. "However, Mr. Davis morphs into a sphinx when questioned about the incident, eliciting unease among the practitioners regarding his potential entanglement in nefarious affairs."
>
> Recognizing the need for swift action, Detective Martinez responded, "Consider me en route to the hospital, Detective Wilson."
>
> Upon reaching the hospital, a chaotic harmony of healthcare urgency greeted Detective Martinez. After obtaining a thorough briefing from the assembled healthcare squadron, she attempted to peel back the layers of Mr. Davis' silence. However, he remained an impenetrable citadel of secrecy, leading to a tussle between patient confidentiality and the dire need for justice.
>
> Unfazed, Detective Martinez trudged through an intricate web of leads and dead ends. It soon unraveled that Mr. Davis was the byproduct of a brutal gang skirmish, his injury serving as a chilling memento. In strict compliance with the legal requirements, the practitioners notified the authorities while ensuring that only the necessary crumbs of information were shared.
>
> Detective Martinez found herself admiring the practitioners' deft navigation between legal obligations and healthcare ethics. Without mandatory reporting, the societal shield of law and order would falter in protecting the public from criminal undertows.
>
> With the hospital's astute reporting, the law's long arm rapidly clamped down on the perpetrators of the gang conflict. Meanwhile, Mr. Davis, under the watchful care of the healthcare team, commenced his journey toward recovery, far removed from his perilous past.
>
> This enthralling case shed light on the delicate equilibrium between mandatory reporting of gunshot wounds and safeguarding patient confidentiality at EthicalResponse Emergency Hospital. It underscored the vital role of practitioners as vigilant whistleblowers while respecting the sanctity of patient trust. Detective Martinez came away with renewed admiration for the practitioners, who had steered through this ethical maze with commendable integrity, ensuring that justice was served while upholding their professional ethics.

* * *

THINK

Gunshot wounds and child and elder abuse are examples of national state-mandated reporting. Failure to comply with mandatory reporting is a Class A misdemeanor.

Statutory laws vary from state to state regarding what evidence is required to report to law enforcement for public safety. Motor vehicle collisions involving serious injury or death, unconsciousness, violent injuries such as knife injury, gunshot wounds such as powder burns, or other injuries resulting from a gun or firearm discharge are frequently, but not always, on the list.

Breaching the duty of patient confidentiality for the greater good of public safety threatens to diminish the strength of the patient–practitioner relationship by setting limits as to the confidentiality and privacy of protected health information (PHI) attained by the practitioner. Care must be taken to avoid the unintended consequence of diminishing the value and strength of the patient–practitioner relationship.

Ensuring that there is no breach of the duty of patient confidentiality has been accomplished by including the categories of mandated reporting in the patient–practitioner social contract.

ASSESS

Patient: Autonomy

Concerning autonomy (informed consent), mandatory reporting laws are statutes that require, under the threat of law, that the practitioner reports the minimal amount of information necessary to the proper authorities regardless of whether or not the patient consents to the reporting. Therefore, the practitioner must disclose such requirements to the patient as soon as possible. The goal is for the patient to understand and voluntarily consent to the reporting without ill effects on the patient–practitioner relationship.

If the patient does not consent to the reporting, even after communicating that the practitioner has no legal choice but to report, then the practitioner has a professional and legal obligation to report. Mandatory reporting is not a breach of patient–practitioner confidentiality because, by law and standards of care, the categories of mandatory reporting are part of the patient–practitioner social contract.

Practitioner: Beneficence and Nonmaleficence

Professionally, the patient–practitioner relationship with confidentiality and privacy is central to maximizing the patient's best interests, as determined by the patient's reasonable goals, values, and priorities. Beneficence (do good) and nonmaleficence (do no harm) are the central professional moral principles. Confidentiality is the practitioner's duty to ensure that all PHI is kept secret from unauthorized personnel. Privacy is the right of the patient to access and control the dissemination of their personal information.

Mandated reporting, such as for gunshot wounds, does not violate the practitioner's professional covenant of patient confidentiality when the reporting is given to authorized personnel and the information provided is the minimum amount required. Mandatory reporting is a legal and professional requirement in the patient–practitioner relationship. Not reporting would be the practitioner's violation of the patient–practitioner relationship—a form of professional misconduct.

Public Policy: Justice

Justice is the fair distribution of benefits and burdens throughout society. The judicial system has determined that gunshot wounds must be reported for public safety and justice (be fair). Gunshot wounds legislation has more weight than patient autonomy (informed consent), professional beneficence (do good), and nonmaleficence (do no harm) combined. Unintended consequences are that individuals with gunshot wounds who are afraid of being reported will avoid seeking healthcare due to the fear of harmful social consequences. Some providers think healthcare services should be an independent environment of security and health without mandatory reporting laws.

Public policy and the courts have determined that artificially separating healthcare from mandatory reporting laws would come at the cost of public health and security and would not be justifiable or fair for other individuals in the community. In this view, the personal liberty of the patient ends when exercising that liberty negatively affects other citizens' constitutional rights and liberties to live in a safe and secure environment. Mandatory reporting does not violate the patient–practitioner relationship because state mandatory reporting categories are part of the patient–practitioner social contract.

CONCLUDE

At no time is it permissible for law enforcement to interfere in treating a patient in a healthcare emergency, and all mandatory reporting needs to be done to only authorized personnel and then only providing the minimum amount of information required.

In summary, mandatory reporting laws are crucial in maintaining public safety and justice despite potential unintended consequences, such as deterring individuals from seeking healthcare. Practitioners must disclose their reporting obligations to patients and provide only the minimum amount of information necessary to authorized personnel. Law enforcement interference in treating a patient in a healthcare emergency is impermissible, and healthcare professionals must balance their legal obligations with preserving the patient–practitioner relationship. (See also Chapter 6.)

* * *

REVIEW QUESTIONS

1. Because reporting of gunshot wounds is mandatory per state laws, the practitioner is exempt from legal accountability for failure to report.
   A. True
   B. False

2. Mandatory reporting of gunshot wounds breaches the patient–practitioner relationship.
   A. True
   B. False

3. Whenever a practitioner must report a gunshot wound, disclosing the reporting requirement and attempting to get the patient's informed consent are essential.
   A. True
   B. False

4. If the gunshot wound patient refuses to provide informed consent for reporting, the practitioner is obliged not to report because the professional obligation is never to breach the patient–practitioner relationship.
   A. True
   B. False

5. Mandatory reporting of a gunshot wound is a legal and professional requirement in the patient–practitioner relationship. Not reporting would be the practitioner's violation of the patient–practitioner relationship, a form of professional misconduct.
   A. True
   B. False

6. When specifying, weighing, and balancing the four principles of bioethics related to mandatory reporting, the principle of justice has more weight than the principles of autonomy, beneficence, and nonmaleficence combined.
   A. True
   B. False

7. Personal liberty of the patient ends when exercising that liberty negatively affects other citizens' constitutional rights and liberties to live in a safe and secure environment.
   A. True
   B. False

Answers: 1B, 2B, 3A, 4B, 5A, 6A, 7A

* * *

CLINICAL VIGNETTES

1. Mr. Nico Anderson is a 27-year-old construction worker who was brought to the emergency room with a gunshot wound to the leg. He reported that he accidentally shot himself while cleaning his gun at home. The differential diagnosis includes soft tissue injury and nerve damage. What is the ethical obligation of the practitioner in this case?
   A. The practitioner should report the case to law enforcement without informing Mr. Anderson.
   B. The practitioner should not report the case to law enforcement.
   C. The practitioner should obtain informed consent from Mr. Anderson before reporting the case to law enforcement.
   D. The practitioner should report the case to law enforcement and inform Mr. Anderson of the requirement.

   Explanation: Gunshot wounds are examples of national state-mandated reporting. Failure to comply with mandatory reporting is a Class A misdemeanor. In this case, although the gunshot wound is not the result of a crime, the practitioner must still report the case to law enforcement as required by law. The practitioner must inform the patient of the requirement to report and only provide the minimum amount of information necessary to authorized personnel. In this case, the practitioner should report the case to law enforcement and inform Mr. Anderson of the requirement to report. Mandatory reporting raises ethical concerns, such as patient confidentiality, autonomy, and the practitioner's duty to act in the patient's best interest. However, reporting cases of gunshot wounds to law enforcement is necessary for public safety and justice, and it is part of the social contract between healthcare providers and their patients. Practitioners must balance their legal obligations with their ethical responsibilities to their patients. *Answer*: D

2. Mr. Nathanael Johnson, a 32-year-old gang member, was brought to the emergency department (ED) after sustaining a gunshot wound to the abdomen. He is conscious and in severe pain. As the ED staff begins to triage Mr. Johnson, a police officer demands information about the incident and attempts to question him about any potential suspects. The ED staff informs the officer that they cannot disclose any information due to patient privacy laws and that it is not permissible for law enforcement to interfere in treating a patient in a healthcare emergency. What is the proper action of the ED staff after this encounter?
   A. The ED staff should provide the police officer with all information it has about the incident, including Mr. Johnson's healthcare condition.
   B. The ED staff should comply with the police officer's request for information to ensure justice is served.
   C. The ED staff should delay notifying the proper authorities of the incident until Mr. Johnson is stable and able to provide information to law enforcement.
   D. The ED staff should follow the mandatory reporting requirements and provide the minimum amount of information required to authorized personnel only.

   Explanation: At no time is it permissible for law enforcement to interfere in treating a patient in a healthcare emergency, and all mandatory reporting needs to be done to only authorized personnel and then only providing the minimum amount of information required. In this case, the ED staff is ethically obligated to prioritize Mr. Johnson's healthcare and privacy over any potential law enforcement inquiries. Delaying the reporting or providing excessive information to law enforcement can compromise the trust and confidentiality of the patient–practitioner relationship and violate Mr. Johnson's autonomy and right to privacy. The proper authorities should be notified with the minimum information necessary for public safety and justice. *Answer*: D

3. A middle-aged woman was brought to the ED with a gunshot wound to the abdomen. She is conscious but in severe pain and cannot provide identifying information. The emergency healthcare team urgently needs to treat her, but the team must also report her injury to the authorities due to mandatory reporting laws. How should the team proceed?
   A. Provide treatment without reporting the injury
   B. Delay treatment until the patient can provide identifying information
   C. Report the injury without providing any identifying information
   D. Provide treatment and report the injury as soon as possible

   Explanation: Gunshot wounds exemplify a state-mandated reporting requirement. Failure to comply with mandatory reporting is a Class A misdemeanor. In emergency situations in which a patient cannot provide identifying information, healthcare providers have a legal and ethical obligation to provide healthcare first, even if identifying information must be obtained later. Therefore, in this situation, the healthcare team should provide immediate treatment to the patient without delay and report the injury as soon as possible while providing only the minimum amount of identifying information required. This approach balances the patient's immediate healthcare needs with the legal and ethical obligation to comply with mandatory reporting laws. Delaying treatment or failing to report the injury would violate the patient–practitioner relationship and may harm the patient and others in the community. *Answer*: D

4. A middle-aged man is brought to the ED after being involved in a shooting incident. He has sustained a gunshot wound to his leg and is bleeding profusely. The practitioner starts assessing and treating the patient, but the patient refuses to provide any identification and insists he does not want to be reported to the authorities. What should the practitioner do in this situation?
   A. Continue treating the patient without reporting the incident to the authorities
   B. Stop treating the patient until he provides identification and consents to the reporting
   C. Report the incident to the authorities without informing the patient
   D. Continue treating and disclose the mandatory reporting requirements to the patient and seek his consent

   Explanation: As per mandatory reporting laws, the practitioner must report gunshot wounds to the authorities, but it is essential to disclose these requirements to the patient as soon as possible. The goal is to make the patient understand the requirements and voluntarily consent to the reporting without ill effects on the patient–practitioner relationship. In this case, the practitioner should explain the legal requirement to report the gunshot wound and seek the patient's consent without disclosing any unnecessary information that might harm the patient's privacy. If the patient still refuses to provide identification or consent to the reporting, then the practitioner must continue treating the patient in the best interest of his health while balancing the obligations of justice and privacy. *Answer*: D

4. Mr. Henry Johnson is a 45-year-old long-term patient of Practitioner Tony Ramirez, who has been treating him for chronic lower back pain. During a routine appointment, Practitioner Ramirez notices a gunshot wound on Mr. Johnson's arm. When asked about it, Mr. Johnson reveals that he was shot during a mugging incident but refuses to provide further information, stating that he does not want the police involved. Practitioner Ramirez

explains to Mr. Johnson that mandatory reporting laws require healthcare providers to report gunshot wounds to law enforcement in the interest of public safety and justice, even if the patient does not consent. Mr. Johnson expresses his belief that reporting his injury is a violation of his confidentiality and privacy. What should Practitioner Ramirez do?
A. Report the gunshot wound to law enforcement without informing Mr. Johnson
B. Respect Mr. Johnson's wishes and not report the gunshot wound
C. Try to persuade Mr. Johnson to report the gunshot wound himself
D. Explain to Mr. Johnson that it is mandated for him to report all gunshot wounds and encourage Mr. Johnson to consent

Explanation: According to the principle of autonomy, patients have the right to make their own healthcare decisions. However, mandatory reporting laws require healthcare providers to report certain injuries, including gunshot wounds, to law enforcement for public safety and justice. It is the practitioner's duty to disclose such requirements to the patient as soon as possible and for the patient to understand and voluntarily consent to the reporting. Practitioner Ramirez must explain to Mr. Johnson the importance of mandatory reporting laws and encourage him to reconsider, emphasizing that mandatory reporting does not violate the patient–practitioner relationship because reporting is part of the patient–practitioner social contract. If Mr. Johnson refuses to allow the reporting, Practitioner Ramirez must report the gunshot wound to law enforcement while providing the minimum amount of information required, as mandated by law. *Answer*: D

REFLECTION VIGNETTES

1. Mr. James Smith, a 32-year-old construction worker, presents to the practitioner's clinic with a gunshot wound to the leg. The practitioner assesses the wound and explains to Mr. Smith that all gunshot wounds must be reported to law enforcement. However, Mr. Smith pleads with the practitioner not to report the case, stating that he was involved in an altercation and feared retaliation. When the practitioner explains that reporting the wound is mandatory, Mr. Smith becomes agitated and demands immediate discharge.

   What is the ethical issue(s)?

   How should the ethical issue(s) be addressed?

2. Mr. John Williams, a 35-year-old bartender, is brought to the ED handcuffed to a gurney with multiple gunshot wounds. Police officers request that the practitioner and staff get a blood alcohol level and drug screen test because Mr. Williams was involved in a high-speed chase that ended with him getting shot. The emergency healthcare protocol does not include a blood alcohol test.

   What is the ethical issue(s)?

   How should the ethical issue(s) be addressed?

* * *

WEB LINKS

1. *Code of Medical Ethics*. American Medical Association. https://code-medical-ethics.ama-assn.org
   The AMA *Code of Medical Ethics*, established by the American Medical Association, is a comprehensive guide for healthcare practitioners. It addresses issues and challenges, promotes adherence to standards of care, and is continuously updated to reflect contemporary practices and challenges.

2. "Doctors Aren't Cops: We Need to Change Gunshot Reporting." Shane Collins. May 4, 2017. https://www.kevinmd.com/blog/2017/05/doctors-arent-cops-need-change-gunshot-reporting.html
A fourth-year healthcare student recounts treating a young man with a gunshot wound, initially claimed as a nail gun injury, in the emergency department. The case illustrates the tension between patient care and legal obligations to report gunshot wounds, highlighting issues of trust in patient–physician confidentiality; the overlap of healthcare and law enforcement; and the potential deterrent effect on victims seeking emergency care, especially among vulnerable populations such as the homeless or those with substance use issues.
3. "Should a Physician Ever Violate SWAT or TEMS Protocol in a Mass Casualty Incident?" Brandon Morshedi and Faroukh Mehkri. 2022. https://journalofethics.ama-assn.org/article/should-physician-ever-violate-swat-or-tems-protocol-mass-casualty-incident/2022-02
This source discusses the standard of care in emergency healthcare services during mass casualty incidents, including gunshot wounds.
4. "Principles of Clinical Ethics and Their Application to Practice." Basil Varkey. 2021. https://karger.com/mpp/article/30/1/17/204816/Principles-of-Clinical-Ethics-and-Their
This source discusses the principles of biomedical ethics and includes legally required reporting of gunshot wounds.
5. "Gunshot Wounds: Ballistics, Pathology, and Treatment Recommendations, with a Focus on Retained Bullets." Gracie R. Baum, Jaxon T. Baum, Dan Hayward, and Brendan J. MacKay. 2022. https://www.ncbi.nlm.nih.gov/pmc/articles/PMC9462949
This source discusses the common cause of traumatic injury due to civilian use of firearms in the United States and the management of gunshot wound injuries.
6. "Reporting of Gunshot Wounds by Doctors in Emergency Departments: A Duty or a Right?" A. Frampton. 2005. https://emj.bmj.com/content/22/2/84
This source discusses the legal and ethical issues surrounding breaking patient confidentiality concerning gunshot wounds.
7. "Reporting Gunshot Wounds." International Committee of the Red Cross. 2019. https://www.icrc.org/en/document/report-gunshot-wounds-study-gives-fresh-insights-obligation-health-staff
This source discusses the legal obligation of healthcare professionals to report gunshot wounds.
8. "Self-Inflicted Gunshot Wound as a Consideration in the Patient Selection Process for Facial Transplantation." Michelle W. McQuinn, Laura L. Kimberly, Brendan Parent, et al. 2019. https://www.cambridge.org/core/journals/cambridge-quarterly-of-healthcare-ethics/article/selfinflicted-gunshot-wound-as-a-consideration-in-the-patient-selection-process-for-facial-transplantation/4597C45DCEB0AD01C1E011C0C0F04133
This source discusses the consideration of self-inflicted gunshot wounds in the patient selection process for facial transplantation.
9. "Gunshot Wounds: The New Public Health Issue." Bradford Mackay. 2004. https://pubmed.ncbi.nlm.nih.gov/14993167
This source discusses gunshot wounds as a new public health issue and the legal enforcement of patient rights.
10. "Medical Ethics." William Ruddick. https://as.nyu.edu/content/dam/nyu-as/philosophy/documents/faculty-documents/ruddick/medethics.html
This source discusses physicians' social and political responsibilities, including the reporting of patients' gunshot wounds as exceptions to traditional pledges of confidentiality.
11. "Initial Evaluation and Management of Abdominal Gunshot Wounds in Adults." Christopher Colwell and Ernest E. Moore. https://www.uptodate.com/contents/initial-evaluation-and-management-of-abdominal-gunshot-wounds-in-adults
This source discusses the initial evaluation and management of abdominal gunshot wounds in adults.

HOSPICE

27

The ultimate value of life depends upon awareness and the power of contemplation rather than upon mere survival.
—Aristotle

* * *

MYSTERY STORY

THE ETHEREAL SPECTER OF MERCY: HOSPICE CARE WALKING THE FINE LINE OF ETHICAL IMPERATIVES

Beneath the seemingly tranquil facade of the Sunset Meadows Hospice Center, an uncanny specter loomed large as Detective Amelia Thompson was summoned to investigate an enigmatic death. The somber whispers of the incident murmured about a certain 72-year-old resident, Mr. Harold Jenkins, whose life had reached its terminus amidst the palliative solace of the hospice, raising a myriad of questions.

With her seasoned acumen, Detective Thompson set about peeling back the layers of ambiguity. Through the labyrinth of staff interviews and patient records, she charted the trajectory of Mr. Jenkins' illness, one that had found a constant in the hospice for the past 3 months. His condition, though terminal, had held steadfast in recent weeks under the vigilant oversight of the hospice caregivers.

Yet, a discordant note resonated in the symphony of Mr. Jenkins' palliative care—his pain medication. A marked aberration in the prescribed dosage sent ripples of suspicion through Detective Thompson's seasoned instincts, stirring the winds of possible foul play.

Detective Thompson narrowed her gaze on Nurse Emma Jones, a crucial cog in the wheel of Mr. Jenkins' multidisciplinary treatment team, whose duty was the administration of his medication. As Detective Thompson unfurled the fabric of discrepancies, Nurse Jones, caught in the glare of truth, crumbled, letting slip a confession that was both tragic and ethically fraught.

Nurse Emma recounted the dire tableau of Mr. Jenkins' agony—a battleground of relentless pain and creeping depression. Driven by compassion, she had overstepped her bounds, believing that an escalated dosage of his pain medication could offer a sliver of respite from his torment.

Patiently, Detective Thompson delineated the grave transgression of Nurse Emma's well-intentioned but misguided act. In her quest for mercy, Nurse Jones had breached the sacrosanct protocol of hospice care, potentially catalyzing harm to Mr. Jenkins. Detective Thompson emphasized the hospice's primary mandate—to bathe the final days of patients in comfort and dignity, not to gamble with their fragile existence.

In the shadow of her actions, Nurse Emma was arrested and indicted with professional misconduct and involuntary manslaughter. This unfortunate event echoed as a potent reminder to healthcare providers across the spectrum: The essence of hospice care lies in adhering to the ethical pillars of beneficence (doing good) and nonmaleficence (doing no harm) while championing patient autonomy.

Hospice care, the harbor at the twilight of life, must focus on an integrative approach that fosters comfort, support, and dignity during the end-of-life process. Any deviation from this path can invite dire repercussions. Thus, healthcare providers, such as those at Sunset Meadows Hospice Center, are reminded to tread this delicate balance with care and vigilance, ensuring they serve as beacons of compassion while upholding the inviolable boundaries of their ethical obligations.

* * *

THINK

The word "hospice" comes from the Latin word *hospitium*, which means hospitality or place of rest and protection for the ill and weary. The first hospice came to the United States from London in 1974. It developed outside of mainstream medicine in response to the excessive use of technology for extending the quantity of life at the expense of quality of life. Technology has had the unintended consequence of creating chronic illnesses, extended hospitalizations, and dying alone in the hospital without family and friends.

Hospice benefits include access to a multidisciplinary treatment team specialized in end-of-life care and can that it can be accessed in the home, long-term care facility, or the hospital. In 1982, federal Medicare started funding Medicare beneficiaries who had a prognosis of less than 6 months to live, as verified by two practitioners. State Medicaid and private insurance soon followed suit. After 6 months, the patient can continue to receive hospice care if the hospice healthcare director or doctor certifies (at a face-to-face meeting) that the patient is still terminally ill with a prognosis of less than 6 months to live.

Although hospice is known for palliating a terminally ill patient's pain and symptoms, hospice care also provides for many other needs, such as a hospital bed for the home, psychological counseling, family bereavement counseling, and other quality-of-life products and activities.

The practice of practitioners having end-of-life discussions with their Medicare patients about end-of-life care, living wills, and advanced directives was politicized in 2009 as "death panels." The provision for paying practitioners for end-of-life practitioner counseling was, as a consequence, removed from the 2010 Patient Protection and Affordable Care Act, also known as Obamacare. A randomized trial published in 2010 in the *New England Journal of Medicine* showed that hospice increased patients' quality of life and had the unexpected result of increasing the quantity of life, over and above traditional patient care. This debunked the notion that patients who choose hospice die sooner, albeit more comfortably, than those who choose more traditional care.

ASSESS

Patient: Autonomy

Hospice is an autonomous option for patients with a terminal prognosis of less than 6 months to live. Any treatment toward regaining of health has been determined to be highly unlikely and, therefore, futile. This is the opportunity for the patient to freely choose quality-of-life measures that align with the standards of care. It is also the opportunity not to die alone but, rather, in the presence of family, friends, and loved ones.

Practitioner: Beneficence and Nonmaleficence

Professionally, the principles of beneficence (do good) and nonmaleficence (do no harm) are also satisfied with the healthcare option of hospice. Hospice respects the patient's reasonable goals, values, and priorities and attempts to maximize the patient's best interests during this challenging end-of-life time. The patient's best interests almost universally include focusing on the quantity and quality of life.

Public Policy: Justice

Social justice (fair distribution of benefits and burdens) recognizes the value of the patient's liberty not to pursue ineffective and futile treatments and not to die in isolation. Hospice focuses on the palliation of pain and symptoms and of dying in the presence of family and friends, along with bereavement counseling.

CONCLUDE

A do not resuscitate (DNR) order is never a condition for hospice. The only requirement for Medicare hospice is a prognosis of less than 6 months of life. Hospice attempts to focus on not just the quantity of life but also the quality of life.

In summary, hospice care is an essential healthcare option for terminally ill patients with a prognosis of less than 6 months of life. It focuses on both the quantity and quality of life, providing comfort, support, and dignity during the end-of-life process. By offering palliative care, psychological counseling, and bereavement support, hospice care respects patient autonomy, adheres to the ethical principles of beneficence and nonmaleficence, and promotes social justice. A DNR order is not a requirement for hospice care, emphasizing the focus on providing quality of life for patients and their families. (See also Chapter 39.)

* * *

REVIEW QUESTIONS

1. The word "hospice" comes from the Latin word *hospitium*, which means
   A. Hospital
   B. Death
   C. Hospitality
   D. Do not resuscitate

2. Technology has had the unintended consequence of creating chronic illnesses, extended hospitalizations, and dying alone in the hospital without family and friends.
   A. True
   B. False

3. Although hospice is known for palliating a terminally ill patient's pain and symptoms, hospice care also provides for many other needs, such as
   A. Hospital bed for the home
   B. Psychological counseling
   C. Family bereavement counseling
   D. Interprofessional team for healthcare
   E. All of the above

4. A randomized trial published in 2010 in the *New England Journal of Medicine* showed that hospice increased patients' quality of life and had the unexpected result of increasing the quantity of life, over and above traditional patient care.
   A. True
   B. False

5. A requirement for hospice is a prognosis of less than 6 months to live and a DNR in the healthcare record.
   A. True
   B. False

6. Social justice (fair distribution of benefits and burdens) recognizes the value of the patient's liberty not to pursue ineffective and futile treatments and not to die in isolation.
   A. True
   B. False

Answers: 1C, 2A, 3E, 4A, 5B, 6A

* * *

CLINICAL VIGNETTES

1. Mrs. Esther Patel is a 70-year-old retired schoolteacher with advanced pancreatic cancer for several months. Her oncologist has recommended considering hospice care, and she has

expressed an interest in learning more about the end-of-life care options available to her. However, she is hesitant to discuss these options with her practitioner because she has heard about the politicization of end-of-life discussions and fears being pressured into making decisions that she is uncomfortable with. It is important to note that healthcare professionals having end-of-life discussions with their Medicare patients about end-of-life care, living wills, and advanced directives were politicized in 2009 as "death panels." Consequently, the provision for paying practitioners for end-of-life practitioner counseling was removed from the 2010 Patient Protection and Affordable Care Act, also known as Obamacare. This has led to many patients, like Mrs. Patel, feeling apprehensive about discussing their end-of-life care preferences with their healthcare providers. Which of the following is true?

A. End-of-life discussions should be avoided because they can lead to patients feeling pressured into making decisions they are not comfortable with.
B. The politicization of end-of-life discussions has had a positive impact on patient care.
C. End-of-life discussions are an important aspect of patient care and should be encouraged.
D. Patients who choose hospice care are more likely to die sooner than those who choose traditional care.

Explanation: The correct answer is that end-of-life discussions are important to patient care. They can help patients make informed decisions that align with their values and preferences. The option of avoiding end-of-life discussions is wrong because doing so can lead to patients feeling confused and uncertain about their options and can also lead to healthcare providers making assumptions about patients' preferences. The politicization of end-of-life discussions has had a negative, not a positive, impact on patient care, causing many patients to feel hesitant about discussing their preferences and limiting their access to important information. Despite the politicization of end-of-life discussions, healthcare providers should encourage patients to discuss their preferences and should provide patients with the information they need to make informed decisions. Last, it is incorrect that hospice patients die sooner. A randomized trial published in 2010 in the *New England Journal of Medicine* showed that hospice increased patients' quality of life and had the unexpected result of increasing the quantity of life, over and above traditional patient care. This study debunked the notion that patients who choose hospice care are more likely to die sooner than those who choose traditional care. *Answer*: C

2. Ms. Sabrina Rodriguez is a 78-year-old woman with advanced pancreatic cancer. She has been experiencing severe pain, fatigue, and weight loss despite undergoing several rounds of chemotherapy. Her oncologist has suggested considering hospice care, which can provide palliative care for her pain and symptoms and other services to improve her quality of life. These services include a hospital bed for her home, psychological counseling to help her cope with the emotional toll of her illness, family bereavement counseling to prepare her loved ones for her eventual passing, and other quality-of-life products and activities. Which of the following is true?

A. Hospice care can only be provided in the hospital setting.
B. Ms. Rodriguez must have a DNR order in place to be eligible for hospice care.
C. Hospice care can provide relief from pain and symptoms as well as a range of other services to improve quality of life, such as a hospital bed for the home, psychological counseling, and family bereavement counseling.
D. Ms. Rodriguez should continue with aggressive chemotherapy despite the complications.

Explanation: Hospice care can be accessed in the home, long-term care facility, or hospital, and a DNR order is not a requirement for hospice; the only requirement for Medicare hospice is a prognosis of less than 6 months of life. Continuing aggressive chemotherapy is also unlikely to be effective and may cause more harm than benefit, especially considering the complications Ms. Rodriguez is experiencing. The correct answer is that hospice care can provide relief from pain and symptoms and a range of other services to improve quality of life, such as a hospital bed for the home, psychological counseling, and family bereavement counseling. Hospice is an autonomous option for patients with a terminal prognosis, and it respects the patient's reasonable goals, values, and priorities, maximizing their best interests during the end-of-life stage. Hospice care promotes social justice by focusing on palliation and bereavement counseling. *Answer*: C

3. Mrs. Cynthia Thompson is an 85-year-old retired librarian diagnosed with end-stage lung cancer. She has been experiencing severe pain, fatigue, and shortness of breath despite undergoing multiple rounds of chemotherapy and radiation therapy. Her oncologist has recommended considering hospice care because any treatment toward regaining health has been determined to be highly unlikely and, therefore, futile. Hospice care will allow Mrs. Thompson to freely choose quality-of-life measures that align with the standards of care. This includes palliative care to manage her pain and symptoms, psychological counseling to help her cope with the emotional toll of her illness, and family bereavement counseling to prepare her loved ones for her eventual passing. Which of the following is true?
   A. Hospice care is only provided in the hospital setting.
   B. Mrs. Thompson must have a DNR order in place to be eligible for hospice care.
   C. Hospice care can provide Mrs. Thompson with the opportunity to freely choose quality of life measures that are in line with the standards of care and to die in the presence of family, friends, and loved ones.
   D. Mrs. Thompson should continue with aggressive treatment to manage her lung cancer symptoms.

   Explanation: Hospice care can be accessed in the home, long-term care facility, or hospital setting, and a DNR order is not a requirement for hospice care; the only requirement for Medicare hospice is a prognosis of less than 6 months of life. Continuing aggressive treatment is unlikely to be effective and may cause more harm than benefit, especially considering Mrs. Thompson's advanced age and declining condition. The correct answer is that hospice care will allow Mrs. Thompson to freely choose the quality of life measures that align with the standards of care and to die in the presence of family, friends, and loved ones. Hospice is an autonomous option for patients with a terminal prognosis, and it respects the patient's reasonable goals, values, and priorities, maximizing their best interests during the end-of-life stage. Hospice care promotes social justice by focusing on palliation and bereavement counseling. *Answer*: C

4. Mr. Kurt Jackson is a 65-year-old retired teacher diagnosed with end-stage lung cancer. His condition has continued to deteriorate despite undergoing multiple treatments, and his oncologist has recommended hospice care. Hospice care offers Mr. Jackson the opportunity to manage his pain and symptoms through palliative care, receive emotional support through psychological counseling, and prepare his loved ones for his eventual passing through family bereavement counseling. Hospice care also allows Mr. Jackson to die in the presence of family and friends, providing comfort and peace in his final days. The differential diagnosis includes lung cancer, treatment-related side effects, and infections. Which of the following is true?
   A. Hospice care should only be considered as a last resort when all other treatment options have been exhausted.
   B. Mr. Jackson must sign a DNR order to be eligible for hospice care.
   C. Hospice care only focuses on the palliation of pain and symptoms in the hospital setting.
   D. Hospice care recognizes the value of the patient's liberty not to pursue ineffective and futile treatments, promoting social justice through a fair distribution of benefits and burdens, and provides comfort and peace in the presence of family and friends.

   Explanation: Hospice care recognizes the value of the patient's liberty not to pursue ineffective and futile treatments and not to die in isolation, promoting social justice through a fair distribution of benefits and burdens. Hospice care allows patients to manage their pain and symptoms, receive emotional support, and die in the presence of family and friends, providing comfort and peace during a challenging time. Hospice care satisfies the principles of beneficence (do good) and nonmaleficence (do no harm) and attempts to maximize the patient's best interests during the end-of-life stage. The other options are incorrect because hospice care can be a beneficial option for patients with a terminal prognosis, as it allows them to focus on the quantity and quality of life rather than pursuing ineffective and futile treatments; a DNR order is not a requirement for hospice care—the only requirement for Medicare hospice is a prognosis of less than 6 months of life; hospice care can be accessed in the home, long-term care facility, or hospital setting; and it provides not only pain and symptom relief but also psychological and bereavement support. *Answer*: D

5. Ms. Clarance Gutierrez is a 75-year-old retired nurse with advanced heart failure for several years. Despite undergoing several treatments and lifestyle modifications, her condition has continued to deteriorate, and her cardiologist has recommended hospice care. Hospice care provides patients with multidisciplinary care and support tailored to the patient's individual needs. Hospice care will offer Ms. Gutierrez the opportunity to manage her symptoms through palliative care, receive emotional support through psychological counseling, and prepare her loved ones for her eventual passing through family bereavement counseling. Hospice care also offers Ms. Gutierrez the opportunity to die in the presence of family and friends, providing comfort and peace in her final days. Which of the following is true?
   A. A DNR order is a requirement for hospice care.
   B. Hospice care is only available in the hospital setting.
   C. Hospice care is only for patients with cancer.
   D. Hospice care attempts to focus not only on the quantity of life but also on the quality of life.

   Explanation: Hospice care attempts to focus not only on the quantity of life but also on the quality of life, providing patients with multidisciplinary care and support tailored to their individual needs. Hospice care allows patients to manage their symptoms, receive emotional support, and die in the presence of family and friends, providing comfort and peace during a challenging time. Hospice care satisfies the principles of beneficence (do good) and nonmaleficence (do no harm) and attempts to maximize the patient's best interests during the end-of-life stage. The other options are incorrect because a DNR order is not a requirement for hospice care, as hospice care attempts to focus not only on the quantity of life but also on the quality of life; hospice care can be accessed in the home, long-term care facility, or hospital setting; and hospice care is not just for patients with cancer as it can be a beneficial option for patients with a terminal prognosis, regardless of the underlying disease or condition. *Answer*: D

REFLECTION VIGNETTES

1. Ms. Elizabeth Davis, a 70-year-old retired teacher, receives palliative care for her terminal illness. During a routine checkup, the practitioner asks whether she has considered enrolling in hospice care. Ms. Davis becomes upset and expresses that hospice is a place of death and where there is no hope. She believes the suggestion is like being thrown out with the trash and questions the patient–practitioner relationship.

   What is the ethical issue(s)?

   How should the ethical issue(s) be addressed?

2. Mr. Robert Johnson, a 65-year-old retired mechanic, has been certified for hospice care due to his terminal cancer. He is in the last stages of the illness, experiencing fecal and urinary incontinence, and dependent on others for all activities of daily living. Mr. Johnson desires

to pass away at home surrounded by family and friends. Still, he and his spouse are concerned about proper care and the potential disruption to family dynamics caused by caregiving role changes.

What is the ethical issue(s)?

How should the ethical issue(s) be addressed?

3. Mr. Charles Thompson, a 70-year-old retired businessman, is admitted to the hospital with end-stage lung cancer. The patient refuses to have a do not attempt resuscitation order and requests that no heroic measures be performed.

What is the ethical issue(s)?

How should the ethical issue(s) be addressed?

* * *

WEB LINKS

1. *Code of Medical Ethics*. American Medical Association. https://code-medical-ethics.ama-assn.org
   The AMA *Code of Medical Ethics*, established by the American Medical Association, is a comprehensive guide for healthcare practitioners. It addresses issues and challenges, promotes adherence to standards of care, and is continuously updated to reflect contemporary practices and challenges.
2. "Hospice Care." Medicare.gov. https://www.medicare.gov/coverage/hospice-care
   Hospice care under Medicare Part A is available for terminally ill patients with a life expectancy of 6 months or less, focusing on comfort rather than curative treatment. It covers various services, including doctors, nursing, healthcare equipment, and counseling, with some costs for medications, inpatient respite care, and potentially room and board in certain facilities. The care plan is created by a hospice team, and although patients can continue hospice care beyond 6 months with recertification, Medicare does not cover treatments intended to cure the terminal illness, non-hospice arranged care, or certain inpatient and emergency services.
3. "Ethical Considerations at the End-of-Life Care." Melahat Akdeniz, Bülent Yardimci, and Ethem Kavukcu. 2021. https://www.ncbi.nlm.nih.gov/pmc/articles/PMC7958189
   This source discusses the ethical difficulties for healthcare professionals in end-of-life care, including decisions regarding resuscitation, mechanical ventilation, artificial nutrition and hydration, terminal sedation, withholding and withdrawing treatments, euthanasia, and physician-assisted suicide.
4. "Palliative Care, Hospice Care, and Bioethics: A Natural Fit." Nessa Coyle. 2014. https://journals.lww.com/jhpn/Fulltext/2014/02000/Palliative_Care,_Hospice_Care,_and_Bioethics__A.3.aspx
   This source discusses ethical issues at the end of life from a mainly Western perspective, including social, legal, religious, and cultural aspects.
5. "Home Health and Hospice Patient Rights." National Association for Home Care and Hospice. https://nahc.org/patient-family-providers/home-health-and-hospice-patient-rights/
   This source outlines the rights of hospice patients to receive care of the highest quality and to have relationships with hospice organizations based on ethical standards of conduct, honesty, dignity, and respect.
6. "Brief Guide to Ethics Committees and Consultation for Hospice Links." National Hospice and Palliative Care Organization. 2021. https://www.nhpco.org/wp-content/uploads/MAID-Brief-Guide-Ethics-Committees-for-Hospice.pdf
   This source provides a guide to addressing ethical dilemmas in palliative and end-of-life care, including questions about the continuation of life-sustaining treatment and scope of patient autonomy in decision-making.

7. "Hospice Patient's Rights." National Hospice and Palliative Care Organization. https://nhpco.rd.net/wp-content/uploads/2019/04/hospice_patient_rights.pdf
   This source provides a comprehensive list of hospice patient rights, including the right to receive quality end-of-life care and information on advance directives.
8. "Code of Ethics." National Association for Home Care and Hospice. https://www.nahc.org/about/code-of-ethics
   This source outlines the code of ethics for the National Association for Home Care and Hospice.

IMMUNIZATION—VACCINE HESITANCY

28

As practitioners say . . . at the beginning of a malady, it is easy to cure but difficult to detect, but over time, not having been either detected or treated in the beginning, it becomes easy to detect but difficult to cure.
—Niccolo Machiavelli

* * *

MYSTERY STORY

THE SERPENT IN EDEN: A TALE OF IMMUNOLOGICAL DEFIANCE

Detective Leopold Sterling could feel a cold coil of urgency gripping his senses as he picked up the somber call. The nerve center of the city's healthcare, Immunity Fortress Hospital—an institution of survival and recuperation—had been infiltrated by an insidious, highly contagious affliction. The specter of illness loomed, casting its malignant shadow over both patients and healthcare practitioners alike, highlighting the urgency of containment and swift justice.

A seasoned investigator, known for his meticulous scrutiny, Detective Sterling armored himself with determination, ready to plunge into the vortex of fear, infectious chaos, and unpredictability. His journey into this healthcare maelstrom began with a trek down the sterile corridors of the hospital, his footsteps echoing ominously against a backdrop of panic-filled whispers and apprehensive glances.

In this turbulent landscape, a single figure presented a paradoxical tranquility amidst the storm—Nurse Eloise Moretti. A beacon of resilience in the face of fear, she embodied an unwavering belief in personal freedom and autonomy. Yet, her conviction stood as an affront to the collective well-being, her unvaccinated status transforming her into a ticking time bomb in a sea of vulnerability.

In the tapestry of healthcare, Nurse Moretti represented a rare deviation, her strong adherence to personal freedoms shaping her professional philosophy. This commitment, however, had resulted in a jarring violation of the patient–practitioner social contract. Her decision not to vaccinate jeopardized the equilibrium of bioethical principles that form the bedrock of healthcare, thereby distorting the trust-filled mirror of care.

By consciously opting out of vaccination, Nurse Moretti breached the ramparts of patient safety, thereby contravening the ethical principle of nonmaleficence (do no harm). Her decision also strained the principle of justice (be fair), casting an ominous pall of disproportionate risk over those in her care.

As Detective Sterling delved into the cryptic labyrinth of his investigation, he unearthed a disconcerting pattern. Documentation revealed that Nurse Moretti had been the recipient of multiple warnings regarding her precarious stance. She had been offered educational counseling about the crucial role of vaccines, yet her obstinate refusal remained unbroken, casting a specter of danger over her patients and colleagues alike.

When Detective Sterling stitched together the scattered shards of his findings, the mosaic of neglect and disregard painted a chilling portrait. Nurse Moretti, with her stubborn persistence, had inadvertently ignited the wildfire of disease outbreak at Immunity Fortress Hospital. Her subsequent indictment for professional misconduct reverberated

> through the corridors of the healthcare system, etching a grim reminder of the uncompromising ethical expectations tethered to the profession.
>
> The foundational responsibility of healthcare practitioners is to uphold patient-centric care by unwaveringly adhering to bioethical principles. When this commitment falters, catastrophic consequences loom. Nurse Moretti's tale serves as a stark reminder of this pivotal duty, illustrating the devastating fallout when individualistic ideologies override professional responsibilities and societal trust.

* * *

THINK

Practitioners have a social contract and a professional obligation to maximize their patients' best interests and not just reduce harm but also reduce even the risk of harm. It is expected that practitioners, as stewards of public health, take reasonable measures for reducing and preventing any spread of infectious diseases in the healthcare setting. Such activities include, but are not limited to, washing hands, wearing personal protective equipment (PPE), appropriate isolation precautions, and being vaccinated.

The American Medical Association (AMA), "8.7 Routine Universal Immunization of Practitioners," states,

> In the context of a highly transmissible disease that poses a significant medical risk for vulnerable patients or colleagues or threatens the availability of the healthcare workforce, particularly a disease that has the potential to become epidemic or pandemic, and for which there is an available, safe, and effective vaccine, practitioners should: (a) Accept immunization.

ASSESS

Patient: Autonomy

Autonomy generally focuses on the patient's choice of informed consent. Part of this "informing" necessarily includes disclosing risks of harm the practitioner presents to the patient due to not being vaccinated. 45 Code of Federal Regulations 46 (45CFR46), also known as the Common Rule, is a federal law that forbids research on human subjects that increases even the risks of harm based on the bioethical principle of nonmaleficence (do no harm) as expressed in the Belmont Report. If healthcare reflects human research standards, then unvaccinated practitioners should not have contact with patients if doing so increases the risk of harm to the patients, and the patients should be made aware of such risks. Healthcare professional vaccine hesitancy is a failure of the professional to recognize the patient's right to be informed about their exposure, increased risks of harm to the patient, and the patient's right not to be exposed to unvaccinated practitioners.

Practitioner: Beneficence and Nonmaleficence

Based on the principle of beneficence (do good), practitioners must pursue the patient's best interests. This patient–practitioner relationship is a patient-centered professional relationship. Beneficence (do good) requires washing hands, taking respiratory precautions, and getting vaccinated.

Nonmaleficence (do no harm) requires not exposing patients and their colleagues to an increased risk of harm due to a failure to get vaccinated. These professional obligations hold more weight than the practitioner's personal beliefs and convictions of individual liberty. If getting immunized is not acceptable to a practitioner, and if not getting vaccinated increases the risk of harm to the patient, then the practitioner should not have patient contact until the practitioner is vaccinated.

Vaccine hesitancy, combined with patient contact, is a self-centered approach to healthcare rather than an altruistic patient-centered approach. Vaccine hesitancy, at the expense of patient safety, is a form of professional misconduct.

Public Policy: Justice

The principle of justice is the fair distribution of benefits and burdens. The patient–practitioner relationship is a social contract that requires that the practitioner's role be a healer, not a harmer. Practitioners who are not vaccinated at the patient's peril violate the social principle of justice. Socially requiring, through the legislative process, that practitioners are vaccinated to prevent social harm is consistent with the role and function of the government toward public safety and the principle of justice. Making vaccinations mandatory is also well within the authority of healthcare institutions. Patients are already in a vulnerable state, and exposing patients to an avoidable pathogen because of vaccine hesitancy of a practitioner would be an unfair and unjust distribution of benefits and burdens and a violation of patients' constitutional right to liberty. Practitioners also have liberties but do not have the liberty to willfully violate their patients' liberty if one or more members of a healthcare team are unvaccinated. Vaccine hesitancy by healthcare professionals is a violation of the patient–practitioner social contract and, therefore, a violation of justice.

Vaccine hesitancy based on individual liberty violates all four principles of autonomy (informed consent), beneficence (do good), nonmaleficence (do no harm), and justice (be fair). Altruistic patient-centered healthcare is central in the patient–practitioner social contract relationship.

CONCLUDE

Practitioners have a professional and social obligation to stay up-to-date on all vaccinations, and not doing so is professional misconduct. It is never acceptable for a practitioner to increase the risk of harm to patients, other healthcare providers, and staff because of personal liberty and beliefs inconsistent with evidence-based care. Patient-centered healthcare is paramount for maximizing the patient's best interests as determined by the patient's reasonable goals, values, and priorities.

In summary, practitioners must uphold their professional and social obligation to stay up-to-date on all vaccinations because not doing so is considered professional misconduct. Vaccine hesitancy based on personal beliefs is unacceptable when it increases the risk of harm to patients and other healthcare providers. Ultimately, the focus should be on providing patient-centered healthcare, ensuring that the patient's best interests are always at the forefront of healthcare decision-making.

* * *

REVIEW QUESTIONS

1. Practitioners have a social contract and a professional obligation to maximize their patients' best interests, reduce harm, and even reduce the risk of harm.
    A. True
    B. False

2. Practitioners have a professional, legal, and moral responsibility to prevent the spread of infectious diseases by
    A. Washing hands
    B. Wearing PPE
    C. Isolation precautions
    D. Being vaccinated
    E. All of the above

3. Patients have the right to be informed about infectious disease exposure risks and not to be exposed to unvaccinated practitioners.
   A. True
   B. False

4. The practitioner's personal beliefs and convictions of individual liberty carry more moral weight than the professional obligation to get vaccinated.
   A. True
   B. False

5. Vaccine hesitancy, combined with patient contact, is a self-centered approach to healthcare rather than an altruistic patient-centered approach.
   A. True
   B. False

6. Patients are vulnerable, and to expose them to an avoidable pathogen because of vaccine hesitancy of a practitioner would be an unfair and unjust distribution of benefits and burdens and a violation of the patient's constitutional right to liberty.
   A. True
   B. False

7. Vaccine hesitancy based on individual liberty violates all four principles of autonomy (informed consent), beneficence (do good), nonmaleficence (do no harm), and justice (be fair).
   A. True
   B. False

Answers: 1A, 2E, 3A, 4B, 5A, 6A, 7A

* * *

CLINICAL VIGNETTES

1. Sandra Roberson is a 32-year-old pharmacist who works at a local hospital. She presents with mild flu-like symptoms and is concerned that she may have contracted the flu. Her differential diagnosis includes influenza, a common cold, or COVID-19. As part of the hospital's policy, Roberson must be up-to-date on her immunizations, including the flu vaccine. An ethical question arises: Should Roberson continue to work until her diagnosis is confirmed, and should she inform her patients and colleagues that she may be infectious?
   A. Sandra Roberson should continue to work until her diagnosis is confirmed and not disclose her symptoms to patients or colleagues.
   B. Sandra Roberson should continue to work but inform her colleagues of her symptoms so they can take precautions.
   C. Sandra Roberson should stay home until her diagnosis is confirmed and inform her colleagues and patients that she may be infectious.
   D. Sandra Roberson should stay home until her diagnosis is confirmed and not inform her colleagues or patients of her symptoms.

   Explanation: Sandra Roberson's ethical obligation is to ensure that she does not expose her patients to any infectious disease she may have contracted. She should stay home until her diagnosis is confirmed and inform her colleagues and patients that she may be infectious. This is in line with the principles of nonmaleficence and beneficence. The principle of nonmaleficence requires that Roberson does not cause harm to her patients, and the principle of beneficence requires that she act in the best interest of her patients. By staying home and informing her colleagues and patients of her symptoms, she is taking appropriate measures to prevent the spread of infectious disease, and this is consistent with the AMA's guidelines for routine universal immunization of practitioners. The other options are inappropriate because they violate the principles of nonmaleficence and beneficence by exposing patients to the risk of harm. It is inappropriate for Roberson to work close to patients or colleagues if she may be infectious, as this violates the principle of nonmaleficence. *Answer*: C

2. Fabian Johnson is a 45-year-old practitioner working at a busy urban hospital. He has hesitated to get vaccinated, violating the hospital's mandatory vaccination policy for healthcare workers. Despite multiple efforts from the hospital administration to encourage Practitioner Johnson to get vaccinated, he still refuses to do so. Clinical symptoms: none. Differential diagnosis: none. What steps should the hospital take to ensure that Practitioner Johnson is taking reasonable measures to reduce and prevent the spread of infectious diseases in the healthcare setting?
    A. The hospital should allow Practitioner Johnson to continue working without getting vaccinated because it is his personal liberty and choice.
    B. The hospital should impose a fine on Practitioner Johnson for not getting vaccinated.
    C. The hospital should reprimand Practitioner Johnson for not getting vaccinated and put him on probation.
    D. The hospital should reassign Practitioner Johnson to an isolating role, such as telemedicine, until he complies with the policy on mandatory vaccination.

    Explanation: Healthcare workers must take reasonable measures to reduce and prevent the spread of infectious diseases in the healthcare setting, including being vaccinated. Practitioner Johnson's refusal to get vaccinated violates the hospital's mandatory vaccination policy for healthcare workers and puts his patients and colleagues at risk. The hospital has a duty to ensure that its healthcare workers are taking reasonable measures to reduce and prevent the spread of infectious diseases, and it cannot allow Practitioner Johnson to continue working directly with patients without complying with its policy on mandatory vaccination. Reassigning Practitioner Johnson to an isolating role until he complies with the policy is an appropriate option because it ensures that the hospital is taking reasonable measures to reduce and prevent the spread of infectious diseases while also providing an opportunity for Practitioner Johnson to comply with the policy and return to his usual patient-facing role. This approach is also more appropriate than the other options because it still allows Practitioner Johnson to work as a healthcare professional and participate in his professional duties while ensuring the safety of patients and colleagues. It is important to note that although the hospital has a duty to ensure that its healthcare workers are taking reasonable measures to reduce and prevent the spread of infectious diseases, it should also consider its employees' concerns and beliefs. By providing education and resources to Practitioner Johnson and other hesitant employees, the hospital can work toward a culture of safety and respect that benefits its employees and patients. *Answer*: D

3. Practitioner Bertha Davis is a 50-year-old oncologist at a hospital specializing in cancer treatment. Despite being informed about the risk of hepatitis B transmission and the availability of a safe and effective vaccine, Practitioner Davis has not been vaccinated against hepatitis B. She has expressed concerns about the vaccine's safety and has chosen not to get vaccinated. Clinical symptoms: none. Differential diagnosis: none. What are the implications of Practitioner Davis's vaccine hesitancy for her patients and colleagues, and what steps should be taken to ensure she is not exposing them to increased risk of harm?
    A. Practitioner Davis should continue working without getting vaccinated because it is her personal choice.
    B. Practitioner Davis should inform her patients and colleagues that she is not vaccinated and let them decide whether they want to be seen by or work with her.
    C. Practitioner Davis should be required to wear additional PPE while working with patients and colleagues instead of getting vaccinated.

D. Practitioner Davis should be required to get vaccinated and not have patient contact until she complies with the policy on mandatory vaccination.

Explanation: Healthcare professionals have a professional and ethical obligation not to expose their patients and colleagues to increased risk of harm due to a failure to get vaccinated. Practitioner Davis' vaccine hesitancy puts her high-risk cancer patients and colleagues at risk of hepatitis B transmission and fails her professional and ethical obligations. Requiring Practitioner Davis to get vaccinated and not have patient contact until she complies with the policy on mandatory vaccination is the most appropriate course of action. This ensures that Practitioner Davis takes reasonable measures to reduce and prevent the transmission of infectious diseases in the healthcare setting while also recognizing her professional and ethical obligations to her patients and colleagues. The other options are inappropriate because they do not adequately address Practitioner Davis' professional and ethical obligations, and requiring Practitioner Davis to wear additional PPE may not be sufficient in reducing the risk of transmission. It is important to note that although healthcare professionals have the right to make personal treatment decisions, they also have a professional and ethical obligation to ensure that their decisions do not put their patients and colleagues at risk. Practitioner Davis fulfills her professional and ethical obligations to her patients and colleagues by getting vaccinated. The hospital should provide education and resources to healthcare professionals who are hesitant to get vaccinated in order to promote a culture of safety and respect that benefits its employees and patients. *Answer*: D

4. Practitioner Hayden Allen is a 60-year-old practitioner who works in a rural hospital with a limited number of other healthcare professionals. He has worked for the hospital for 25 years and is highly respected by his colleagues and patients. However, Practitioner Allen hesitates to get vaccinated against preventable diseases, including influenza, measles, and hepatitis B. Clinical symptoms: none. Differential diagnosis: none. What are the implications of Practitioner Allen's vaccine hesitancy for his patients and the principle of justice, and what can be done to ensure that he fulfills his professional and ethical obligations?
   A. The hospital should allow Practitioner Allen to continue working without being vaccinated because he has a right to make his own treatment decisions.
   B. The hospital should require Practitioner Allen to wear additional PPE while working with patients instead of getting vaccinated.
   C. The hospital should reassign Practitioner Allen to non-patient-facing activities until he complies with the policy on mandatory vaccination.
   D. The hospital should require Practitioner Allen to get vaccinated or take a leave of absence until he complies with the policy on mandatory vaccination.

Explanation: Healthcare professionals have a professional and ethical obligation to maximize their patients' best interests and not expose them to increased risk of harm due to a failure to get vaccinated. Furthermore, healthcare providers who do not get vaccinated put their patients at risk and violate the social principle of justice. Practitioner Allen's vaccine hesitancy jeopardizes the health and safety of his patients and goes against his role as a healer. Requiring Practitioner Allen to get vaccinated or take a leave of absence until he complies with the policy on mandatory vaccination is the most appropriate course of action. This ensures that Practitioner Allen takes reasonable measures to reduce and prevent the transmission of preventable diseases in the healthcare setting while also recognizing his professional and ethical obligations to his patients and the principle of justice. Reassigning Practitioner Allen to non-patient-facing activities may not be feasible due to the limited number of rural practitioners and the hospital's and its patients' needs. The other options are inappropriate because they do not adequately address Practitioner Allen's professional and ethical obligations to his patients and the principle of justice. It is important to note that although healthcare professionals have the right to make personal treatment decisions, they also have a professional and ethical obligation to ensure that their decisions do not put their patients at risk and violate the principle of justice. By getting vaccinated or taking a leave of absence until he complies with the policy on mandatory vaccination, Practitioner Allen is fulfilling his professional and ethical obligations to his patients and the community. The hospital should provide education and resources to healthcare professionals who are hesitant to get vaccinated in order to promote a culture of safety and respect that benefits its employees and patients. *Answer*: D

5. Ms. Chloe Williams is a 35-year-old woman concerned about vaccine safety for her young daughter. She has read about possible links between vaccines and autism, and she is hesitant to vaccinate her daughter. Her practitioner has recommended that her daughter receive all the recommended vaccines to protect her from serious diseases. Which of the following statements most accurately reflects the parental authority of a minor child?
   A. Ms. Williams does not have the parental right to refuse consent for the vaccination of her daughter.
   B. Ms. Williams does not have the moral obligation to become informed of peer-reviewed evidence-based care and the resultant standards of care regarding vaccinations.
   C. Ms. Williams should be allowed to make her own decision regarding vaccination for her daughter but should be informed of the risks and benefits of vaccination and the potential consequences of not vaccinating.
   D. Ms. Williams' decision not to provide consent for her daughter's vaccination should not be respected because it puts her daughter at greater risk for serious diseases.

   Explanation: Ms. Williams' concerns about the safety of vaccines are understandable, but it is important to consider the potential risks and benefits of vaccination in the context of her daughter's health. Although Ms. Williams has the right to make decisions regarding her daughter's healthcare, she is also responsible for reducing the risk of harm to her daughter and others. Vaccines are important for protecting individuals from serious diseases and preventing outbreaks of infectious diseases in the community. Ms. Williams should be allowed to decide on vaccination for her daughter. However, it is important that she is informed of the risks and benefits of vaccination and the potential consequences of not vaccinating. Her practitioner must address her concerns about vaccine safety and provide accurate information to help her make an informed decision. Ultimately, it is Ms. Williams' responsibility to make a decision that is in her daughter's best interest and the best interest of public health. *Answer*: C

REFLECTION VIGNETTES

1. Ms. Samantha Lee, a 35-year-old nurse, has decided not to get vaccinated. She believes that being forced to be vaccinated would violate her constitutional rights and liberties.

   What is the ethical issue(s)?

   How should the ethical issue(s) be addressed?

2. Mr. John Smith, a 45-year-old practitioner, has decided not to get vaccinated. He believes there is a government conspiracy for vaccination and that vaccine safety and efficacy are fraudulent.

   What is the ethical issue(s)?

   How should the ethical issue(s) be addressed?

* * *

WEB LINKS

1. *Code of Medical Ethics*. American Medical Association. https://code-medical-ethics.ama-assn.org

   The AMA *Code of Medical Ethics*, established by the American Medical Association, is a comprehensive guide for healthcare practitioners. It addresses issues and challenges, promotes adherence to standards of care, and is continuously updated to reflect contemporary practices and challenges.

2. "Ethical Issues in Mandating COVID-19 Vaccination for Health Care Personnel." Robert Olick, Jana Shaw, and Y. Tony Yang. 2021. https://www.mayoclinicproceedings.org/article/S0025-6196(21)00804-1/fulltext

   This source discusses the ethical debate surrounding compulsory vaccination for healthcare professionals, focusing on the duty to protect and promote the greater good versus individual rights of autonomy and informed consent.

3. "Ethics of Vaccination in Childhood—A Framework Based on the Four Principles of Biomedical Ethics." Meta Rus and Urh Groselj. 2021. https://www.ncbi.nlm.nih.gov/pmc/articles/PMC7913000

   This source applies a framework for ethical analysis of vaccination in childhood based on the four principles of biomedical ethics (respect for autonomy, nonmaleficence, beneficence, and justice).

4. "Vaccination Refusal: Ethics, Individual Rights, and the Common Good." Jason L. Schwartz and Arthur L. Caplan. 2011. https://pubmed.ncbi.nlm.nih.gov/22094142

   This source discusses the ethical implications of vaccination refusal, focusing on balancing individual rights and the common good.

5. "No COVID-19 Vaccination, No Care? Why That's the Wrong Path." Timothy M. Smith. 2021. https://www.ama-assn.org/delivering-care/ethics/no-covid-19-vaccination-no-care-why-s-wrong-path

   This source discusses the ethical implications of refusing care to patients who have forgone SARS-CoV-2 vaccination.

6. "Addressing Vaccine Hesitancy Requires an Ethically Consistent Health Strategy." Laura Williamson and Hannah Glaab. 2018. https://pubmed.ncbi.nlm.nih.gov/30355355

   This source discusses the growing threat of vaccine hesitancy to public health and its connection to a lack of trust in vaccines, expertise, and traditional sources of authority.

7. "COVID-19 Vaccination: Ethical Issues Regarding Mandatory Vaccination for Healthcare Providers." Alireza Hamidian Jahromi, Jenna Rose Stoehr, and Clayton Thomason. 2021. https://www.ncbi.nlm.nih.gov/pmc/articles/PMC8547809

   This source discusses the ethical dilemma of whether COVID-19 vaccination for healthcare providers should be mandated, considering their critical role in patient care and potential role in the spread of the virus.

IMPAIRED DRIVER

29

The practitioner who knows only medicine knows not even medicine.
—Mark Twain

* * *

MYSTERY STORY

THE REAPER'S EDICT: A DEADLY CONSEQUENCE OF UPHOLDING HEALTHCARE RESPONSIBILITY

Renowned neurologist, Practitioner Elizabeth White, was a beacon of hope and healing at SafeJourney Recovery Hospital in the quaint town of Millville. For more than two decades, her dedicated service had won the hearts of the townsfolk. Her intuitive diagnostic prowess and gentle care often offered solace to those grappling with the complexities of neurological conditions.

One day, Practitioner White was startled by a call from the local constabulary. An unfortunate automobile incident had occurred, and the driver, Mr. Matthew Taylor, had convulsed into a seizure behind the wheel, catapulting his vehicle into a menacing tree. Although he escaped with minor injuries, the incident raised concerns about his healthcare condition's role in causing the crash.

A bolt of realization struck Practitioner White. Mr. Taylor was no stranger but one of her regular patients, grappling with epilepsy for many years. She was aware of the legal mandate for epileptic patients to report their condition to the Department of Motor Vehicles (DMV). However, she was uncertain if Mr. Taylor had adhered to this obligation.

Driven by her professional ethics, Practitioner White rushed to the hospital, only to discover a harsh reality. Mr. Taylor had remained tight-lipped about his condition to the DMV. She impressed upon him the legal and moral gravity of his oversight, warning him of the potentially catastrophic consequences. In addition, she reinforced her duty as a healthcare professional to report the incident to the DMV as an essential measure to ensure public safety.

Mr. Taylor's plea for mercy, citing his economic necessities and family commitments, left Practitioner White in a quandary. Despite her sympathy for his predicament, she held steadfast to her ethical obligations. Assuring Mr. Taylor of finding alternate transportation solutions and continuing his healthcare, she reported his condition to the DMV.

The DMV's subsequent letter of acknowledgment brought both relief and a sense of achievement to Practitioner White. Her decision had ensured that Mr. Taylor was no longer a threat on the road, and she had upheld her sacred oath of prioritizing public safety.

However, a grim twist awaited at SafeJourney Recovery Hospital. A few weeks later, Practitioner White was found lifeless in her office, a chilling testament to the potential backlash of stringent professional ethics. Police investigations revealed threatening communications against Practitioner White, expressing venomous outrage for her decision to report Mr. Taylor. The digital breadcrumbs led to Mr. Taylor's wife, her wrath against Practitioner White for jeopardizing her family's livelihood manifesting in the ultimate act of violence.

> The arrest and subsequent charging of Mr. Taylor's wife for Practitioner White's tragic demise underscored a sobering paradox. Although reporting to the DMV serves to protect public safety, it can sometimes catalyze devastating consequences for the upholders of such responsibility. The tale of Practitioner White is a stark reminder of the precarious balance healthcare practitioners must strike between professional duty and personal relationships, a lesson that resonates deeply within the halls of SafeJourney Recovery Hospital.

* * *

THINK

A wide array of healthcare conditions can compromise a driver's ability to manage a vehicle securely. These conditions can affect anyone operating cars, trucks, motorcycles, buses, trains, planes, and other modes of transportation. In the United States, the requirements for mandatory reporting of medically unfit drivers, including those with visual impairments, vary greatly across states. For instance, in California and Pennsylvania, practitioners are legally required to report patients with certain conditions, whereas in Oregon, reporting remains voluntary. Commercial vehicles are typically subjected to stricter governmental healthcare and healthcare guidelines to ensure driver competence compared to personal vehicles. A pivotal question in this context pertains to the practitioner's role in identifying and assessing these impaired drivers.

Every state requires patients to self-disclose any seizure disorders to the DMV. However, the responsibility for practitioners to report these conditions needs to be uniformly regulated, remaining voluntary in most states. Despite practitioners' role in diagnosing and managing these conditions, they do not possess the authority to suspend or revoke driving privileges—the DMV holds this power. Nevertheless, physicians are always responsible for discussing with patients how their healthcare conditions may affect their ability to drive safely. In states where reporting is not required, physicians choosing to report a patient to the DMV may be protected by "good faith" or "Good Samaritan" laws. These laws offer legal protection to practitioners who choose to report a patient's condition to the DMV in the interest of public safety. The extent and nature of this protection can differ significantly across states, adding another layer of complexity to the decision-making process.

ASSESS

Patient: Autonomy

Mandatory patient reporting can infringe on patient autonomy (informed consent) and restrict the liberties of both the patient and the practitioner. In cases in which a practitioner is legally bound to report an impaired driver, it is paramount to relay this obligation to the patient, ideally before but certainly following the examination. The relationship between a patient and a practitioner is built on trust, confidentiality, and privacy. Any deviations from these principles should be explicitly communicated to patients to ensure their understanding and avoid any perceived violation of confidentiality. When reporting is compulsory, the practitioner and the patient must report. This obligation must be communicated unequivocally to the patient. Best practice in such scenarios entails explicating the circumstances so that the patient understands and consents to the healthcare risk reporting mandate. However, even with clear communication, patients may still believe their privacy is being invaded, representing a significant psychological impact that needs to be factored into the decision-making process.

Mandatory reporting, although potentially intrusive, does not violate the patient–practitioner relationship. This is because these obligations of mandatory reporting are integral to the mutual agreement and understanding between the patient and the practitioner, a condition typically discussed and agreed upon at the outset of care.

Practitioner: Beneficence and Nonmaleficence

From a professional standpoint, healthcare practitioners must prioritize the patient's best interests, which are often guided by the patient's reasonable goals, values, and priorities—the beneficence (do good) principle. Such interests naturally extend to include safety from potential harm caused by an impaired driver, especially if the impaired driver is the patient themself. This represents an implementation of the principle of nonmaleficence (do no harm).

As a patient educator, practitioners should openly and sensitively discuss the patient's options to mitigate driving risks. This includes exploring alternative modes of transportation such as public transportation, rideshare services, or even assistance from family members.

Public Policy: Justice

From a societal standpoint, allowing an impaired driver to operate a vehicle represents an unjust distribution of benefits and burdens. The practitioner is putting the driver and other road users in danger by granting the ability to drive to someone who could pose a risk.

If a state imposes mandatory reporting laws, practitioners must adhere to them. Even when a patient's autonomous choice opposes such reporting, this preference would be primarily overridden by the principle of justice (be fair), emphasizing public safety. This is combined with the practitioner's ethical commitment to nonmaleficence or avoiding harm to others. Notably, states that enforce mandatory reporting typically have systems for evaluating the reported information and determining the appropriate course of action, ensuring that decisions are made based on accurate risk assessments.

Mandatory reporting is a legal and professional requirement within the patient–practitioner relationship. The failure to report when required is viewed as a violation of this relationship and is classified as professional misconduct.

CONCLUDE

Federal law obligates practitioners to encourage patients with seizure disorders to self-report to the DMV. Whereas the DMV wields the authority to revoke, suspend, or limit an individual's driving license based on health status, practitioners do not hold such legal powers.

In summary, healthcare practitioners are expected to comply with mandatory reporting laws, delicately balancing patient autonomy (informed consent) and their professional obligations of beneficence (do good), nonmaleficence (do no harm), and justice (be fair). Urging patients with seizure disorders to self-report to the DMV is integral to the practitioner's duty in promoting public safety. Any deviation or noncompliance with these regulations is viewed as a breach of the patient–practitioner relationship and can be deemed professional misconduct. The consequences of this misconduct can be severe and include sanctions, reputational damage, or legal action, emphasizing the gravity of this responsibility. Ultimately, practitioners must prioritize their patients' well-being and public safety when handling issues related to impaired driving.

* * *

REVIEW QUESTIONS

1. The practitioner must report seizure disorders to the DMV.
    A. True
    B. False

2. If the practitioner honestly believes that a patient might be a driving risk, then the practitioner can suspend or revoke the patient's driving privileges.
    A. True
    B. False

3. The ideal is for the practitioner to explain the situation so the patient understands and consents to the medically at-risk reporting mandate. This approach attempts to keep intact the patient–practitioner relationship and minimize any perception of a breach of confidentiality and privacy.
   A. True
   B. False

4. As a patient educator, the practitioner should have a tactful but candid conversation about reducing driving risks and devising alternative forms of transportation, such as public transportation, Uber or Lift, and family members.
   A. True
   B. False

5. The government has mandated practitioners to report various impairments to various authorities, and failure to comply with mandatory reporting is a Class A misdemeanor.
   A. True
   B. False

6. Mandatory reporting is a legal and professional requirement in the patient–practitioner relationship. Not reporting would be the practitioner's violation of the patient–practitioner relationship, a form of professional misconduct.
   A. True
   B. False

Answers: 1B, 2B, 3A, 4A, 5A, 6A

* * *

CLINICAL VIGNETTES

1. Mr. Solomon Thompson, a 45-year-old truck driver, was brought to the emergency room after a motor vehicle accident. He was found to have slurred speech, delayed reaction time, and an unsteady gait. Blood alcohol content (BAC) was measured to be 0.14%, above the legal limit for driving. The healthcare team suspected Mr. Thompson was under the influence of alcohol, which may have contributed to the accident. The differential diagnosis included alcohol intoxication, brain injury, and stroke. Which of the following actions should the healthcare team take?
   A. Report Mr. Thompson's BAC to the authorities for a possible driving under the influence (DUI) charge
   B. Report Mr. Thompson's BAC to his employer for potential employment consequences
   C. Discharge Mr. Thompson without reporting his BAC to anyone
   D. Report Mr. Thompson's BAC to the DMV for possible license suspension

   Explanation: Based on the scenario, the most suitable action is to discharge Mr. Thompson without reporting his BAC to anyone. Typically, it is not the role of physicians to report a patient's BAC to law enforcement, the DMV, or an employer because it may infringe upon the patient's privacy rights and potentially violate Health Insurance Portability and Accountability Act (HIPAA) laws. Instead, the responsibility to report BAC levels for potential legal action often lies with police officers at the scene of an accident. The authority to suspend or revoke driving privileges rests with the DMV, not physicians. Although physicians should discuss the implications of drinking and driving with patients and recommend appropriate treatments, respecting patient privacy and autonomy in these discussions is crucial. It is important to remember that these are general guidelines, and the exact approach may vary based on local laws, hospital policies, and specific circumstances. *Answer*: C

2. Mr. Augustus Francis, a 55-year-old construction worker, was brought to the emergency department after a car accident. Upon examination, the healthcare team noticed Mr. Francis had slurred speech, difficulty walking, and delayed reaction time. Mr. Francis admitted to taking his prescription pain medication before driving. The differential diagnosis included medication side effects, interactions, and substance abuse. Which of the following actions should the healthcare team take?
    A. Report Mr. Francis to the authorities for possible DUI charge
    B. Discharge Mr. Francis without reporting his medication use to anyone
    C. Report Mr. Francis' medication use to his employer
    D. Report Mr. Francis' medication use to the prescribing practitioner

    Explanation: Based on the given scenario and information, the most appropriate response seems to be option D, report Mr. Francis' medication use to the prescribing practitioner. In this case, there is a clear health issue related to prescribed medication, and the physician should discuss this with the prescribing practitioner to potentially adjust the medication and the dosage or discuss alternative treatment options that would not impair Mr. Francis' ability to drive. Reporting Mr. Francis to authorities for a DUI charge (option A) typically is not the role of a healthcare provider. As for option C, reporting Mr. Francis' medication use to his employer could violate his privacy rights under HIPAA unless Mr. Francis provides explicit consent. Option B, discharging Mr. Francis without addressing the issue, could potentially endanger him and others if the side effects of the medication continue to impair his driving. As such, communicating with the prescribing practitioner (option D) offers the best chance to resolve the issue in a way that respects Mr. Francis' rights and protects public safety. *Answer*: D

3. Ms. Abigail Rodriguez, a 30-year-old teacher, visited her primary care practitioner complaining of frequent headaches and fainting spells. After further examination, the practitioner diagnosed her with a seizure disorder and informed her that it is mandatory for patients to self-report any seizure disorders to the DMV. The differential diagnosis included epilepsy, medication side effects, and brain injury. Which of the following actions should the practitioner take?
    A. Advise Ms. Rodriguez to self-report her seizure disorder to the DMV
    B. Report Ms. Rodriguez's seizure disorder to the DMV without her consent
    C. Discharge Ms. Rodriguez without discussing DMV reporting requirements
    D. Report Ms. Rodriguez's seizure disorder to her employer

    Explanation: The practitioner should advise Ms. Rodriguez to self-report her seizure disorder to the DMV. It is mandatory for patients to self-report any seizure disorders to the DMV to ensure public safety on the roads. Reporting Ms. Rodriguez's seizure disorder to the DMV without her consent is inappropriate because it violates patient autonomy and confidentiality. Discharging Ms. Rodriguez without discussing DMV reporting requirements is not an ethical option because it fails to ensure that Ms. Rodriguez receives appropriate care and may put others at risk. Reporting Ms. Rodriguez's seizure disorder to her employer is unnecessary and could compromise patient confidentiality. Therefore, the correct answer is to advise Ms. Rodriguez to self-report her seizure disorder to the DMV to ensure public safety. The practitioner should also work with Ms. Rodriguez to develop a treatment plan to manage her seizures and minimize the risk of harm to herself and others. *Answer*: A

4. Mrs. Helen Kim, a 50-year-old real estate agent, visited her primary care practitioner for a routine checkup. During the examination, the practitioner noticed that Mrs. Kim had uncontrolled blood sugar levels and neuropathy in her legs. After further examination and tests, the practitioner diagnosed Mrs. Kim with poorly controlled diabetes. The differential diagnosis included peripheral artery disease, medication side effects, and hypothyroidism. Which of the following actions should the practitioner take?
    A. Report Mrs. Kim's condition to the DMV for a possible license suspension
    B. Discharge Mrs. Kim without discussing the impact of her condition on her ability to drive

C. Prescribe medication to manage Mrs. Kim's blood sugar levels and allow her to continue driving
D. Advise Mrs. Kim to self-report her condition to the DMV

Explanation: Given the scenario, the most appropriate answer is option D, advise Mrs. Kim to self-report her condition to the DMV. Because Mrs. Kim's condition could impair her ability to drive, it is within her primary care practitioner's ethical and professional responsibilities to discuss these implications with her. Although a physician can prescribe medication to manage Mrs. Kim's blood sugar levels (option C), she must still understand her condition's potential effects on her driving abilities. The decision to report to the DMV ultimately lies with the patient in most states. Option B, discharging Mrs. Kim without discussing the impact of her condition, would be against the physician's duty to promote patient welfare. Option A, reporting Mrs. Kim's condition to the DMV for a possible license suspension, usually is not the physician's duty and could infringe upon her privacy rights. Therefore, the most ethical and appropriate action for the physician would be to advise Mrs. Kim to self-report her condition to the DMV (option D) and to help her manage her condition in a way that maximizes her health and safety. *Answer*: D

5. Mr. Trey Hernandez, a 35-year-old Uber driver, visited his primary care practitioner for a routine checkup. During the examination, the practitioner noticed that Mr. Hernandez had high blood pressure and was taking medication to manage it. After further examination and tests, the practitioner diagnosed Mr. Hernandez with hypertension. The differential diagnosis included kidney disease, medication side effects, and sleep apnea. Which of the following actions should the practitioner take?
   A. Report Mr. Hernandez's condition to the DMV for a possible license suspension
   B. Discharge Mr. Hernandez without discussing the impact of his condition on his ability to drive
   C. Prescribe medication to manage Mr. Hernandez's hypertension and allow him to continue driving
   D. Advise Mr. Hernandez to self-report his condition to the DMV

Explanation: In this case, the practitioner may discharge Mr. Hernandez without reporting his condition to the DMV or discussing the impact of his condition on his ability to drive, as hypertension alone is not a healthcare condition that impairs a driver's ability to operate a vehicle safely. A wide range of healthcare conditions can impair a driver's ability to operate a vehicle safely. However, hypertension is not typically one of them. Therefore, reporting Mr. Hernandez's condition to the DMV for a possible license suspension is unnecessary. Prescribing medication to manage Mr. Hernandez's hypertension and allowing him to continue driving may be appropriate. Advising Mr. Hernandez to self-report his condition to the DMV is also unnecessary because hypertension is not typically a healthcare condition requiring reporting. Therefore, the correct answer is to discharge Mr. Hernandez without discussing the impact of his condition on his ability to drive, as it is unnecessary in this case. The practitioner should still work with Mr. Hernandez to manage his hypertension and minimize the risk of harm to himself and others. *Answer*: B

REFLECTION VIGNETTES
1. Ms. Sarah Johnson, a 65-year-old retired accountant, presents for her yearly physical. She reports being in good health but has age-related macular degeneration, and her visual acuity has decreased to 20/200 OU. The differential diagnosis includes age-related macular degeneration, cataracts, glaucoma, and diabetic retinopathy.

   What is the ethical issue(s)?

   How should the ethical issue(s) be addressed?

2. Mr. Edward Phillips, a 35-year-old construction worker, is brought into the emergency department handcuffed to a stretcher by police officers. The officers suspect that Mr. Phillips was driving under the influence of alcohol, causing a motor vehicle accident that resulted in three deaths. On examination, the practitioner notes that Mr. Phillips is slurring words, confused, unable to identify himself, and unaware of what has happened. The differential diagnosis includes alcohol intoxication, head trauma, and underlying healthcare conditions.

What is the ethical issue(s)?

How should the ethical issue(s) be addressed?

* * *

WEB LINKS

1. *Code of Medical Ethics*. American Medical Association. https://code-medical-ethics.ama-assn.org
   The AMA *Code of Medical Ethics*, established by the American Medical Association, is a comprehensive guide for healthcare practitioners. It addresses issues and challenges, promotes adherence to standards of care, and is continuously updated to reflect contemporary practices and challenges.
2. "Driverless Cars, Impaired Drivers and the Code of Medical Ethics." Timothy M. Smith. 2017. https://www.ama-assn.org/delivering-care/ethics/driverless-cars-impaired-drivers-and-code-medical-ethics
   This source discusses the ethical implications of impaired driving and the role of physicians in recognizing and addressing this issue.
3. "Physicians & the Health of the Community." American Medical Association. https://www.ama-assn.org/system/files/2020-12/code-of-medical-ethics-chapter-8.pdf
   This source includes opinions on physicians' responsibilities regarding impaired drivers.
4. "Who Makes Decisions for Incapacitated Patients Who Have No Surrogate or Advance Directive?" Scott J. Schweikart. 2019. https://journalofethics.ama-assn.org/article/who-makes-decisions-incapacitated-patients-who-have-no-surrogate-or-advance-directive/2019-07
   This source discusses the ethical considerations when making decisions for incapacitated patients, including those who may be impaired drivers.
5. "Physicians' Legal Responsibility to Report Impaired Drivers." Lee Black. 2008. https://journalofethics.ama-assn.org/article/physicians-legal-responsibility-report-impaired-drivers/2008-06
   This source discusses the legal responsibilities of physicians to report impaired drivers, focusing on the balance between patient confidentiality and public safety.
6. "AMA *Code of Medical Ethics*' Opinion on Reporting Impaired Drivers." AMA Council on Ethical and Judicial Affairs. 2010. https://journalofethics.ama-assn.org/article/ama-code-medical-ethics-opinion-reporting-impaired-drivers/2010-12
   This source provides the American Medical Association's opinion on reporting impaired drivers, discussing the dual responsibilities of physicians to promote patient welfare and protect public safety.
7. "Impaired Drivers & Their Physicians." American Medical Association. https://code-medical-ethics.ama-assn.org/ethics-opinions/impaired-drivers-their-physicians
   This source outlines the responsibilities of physicians when dealing with impaired drivers, focusing on the balance between patient confidentiality and public safety.

IMPAIRED PRACTITIONER

30

The aim of medicine is to prevent disease and prolong life; the ideal of medicine is to eliminate the need for a practitioner.
—William James Mayo

* * *

MYSTERY STORY

FALLING FAÇADE: THE PUZZLING MYSTERY OF AN AILING PRACTITIONER

Practitioner Sarah Jones was the beacon of Healing Hands Hospital, a compassionate and deft healthcare practitioner revered by patients and colleagues alike. However, an unsettling transformation had begun to take root in her character over recent months. The once tranquil and attentive healer had morphed into a frequently irritable and forgetful persona. The troubling metamorphosis had not gone unnoticed; errors began to creep into her diagnoses and treatment strategies, and patients were often left feeling neglected and unheard.

Practitioner Jones had always championed her patients' autonomy and rights, a facet of her profession that she held dear. But the shift in her demeanor ignited alarm among her colleagues, spurring whispers of concern. They wrestled with a distressing suspicion: Was she struggling with a concealed impairment?

The healthcare fraternity of Healing Hands Hospital faced an ethical quagmire. While they acknowledged their professional obligation to report Practitioner Jones, they were also acutely aware of the profound consequences that could ensue. Yet, the predicament took a perilous turn when a glaring misdiagnosis culminated in severe complications for a patient. Although the patient made a fortuitous recovery, the incident sent shock waves through the hospital staff, highlighting the pressing need for intervention.

The corridors of Healing Hands Hospital reverberated with hushed consultations and confabulations as Practitioner Jones' colleagues sought guidance. They turned to their superiors, pored over legal intricacies, and wrestled with the ethical conundrum of reporting her impairment. Balancing a desire for fairness to Practitioner Jones against their paramount duty toward patient safety, they reached a collective decision: a direct confrontation expressing their worries and extending their support.

Practitioner Jones' initial response was an eruption of defensiveness, as the biting sting of perceived betrayal bit deep. But as the sincerity of their concern emerged, she began to understand their intentions were not to question her abilities but, rather, to uphold the sanctity of patient safety. Reluctantly, she agreed to a comprehensive healthcare evaluation and agreed to treatment, with her colleagues providing an unwavering support system.

The thorough examination peeled back the layers of Practitioner Jones' guarded facade, unearthing a struggle with substance use disorder, depression, and anxiety. Surrounded by her colleagues' empathy and the unwavering dedication of healthcare professionals, Practitioner Jones embarked on a journey to recovery. Her path was closely monitored, ensuring her fitness to resume her sacred responsibility as a healthcare practitioner.

The bewildering case of Practitioner Jones served as a poignant reminder to healthcare practitioners about the importance of recognizing, acknowledging, and

> addressing impairments, both in themselves and in their peers. Navigating through the heart-wrenching challenges, the incident unfolded into a beacon of hope for Practitioner Jones and sparked a renaissance of commitment toward patient safety and professional excellence at Healing Hands Hospital.

* * *

THINK

An impaired practitioner is one whose healthcare judgment and practice have been compromised such that the impaired practitioner

- violates the patient–practitioner relationship by exposing the patient to increased risk of harm;
- violates the professional duty of maximizing the patient's best interests; and
- diminishes the community's trust in the healthcare profession.

Impairment can have various causes, including psychoactive agents, prescription medications, aging, substance abuse, Alzheimer's disease, psychiatric disorders, emotional disorders, and fatigue, to name a few. It is essential to recognize that substance use disorders occur at the same rate among practitioners as the general population.

Practitioners, as stewards, have a mandatory duty to report an impaired practitioner legally, professionally, and morally. Reporting an impaired practitioner protects patients from harm and the impaired practitioner from malpractice lawsuits. However, because of the sensitivity and potential hostility that can arise from reporting, it is advisable to seek counsel from colleagues and a designated institutional official as a matter of due diligence when there is no immediate threat to a patient. The same legal protections for liability granted for reporting child and elder abuse will also apply to reporting an impaired practitioner as long as the reporting practitioner is doing so in good faith, sincerely, honestly, and without malice.

Practitioners are obligated to protect patients from an impaired practitioner and help the impaired practitioner get the healthcare and recovery needed to eliminate the impairment. Reporting should be made first to the impaired practitioner's local supervisor or superior, such as the program director, department chair, or division head. If that is not possible, then reported should be made to the healthcare board of healthcare conduct or the state health or education department.

If a practitioner is in treatment, that does not mean the practitioner is impaired. Patient confidentiality and privacy are the same as for any other patient–practitioner relationship, except if confidentiality and privacy compromise patient safety and if there is a legal, professional, and moral duty to divulge the protected health information. After successful treatment and rehabilitation, someone other than the treating practitioner should assist and monitor the recovered colleague while they resume safe and effective patient care.

A practitioner's health and wellness are an essential part of what it means to be a healthcare professional because without them, the practitioner will not be able to

- honor the patient–practitioner relationship;
- fulfill professional obligations and expectations necessary for maximizing the patient's best interests; and
- promote positive public trust in the healthcare profession.

Every healthcare institution needs a supportive environment that promotes, maintains, and restores health and wellness. Proper diet, enough sleep and exercise, and exposure to art and the humanities are necessary for health and wellness.

ASSESS

Patient: Autonomy

The mandatory duty to report an impaired practitioner reduces individual liberty for the reporter and the reported person. Mandatory reporting promotes patient autonomy (informed consent) because patients can only make an informed practitioner authorization if the practitioner's judgments are accurately grounded on evidence-based care. The impaired practitioner's diagnosis, prognosis, risks and benefits assessments of various treatments (including no treatment), answers to the patient's questions, and patient's management are all called into question. Therefore, active mandatory reporting measures must be taken to protect the patient's autonomous decision-making.

Practitioner: Beneficence and Nonmaleficence

Practitioners have a professional obligation to report an impaired practitioner, primarily to maximize the patient's best interests as determined by the patient's reasonable goals, values, and priorities. This professional obligation to report an impaired practitioner is also a collegial duty for the benefit of the impaired practitioner so that the impaired practitioner can be helped and treated to return and practice medicine effectively.

Public Policy: Justice

Justice is the public policy principle of fairness. Legally, practitioners are mandated to report an impaired practitioner to the proper authorities. Failure to report is a Class A misdemeanor, and the nonreporting practitioner is legally liable for patient harm caused by nonreporting. Public knowledge of impaired practitioners diminishes the community's trust in the healthcare profession, which increases the importance of timely identification of impaired practitioners so that they can be reintegrated through treatment and recovery.

CONCLUDE

Practitioners must recognize the importance of

- not being impaired;
- reporting of impaired practitioners;
- patient safety; and
- treatment and recovery.

There are legal protections for reporting as long as the reporting practitioner sincerely and honestly suspects the practitioner is impaired while treating patients. If the practitioner believes that a colleague is impaired while treating patients, then there is a legal, professional, and moral duty to report the incident regardless of explanations, justifications, or promises.

In summary, practitioners must recognize the importance of not being impaired, reporting impaired colleagues, protecting patient safety, and promoting treatment and recovery. Legal protections exist for reporting practitioners acting in good faith, honestly, and without malice. Timely identification of impaired practitioners is critical to protect patients, help the impaired practitioner, and maintain public trust in the healthcare profession. Practitioners have a legal, professional, and moral duty to report incidents involving impaired colleagues, ensuring the safety and well-being of all involved.

* * *

REVIEW QUESTIONS

1. An impaired practitioner is one whose healthcare judgment and practice have been compromised such that the impaired practitioner
    A. Violates the patient–practitioner relationship by exposing the patient to increased risks of harm
    B. Violates the professional duty of maximizing the patient's best interests
    C. Diminishes the community's trust in the healthcare profession
    D. All of the above

2. Proper diet, enough sleep and exercise, and exposure to art and the humanities are necessary for health and wellness.
    A. True
    B. False

3. Mandatory reporting of an impaired practitioner promotes patient autonomy (informed consent) because patients can only make an informed practitioner authorization if the practitioner's judgments are accurately grounded on evidence-based care.
    A. True
    B. False

4. Professional colleagues are not obligated to report an impaired practitioner to benefit patients and the practitioner. Rather, the obligation is primarily a legal mandate.
    A. True
    B. False

5. Failure to report an impaired practitioner is a Class A misdemeanor, and the nonreporting practitioner is legally liable for patient harm caused by nonreporting.
    A. True
    B. False

Answers: 1D, 2A, 3A, 4B, 5A

* * *

CLINICAL VIGNETTES

1. Practitioner David O'Brien is a 45-year-old pharmacist who has practiced for more than 10 years. He has recently been exhibiting symptoms of impairment, including forgetfulness, fatigue, and difficulty concentrating during procedures. He has also been observed with slurred speech and difficulty walking. Clinical differential diagnosis includes substance abuse, psychiatric disorders, emotional disorders, and fatigue. An ethical question arises regarding whether Practitioner O'Brien's colleagues are obligated to report him for impaired practice.
    A. No, it is not necessary to report Practitioner O'Brien because he may be going through a difficult time.
    B. Yes, colleagues have a duty to report Practitioner O'Brien to protect patients and maintain public trust in the healthcare profession.
    C. No, it is not the responsibility of colleagues to report Practitioner O'Brien. It should be left to the hospital administration.
    D. Yes, but only after offering to help Practitioner O'Brien with his condition.

    Explanation: Colleagues have a professional obligation to report an impaired practitioner to protect patients from harm and maintain public trust in the healthcare profession. The duty to report is a matter of public justice to maintain social trust in the healthcare profession. It is not enough to ignore Practitioner O'Brien's impairment and hope it will pass, as this violates the professional duty of maximizing the patient's best interests. It is advisable to seek counsel from other colleagues and institutional officials before reporting, but the reporting practitioner is protected from liability if they are acting in good faith, honestly, and without malice. The reporting responsibility lies with the colleagues who observe Practitioner O'Brien's impairment, and it is not enough to leave it to the hospital administration. Offering to help Practitioner O'Brien is a professional obligation but does not preclude the need to report him for impaired practice. *Answer*: B

2. Bianca Jenkins is a 38-year-old family medicine practitioner who has practiced for more than 10 years. Her colleagues have noticed that her personal hygiene and physical appearance have declined. Practitioner Jenkins has a history of a high-stress lifestyle and has neglected self-care, including proper diet, enough sleep, and exercise. An ethical question arises whether Practitioner Jenkins' colleagues are obligated to report her for impaired practice.
    A. No, it is not necessary to report Practitioner Jenkins because her impairment is a result of a high-stress lifestyle and a lack of self-care.
    B. Yes, colleagues have a duty to report Practitioner Jenkins to protect patients and maintain public trust in the healthcare profession.
    C. No, it is not the responsibility of colleagues to report Practitioner Jenkins. She should be encouraged to seek help and take time off to care for herself.
    D. Yes, but only after consulting with a mental health professional to determine the extent of her impairment.

    Explanation: The correct answer is "No, it is not the responsibility of colleagues to report Practitioner Jenkins. She should be encouraged to seek help and take time off to care for herself." Although maintaining patient safety and public trust is paramount, there is no explicit evidence of Practitioner Jenkins' ability to provide patient care being compromised. However, it is also important not to underestimate the potential impact of a high-stress lifestyle and lack of self-care on professional performance. Although involving a mental health professional is not wrong, a primary step is to address the issue directly with Practitioner Jenkins. Therefore, reporting her is unnecessary, and it is good to advocate for Jenkins' welfare by recognizing signs of possible burnout. *Answer*: C

3. Practitioner Paloma Patterson is a 35-year-old family medicine practitioner who has been practicing for 8 years. Recently, she has been exhibiting signs of impairment, including poor communication with patients, poor clinical judgment, and difficulty maintaining her schedule. Her patients have complained about her poor bedside manner and inappropriate prescriptions. Her colleagues have also noticed her poor performance and behavior. The clinical differential diagnosis includes psychiatric disorders, substance abuse, and emotional disorders. An ethical question arises whether Practitioner Patterson's colleagues are obligated to report her for impaired practice.
    A. No, it is not necessary to report Practitioner Patterson because she may improve on her own.
    B. Yes, colleagues have a duty to report Practitioner Patterson to protect patients and maintain public trust in the healthcare profession.
    C. No, it is not the responsibility of colleagues to report Practitioner Patterson. It should be left to the hospital administration.
    D. Yes, but only after confronting Practitioner Patterson directly and giving her a chance to seek help.

    Explanation: Colleagues have a professional obligation to report an impaired practitioner to protect patients from harm and maintain public trust in the healthcare profession. The duty to report is a matter of public justice to maintain social trust in the healthcare profession. It is not enough to ignore Practitioner Patterson's impairment and hope it will pass, as this violates the professional duty of maximizing the patient's best interests. It is advisable to seek counsel from other colleagues and institutional officials before reporting, but the reporting practitioner is protected from liability if they are acting in good faith, honestly, and without malice. The reporting responsibility lies with the colleagues who observe Practitioner Patterson's impairment, and it is not enough to leave it to the hospital administration. Confronting Practitioner Patterson directly and giving her a chance to seek help is a professional obligation but does not preclude the need to report her for impaired practice. *Answer*: B

4. Practitioner Aurora Dubber is a 45-year-old pediatrician who has been practicing for more than 20 years. Recently, her colleagues have noticed that she has become increasingly forgetful and disorganized. She has been misplacing important patient files and forgetting scheduled appointments. Practitioner Dubber slurs her speech and appears drowsy during patient consultations. Her colleagues have observed that her behavior is inconsistent with her past performance, and they suspect she may have a substance use disorder. The differential diagnosis includes depression, anxiety, and other psychiatric disorders. What is the ethical obligation of Practitioner Dubber's colleagues in this situation?
   A. They should confront her about her suspected substance use and encourage her to seek treatment.
   B. They should keep quiet about their suspicions until there is clear evidence of substance abuse.
   C. They should anonymously report her behavior to the relevant authorities.
   D. They should ignore her behavior and let her continue to practice medicine as she sees fit.

   Explanation: The best answer is option A, "They should confront her about her suspected substance use and encourage her to seek treatment." Given Practitioner Dubber's observed behaviors that could indicate a substance use disorder, the priority is ensuring patient safety and advocating for the well-being of a fellow healthcare provider. Ignoring the behavior (option D) or keeping quiet until there is clear evidence (option B) could harm patients and further deteriorate Practitioner Dubber's health. An anonymous report (option C) may be warranted in some cases, but it can also result in punitive actions without allowing Dubber to self-correct. Confronting her directly and encouraging her to seek treatment (option A) aligns with professional principles of collegiality, nonmaleficence (do no harm), and beneficence (do good) by addressing the issue compassionately while prioritizing patient safety. *Answer*: A

5. A practitioner's colleagues are starting to be concerned about whether a fellow practitioner is becoming impaired. Which of the following are signs that a practitioner's healthcare judgment and practice have been compromised, leading to impairment?
   A. Making frequent errors in documentation and healthcare records
   B. Being absent from work due to a chronic healthcare condition
   C. Failing to attend mandatory professional development training sessions
   D. Prescribing medication outside the scope of their specialty

   Explanation: An impaired practitioner is one whose healthcare judgment and practice have been compromised, leading to a violation of the patient–practitioner relationship by exposing the patient to increased risk of harm, violating the professional duty of maximizing the patient's best interests, and diminishing the community's trust in the healthcare profession. Making frequent errors in documentation and healthcare records can be considered a sign of impaired judgment and practice, and it can compromise patient safety and overall trust in the healthcare profession. Being absent from work due to a chronic healthcare condition does not necessarily indicate impaired judgment or practice, and it may be a protected healthcare condition under the Americans with Disabilities Act. Failing to attend mandatory professional development training sessions may indicate unprofessional behavior but does not necessarily indicate impaired judgment or practice. Prescribing medication outside the scope of one's specialty violates professional practice, but it does not necessarily indicate impaired judgment or practice. *Answer*: A

REFLECTION VIGNETTES

1. Practitioner Jane Smith, a 40-year-old practitioner, is out for a night on the town with several colleagues. As the evening progresses, one colleague on call gets particularly tipsy. Later in the evening, the colleague receives a call about a patient care issue, and Practitioner Smith notices that the tipsy practitioner is providing medication orders over the phone.

   What is the ethical issue(s)?

   How should the ethical issue(s) be addressed?

2. Mr. James Lee, a 69-year-old practitioner, works at a local teaching hospital and seems to treat patients effectively but has difficulty remembering patients' names and healthcare conditions. The differential diagnosis includes age-related memory decline, attention-deficit disorder, and other cognitive impairments. Despite these challenges, Mr. Lee compensates for his memory difficulties by reviewing each patient's healthcare records before each visit. This enables him to provide effective and comprehensive care to his patients despite his memory limitations. The practitioner also ensures that he takes detailed notes during patient visits to help him remember important details for future reference.

   What is the ethical issue(s)?

   How should the ethical issue(s) be addressed?

* * *

WEB LINKS

1. *Code of Medical Ethics*. American Medical Association. https://code-medical-ethics.ama-assn.org
   The AMA *Code of Medical Ethics*, established by the American Medical Association, is a comprehensive guide for healthcare practitioners. It addresses issues and challenges, promotes adherence to standards of care, and is continuously updated to reflect contemporary practices and challenges.
2. "Identifying an Impaired Practitioner." Stephen Ross. 2003. https://journalofethics.ama-assn.org/article/identifying-impaired-physician/2003-12
   This article highlights the critical issue of substance use disorders among physicians, emphasizing the importance of early detection, intervention through physician health programs, and the ethical duty of colleagues to report, focusing on treatment and long-term recovery to ensure patient safety and physician well-being.
3. "Patient Rights and Ethics." Jacob P. Olejarczyk and Michael Young. https://www.ncbi.nlm.nih.gov/books/NBK538279
   This source discusses patient rights in the United States, including the right to informed consent, protected by state and federal legislation.
4. "Principles of Clinical Ethics and Their Application to Practice." Basil Varkey. 2021. https://www.ncbi.nlm.nih.gov/pmc/articles/PMC7923912
   This source presents a model for patient care that integrates ethical aspects with clinical and technical expertise.
5. "Everyday Ethics: What Should You Do When a Colleague Comes to Work Impaired." Donna Euben. 2020. https://leader.pubs.asha.org/do/10.1044/2020-0713-ethics-impaired-practitioner/full
   This source discusses the ethical obligations of healthcare professionals when dealing with an impaired colleague, emphasizing the importance of protecting the welfare of patients.
6. "Who Makes Decisions for Incapacitated Patients Who Have No Surrogate or Advance Directive?" Scott J. Schweikart. 2019. https://journalofethics.ama-assn.org/article/

who-makes-decisions-incapacitated-patients-who-have-no-surrogate-or-advance-directive/2019-07

This source discusses the ethical considerations when making decisions for incapacitated patients, including those who may be impaired practitioners.

7. "Patient Rights." American Medical Association. https://code-medical-ethics.ama-assn.org/ethics-opinions/patient-rights

This source provides the American Medical Association's opinions on patient rights, emphasizing the importance of a collaborative effort between patient and physician in a mutually respectful alliance.

8. "Physician Responsibilities to Colleagues with Illness, Disability or Impairment." American Medical Association.https://code-medical-ethics.ama-assn.org/ethics-opinions/physician-responsibilities-colleagues-illness-disability-or-impairment

This source outlines the responsibilities of physicians when dealing with colleagues with illness, disability, or impairment, focusing on the balance between patient safety and practice competency.

INTERROGATION

31

A person may cause evil to others not only by his actions but by his inaction, and in either case, he is justly accountable to them for the injury.
—John Stuart Mill

* * *

MYSTERY STORY

SHADOWS OF COERCION: THE REPERCUSSIONS OF UNLAWFUL INTERROGATION

In the pulsating veins of SecureSanctuary Hospital, esteemed Practitioner Maria Sanchez was a beacon of steadfast dedication and unwavering ethics. The corridors of the hospital echoed tales of her tireless service, and she was revered as a champion of patient autonomy. As the twilight hours of her shift approached, she looked forward to returning home to the comforts of family and rest. However, an urgent call soon shattered her tranquility, pulling her into an unprecedented gathering that would test her moral compass.

Seated across from her were stern-faced law enforcement officials, their grim expressions casting long shadows in the dimly lit meeting room. The most imposing of them all was Detective Terry Smith, a seasoned veteran with a reputation for bending the rules in the name of justice. He unveiled a disconcerting request: They sought critical information from a severely compromised patient, a tangled nexus between life, law, and ethics.

The proposition sent a chill down Practitioner Sanchez's spine. Healthcare ethics were sacred to her, serving as the guiding light in her healing mission. Recoiling at the proposal, she swiftly rebuffed Detective Braxton, standing as the guardian of her patient's autonomy (informed consent). Her role, she explained with unwavering conviction, was not of an interrogator but a healer, an advocate for the vulnerable, and the custodian of trust.

Despite the simmering tension, Detective Braxton pressed on, his pleas of urgency underlining the high stakes. He painted a picture of a looming public safety threat, but Practitioner Sanchez's resolve was an unyielding fortress. She iterated the principles of her profession: patient-centric care that focuses on beneficence (do good) and nonmaleficence (do no harm), emphasizing that these ideals could not be compromised under any circumstances.

The confrontation did not pass without consequences. A chilling phone call punctuated Practitioner Sanchez's night, an anonymous voice threatening her into silence over the rejected interrogation request. Although deeply shaken, her determination remained unscathed; her commitment to professional principles was unassailable.

The dawn broke with a shocking twist: SecureSanctuary Hospital plunged into chaos, mourning the murder of the contentious patient and the grim discovery of Detective Braxton's lifeless body in his office. The ensuing investigation unearthed a sinister conspiracy, revealing law enforcement's deep-seated corruption and the ghastly abuse of power through illegal interrogation practices that violated the sanctity of human rights.

Practitioner Sanchez found herself in the eye of the storm, her testimony becoming a lynchpin in the investigation. Her bold revelations of the attempted transgression of bioethics facilitated the unmasking of the perpetrators, leading to their eventual arrest and conviction. Her harrowing experience served as a reminder of the crucial role healthcare professionals play in upholding the bedrock of ethical standards in society and the perilous consequences that could ensue when these boundaries are violated.

* * *

THINK

Interrogation is the questioning of a person to gain information or intelligence for law enforcement, military, or national security and is generally limited to no more than 8 hours per day. However, one study of 44 confession cases noted that the average length of interrogation was 16.3 hours.

The interrogation method used by police throughout the United States is the Reid technique. It was developed in the 1950s by John Reid, a psychologist, polygraph expert, and Chicago police officer. With the Reid technique, the investigator introduces fictitious evidence and lies to the suspect being interrogated, saying that the investigation has turned up direct evidence of the suspect committing the crime in question, such as fingerprints, and witnesses who have identified them to the crime, thus creating a very stressful environment. The interrogator then follows up, showing sympathy, understanding, and promising leniency and help, but only if the suspect confesses. The Reid technique has resulted in many false confessions, especially among youth and the mentally vulnerable, giving Illinois the name "False Confession Capital of the United States" by the Innocence Project. Because of Illinois' large number of wrongful convictions predicated on false confessions, the damage to public perception, and the high costs associated with incarcerations and exonerations, in July 2021, Illinois became the first state in the nation to make it illegal for police officers to lie to children during interrogations. However, the law still permits interrogators to use fictitious evidence and to lie when interrogating adults.

The United States Central Intelligence Agency (CIA), in addition to using fictitious evidence and lying, took the Reid technique of creating a stressful environment up a few notches by including waterboarding and defined it as "enhanced interrogation." Waterboarding is strapping a victim to a board on their back with the head lower than the rest of their body. Water is poured over a cloth covering the face, giving the person being interrogated the drowning experience. Using this enhanced interrogation method, the interrogator can extract nearly any confession they want from the person. This technique has had little success other than creating a plethora of false confessions and false intelligence, as the victim will say nearly anything to stop the abuse. The false confessions of weapons of mass destruction in Iraq were one of many consequences of such interrogation methods, resulting in 800,000 to 1.3 million Iraqi deaths, the destruction of a civilization, and the destabilization of the Middle East.

ASSESS

Patient: Autonomy

Concerning the principle of autonomy, the person being interrogated does not have a decisional choice and is therefore not acting autonomously. The information given to the interrogated person is often lies or false criminal accusations combined with the false hope of leniency in exchange for confessing. Coercion and manipulation of information make autonomy impossible because every element of the necessary conditions for informed consent is violated.

Practitioner: Beneficence and Nonmaleficence

The practitioner, as a healthcare professional, must not conduct, participate, monitor, or even be present during any interrogation as a matter of nonmaleficence (do no harm) and the public perception of the profession. The practitioner's role is to pursue the patient's best interests, as determined by the patient's reasonable goals, values, and priorities. This is certainly not what is happening with involuntary interrogation, where there is no relationship toward healing, nor does interrogation focus on maximizing the person's best interests. Because of this, it is incumbent upon the practitioner not to participate in interrogation activities that violate the healthcare profession's patient-centered focus of maximizing the patient's best interests as determined by the patient's reasonable goals, values, and priorities. The practitioner also has a professional, legal, and moral obligation to oppose, expose, and report abusive or coercive interrogations to authorities that can intervene, investigate, and adjudicate.

Public Policy: Justice

As a matter of justice (be fair), it is difficult to understand why the law permits interrogation strategies and techniques that violate human dignity and human rights by allowing outright lying, fabrication of evidence, and physically and mentally threatening environments. One might try to justify such interrogation techniques by their positive outcomes with true confessions and valuable intelligence necessary to protect individuals and society. However, evidence shows that torture and enhanced interrogation are less effective than noncoercive interrogation methods because the victims will confess to anything they believe the interrogators want to hear in order to stop the torture and interrogation.

CONCLUDE

The practitioner's and the healthcare profession's prime directive is to be patient-centered by maximizing the patient's best interests according to the patient's reasonable goals, values, and priorities. Because of this prime directive, practitioners are to have no association with any activity promoting physical or psychological harm to a person and no association with any activity compromising the public's perception of the healthcare profession as a healing art. Practitioners have a professional, legal, and moral obligation to oppose, expose, and report any abusive or coercive interrogations to authorities that have the power to intervene, investigate, and adjudicate.

In summary, practitioners must prioritize patient-centered care and maximize the patient's best interests. As such, they should not participate in or associate with activities that promote physical or psychological harm or compromise public perception of the healthcare profession. Practitioners have a professional, legal, and moral obligation to oppose, expose, and report any abusive or coercive interrogations to authorities with the power to intervene, investigate, and adjudicate. Upholding these obligations is essential for maintaining the integrity of the healthcare profession and protecting the well-being of individuals subjected to interrogation. (See also Chapters 9, 10, 15, 21, and 62.)

* * *

REVIEW QUESTIONS

1. With the Reid technique, the investigator introduces fictitious evidence and lies to the suspect being interrogated, saying that the investigation has turned up direct evidence of the suspect committing the crime in question, such as fingerprints, witnesses who have identified them to the crime, and creating a very stressful environment.
   A. True
   B. False

2. The Reid technique has resulted in many false confessions, especially among youth and the mentally vulnerable, giving Illinois the name "False Confession Capital of the United States" by the Innocence Project.
   A. True
   B. False

3. The information given to the interrogated person is often lies or false criminal accusations combined with the false hope of leniency in exchange for confessing. Coercion and manipulation of information make autonomy impossible because every element of the necessary conditions for informed consent is violated.
   A. True
   B. False

4. As a healthcare professional, the practitioner can be present during interrogation as long as it is for beneficence (do good).
   A. True
   B. False

5. The practitioner has a professional, legal, and moral obligation to report abusive or coercive interrogations to authorities that can intervene, investigate, and adjudicate such activities.
   A. True
   B. False

6. Because torture and enhanced interrogation are more effective at getting true confessions than noncoercive methods, it is morally permissible for practitioners to participate.
   A. True
   B. False

Answers: 1A, 2A, 3A, 4B, 5A, 6B

* * *

CLINICAL VIGNETTES

1. Ms. Bridget Thompson is a 24-year-old college student brought to the police station for questioning concerning a robbery in a nearby store. During the interrogation, the investigator used the Reid technique, telling Ms. Thompson they had evidence linking her to the crime and pressuring her to confess. Ms. Thompson eventually confessed to the crime but later recanted, stating that she only confessed because she was afraid and believed she had no other choice. Clinical differential diagnoses for Ms. Thompson's symptoms may include anxiety, post-traumatic stress disorder (PTSD), or acute stress disorder (ASD). The ethical question arising from this clinical vignette is whether the investigator's use of the Reid technique in interrogating Ms. Thompson violated her autonomy and dignity as a person.
   A. Yes, the investigator's use of the Reid technique violated Ms. Thompson's autonomy and dignity.
   B. No, the investigator's use of the Reid technique was necessary to get a confession and solve the crime.
   C. It depends on whether Ms. Thompson was guilty of the crime or not.
   D. It depends on whether the investigator followed proper protocol and did not use any coercive or abusive tactics during the interrogation.

   Explanation: Using the Reid technique in interrogation violates the principle of patient autonomy because the person being interrogated does not have a decisional choice and is therefore not acting autonomously. In addition, using lies, false evidence, and manipulation in interrogation violates human dignity and rights, calling into question the justice of such practices. In this clinical vignette, Ms. Thompson only confessed due to the pressure put on her by the investigator and later recanted her confession, suggesting that her autonomy was violated. As a healthcare professional, the practitioner must not participate, monitor, or even be present during any interrogation as a matter of nonmaleficence (do no harm) and the public perception of the profession. The practitioner also has a professional, legal, and moral obligation to oppose, expose, and report abusive or coercive interrogations to authorities that can intervene, investigate, and adjudicate. *Answer*: A

2. Mr. Edmond Lee is a 38-year-old former CIA agent involved in waterboarding during interrogations of suspected terrorists. Since retiring from the agency, Mr. Lee has experienced symptoms of anxiety, depression, and PTSD. Clinical differential diagnoses may include

PTSD, anxiety, or major depressive disorder. The ethical question arising from this clinical vignette is whether using waterboarding during interrogation violates the principle of nonmaleficence and the dignity of human beings.
- A. Yes, the use of waterboarding during interrogation violates the principle of nonmaleficence and the dignity of human beings.
- B. No, the use of waterboarding during interrogation is necessary to extract information and protect national security.
- C. It depends on the specific circumstances of the interrogation and the information that was obtained.
- D. It depends on whether the interrogator followed proper protocol and did not use any abusive or coercive tactics during the interrogation.

Explanation: Using waterboarding during interrogation violates the principle of nonmaleficence, which requires that healthcare professionals do no harm to their patients. Using such enhanced interrogation techniques promotes physical and psychological harm to individuals and violates the dignity of human beings. The evidence shows that torture and enhanced interrogation are less effective than noncoercive interrogation methods because the victims will confess to anything they believe the interrogators want to hear in order to stop the torture and interrogation. Mr. Lee's symptoms of anxiety, depression, and PTSD may have resulted from his participation in these interrogation practices. As a healthcare professional, it is incumbent upon the practitioner not to participate in interrogation activities that violate the healthcare profession's patient-centered focus. *Answer*: A

3. Practitioner Nathanial Park is a 30-year-old practitioner who received a request from a government official to provide healthcare assistance during an interrogation of a suspected criminal. The official tells him that the suspect has been injured during his arrest and needs healthcare attention and that his role would be to provide healthcare to the suspect during the interrogation. What ethical obligation does Practitioner Park have as a healthcare professional in this situation?
   - A. To provide healthcare assistance during the interrogation
   - B. To participate as a public servant
   - C. To assist the government official in any way possible
   - D. To refuse the request and report to the appropriate authorities

   Explanation: Practitioner Park's ethical obligation as a healthcare professional is to pursue the patient's best interests, as determined by the patient's reasonable goals, values, and priorities. In this case, he is being asked to participate in an interrogation that does not focus on maximizing the person's best interests but is aimed at obtaining information through coercion and manipulation. As a matter of nonmaleficence and the public perception of the profession, he must not conduct, participate, monitor, or even be present during any interrogation that violates the healthcare profession's patient-centered focus. Practitioner Park has a professional, legal, and moral obligation to oppose, expose, and report any abusive or coercive interrogations to authorities that have the power to intervene, investigate, and adjudicate. Therefore, his ethical obligation in this situation is to report the request to his supervisor and the relevant authorities. *Answer*: D

4. Practitioner Penelope Woods is a practitioner who works at a police department as a healthcare examiner. One day, she is asked to conduct a healthcare examination on a suspect arrested for a serious crime. During the examination, the suspect tells Practitioner Woods that the arresting officers physically abused him. What is Practitioner Woods's ethical obligation as a healthcare professional?
   - A. To dismiss the suspect's claims and proceed with the examination
   - B. To conduct the examination and keep the suspect's claims confidential

C. To report the suspect's claims to her supervisor and the relevant authorities
D. To confront the arresting officers and demand an explanation for their behavior

Explanation: Practitioner Woods's ethical obligation as a healthcare professional is to oppose, expose, and report any abusive or coercive interrogations to authorities that have the power to intervene, investigate, and adjudicate. In this case, she is being made aware of allegations of physical abuse by the arresting officers, which constitutes abusive or coercive behavior. As a healthcare examiner, Practitioner Woods has a professional, legal, and moral obligation to report such claims to her supervisor and the relevant authorities, such as internal affairs or the police department's oversight board. Failure to report such claims could result in the continuation of abusive or coercive behavior, which directly violates the healthcare profession's commitment to patient-centered care and the obligation to do no harm. Furthermore, reporting such claims is in line with the principle of justice because it ensures that those responsible for abusive or coercive behavior are held accountable for their actions. Therefore, Practitioner Woods's ethical obligation is to report the suspect's claims to her supervisor and the relevant authorities. *Answer*: C

5. Practitioner Heather Nguyen is a psychiatrist at a detention center that houses immigrants awaiting deportation. One day, she is approached by an immigration officer who asks her to interrogate a detainee suspected of having mental health issues. The officer tells Practitioner Nguyen that her role would be to assess the detainee's mental health and provide recommendations for interrogation techniques that would not exacerbate the detainee's symptoms. What is Practitioner Nguyen's ethical obligation as a healthcare professional in this situation?
    A. To participate in the interrogation and provide recommendations for interrogation techniques
    B. To participate as a public servant
    C. To assist the immigration officer in any way possible
    D. To refuse the request and report to the appropriate authorities

Explanation: Practitioner Nguyen's ethical obligation as a healthcare professional is to pursue the patient's best interests, as determined by the patient's reasonable goals, values, and priorities. In this case, she is being asked to participate in an interrogation that does not focus on maximizing the person's best interests but, rather, is aimed at obtaining information through coercion and manipulation. The involvement of practitioners in interrogation practices also violates the principle of patient autonomy because it is not a decisional choice for the person being questioned. Furthermore, using mental health assessments to facilitate interrogation techniques raises concerns about professional boundaries and the ethical use of psychiatric expertise. Therefore, as a matter of nonmaleficence, patient confidentiality, and the public perception of the profession, Practitioner Nguyen must not conduct, participate, monitor, or even be present during any interrogation that violates the healthcare profession's patient-centered focus. Practitioner Nguyen has a professional, legal, and moral obligation to oppose, expose, and report abusive or coercive interrogations to authorities that can intervene, investigate, and adjudicate. Her ethical obligation in this situation is to refuse the request and not be involved in the interrogation. *Answer*: D

REFLECTION VIGNETTES

1. Practitioner Sarah Patel, a 35-year-old practitioner, is serving in the military. One day, her commanding officer asks her to provide a blood transfusion to a prisoner being interrogated so that the interrogation can continue.

    What is the ethical issue(s)?

    How should the ethical issue(s) be addressed?

2. Mr. John Kim, a 40-year-old practitioner, is asked to be present during an enhanced interrogation procedure to ensure the prisoner does not die.

   What is the ethical issue(s)?

   How should the ethical issue(s) be addressed?

   * * *

WEB LINKS

1. *Code of Medical Ethics*. American Medical Association. https://code-medical-ethics.ama-assn.org
   The AMA *Code of Medical Ethics*, established by the American Medical Association, is a comprehensive guide for healthcare practitioners. It addresses issues and challenges, promotes adherence to standards of care, and is continuously updated to reflect contemporary practices and challenges.
2. "Torture, Coercive Interrogations and Physicians." American Medical Association. 2014. https://www.ama-assn.org/delivering-care/ethics/torture-coercive-interrogations-and-physicians
   This source discusses the ethical obligations of physicians regarding torture and interrogation.
3. "Physician Participation in Interrogation." American Medical Association. https://code-medical-ethics.ama-assn.org/ethics-opinions/physician-participation-interrogation
   This source provides the American Medical Association's opinion on physician participation in interrogation, discussing the balance between law enforcement and patient rights.
4. "Principles of Clinical Ethics and Their Application to Practice." Basil Varkey. 2021. https://www.ncbi.nlm.nih.gov/pmc/articles/PMC7923912
   This source presents a model for patient care that integrates ethical aspects with clinical and technical expertise.
5. "Patient Rights and Ethics." Jacob P. Olejarczyk and Michael Young. https://www.ncbi.nlm.nih.gov/books/NBK538279
   This source discusses patient rights in the United States, including the right to informed consent, which is protected by state and federal legislation.
6. "The Four Principles of Biomedical Ethics." Healthcare Ethics and Law. https://www.healthcareethicsandlaw.co.uk/intro-healthcare-ethics-law/principlesofbiomedethics
   This source discusses the four principles of biomedical ethics: autonomy, nonmaleficence, beneficence, and justice.
7. "Ethical Conflicts in Patient-Centered Care." Sven Ove Hansson and Barbro Fröding. 2021. https://journals.sagepub.com/doi/full/10.1177/1477750920962356
   This source discusses the ethical conflicts in patient-centered care, including the balance between patient rights and practitioner responsibilities.
8. "Medical Ethics." American Medical Association. https://www.ama-assn.org/topics/medical-ethics
   This source provides an overview of bioethics, including the moral framework for the practice of clinical medicine.
9. "Issues in Biomedical Ethics." J. R. Vevaina, L. M. Nora, and R. C. Bone. 1993. https://pubmed.ncbi.nlm.nih.gov/8243220
   This source discusses various issues in biomedical ethics, including moral problems arising in interrogation.

LIFELONG LEARNING

32

The doctor of the future will be a teacher as well as a practitioner. The practitioner's real job will be to teach people how to be healthy.
—D. C. Jarvis

* * *

MYSTERY STORY

> **ACADEMIC ASSASSIN: THE IMPORTANCE OF LIFELONG LEARNING IN HEALTHCARE**
>
> Bathed in the brilliant glow of intellectual pursuit, a distinguished healthcare symposium buzzed with anticipation at Wisdom Tree Medical Institute. It was an assembly of scholarly gladiators, each a titan in their respective fields, convened to delve into the riveting depths of modern healthcare research and practice.
>
> Among these academic giants, one figure towered above the rest—Practitioner Elizabeth Moore, an illustrious cardiac surgeon. Revered for her breakthroughs in heart disease research and her mentorship to a generation of healthcare acolytes, she was the chosen torchbearer, set to deliver a compelling keynote on the imperative of lifelong learning in healthcare.
>
> But as the clock ticked relentlessly toward her appointed hour, Practitioner Moore was conspicuous by her absence. Disquiet rippled through the crowd, mutating into alarm as it was discovered that the esteemed surgeon had been brutally murdered in the sanctuary of her hotel suite. The gruesome discovery was swiftly escalated to the police, and the mysterious case landed on the weathered desk of Detective John Wilson.
>
> A seasoned sleuth, Wilson soon discerned that Moore's demise was inexplicably intertwined with her pioneering research on a revolutionary cardiac procedure. This cutting-edge technique, while promising to redefine the boundaries of cardiac surgery, was shrouded in controversy, attracting both high praise and harsh criticism within the healthcare community.
>
> The detective dived into a sea of potential suspects. They were a motley crew composed of Moore's disciples, colleagues, and even the naysayers of her work, each harboring motives steeped in professional envy or personal grudges.
>
> As Wilson delved deeper, he realized the significance of lifelong learning in healthcare as a pivotal aspect of the case. His inquiry revealed a trail of resentment among suspects who had failed to keep pace with the relentless march of healthcare progress, resulting in professional stagnation and personal dissatisfaction.
>
> Finally, the smoky veil of mystery lifted, and Detective Wilson unmasked the assassin. The perpetrator was a fellow practitioner, poisoned by jealousy and tormented by the fear of their work being eclipsed by Moore's groundbreaking research.
>
> The chilling case became a stark reminder for healthcare professionals, including those at Wisdom Tree Medical Institute, of the need to remain lifelong learners, to swim in the tide of progress rather than be swept away by it. It underscored the profound influence of educators like Moore and emphasized the need to nurture an environment of unending learning and collaboration within the healthcare fraternity. The tragedy served as a poignant lesson, etched in blood and loss, illuminating the dangerous implications of stagnation in a field that thrives on innovation.

* * *

THINK

The patient–practitioner relationship is a unique social contract between the patient and practitioner that establishes confidentiality and privacy of protected health information (PHI). This is based on the patient's trust, professional obligations, and social expectations that practitioners will be engaged in lifelong learning to stay up-to-date on evidence-based care's ever-changing knowledge, skills, and ethical values.

The term *doctor* comes from the Latin *docere*, which means "to teach." Doctor is a term of formal address associated with being a practitioner because in addition to being engaged in lifelong learning as a lifetime student, practitioners are also responsible for being lifetime educators. The term *preceptor* comes from the Latin *praecipere*, which means "instructor." Doctors educate each patient regarding prognosis, diagnosis, treatment options, and activities of daily living. As preceptors, licensed practitioners instruct and supervise the next generation of practitioners, passing on clinical knowledge, skills, and ethical values to students.

ASSESS

Patient: Autonomy

Patient autonomy, exercised through informed consent, is grounded in the patient–practitioner social contract. The patient trusts that the clinical diagnosis, prognosis, treatment options, risk assessments and benefits, and the answers will be up-to-date, evidence-based medicine and care. Anything less than that questions the practitioner's clinical competence and the legitimacy of the patient's informed consent, especially if the patient's consent is based on outdated and no longer relevant information. Informed consent as a shared decision-making process requires that the information provided by the practitioner is current and accurate. These conditions can only be met if the practitioner is engaged in lifelong learning.

Practitioner: Beneficence and Nonmaleficence

Professionally, healthcare recognizes that if the practitioner is to maximize the patient's best interests according to the patient's reasonable goals, values, and priorities—beneficence (do good)—and if the practitioner is to reduce even the risks of harm to the patient—nonmaleficence (do no harm)—then that can be realized only if the practitioner is engaged in lifelong learning. The patient–practitioner relationship exists with the assumption that the healthcare profession will take the necessary steps to ensure that its responsibility of lifelong learning will occur universally among its members.

States require practitioners to earn a minimum number of lifelong learning and continuing professional development credits per year. This includes, but is not limited to, continuing medical education (CME) credits, continuing nursing education (CNE) credits, continuing pharmacy education (CPE) credits, continuing dental education (CDE) credits, continuing education units (CEUs), maintenance of certification (MOC) points, performance improvement continuing medical education (PI CME), continuing professional development (CPD) points, and quality improvement (QI) points. The number of required continuing education credits varies widely depending on the profession, specialty, licensing board, and state or country of practice. However, the following is a general guideline:

- Physicians usually need to earn between 20 and 50 CME credits per year. This can depend on the specific requirements of their state healthcare board and specialty board. For example, the American Board of Internal Medicine requires at least 20 CME credits per year for MOC.
- Registered nurses typically need to earn between 20 and 30 CNE credits per 2-year period, depending on their state board of nursing.
- Pharmacists are generally required to earn 15–30 CPE credits per year, but the specific requirement can vary by state.

- Dentists typically need to earn between 20 and 40 CDE credits every 1 to 2 years, depending on the requirements of their state dental board.

Several accrediting bodies oversee the provision of lifelong learning and continuing professional development credits in the healthcare field. They vary depending on the specific profession:

- Physicians (CME credits): The Accreditation Council for Continuing Medical Education is the primary accreditor for CME providers in the United States. There are also state-level accrediting bodies.
- Nurses (CNE credits): The American Nurses Credentialing Center is a major accreditor of CNE activities, but there are also state nursing boards and other organizations that accredit CNE providers.
- Pharmacists (CPE credits): The Accreditation Council for Pharmacy Education accredits CPE providers.
- Dentists (CDE credits): The American Dental Association's Continuing Education Recognition Program recognizes providers of CDE.
- CEUs: These are not limited to healthcare and are accredited by various organizations depending on the field. In healthcare, the aforementioned organizations often also accredit CEU providers.
- MOC points: These are overseen by the various specialty boards of the American Board of Medical Specialties.
- PI CME: These are recognized by organizations such as the American Medical Association.
- CPD points: These can be accredited by various organizations, depending on the profession and specialty.
- QI points: These are often required for hospital staff and administrators and are typically overseen by hospital accreditation organizations such as The Joint Commission in the United States.

Accreditation for lifelong learning and professional development is not for each individual educational activity; rather, accreditation is for the organization or institution that awards the credits and/or points. Each organization and institution must implement all the required quality control criteria to ensure that the educational and professional development activities meet the accreditation standards of the accrediting body.

Public Policy: Justice

The healthcare profession has a social contract with the community in exchange for more professional liberty and fewer federal, state, and institutional regulations. These freedoms depend on the populace trusting that their healthcare providers are staying up-to-date with the knowledge, skills, and clinical practice of evidence-based care and ethical values, which can only occur through the practice and implementation of lifelong learning and professional development.

Lifelong learning is a healthcare and legal licensure requirement. Failure to attain the necessary credit hours and/or points results in failure to renew the state healthcare board license and/or institutional privileges. Practicing healthcare without a license is a prosecutable offense.

CONCLUDE

Although continuing education and professional development credits and points vary widely depending on the profession, specialty, licensing board, and state or country of practice, they are necessary qualifications for healthcare licensure. Practicing medicine without a proper healthcare license is a prosecutable offense. It is imperative that practitioners have internalized the value and importance of lifelong learning and that practitioners view themselves as learners, preceptors, and teachers.

In summary, lifelong learning is essential for practitioners to fulfill their responsibilities as educators, learners, and providers of quality healthcare. Practitioners must internalize the value and importance of continuous learning to maintain their healthcare license and uphold the trust placed in them by their patients and society. By doing so, they can ensure that they provide the most up-to-date, evidence-based healthcare, care, and ethical values, ultimately improving patient outcomes and preserving the integrity of the healthcare profession.

* * *

REVIEW QUESTIONS

1. The term doctor comes from the Latin *docere*, which means "practitioner."
   A. True
   B. False

2. The term preceptor comes from the Latin *praecipere*, which means "advisor."
   A. True
   B. False

3. Informed consent as a shared decision-making process requires that the information provided by the practitioner is current and accurate. These conditions can only be met if the practitioner is engaged in lifelong learning.
   A. True
   B. False

4. Accreditation for lifelong learning and professional development is for each individual educational activity, not the organizations and institutions that award the credits and points.
   A. True
   B. False

5. Lifelong learning is a healthcare and legal licensure requirement. Failure to attain the necessary credit hours will result in the failure to renew the state healthcare board license.
   A. True
   B. False

Answers: 1B, 2B, 3A, 4B, 5A

* * *

CLINICAL VIGNETTES

1. Practitioner Wilma Jenkins is a 35-year-old oncologist practicing medicine for 10 years. She is committed to providing the best possible care to her patients with cancer and wants to ensure that she is up-to-date on the latest treatment options. Which of the following approaches should Practitioner Jenkins use to stay up-to-date on the latest advancements in cancer treatment and ensure patient safety?
   A. Attend a healthcare conference focused on cancer treatment
   B. Engage in self-directed learning and research to stay up-to-date with the latest advancements in cancer treatment
   C. Participate in a formal CME program that offers credit hours related to cancer treatment
   D. Consult with a colleague who has expertise in cancer treatment
   E. All of the above

   Explanation: The correct answer is option E, all of the above. To stay current on the latest advancements in cancer treatment, Practitioner Jenkins should consider a multipronged approach. This includes attending healthcare conferences focused on cancer treatment, which will offer insights into recent research and emerging therapies. Self-directed learning and research are also crucial because they allow her to delve into specific areas of interest and keep up with new publications. Participating in a formal CME program can provide structured learning experiences and often cover the latest best practices. Finally, consulting with colleagues with cancer treatment expertise can offer practical insights and shared experiences. Therefore, by engaging in all these strategies, Practitioner Jenkins will be best positioned to provide safe and cutting-edge care to her patients. *Answer*: E

2. Practitioner Cedric Wallace is a 45-year-old practitioner working in a busy emergency department for more than a decade. One day, he is presented with a patient with a rare and complex healthcare condition he has never encountered before. He is still determining the best course of action and how to treat the patient effectively. Which of the following is the best option for Practitioner Wallace?
   A. Attend a relevant healthcare conference to learn more about the condition
   B. Consult with a specialist in the field
   C. Conduct a literature review to learn more about the condition
   D. Wait for another practitioner to take over the patient's care

   Explanation: Consult with a specialist in the field is the best choice for Practitioner Wallace when encountering a patient with a rare and complex condition in an emergency department setting. This choice is the most beneficial due to the urgency and complexity of the situation. A specialist can provide immediate, accurate advice on managing the patient's condition, capitalizing on their extensive knowledge and experience. The other options, such as attending a healthcare conference or conducting a literature review, although informative, are not feasible due to time constraints inherent in emergency medicine. There are better choices than waiting for another practitioner to take over, as it could result in harmful delays and does not guarantee the incoming practitioner would have more knowledge about the condition. Therefore, consulting a specialist ensures timely, appropriate care for the patient while enhancing Practitioner Wallace's understanding of the condition for future cases. *Answer*: B

3. Practitioner Dakota Henderson is a 40-year-old OB/GYN practicing for more than 10 years. She has recently noticed that her patients are increasingly asking her questions about alternative and complementary healthcare, and she feels ill-equipped to provide them with accurate information. How can Practitioner Henderson provide her patients with the most up-to-date and accurate information regarding alternative and complementary medicine?
   A. Attend a healthcare conference focused on alternative and complementary medicine
   B. Conduct a literature review on alternative and complementary medicine
   C. Consult with a colleague who has experience with alternative and complementary medicine
   D. Engage in self-directed learning and research to better understand alternative and complementary medicine
   E. All of the above

   Explanation: The correct answer is option E, all of the above. Practitioner Dakota Henderson can use all the listed methods to enhance her understanding of alternative and complementary medicine. Attending specialized healthcare conferences, conducting a literature review, consulting with experienced colleagues, and engaging in self-directed learning and research are all effective approaches. Each strategy offers unique benefits, from getting a comprehensive view of the field and access to current research and trends to benefiting from practical experiences and tailored knowledge acquisition. All these methods collectively will equip her with a broad and in-depth understanding, enabling her to provide her patients with the most accurate and up-to-date information. *Answer*: E

4. Practitioner Oskar Zabrousky is a 55-year-old primary care practitioner who has been in practice for 25 years. He has noticed that his patients increasingly ask him about healthcare marijuana and its potential benefits and risks. Practitioner Zabrousky received his healthcare degree before healthcare marijuana was legalized in his state, and he feels ill-equipped to provide his patients with accurate and up-to-date information. What is the most appropriate approach for Dr. Zabrousky to ensure that he provides his patients with accurate and up-to-date information regarding healthcare marijuana?
    A. Consult with colleagues who have experience with healthcare marijuana and engage in self-directed learning and research, including attending relevant healthcare conferences and conducting a literature review on the benefits and risks of healthcare marijuana
    B. Rely solely on his existing knowledge from his healthcare training, regardless of the changes in the law and emerging research on healthcare marijuana
    C. Suggest that his patients research healthcare marijuana independently and make their own judgments on its potential benefits and risks
    D. Avoid discussing healthcare marijuana with his patients due to his lack of knowledge and training

    Explanation: The correct answer is that Practitioner Zabrousky should consult with colleagues who have experience with healthcare marijuana and engage in self-directed learning and research, including attending relevant healthcare conferences and conducting a literature review on the benefits and risks of healthcare marijuana. This approach provides a comprehensive and reliable way for him to acquire up-to-date, evidence-based information on healthcare marijuana. It allows him to accurately respond to his patients' questions and make informed decisions when recommending treatments. By doing so, he upholds the principles of continual professional development and the duty to provide the best possible care for his patients. The other options are insufficient or inappropriate because they may result in misinformation, negligence, or a lack of patient engagement and education. *Answer*: A

5. Practitioner Darian O'Brian is a 40-year-old emergency department practitioner. She has noticed that her colleagues often rely on clinical intuition and prior experience rather than keeping up with the latest research in the field. Practitioner O'Brian believes this approach can be dangerous, particularly in emergency medicine, in which decisions must be made quickly and accurately. She is concerned that her colleagues are not engaging in enough lifelong learning. What is the most appropriate approach for Practitioner O'Brian to promote lifelong learning among her colleagues?
    A. Encourage her colleagues to adopt evidence-based practice by demonstrating how it improves patient outcomes and promote activities such as journal clubs, case discussions, and CME programs within her department
    B. Ignore the issue, assuming her colleagues' experiences are sufficient to make accurate decisions in emergencies
    C. Publicly criticize her colleagues for lacking engagement in ongoing learning during departmental meetings
    D. Focus solely on her own learning and disregard what her colleagues do

    Explanation: The correct answer is that Practitioner O'Brian should encourage her colleagues to adopt evidence-based practice by demonstrating how it improves patient outcomes and promote activities such as journal clubs, case discussions, and CME programs within her department. These strategies facilitate ongoing learning and ensure that practitioners stay up-to-date with recent advancements in the field. They also foster a culture of continuous learning and improvement. The other options are not conducive to fostering a learning environment. Ignoring the issue or focusing solely on her own learning does not address the problem, and publicly criticizing colleagues could create a hostile work environment, which is not conducive to promoting a culture of learning and improvement. *Answer*: A

REFLECTION VIGNETTES

1. Practitioner Sarah Dabrowski, a 35-year-old practitioner, struggles to meet the state requirement of earning 50 American Medical Associations Physician's Recognition Award Category 1 Credits biennially. Practitioner Dabrowski finds this requirement to be time-consuming, inconvenient, and expensive. She is a busy practitioner with a hectic practice and finds it challenging to take time away from her patients to attend conferences and workshops. She is also frustrated with the high costs of attending these events and believes that the financial burden falls solely on practitioners. The differential diagnosis includes the importance of continuing healthcare education for maintaining competence, staying up-to-date on the latest research and treatments, and improving patient outcomes. Practitioner Dabrowski recognizes the importance of continuing education but believes that the current system may be too burdensome and potentially discriminatory against practitioners who cannot take time away from their practice or afford the high costs of attending these events.

    What is the ethical issue(s)?

    How should the ethical issue(s) be addressed?

2. Practitioner John Smith, a 42-year-old practitioner, is dedicated to pursuing lifelong learning. In addition to meeting his state's CME requirements, Practitioner Smith engages in various other learning activities to keep his knowledge and skills current. He regularly attends local healthcare conferences and grand rounds, reads peer-reviewed journals, participates in online healthcare forums, and engages in case-based discussions with his colleagues. He also mentors students and residents, stays updated with clinical practice guidelines, and attends webinars and podcasts on emerging healthcare topics. By engaging in these lifelong learning activities, Practitioner Smith ensures that he provides the highest quality care to his patients and keeps pace with the ever-evolving healthcare field.

    What is the ethical issue(s)?

    How should the ethical issue(s) be addressed?

* * *

WEB LINKS

1. *Code of Medical Ethics*. American Medical Association. https://code-medical-ethics.ama-assn.org
   The AMA *Code of Medical Ethics*, established by the American Medical Association, is a comprehensive guide for healthcare practitioners. It addresses issues and challenges, promotes adherence to standards of care, and is continuously updated to reflect contemporary practices and challenges.
2. "Principles of Clinical Ethics and Their Application to Practice." Basil Varkey. 2021. https://www.ncbi.nlm.nih.gov/pmc/articles/PMC7923912
   This source discusses the ethical obligations of physicians and how they can improve their ethical skills through lifelong learning.
3. "What Is Ethically Informed Risk Management?" Alan J. Card. 2020. https://journalofethics.ama-assn.org/article/what-ethically-informed-risk-management/2020-11
   This source frames ethically informed risk management as a patient-centered and evidence-based practice, aligning its scope with biomedical ethics and proposing specific ethical duties to guide risk management practice.
4. "Patient Rights and Ethics." Jacob P. Olejarczyk and Michael Young. https://www.ncbi.nlm.nih.gov/books/NBK538279
   This source discusses the right to informed consent in the United States, which is protected to some degree by state and federal legislation.

5. "Reflection-Based Learning for Professional Ethical Formation." William T. Branch, Jr., and Maura George. 2017. https://journalofethics.ama-assn.org/article/reflection-based-learning-professional-ethical-formation/2017-04
This source discusses the importance of reflection-based learning for forming a professional identity as a humanistic practitioner.
6. "Module 12: Biomedical Ethics." Quizlet. https://quizlet.com/181201323/module-12-biomedical-ethics-flash-cards
This source discusses the history of physicians ignoring the wishes of patients and of abusing research subjects.
7. "AARC Statement of Ethics and Professional Conduct." American Association for Respiratory Care. https://c.aarc.org/resources/position_statements/ethics_detailed.html
This source provides a statement of ethics and professional conduct for respiratory therapists, including the importance of lifelong learning.
8. "Medical Ethics." American Medical Association. https://www.ama-assn.org/topics/medical-ethics
This source provides an overview of bioethics, including the importance of lifelong learning for practitioners.

MALPRACTICE 33

When a doctor does go wrong, they are the best of criminals. The doctor has nerve and knowledge.
—Arthur Conan Doyle

* * *

MYSTERY STORY

OVERSIGHT REDEMPTION

In the heart of Arborville, Grace Hospital had long been recognized for its exceptional care. Practitioner Patrick Warren, a seasoned neurologist, was among the most respected on the staff. His unwavering commitment to his patients made him a beacon of trust in the community.

One morning, a 30-year-old journalist, Lydia Carter, was admitted with chronic headaches. Initial magnetic resonance imaging (MRI) revealed an anomaly. Practitioner Warren, juggling multiple patients that day, made a note to order a more detailed scan. However, amidst the chaos, he inadvertently failed to communicate this to Lydia or add it to her electronic health records.

The oversight became evident when Lydia was discharged with a painkiller prescription but no follow-up for the MRI. Her headaches persisted, growing in intensity. When Lydia reviewed her records online, she noticed the absence of any mention of a further scan. Confused and concerned, she reached out to Grace Hospital's patient ombudsman.

The ombudsman conducted a preliminary review, uncovering the missed note in Practitioner Warren's personal digital assistant but not in Lydia's official records. The hospital swiftly arranged the necessary scan for Lydia, which confirmed a benign growth requiring minor surgery. Although relieved at the diagnosis, Lydia could not help but ponder the unintended negligence.

The incident was a wake-up call for Grace Hospital. Realizing the gaps in the hospital's electronic record synchronization and communication protocols, the administration took proactive measures. It introduced a two-step verification system for critical healthcare notes. It made it mandatory for immediate peer reviews in complex cases, ensuring that no single practitioner's oversight could jeopardize patient care.

Practitioner Warren, distressed by his rare oversight, personally apologized to Lydia. In a surprising turn, Lydia chose not to pursue a malpractice suit, realizing the genuine error amid human endeavors to provide care. Instead, she wrote an insightful article on the importance of patient awareness and advocacy, emphasizing the need for open communication lines between patients and healthcare professionals.

Grace Hospital's proactive response became a case study, emphasizing continuous improvement in healthcare processes. Although rooted in an honest mistake, the incident highlighted the fragile balance between healthcare expertise and the ever-present potential for human error. Through understanding, reform, and commitment to better care, Grace Hospital continued to serve as a beacon of trust in Arborville.

* * *

THINK

Within the context of a patient–practitioner relationship, healthcare malpractice occurs when

- a practitioner's action or omission sufficiently deviates from the standards of care to be considered negligent and
- it results in an injury to the patient.

Both criteria must be met as necessary conditions to satisfy the definition of healthcare malpractice; thus, practitioner negligence without patient injury or patient injury without practitioner negligence is not grounds for healthcare malpractice. In addition, the statute of limitations varies from state to state but is usually 2 or 3 years for adults and 8 years for minors before their 22nd birthday.

The first part of the healthcare malpractice equation defines a negligent act or omission. Negligent acts or omissions occur when a standard of care is breached. Standards of care are a legal definition based on the hypothetical practice of a reasonably competent practitioner in the same or similar community that addresses the same healthcare condition of the malpractice claim. A reasonably competent practitioner is a healthcare professional who provides the patient with treatment options based on up-to-date, peer-reviewed, evidence-based care that the healthcare community has generally recognized. Using these definitions, it can be determined if a deviation from the accepted standard was a negligent act or omission.

Examples of negligent acts include, but are not limited to, making a diagnostic error, providing a medication error, or committing a surgical error.

Examples of negligent omissions include, but are not limited to, failure to order proper testing; failure to arrange appropriate follow-up for abnormal test results; failure to provide appropriate treatment options to the patient; failure to inform the patient of the risks involved in each treatment option, including no treatment; or not fixing defective healthcare equipment.

The second part of the malpractice equation requires defining injury. Examples of injury include more than mere physical damages: disability; unusual pain, suffering, and hardship; significant past and future healthcare bills; and other adverse outcomes.

Healthcare malpractice involves the practitioner's negligent act or omission resulting in patient injury. Healthcare malpractice is a practitioner's violation of the patient–practitioner relationship in which patients trust their practitioner will practice due diligence. It is also a practitioner's violation of society's trust in the healthcare profession's ability to self-regulate.

ASSESS

Patient: Autonomy

A negligent act or omission by a practitioner is not something that a reasonable patient would autonomously consent to. Therefore, practitioner negligence violates patient autonomy (informed consent). Because a patient's informed consent is the authorization of a specific treatment option, it follows that the practitioner who negligently does something different from what was authorized, resulting in patient injury, could be committing a form of battery, which in legal terminology is an unauthorized touching of a person without their consent. The patient's informed consent was violated because the patient did not autonomously choose the negligent act or omission, resulting in patient injury.

Practitioner: Beneficence and Nonmaleficence

Professionally, if the practitioner fails to maximize the patient's best interests according to the patient's reasonable goals, values, and priorities because of negligently deviating from standard healthcare practices, that would violate beneficence (do good). If the practitioner injures the patient because of negligently violating the standards of care, then nonmaleficence (do no harm) is violated. Because the negligent act or omission violates the professional principle of beneficence (do good) that then causes a violation of the professional principle of nonmaleficence (do no harm), it logically follows that both necessary conditions of negligence and harm have been met for justifying a healthcare malpractice suit.

Unlike the doctrine of double effect, the practitioner's intention is not part of this healthcare malpractice equation. Good intentions and bad intentions indicate a lot about the practitioner's moral character, but intentions are independent of the legal status of a healthcare malpractice suit. A practitioner may have the best of intentions when they deviate from standard healthcare practice, causing harm to the patient; however, good intentions will not absolve the practitioner's accountability for violating the standards of care and the patient's subsequent harm incurred because of that violation.

Public Policy: Justice

Justice or public policies for the fair distribution of benefits and burdens try to balance the weight of benefits and burdens. The healthcare malpractice judge compensates the harmed patient to mitigate the unfair distribution of the patient's burdens caused by the negligent injury.

Healthcare malpractice compensation is awarded to the patient for two main reasons:

1. To even out the unfair balance of burdens caused by injury damages
2. To motivate the healthcare profession and the institution to perform a root cause analysis; identify what, how, and why a negligent adverse event occurred; and implement proactive measures to avert any potential future negligence

Examples of practical measures developed to avoid negligent acts and omissions are surgical checklists, wristband identifiers, having the patient point and mark the exact body part that needs surgery, and other risk mitigation procedures. Sometimes, healthcare malpractice settlements are awarded to the patient to make treatment options accessible to the patient, rather than solely on the culpability of the negligent practitioner, so that social burdens are more fairly distributed. This is mainly the case when a healthcare system causes patient burdens because of systemic healthcare disparities, resulting in unequal healthcare benefits and burdens.

CONCLUDE

Malpractice is a central concern for practitioners, the healthcare profession, institutions, and, most important, patients. The Federation of State Medical Boards and the National Board of Medical Examiners want to ensure that the practitioner understands that intention is not one of the necessary conditions for healthcare malpractice. Within the context of a patient–practitioner relationship, healthcare malpractice occurs when there is (a) a negligent act or omission that (b) results in patient injury. Practitioner negligence without injury and patient injury without practitioner negligence are not grounds for healthcare malpractice.

Healthcare malpractice is a significant concern for practitioners, the healthcare profession, institutions, and patients. It occurs when a negligent act or omission results in patient injury. Healthcare malpractice suits compensate the harmed patient and motivate the healthcare profession and institutions to prevent future negligent events by identifying root causes and implementing practical measures. Understanding the legal implications of healthcare malpractice is essential for practitioners to fulfill their professional obligations and maintain the trust of their patients and society as a whole. (For more discussion on the issue of intent, see Chapter 14.)

* * *

REVIEW QUESTIONS
1. Healthcare malpractice occurs when
    A. A practitioner's action or omission sufficiently deviates from the standards of care to be negligent
    B. A practitioner injures a patient
    C. Both A and B

2. Practitioner negligence without patient injury or patient injury without practitioner negligence are not grounds for healthcare malpractice.
    A. True
    B. False

3. Standards of care are based on the hypothetical practices of a reasonably competent practitioner in the same or similar community that address the same healthcare condition of the malpractice claim.
    A. True
    B. False

4. A negligent act or omission by a practitioner is not something that a reasonable patient would autonomously consent to. Therefore, practitioner negligence violates patient autonomy (informed consent).
    A. True
    B. False

5. The practitioner's intentions are significant in determining healthcare malpractice.
    A. True
    B. False

Answers: 1C, 2A, 3A, 4A, 5B

* * *

CLINICAL VIGNETTES

1. Ms. Angela Johnson, a 35-year-old accountant, presents to the emergency department with severe abdominal pain and vomiting. She reports having had diarrhea for the past week, and the symptoms have progressively worsened. Her clinical symptoms include abdominal tenderness, distention, and guarding. The differential diagnosis includes appendicitis, diverticulitis, and inflammatory bowel disease. A computed tomography (CT) scan is ordered, but the attending practitioner misinterprets the radiologist's report as showing normal results. Ms. Johnson was discharged with a diagnosis of gastroenteritis and prescribed over-the-counter medications. Her symptoms worsened, and she returned to the hospital within 2 days when she was diagnosed with a perforated appendicitis. She requires surgery and spends several weeks in the hospital. Which of the following is correct?
    A. Ms. Johnson cannot sue the hospital for healthcare malpractice because the attending practitioner did not intend to misinterpret the radiologist's report.
    B. Ms. Johnson can sue the hospital for healthcare malpractice because the attending practitioner was negligent in misinterpreting the radiologist's report, and she suffered harm.
    C. Ms. Johnson cannot sue the hospital for healthcare malpractice because the patient had the responsibility of coming back sooner once they realized that they were not getting better.
    D. Ms. Johnson cannot sue the attending practitioner for healthcare malpractice because the CT scan was a difficult case, and their interpretation of the results was within the accepted standards of care.

    Explanation: The attending practitioner was negligent in misinterpreting the radiologist's report, and Ms. Johnson suffered harm. In this case, the attending practitioner breached the standards of care by failing to identify and diagnose Ms. Johnson's appendicitis, which resulted in a delayed diagnosis and harm to the patient. The other options are wrong because the intention of the attending practitioner is not relevant in a healthcare malpractice case, as the necessary conditions are practitioner negligence and patient harm; the patient did not have the responsibility of coming back sooner; and difficulty of the case is not the definition of negligence, which is a breach of the standards of care. *Answer*: B

2. Mr. Nico Lopez is a 50-year-old salesman who presented to the emergency department with complaints of left arm pain and shortness of breath. He had a history of smoking and hypertension. The attending practitioner ordered an electrocardiogram, which was read as normal, and discharged Mr. Lopez with a diagnosis of musculoskeletal chest pain. Mr. Lopez had a myocardial infarction 2 days later and required emergency cardiac catheterization. The differential diagnosis for Mr. Lopez's initial presentation included acute coronary syndrome, pulmonary embolism, and musculoskeletal chest pain. The two necessary conditions for healthcare malpractice are a negligent act or omission and patient injury. Which of the following is correct?
   A. The attending practitioner followed the appropriate standards of care and is not liable for healthcare malpractice.
   B. The attending practitioner deviated from the standards of care, but Mr. Lopez's injury was not a result of the deviation, so the attending practitioner is not liable for healthcare malpractice.
   C. The attending practitioner deviated from the standards of care and is liable for healthcare malpractice because the deviation resulted in Mr. Lopez's injury.
   D. The attending practitioner deviated from the standards of care but is not liable for healthcare malpractice because the electrocardiogram was read as normal.

   Explanation: In this clinical vignette, the attending practitioner deviated from the standards of care by failing to consider other possible diagnoses, order additional testing, or admit Mr. Lopez for observation, given his healthcare history and presentation. The attending practitioner's negligence resulted in Mr. Lopez's injury, a myocardial infarction requiring emergency cardiac catheterization. Therefore, the attending practitioner is liable for healthcare malpractice because both necessary conditions—a negligent act or omission and patient injury—have been met. The other options are incorrect because the attending practitioner did not follow the appropriate standards of care, Mr. Lopez's injury resulted from the attending practitioner's deviation from the standards of care, and the electrocardiogram is not a definitive test for acute coronary syndrome. The attending practitioner should have considered additional testing or observation, given Mr. Lopez's healthcare history and presentation. *Answer*: C

3. Ms. Gwendolyn Davis, a 55-year-old teacher, presents to her primary care practitioner complaining of shortness of breath and a persistent cough. She reports that her symptoms have been present for several weeks, and over-the-counter medications have not provided relief. Clinical examination reveals decreased breath sounds in the left lower lobe of her lungs, and a chest X-ray is ordered. The radiologist reports the X-ray as showing no abnormalities. Despite the X-ray report, the practitioner suspects pneumonia and orders a CT scan, which confirms the diagnosis. Ms. Davis is treated with antibiotics and recovers well. Which of the following is correct?
   A. The practitioner was negligent in ordering a CT scan, as the radiologist reported normal results.
   B. The radiologist was negligent in misreading the chest X-ray, leading to delayed diagnosis and treatment.
   C. The practitioner was not negligent, as they ordered a CT scan based on their clinical judgment, which was supported by Ms. Davis' symptoms.
   D. The radiologist was not negligent, as the chest X-ray can be a difficult diagnostic tool and can result in false-negative results.

   Explanation: The practitioner was not negligent in ordering a CT scan based on their clinical judgment, supported by Ms. Davis' symptoms. Standards of care are based on the hypothetical practices of a reasonably competent practitioner in the same or similar community that addresses the same healthcare condition of the malpractice claim. In this case, the practitioner met the standards of care by ordering a CT scan to confirm their clinical suspicion of pneumonia, despite the radiologist's report. The practitioner's clinical judgment was supported by Ms. Davis' symptoms, which warranted further investigation to ensure an accurate diagnosis and prompt treatment. The practitioner was not negligent in ordering a CT scan based on their clinical judgment, as it complied with standards of care. The radiologist was negligent in misreading the chest X-ray but cannot be held liable because there was no resultant harm to the patient as there was no significant delay in the diagnosis. *Answer*: C

4. Mr. Alfred Wilson, a 52-year-old construction worker, was referred to Practitioner Tanner Chen, an orthopedic surgeon, for chronic knee pain. Practitioner Chen ordered an MRI, which showed a torn meniscus, and recommended surgery to repair the damage. Before the surgery, Practitioner Chen asked Mr. Wilson to mark the knee with a pen to avoid wrong-site surgery. On the day of the surgery, the operating team confirmed it had the correct patient, and Mr. Wilson's marked knee was visible. The team proceeded with the surgery, but during the procedure, Practitioner Chen realized that he had misread the MRI and operated on the wrong knee. Mr. Wilson was left with two damaged knees and filed a healthcare malpractice suit against Practitioner Chen. Clinical differential diagnosis: Wrong-site surgery resulting in damage to both knees. Ethical question: Should Practitioner Chen be held liable for healthcare malpractice, even though he followed practical measures to avoid wrong-site surgery, such as surgical checklists and patient marking?
    A. No, Practitioner Chen should not be held liable for healthcare malpractice because he took appropriate measures to avoid wrong-site surgery.
    B. Yes, Practitioner Chen should be held liable for healthcare malpractice because he misread the MRI and operated on the wrong knee, despite the patient marking.
    C. No, Practitioner Chen should not be held liable for healthcare malpractice because Mr. Wilson marked the wrong knee.
    D. Yes, Practitioner Chen should be held liable for healthcare malpractice because wrong-site surgery happened, irrespective of the precautions taken.

    Explanation: The correct answer is option B. Despite Practitioner Chen following standard safety procedures such as surgical checklists and patient marking, he still operated on the wrong knee. The fact that he misread the MRI and performed surgery on the wrong knee indicates negligence on his part. It is important to remember that the ultimate responsibility for ensuring the correct surgical site lies with the surgeon. Practitioner Chen's actions damaged both of Mr. Wilson's knees, a harm that would not have occurred without his negligence. Therefore, he should be held liable for healthcare malpractice. The incorrect options A, C, and D fail to correctly identify Practitioner Chen's liability by incorrectly assuming the precautions taken absolve him of responsibility (A), incorrectly suggesting that Mr. Wilson marked the wrong knee (C), or not specifically linking Practitioner Chen's negligence in misreading the MRI to the wrong-site surgery (D). *Answer*: B

5. Mr. Omar Larson, a 45-year-old software engineer, was diagnosed with a rare cancer. His practitioner, Practitioner Johnson, recommended a new and experimental treatment that had only been tried in a handful of patients. Despite knowing the risks, Practitioner Johnson administered the treatment to Mr. Larson, and the patient's condition worsened. After being admitted to the hospital, Mr. Larson's condition continued to deteriorate, and he eventually died. Should the practitioner's intentions be considered when evaluating whether healthcare malpractice occurred?

A. Yes, the practitioner's good intentions should be taken into account when determining if healthcare malpractice occurred.
B. No, the practitioner's intentions are irrelevant in determining if healthcare malpractice occurred.
C. It depends on the outcome of the treatment.
D. It depends on the practitioner's level of experience.

Explanation: A practitioner's intentions are not part of the legal equation for determining whether healthcare malpractice occurred. In other words, even if the practitioner had good intentions when they deviated from standard healthcare practice, this would not absolve them of accountability for violating standards of care and causing harm to the patient. The other options are incorrect because the outcome of the treatment does not change the fact that a deviation from standards of care occurred, and the practitioner's level of experience is not relevant to the issue of whether healthcare malpractice occurred. *Answer*: B

REFLECTION VIGNETTES

1. Ms. Laura Johnson, a 35-year-old elementary school teacher, presents to the emergency department with severe abdominal pain, fever, and chills. She is diagnosed with a ruptured appendix and requires urgent surgery. The practitioner explains the risks and benefits of the surgical procedure to Ms. Johnson, including the risk of complications such as bleeding, infection, and anesthesia reactions. However, the practitioner failed to fully inform Ms. Johnson of the risks of not having the surgery, such as sepsis and death. Ms. Johnson is afraid of surgery and decides to decline the procedure. Unfortunately, her condition deteriorated rapidly, and she passed away due to complications from the ruptured appendix.

    What is the ethical issue(s)?

    How should the ethical issue(s) be addressed?

2. Mr. Jake Anderson, a 35-year-old elementary school teacher, presents to the emergency department with severe abdominal pain, fever, and chills. He is diagnosed with a ruptured appendix and requires urgent surgery. The practitioner explains the risks and benefits of the surgical procedure to Mr. Anderson, including the risk of complications such as bleeding, infection, and anesthesia reactions. However, the practitioner fails to fully inform Mr. Anderson of the risks of not having the surgery, such as sepsis and death. Mr. Anderson is afraid of surgery and decides to decline the procedure. Fortunately, the ruptured appendix was a misdiagnosis, and Mr. Anderson fully recovered.

    What is the ethical issue(s)?

    How should the ethical issue(s) be addressed?

* * *

WEB LINKS

1. *Code of Medical Ethics*. American Medical Association. https://code-medical-ethics.ama-assn.org
   The AMA *Code of Medical Ethics*, established by the American Medical Association, is a comprehensive guide for healthcare practitioners. It addresses issues and challenges, promotes adherence to standards of care, and is continuously updated to reflect contemporary practices and challenges.
2. "Ethical Issues in Medical Malpractice." Robert C. Solomon. 2006. https://pubmed.ncbi.nlm.nih.gov/16877140
   This source discusses the interrelationships between biomedical ethics and the law in healthcare malpractice.

3. "Patient Rights." American Medical Association. https://code-medical-ethics.ama-assn.org/ethics-opinions/patient-rights
   This source discusses the health and well-being of patients depending on a collaborative effort between patient and physician in a mutually respectful alliance.
4. "Health Law and Medical Practice." Arina Evgenievna Chesnokova. 2016. https://journalofethics.ama-assn.org/article/health-law-and-medical-practice/2016-03
   This source discusses the history of health law and healthcare practice, including the area of healthcare malpractice.
5. "What Is Ethically Informed Risk Management?" Alan J. Card. 2020. https://journalofethics.ama-assn.org/article/what-ethically-informed-risk-management/2020-11
   This source discusses ethically informed risk management as a patient-centered and evidence-based practice, aligning its scope with biomedical ethics.
6. "Medical Ethics." American Medical Association. https://www.ama-assn.org/topics/medical-ethics
   This source provides an overview of bioethics, including the moral framework for the practice of clinical medicine.
7. "Medical Beneficence, Nonmaleficence, and Patients' Well-Being." Lynn A Jansen. 2022. https://pubmed.ncbi.nlm.nih.gov/35302515
   This source critically analyzes the principles of beneficence and nonmaleficence in clinical bioethics.
8. "Ethical Malpractice." Nadia Sawicki. 2022. https://houstonlawreview.org/article/36539-ethical-malpractice
   This source discusses why there is no tort cause of action for "ethical malpractice" and the expectations for healthcare providers to comply with professional standards of care.
9. "Ethics and Medical Malpractice." Andres, Berger & Tran. 2013. https://www.andresberger.com/blog/ethics-and-medical-malpractice
   This source notes that when practitioners fail to live up to the standard of care consistent with the best practices in the healthcare community, they may be held responsible for healthcare malpractice.

MEDIA COMMUNICATIONS

34

The practice of medicine is an art, not a trade; a calling, not a business; a calling in which your heart will be exercised equally with your head. Often the best part of your work will have nothing to do with potions and powders, but with the exercise of an influence of the strong upon the weak, of the righteous upon the wicked, of the wise upon the foolish.
—Sir William Osler

* * *

MYSTERY STORY

AN UNINTENDED BREACH: AN INTRIGUING CAUTIONARY TALE FOR THE HEALTHCARE FRATERNITY

The tapestry of healthcare is a complex and delicate blend of professional dedication, patient trust, and ethical boundaries. One such narrative centers on Practitioner Samantha Park, a renowned pediatrician at ConfidentialCare Children's Hospital, recognized for her expertise in treating a multitude of challenging cases. Among the galaxy of her patients was a young boy wrestling with the hardships of a rare genetic disorder.

The boy's parents, strong pillars of support and advocacy, courageously navigated this challenging journey, often utilizing social media to share their son's story, hoping to connect with other families facing similar battles. Practitioner Park, known for her meticulous commitment to patient privacy, made sure to respect confidentiality agreements, revealing healthcare details only to those directly involved in the boy's care.

As fate would have it, a local news station, captivated by the poignant journey of this young warrior, reached out to Practitioner Park. The station wanted to shine a spotlight on the reality of living with rare genetic disorders, potentially driving public interest and more research funding. Torn between her obligation to protect her patient's privacy and the tantalizing prospect of amplifying awareness, Practitioner Park decided to participate in the interview.

As the camera rolled and questions flowed, Practitioner Park, swept up in the moment, unknowingly crossed a line. She shared specifics about the boy's condition, along with his full name—details not authorized for disclosure by the parents. The world of the boy's family collapsed when they encountered the news segment. They immediately vented their outrage by filing a complaint against Practitioner Park.

In the aftermath of the interview, a thorough investigation by ConfidentialCare Children's Hospital revealed that Practitioner Park had unwittingly breached several tenets of the Health Insurance Portability and Accountability Act (HIPAA), revealing protected health information (PHI) without the necessary patient/guardian consent. As the dust settled, the once esteemed Practitioner Park was faced with severe repercussions, including a hefty fine and the looming threat of losing her cherished healthcare license.

The incident echoed far beyond the direct players, inflicting a ripple effect of harm upon the patient, his family, and Practitioner Park's once unblemished reputation at ConfidentialCare Children's Hospital. A lesson harshly learned, Practitioner Park openly regretted her lapse in judgment and vowed to prioritize her loyalty to her patients and to the principles of patient autonomy (informed consent), practitioner beneficence (do good), practitioner nonmaleficence (do no harm), and public policy justice (be fair).

> A poignant tale of unintended consequences, this incident serves as a stark reminder to all healthcare professionals of the critical importance of patient privacy and confidentiality. It underscores the fact that the strictest adherence to HIPAA regulations is not just a guideline but an unwavering commitment, safeguarding against the potential devastation caused by a single moment's oversight.

* * *

THINK

Informing the media about evidence-based care is a public service that can significantly benefit individuals and the community. Addressing epidemiological issues and healthcare disparities is a professional and social responsibility practitioners must pursue. However, practitioners who speak on behalf of an organization must fully comply with their professional and institutional media policies and guidelines. Failure to do so could have legal, institutional, and professional consequences.

PHI is a specific legal term defined by HIPAA in the United States. Practitioners who speak to the media on behalf of a patient must only do so with documented informed consent from the patient that specifically expresses what PHI, which is meant to be confidential and private, is permissible to disclose. No PHI can be released to the media without documented patient authorization and institutional approval. If the authorized information is specific to a diagnosis or prognosis, the information should only be disclosed by the patient's attending practitioner in addition to documented patient consent and authorization. If information is requested about nefarious or criminal activities by court order, then only the minimal amount of patient information should be disclosed to those who have the authority to receive such information, which would not be the media. Confidentiality and privacy are the cornerstones of the patient–practitioner relationship, and trust must never be compromised for the integrity of the patient–practitioner relationship and for the public's trust in the healthcare profession.

When communicating with the media, being addressed as a healthcare professional and/or wearing the "white coat" changes the role of the practitioner into someone who now represents the healthcare profession and the healthcare institution where the practitioner works and has healthcare privileges and where the patient is being treated. All media policies and procedures mandated by the healthcare profession and the associated institutions must not be violated. When wearing the white coat and/or being represented as a healthcare professional, it is imperative that the practitioner minimizes the expression of personal beliefs and political viewpoints and accentuates the demeanor and ethical values of the patient-centered, evidence-based healthcare profession. As a representative of the healthcare profession and the healthcare facility, all media communications must reflect the practitioner's patient–practitioner relationship as always being the practitioner's primary loyalty.

ASSESS

Patient: Autonomy

Autonomy (informed consent) requires that any patient's information may only be disclosed if authorized by the patient. Although the institution may own the healthcare records, the patient owns the information. Any disclosure without patient authorization violates the patient's civil rights of confidentiality and privacy and violates professional and institutional policy. It is also a violation of the HIPAA federal statute that is punishable by fines of up to $250,000 and a jail term of up to 5 years. When addressing the media, extreme care must be taken not to violate the patient-centered patient–practitioner relationship.

Practitioner: Beneficence and Nonmaleficence

Professionally, the practitioner must continuously pursue the patient's best interests as determined by the patient's reasonable goals, values, and priorities. Therefore, it is incumbent upon the practitioner to reveal to the media only the information that the patient has authorized and only the information that will not result in psychological, social, or physical harm to the patient, in accordance with the professional principle of nonmaleficence (do no harm).

Wearing the white coat and being addressed as a practitioner is a professional honor and should never be taken lightly. Every effort should be taken to keep the white coat immaculately clean and unwrinkled because the white coat represents the healthcare profession's clean and ordered environment along with conscientious patient care. All professional and institutional media policies must be followed along with the ethical and moral precepts of the healthcare profession.

Public Policy: Justice

The healthcare profession has a social obligation of justice to keep the media informed of evidence-based care's progress and how communities can better address various issues related to eliminating healthcare disparities. Communicating to the public the importance of the patient–practitioner relationship and the healthcare profession's unwillingness to compromise on the patient's right to confidentiality and privacy will help increase social trust.

CONCLUDE

The practitioner must understand that the patient–practitioner relationship must not be breached when communicating with the media. No PHI can be disclosed to the media without documented patient authorization in the patient's healthcare record and institutional approval. One exception would be if there were to be a court order. Then that information would still not be disclosed to the media but, rather, only to those authorized to receive that information, and then only the minimal amount of patient information required.

In summary, practitioners must balance their responsibilities to their patients, the healthcare profession, and the public when communicating with the media. They must protect patient privacy by obtaining informed consent and adhering to professional and institutional media policies. By maintaining the integrity of the patient–practitioner relationship, practitioners can help build trust in the healthcare profession and contribute to a better informed society on matters related to medicine and healthcare disparities.

* * *

REVIEW QUESTIONS

1. Practitioners who speak to the media on behalf of a patient must have documented informed consent from the patient that specifically expresses what PHI is permissible to disclose along with institutional approval.
    A. True
    B. False

2. When communicating with the media, being addressed as a practitioner and/or wearing the "white coat" changes the person's role into someone who now represents the healthcare profession and the healthcare institution where the practitioner works and has healthcare privileges.
    A. True
    B. False

3. When wearing the white coat and/or being represented as a healthcare professional, it is imperative that the practitioner minimizes the expression of personal beliefs and

political viewpoints and accentuates the demeanor and ethical values of the patient-centered, evidence-based healthcare profession.
A. True
B. False

4. Every effort should be taken to keep the white coat immaculately clean and unwrinkled because the white coat represents the healthcare profession's clean and ordered environment along with conscientious patient care.
A. True
B. False

5. The healthcare profession has a social obligation of justice to keep the media informed of evidence-based care's progress and how communities can better address various issues related to eliminating healthcare disparities.
A. True
B. False

Answers: 1A, 2A, 3A, 4A, 5A

* * *

CLINICAL VIGNETTES

1. Mr. John Smith is a 45-year-old patient diagnosed with stage 3 colon cancer. He has been receiving chemotherapy treatments for the past 6 months and has been closely monitored by his attending oncologist, Practitioner Kaiden Rodriguez. Recently, a local news agency contacted Practitioner Rodriguez requesting an interview to discuss the effectiveness of the chemotherapy treatments Mr. Smith has received. What ethical considerations should Practitioner Rodriguez consider when deciding whether or not to grant an interview to the news agency about Mr. Smith's treatments?
   A. Practitioner Rodriguez should provide the news agency with all of Mr. Smith's healthcare history, including the details of his current chemotherapy treatment.
   B. Practitioner Rodriguez should not grant an interview to the news agency because it would violate Mr. Smith's right to confidentiality and privacy.
   C. Practitioner Rodriguez should grant the interview but withhold information about Mr. Smith's chemotherapy treatments to protect his privacy.
   D. Practitioner Rodriguez must comply with institutional media policies and guidelines and obtain documented informed consent from Mr. Smith before disclosing any PHI to the media.

   Explanation: Practitioner Rodriguez must prioritize Mr. Smith's autonomy and privacy when communicating with the media. No PHI can be released to the media without documented patient authorization and institutional approval. If Mr. Smith has consented, then only the information he has authorized should be disclosed to the news agency. The ethical principles of beneficence and nonmaleficence also require that Practitioner Rodriguez disclose information that will not harm Mr. Smith. Furthermore, Practitioner Rodriguez must follow professional and institutional media policies and guidelines and ensure that he represents the healthcare profession and the healthcare facility in a manner that accentuates the demeanor and ethical values of the patient-centered, evidence-based healthcare profession. In conclusion, Practitioner Rodriguez must obtain Mr. Smith's informed consent and only disclose the information that Mr. Smith has authorized. *Answer*: D

2. Practitioner Emily Johnson is a 30-year-old nurse who works in a local hospital's emergency department. One evening, she witnessed a car accident on her way home from work and

stopped to provide assistance. The local news agency arrived on the scene and requested an interview with Practitioner Johnson about her experience as a nurse and the care she provided to the accident victims. What ethical considerations should Practitioner Johnson consider when deciding whether or not to grant an interview to the news agency about the accident?

   A. Practitioner Johnson should grant the interview and disclose all the details of the care she provided to the accident victims.
   B. Practitioner Johnson should not grant the interview because it would violate the privacy of the accident victims and their families.
   C. Practitioner Johnson should grant the interview but only disclose nonidentifiable details of the care she provided.
   D. Practitioner Johnson should grant the interview but only disclose the information that is in accordance with her professional and institutional media policies and HIPAA regulations.

   Explanation: When communicating with the media, being addressed as a healthcare professional and/or wearing the "white coat" changes the practitioner's role into someone who represents the healthcare profession, the healthcare institution where they work, and the patient they are treating. As such, Practitioner Johnson must follow her professional and institutional media policies and HIPPA regulations to ensure that the information she discloses is in accordance with the principles of beneficence and nonmaleficence and will not result in psychological, social, or physical harm to the patients. Practitioner Johnson must also minimize the expression of personal beliefs and political viewpoints and accentuate the demeanor and ethical values of the patient-centered, evidence-based healthcare profession. Although Practitioner Johnson may grant the interview, she must only disclose the information authorized by her professional and institutional media policies and HIPAA regulations. Practitioner Johnson's loyalty must always be to her patient–practitioner relationship, and her disclosures must not violate the patients' privacy or confidentiality. *Answer*: D

3. Mr. James Wilson, a 52-year-old engineer, presents to the emergency department with severe abdominal pain and vomiting. The differential diagnosis includes acute appendicitis, diverticulitis, and intestinal obstruction. The healthcare team orders several diagnostic tests and administers intravenous fluids and pain medication. While waiting for the test results, Mr. Wilson asks to speak to a hospital's media relations team member to share his positive experience with the care he received so far. He requests to take a photo with one of the practitioners, wearing a white coat, to share on his social media platforms. How should the practitioner respond?
   A. Allow Mr. Wilson to take the photo with the practitioner wearing the white coat to promote positive publicity for the hospital
   B. Inform Mr. Wilson that the healthcare team cannot participate in any media relations activities without prior approval from the hospital's media relations team
   C. Inform Mr. Wilson that the healthcare team cannot participate in any media relations activities due to patient privacy laws and regulations
   D. Inform Mr. Wilson that the healthcare team cannot participate in any media relations activities while Mr. Wilson is a patient in the hospital

   Explanation: All media policies and procedures mandated by the healthcare profession and the associated institutions must not be violated. In addition, the white coat represents the healthcare profession's clean and ordered environment and conscientious patient care, and every effort should be taken to keep it immaculately clean and unwrinkled. The healthcare team cannot participate in any media relations activities without prior approval from the hospital's media relations team. The reason for not participating in media relations activities is not patient privacy laws and regulations but, rather, institutional media policies and guidelines. *Answer*: B

4. Mr. Solomon Yates, a 50-year-old retired construction worker, visits his healthcare provider with concerns about his hypertension and high cholesterol. His recent readings show a blood pressure of 150/90 mmHg and a total cholesterol level of 250 mg/dl. As a healthcare practitioner, you are often called upon to discuss the latest advancements in healthcare for conditions such as hypertension and hypercholesterolemia. In this context, an important ethical question arises: How can you balance your responsibility to inform the public about progress in these areas of healthcare while preserving the confidentiality and privacy of your patients, such as Mr. Yates?
   A. Disclose patient-specific details in discussions with the media to provide real-world context
   B. Use generalized data and anonymized case studies to discuss progress without identifying individual patients
   C. Refrain from engaging with the media about specific healthcare conditions to avoid possible HIPAA violations
   D. Request patients share their personal experiences with the media directly

   Explanation: To uphold their ethical responsibility toward patient privacy and confidentiality, healthcare practitioners should avoid disclosing patient-specific details in public or media discussions. Instead, they can use generalized data and anonymized case studies when discussing advancements in healthcare. This approach allows them to contribute to public awareness and understanding of health issues such as hypertension and hypercholesterolemia without violating the privacy of specific individuals like Mr. Yates. By doing so, practitioners fulfill their social responsibility to keep the public informed about the progress of evidence-based care while adhering to privacy laws and ethical guidelines. *Answer*: B

5. Practitioner Miles Jones is a 55-year-old practitioner and researcher who works at a prestigious healthcare institution. He has been invited to speak on a panel about the importance of evidence-based care in addressing healthcare disparities. During the event, Practitioner Jones wears his white coat and is introduced as a healthcare professional from his work institution. He speaks about the institution's efforts to address healthcare disparities and emphasizes the importance of evidence-based care. What is the ethical obligation of Practitioner Jones when communicating with the media and being addressed as a healthcare professional and/or wearing the white coat?
   A. To reveal all information about the institution's research and patients to the media, regardless of patient consent or institutional policy
   B. To minimize the expression of personal beliefs and political viewpoints when communicating with the media and to accentuate the demeanor and ethical values of the patient-centered, evidence-based healthcare profession
   C. To wear the white coat and represent the institution at all times, regardless of whether he is off-duty or not
   D. To prioritize the media's need for information over the patient's right to confidentiality and privacy

   Explanation: When communicating with the media, being addressed as a healthcare professional and/or wearing the white coat changes the role of the practitioner into someone who now represents the healthcare profession and the healthcare institution where the practitioner works and has healthcare privileges and where the patient is being treated. As a healthcare professional, Practitioner Jones has a professional obligation to pursue the patient's best interests; follow professional and institutional media policies; and maintain the ethical and moral principles of the patient-centered, evidence-based healthcare profession. When speaking to the media, Practitioner Jones must minimize the expression of personal beliefs and political viewpoints and accentuate the demeanor and ethical values of the healthcare profession to maintain the trust of the patient–practitioner relationship and the public's trust in the healthcare profession. The other options are incorrect because it is not acceptable to violate a patient's autonomy and institutional policies, it is not a professional obligation to at all times wear the white coat and represent the institution, and

it is inappropriate to prioritize the media's needs over patient confidentiality and privacy.
Answer: B

REFLECTION VIGNETTES

1. Ms. Samantha Lee, a 45-year-old politician, is admitted to the hospital with severe chest pain, shortness of breath, and dizziness. The practitioner ordered a series of tests and diagnosed Ms. Lee with an acute myocardial infarction. The patient's political status attracts the media, and they request updates on Ms. Lee's condition.

   What is the ethical issue(s)?

   How should the ethical issue(s) be addressed?

2. Mr. Mark Lewis, a 55-year-old politician, is admitted to a hospital with symptoms of a sexually transmitted infection (STI). After testing positive for syphilis, he is treated by a prominent venereologist who maintains an active blog on STI treatments. Unaware of Mr. Lewis' political status, the venereologist posts about treating a patient with syphilis on their blog. News outlets soon pick up on the story and begin speculating about the identity of the high-profile patient.

   What is the ethical issue(s)?

   How should the ethical issue(s) be addressed?

* * *

WEB LINKS

1. *Code of Medical Ethics*. American Medical Association. https://code-medical-ethics.ama-assn.org
   The AMA *Code of Medical Ethics*, established by the American Medical Association, is a comprehensive guide for healthcare practitioners. It addresses issues and challenges, promotes adherence to standards of care, and is continuously updated to reflect contemporary practices and challenges.
2. "Ethical Issues in Medical Malpractice." Robert C. Solomon. 2006. https://pubmed.ncbi.nlm.nih.gov/16877140
   This source discusses the interrelationships between biomedical ethics and the law in the realm of healthcare malpractice.
3. "Patient Rights." American Medical Association. https://code-medical-ethics.ama-assn.org/ethics-opinions/patient-rights
   This source discusses the health and well-being of patients depending on a collaborative effort between patient and physician in a mutually respectful alliance.
4. "Federal Privacy Protections: Ethical Foundations, Sources of Confusion in Clinical Medicine, and Controversies in Biomedical Research." Mary Anderlik Majumder and Christi J. Guerrini. 2016. https://journalofethics.ama-assn.org/article/federal-privacy-protections-ethical-foundations-sources-confusion-clinical-medicine-and/2016-03
   This source discusses the belief that HIPAA forbids appropriate communication with patients' families, friends, and the clergy (potentially isolating patients and depriving them of support).
5. "The Top 10 Most-Read Bioethics Articles in 2021." American Medical Association. https://www.ama-assn.org/delivering-care/ethics/top-10-most-read-medical-ethics-articles-2021
   This source discusses the resurgence of anti-Asian racism and xenophobia during the COVID-19 pandemic in light of American history and personal experience.
6. "6 Ethical Issues in Healthcare in 2020." AdventHealth University. 2020. https://www.ahu.edu/blog/ethical-issues-in-healthcare

This source discusses how the seemingly simple maxim of healthcare proves far more complex when considering rapidly advancing healthcare technology, constant budget constraints, and new health threats.

7. "Medical Ethics." American Medical Association. https://www.ama-assn.org/topics/medical-ethics

    This source provides physicians with a moral framework for the practice of clinical medicine.

8. "The Ethics of Health Communication." T. Strasser and J. Gallagher. 1994. https://pubmed.ncbi.nlm.nih.gov/8018285

    This source discusses the scope of application of ethical principles in health communication with particular reference to the influence of the mass media on people's perceptions of benefit and risk in the field of healthcare.

HEALTHCARE RECORDS

35

The good practitioner treats the disease; the great practitioner treats the patient who has the disease.
—William Osler

* * *

MYSTERY STORY

> **DATA INTRUSION: A CAUTIONARY TALE OF TRUST SHATTERED AND PRIVACY BREACHED IN HEALTHCARE**
>
> In the bustling epicenter of HealthPrivacy General Hospital, Practitioner Catherine Kensington stood as a beacon of exceptional healthcare delivery. An accomplished practitioner with two decades of triumphant service, her name was synonymous with unwavering patient commitment. However, one ominous day, a ripple of uncertainty disrupted this picture of perfection. The hospital's cyber-vigilant information technology department unearthed an anomaly in Practitioner Kensington's electronic medical record (EMR) activities—an enigma that would soon send shock waves across the entire institution.
>
> The aberrations indicated a disconcerting pattern: Practitioner Kensington had been plunging into the confidential depths of EMRs, prying into the healthcare narratives of patients outside her jurisdiction, all without their knowledge or consent. An alarm was triggered, prompting the hospital's compliance crusader to swiftly step in, igniting an exhaustive investigation into the concerning matter.
>
> The inquiry uncovered a chilling truth: Practitioner Kensington had been passing on the intimate patient information to an unauthorized confidante, a nonmedical healthcare individual. This act, a flagrant violation of the sacrosanct Health Insurance Portability and Accountability Act (HIPAA), had been a clandestine operation for several long months. A swift retribution ensued; Practitioner Kensington found herself in the eye of the storm.
>
> Convened under the weight of the situation, a tribunal of her fellow practitioners sat in judgment, determining the fate of Practitioner Kensington's healthcare career. Their verdict echoed the hospital corridors: A grave breach of trust warranted severe retribution—revocation of her healthcare license and immediate termination from the hospital were deemed necessary.
>
> News of Practitioner Kensington's transgressions and her consequent downfall rippled across the community, leaving a trail of disbelief and fear. The once-trusting patients found themselves adrift in a sea of violation, uncertainty gnawing at them about the illicit sharing of their private healthcare information. The hospital engaged all gears to mend the gaping wound of trust, reiterating the sacrosanctity of HIPAA laws and the importance of protected health information (PHI) regulations to its healthcare providers.
>
> The disturbing incident served as a grim reminder to all healthcare providers about the sanctity of healthcare records, shining a light on the repercussions of breaching confidentiality and privacy laws. PHI, although housed within the digital vaults of the institution, belongs to the patients, mandating stringent adherence to HIPAA regulations by healthcare providers.
>
> Ultimately, the incident at HealthPrivacy General Hospital imparted a sobering lesson, a powerful reminder of the unwritten contract between a patient and their practitioner. The core pillars of autonomy (informed consent), beneficence (do good), nonmaleficence (do no harm), and justice (be fair) are paramount in fostering trust and maintaining the integrity of the healthcare profession. The saga of Practitioner Kensington, thus, became a stark illustration of the imperative to treat the patient, not merely the disease.

* * *

THINK

Healthcare records can be either physical or electronic, and although the healthcare records belong to the healthcare facility, the patient owns the content of the healthcare records. HIPAA has defined this patient information as PHI, and the patient has the legal right to access and copy all information that their healthcare records contain. Because the healthcare record's information belongs to the patient, that information cannot be withheld from the patient for any reason, including nonpayment of bills. The patient does not have to provide any justification or reason for attaining access to the information in the healthcare records, as all that information is the property of the patient, even though the physical paper and electronic healthcare records, with the contained information, are the property of the practitioner, clinic, or healthcare institution. The patient has the right to the information but does not have the right to delete or change the information in the healthcare records or remove the healthcare records from the provider of care.

Under no circumstance is the healthcare facility to release any of the patient's healthcare records containing the patient's PHI to anybody without the patient's explicit consent or because of a court order. To do otherwise violates HIPAA, a federal statute that has punishable fines of up to $250,000 and a jail term of up to 5 years.

If there is any need for a correction in a physical healthcare record, the practitioner should draw a line through the error, initial the correction, and provide the current date and time. It is never permissible to whiteout mistakes, remove pages, or back-date notes or corrections. Most healthcare institutions have switched from physical healthcare records to EMRs, where there is much less opportunity or temptation to change or revise past entries. The same notation of author, date, and time applies to emendations in an EMR.

ASSESS

Patient: Autonomy

The bioethical principle of autonomy is not just manifested in the process of a patient providing informed consent for authorizing a treatment option; it is also manifested in the patient's informed consent to disclose confidential and private PHI as found in the healthcare records. Although the interprofessional healthcare team must have readily available access to the patient's healthcare records while treating the patient, all team members must follow strict confidentiality and privacy protocols.

During clinical services, healthcare information must only be provided to the patient and the patient's interprofessional healthcare team unless the patient has explicitly given the authority through consent that others, such as family members, should also receive the healthcare record information. Healthcare record information is the patient's personal property, and violating confidentiality and privacy rules violates the patient–practitioner relationship in which PHI is to be kept confidential and private.

One rare exception to this explicit requirement for the patient's consent before disclosing the healthcare record's information is if there were to be a specific court order for that disclosure. Then, the practitioner should only disclose the minimum amount of information requested and only to those given the authority to receive that information.

Court order is an example of when the principle of justice (be fair) carries more moral weight than the principles of autonomy (informed consent), beneficence (do good), and nonmaleficence (do no harm) because it maximizes the best interests of the community at large.

Practitioner: Beneficence and Nonmaleficence

Practitioners must disclose all healthcare records information to the patient and maintain strict confidentiality and privacy standards toward their patient's PHI. This patient–practitioner relationship mandate is based on the professional moral principles of beneficence (do good) and nonmaleficence (do no harm). Professional patient-centered healthcare occurs as the practitioner makes healthcare judgments based on what will maximize the patient's best interests, as determined by the patient's reasonable goals, values, and priorities as a matter of beneficence (do good).

The patient's best interests allow for circumstances in which, because of the therapeutic privilege, it is permissible not to disclose some information to the patient if disclosing the information would induce the patient to harm themself or others—nonmaleficence (do no harm).

Therapeutic privilege is an example of when the principle of nonmaleficence (do no harm) and the principle of justice (be fair) carry more moral weight than the principle of autonomy (informed consent) because it maximizes the best interests of the patient and the community.

These exceptions are not instances of violating the patient–practitioner relationship. These exceptions are expected and understood as part of the social contract, in which not recognizing these exceptions would violate the patient–practitioner relationship and social contract expectations.

Public Policy: Justice

Society has prohibited healthcare confidentiality and privacy violations based on the public policy principle of justice (be fair). Society and the legal system recognize the importance of the sanctity of the patient–practitioner relationship and, therefore, have legislated patient healthcare record information, confidentiality, and privacy laws. These laws promote the public's trust and professional image of the healthcare profession.

Generally, only the patient and the patient's interprofessional healthcare team have the legal right to access the patient's healthcare record information. However, under extenuating circumstances, such as public safety and a court order, the practitioner is socially mandated to turn over a patient's healthcare record information, even at the objection of the patient, but only by the practitioner delivering the minimal amount of requested information and then only to the appropriate authorities. This is not a violation of the patient–practitioner relationship; rather, it is an expected and understood part of the social contract in which doing anything otherwise would violate the patient–practitioner relationship.

CONCLUDE

Practitioners must recognize patient confidentiality, privacy, access, and ownership of healthcare record information defined by HIPAA as PHI. Providing full patient access to all the patient's healthcare record information is a fundamental right established by society and the healthcare profession.

In summary, the sanctity of healthcare records is essential to the patient–practitioner relationship and the public's trust in the healthcare profession. Practitioners must recognize and respect patients' rights concerning their PHI as found in the healthcare records, including confidentiality, privacy, access, and ownership. By upholding these principles, practitioners can help maintain a strong patient–practitioner relationship and preserve the integrity of the healthcare profession.

* * *

REVIEW QUESTIONS

1. The institution owns both the physical or electronic healthcare records and the content in those records.
    A. True
    B. False

2. It is permissible to deny the patient a copy of their healthcare record due to unpaid healthcare bills.
    A. True
    B. False

3. It is unnecessary to get explicit authority from the patient to share their healthcare information with their family and loved ones.
   A. True
   B. False

4. The principle of autonomy (informed consent) always has more moral weight than the principles of beneficence (do good), nonmaleficence (do no harm), and justice (be fair).
   A. True
   B. False

5. Therapeutic privilege, where information is not disclosed to the patient, and court order, where information is given to others without consent, are not breaches of the patient–practitioner relationship.
   A. True
   B. False

Answers: 1B, 2B, 3B, 4B, 5A

* * *

CLINICAL VIGNETTES

1. Mr. Gage Nicholson, a 40-year-old software developer, requests a copy of his healthcare records from his practitioner. His healthcare records include information about his allergies and past surgeries. Who owns the physical or electronic healthcare records and the content of those records?
   A. The healthcare facility owns the physical or electronic healthcare records and the right to access the content based on need, but the patient owns the access and the content of those records.
   B. The patient owns both the physical or electronic healthcare records and the content in those records.
   C. The healthcare facility owns the content of the records but not the physical or electronic healthcare records.
   D. The patient owns the content of those records and, therefore, has the right to delete or modify them.

   Explanation: The healthcare facility owns both the physical and electronic healthcare records and has the right to access the content of those records based on the need for healthcare. However, the patient has the right to access those records' content and obtain a copy of the information contained in those records under HIPAA law. The patient cannot change the information in those records, which is the legal property of the healthcare facility. The other options are incorrect because the patient does not own the physical or electronic healthcare records; rather, the institution does, and the patient and the institution cannot delete or freely change the information in those records; only added emendations can be made. *Answer*: A

2. Mr. Hayden Garcia is a 55-year-old accountant diagnosed with hypertension and type 2 diabetes. He requests that his practitioner change his healthcare records to reflect that his hypertension and diabetes were pre-existing conditions before he started visiting this healthcare facility. What is the appropriate response of the practitioner?
   A. The practitioner must change the healthcare record as requested by the patient.
   B. The practitioner must explain to the patient that the healthcare record cannot be changed; only added emendations can be made.

C. The practitioner must change the healthcare record to reflect the patient's request but with a note specifying that the change was made at the patient's request.
D. The practitioner must inform the patient that changing the healthcare record would be a violation of the law and bioethics.

Explanation: Under the HIPAA privacy rule, patients can request added emendations to their healthcare records if the information is incorrect or inaccurate. However, the practitioner can only make changes if the information is incorrect. In this case, the practitioner should explain to the patient that the information in the healthcare record cannot be changed. Changing the healthcare record at the patient's request would violate the law and bioethics because healthcare records are legal documents and must be accurate and complete. The other options are incorrect because changing the healthcare record to reflect the patient's request or violating the law and bioethics would not be appropriate responses. *Answer*: B

3. Ms. Violet Sullivan is a 50-year-old teacher who has been seeing a doctor for the past year to manage her high blood pressure. After missing several appointments, she receives a bill for unpaid healthcare bills, including charges for the missed appointments. Ms. Sullivan contacted her doctor's office to request a copy of her healthcare records but was told that the office could not provide her with a copy due to her unpaid healthcare bills. Is it permissible to deny Ms. Sullivan a copy of her healthcare records due to unpaid healthcare bills?
    A. Yes, it is permissible to deny Ms. Sullivan a copy of her healthcare records because of her unpaid healthcare bills.
    B. No, it is not permissible to deny Ms. Sullivan a copy of her healthcare records because of her unpaid healthcare bills.
    C. Yes, it is permissible to deny Ms. Sullivan a copy of her healthcare records until her unpaid healthcare bills are paid in full.
    D. Yes, it is permissible to deny Ms. Sullivan a copy of her healthcare records because the office owns access to the healthcare records.

Explanation: Patients have the legal right to access and copy their healthcare records, even without providing a reason. Therefore, denying Ms. Sullivan a copy of her healthcare records is not permissible due to her unpaid healthcare bills. Furthermore, refusing to provide a copy of healthcare records due to nonpayment of healthcare bills violates the principle of patient autonomy, which mandates patients' informed consent to disclose confidential information, including healthcare record information. In addition, withholding healthcare record information due to unpaid healthcare bills would violate the HIPAA federal statute. All other options are incorrect because denying patients access to their healthcare records is not permissible. *Answer*: B

4. Mr. Alex Cooper, a 68-year-old retired maintenance director, was hospitalized with chest pain. After several tests and a consultation with a cardiologist, Mr. Cooper is diagnosed with a heart attack and requires further treatment. During his hospital stay, Mr. Cooper's family members, who are not directly involved in his care, call the hospital and ask about his condition. Due to patient confidentiality laws, the hospital staff informs them that Mr. Cooper is admitted and receiving treatment but cannot provide specific healthcare information. However, Mr. Cooper has told the practitioner that he does not want his family members to know about his healthcare condition. What should the practitioner do in this situation?
    A. The practitioner should provide Mr. Cooper's family members with specific healthcare information about his condition because they are concerned about his well-being.
    B. The practitioner should inform Mr. Cooper of his family members' concerns and allow him to decide whether to share his healthcare information with them.
    C. The practitioner should ignore Mr. Cooper's wishes and provide his family members with specific healthcare information about his condition.

D. The practitioner should inform Mr. Cooper's family members that they need to talk directly with Mr. Cooper and make him disclose his condition.

Explanation: In this situation, the practitioner should respect Mr. Cooper's right to healthcare privacy and confidentiality, as outlined in HIPAA laws. The ethical principle of autonomy requires that healthcare information only be provided to the patient and the patient's interprofessional healthcare team unless the patient has explicitly given the authority through consent that others, such as family members, should also receive the healthcare record information. Therefore, the practitioner should inform Mr. Cooper of his family members' concerns and allow him to decide whether to share his healthcare information with them. This approach upholds Mr. Cooper's right to privacy and autonomy and addresses his family members' concerns. It is important to note that although the hospital staff cannot provide specific healthcare information without explicit consent, they can acknowledge that the patient is receiving treatment and is admitted to the hospital. *Answer*: B

5. Mr. Steven Williams is a 42-year-old construction worker who comes to the clinic with chest pain, shortness of breath, and fatigue symptoms. After taking a detailed history and performing a physical examination, the differential diagnosis includes a possible myocardial infarction, pneumonia, or pulmonary embolism. As the practitioner, you explain to Mr. Williams the importance of conducting diagnostic tests and procedures to rule out potential health problems, and you recommend hospitalization for further workup. Mr. Williams requests access and a copy of his healthcare records. What ethical obligation do you have as a practitioner regarding Mr. Williams' healthcare records?
   A. Mr. Williams has no legal right to access his healthcare records because the healthcare facility owns them. Therefore, the healthcare records should not be released to Mr. Williams without the explicit consent of the practitioner.
   B. Mr. Williams has the legal right to access and copy his healthcare records, which belong to him. The healthcare records should be released to Mr. Williams without delay or justification, and confidentiality and privacy protocols must be maintained.
   C. Mr. Williams can only access his healthcare records with a court order. Therefore, the healthcare records should not be released to Mr. Williams without the explicit consent of the practitioner or a court order.
   D. Mr. Williams can only access his healthcare records with the explicit consent of the practitioner. Therefore, the healthcare records should be released to Mr. Williams only after getting his consent.

Explanation: Mr. Williams has the legal right to access and copy his healthcare records, which belong to him, as per HIPAA law. The practitioner has an ethical and legal obligation to maintain confidentiality and follow privacy protocols. The patient's right to access and copy their healthcare records is a fundamental patient right established by society and the healthcare profession. The patient's healthcare record information is the patient's personal property, and violating confidentiality and privacy rules violates the patient–practitioner relationship in which PHI is to be kept confidential and private. The practitioner can only disclose the minimal amount of requested information and only to those given the authority to receive that information if there is a specific court order for that disclosure. It is not permissible to withhold healthcare record information from the patient for any reason, including nonpayment of bills. The other options are incorrect because denying patient access to the healthcare records is a violation of the patient's legal right to access, a court order is not required for patient access, and it is not true that the practitioner's consent is required for a patient to access their healthcare records. *Answer*: B

REFLECTION VIGNETTES

1. Ms. Olivia Smith, a 45-year-old accountant, presents to the healthcare office requesting a copy of her healthcare records. Upon reviewing her account, the practitioner noticed that Ms. Smith has an outstanding balance of several thousand dollars for previous healthcare services rendered. The practitioner informs Ms. Smith that the practice does not release healthcare records until the patient's outstanding balance has been resolved. The practitioner advises Ms. Smith to set up a payment plan to pay off the balance and receive her healthcare records once the plan has been agreed upon.

   What is the ethical issue(s)?

   How should the ethical issue(s) be addressed?

2. Mr. Ben Whitmore, a 42-year-old businessman, comes into the clinic and demands that all physical and electronic healthcare records related to him be destroyed and erased.

   What is the ethical issue(s)?

   How should the ethical issue(s) be addressed?

* * *

WEB LINKS

1. *Code of Medical Ethics*. American Medical Association. https://code-medical-ethics.ama-assn.org
   The AMA *Code of Medical Ethics*, established by the American Medical Association, is a comprehensive guide for healthcare practitioners. It addresses issues and challenges, promotes adherence to standards of care, and is continuously updated to reflect contemporary practices and challenges.
2. "The Anatomy of Electronic Patient Record Ethics: A Framework to Guide Design, Development, Implementation, and Use." Tim Jacquemard, Colin P. Doherty, and Mary B. Fitzsimons. 2021. https://bmcmedethics.biomedcentral.com/articles/10.1186/s12910-021-00574-x
   This source discusses an EPR as a digital repository used to collect, store, and display information regarding an individual's healthcare history.
3. "Examination and Diagnosis of Electronic Patient Records and Their Associated Ethics: A Scoping Literature Review." Tim Jacquemard, Colin P. Doherty, and Mary B. Fitzsimons. 2020. https://bmcmedethics.biomedcentral.com/articles/10.1186/s12910-020-00514-1
   This source discusses how EPR technology is a key enabler for improvements to healthcare service and management.
4. "Ethical Issues in Biomedical Research Using Electronic Health Records: A Systematic Review." Jan Piasecki, Ewa Walkiewicz-Żarek, Justyna Figas-Skrzypulec, Anna Kordecka, and Vilius Dranseika. 2021. https://pubmed.ncbi.nlm.nih.gov/34146228
   This source discusses how digitization of a health record changes its accessibility and how multiple authorized users can access EHRs.
5. "Ethical Implications of the Electronic Health Record: In the Service of the Patient." Lois Snyder Sulmasy, Ana Maria López, Carrie A. Horwitch, and American College of Physicians Ethics, Professionalism and Human Rights Committee. 2017. https://pubmed.ncbi.nlm.nih.gov/28321550
   This source discusses the ethical implications of EHRs.
6. "Confidentiality & Electronic Medical Records." American Medical Association. https://code-medical-ethics.ama-assn.org/ethics-opinions/confidentiality-electronic-medical-records
   This source discusses how information gathered and recorded in association with the care of a patient is confidential, regardless of the form in which it is collected or stored.

7. "Ethical Issues in Biomedical Research Using Electronic Health Records: A Systematic Review." Jan Piasecki, Ewa Walkiewicz-Żarek, Justyna Figas-Skrzypulec, Anna Kordecka, and Vilius Dranseika. 2021. https://link.springer.com/article/10.1007/s11019-021-10031-6
This source discusses the ethical issues concerning research using EHRs in the literature.
8. "Privacy, Confidentiality & Medical Records." American Medical Association. https://code-medical-ethics.ama-assn.org/chapters/privacy-confidentiality-medical-records
This source discusses patient privacy, including personal space, personal data, personal choices, and personal relationships.

MINOR PATIENTS

36

It is easier to build strong children than to repair broken men.
—Frederick Douglass

* * *

MYSTERY STORY

CONFLICTING PRINCIPLES: STRUGGLING BETWEEN PARENTAL RIGHTS AND PATIENT AUTONOMY IN THE HEART OF A PEDIATRIC MYSTERY

In the labyrinthine corridors of Little Stars Pediatric Hospital, the morning seemed uneventful until Practitioner Samantha Davis encountered Emily Kilpatrick, a 16-year-old girl stricken with excruciating abdominal pain. Accompanied by her imposing mother, Mrs. Kilpatrick, Emily's anxious eyes hinted at something deeper than the evident discomfort.

Practitioner Davis, an astute and compassionate healthcare professional, sensed Emily's hesitation as she struggled to answer even the simplest of questions. The girl's reluctance to be examined was palpable. An air of mystery cloaked the room, thickening as Practitioner Davis tried to obtain information from Emily's mother.

Mrs. Kilpatrick, a stern and dominant figure, guarded her daughter's healthcare history like a treasure chest of secrets. "I know everything there is to know about my Emily's health," she snapped, her eyes narrowing with a mix of protectiveness and suspicion.

Days turned into a nightmare as Emily's condition deteriorated alarmingly. Her once vibrant eyes became clouded with confusion and pain, and she eventually became unresponsive. Practitioner Davis' concern turned into a desperate plea for action. Her instincts told her that Emily's life was slipping through her fingers like sand.

The young girl's illness remained enigmatic, but the dark shadow of severe infection was looming. Practitioner Davis' experienced mind mapped the possibilities, but Mrs. Kilpatrick's obstructionism turned a healthcare emergency into a battlefield of principles. The mother's refusal to consent to any treatment, justified by her own inflated sense of wisdom, threatened to suffocate Emily's chances of survival.

With the clock ticking and the ethics of healthcare practice under strain, Practitioner Davis reached out to the Little Star's moral compass—the ethics committee. The committee's analysis was clear: Child Protective Services must be called. The hand of suspected healthcare neglect was now in play.

A whirlwind of an investigation unraveled Emily's hidden agony. The girl had harbored a secret too heavy for her young shoulders, fearing her mother's wrath more than the invisible tormentor wracking her body. Emily's sexual activity, her mother's anticipated fury, and the resultant ailment had collided to form a perilous storm.

Mrs. Kilpatrick's hubris gave way to the harsh reality of healthcare neglect charges, and Emily's fragile life was finally pulled back from the brink by the urgently required treatment.

Through this harrowing experience at Little Stars Pediatric Hospital, Practitioner Davis' resolve was fortified. Her understanding of the fragile balance between parental authority and a minor patient's autonomy (informed consent) evolved into a steadfast belief. She had tasted the bitter complexities of human emotions and healthcare ethics, learning that in the labyrinth of human suffering, the compass must always point toward the best interests of the patient, no matter how young, vulnerable, or entangled in familial strife.

* * *

THINK

The legal definition of a minor is an individual younger than age 18 years who is deemed not legally competent to make treatment decisions independently. In this context, only a parent or legal guardian possesses the authority to consent, authorizing a practitioner to provide healthcare treatment. This unique scenario engenders a three-way relationship involving the minor patient, the parents (or legal guardian), and the practitioner. Within this relationship, each party assumes distinct roles and degrees of authority, unified by the common goal of maximizing the child's best interests.

ASSESS

Patient: Autonomy

Medicine prioritizes patient-centered care, and consequently, minor patients capable of assent should be substantially informed about their diagnosis, prognosis, and treatment options tailored to their cognitive and emotional maturity. Parents and practitioners share an ethical duty to foster the minor's autonomous decision-making skills by creating a supportive environment, simplifying complex treatment regimens, and obtaining informed assent from the minor patient. In this context, "assent" denotes an agreement from someone not legally authorized to provide full consent.

Within the three-way relationship involving minor patients, parents must recognize that their decisional authority over healthcare treatment, while acknowledged, is constrained to the child's best interests. Similar to how parents lack the authority to abuse or neglect a child, they cannot demand harmful treatments or reject standard care if it significantly increases the risk of harm or death to the minor. In such instances, the state's interest in protecting vulnerable populations takes precedence, compelling the practitioner to administer the necessary treatment despite parental objections. As a result, parental consent is not legally required for providing a minor with contraception; prenatal care; or treatment for substance abuse, HIV, or sexually transmitted infections (STIs).

The minor's patient–practitioner relationship concerning confidential and private information carries more weight than the parent's wish to be informed and to control healthcare treatment in the circumstances listed in Table 36.1.

Table 36.1 Circumstances in Which Minors Do Not Require Parental Consent

Healthcare	Emergency care
	Sexually transmitted infections
	Contraception
	Substance abuse (most states)
	Prenatal care
Emancipated minor	Homeless
	Parent
	Married
	Military
	Financially independent

When a minor patient requests that their protected health information be kept confidential from their parents, the practitioner is professionally, ethically, and legally obligated to honor this request. Many states have enacted "mature minor" legislation, enabling minors to make treatment decisions if they demonstrate a sufficient understanding of the decision itself and its potential consequences.

Practitioner: Beneficence and Nonmaleficence

Guided by the principle of beneficence (do good), if a minor patient seeks to keep their treatment confidential, the practitioner must professionally inquire why the minor patient does not want their parents informed. The practitioner should promote patient–parent communication and offer assistance in facilitating open dialogue. If the minor continues to object to parental disclosure, the practitioner must ensure the minor is well-informed about available social resources and how to access them.

Adhering to the principle of nonmaleficence (do no harm), the practitioner must notify the patient about information that will be disclosed to the parents through the insurance company's healthcare billing. This includes details such as the practitioner's identity, lab tests performed, and results. Publicly funded social services, which can bypass insurance billing, offer a better option for minor patients wishing to keep aspects of their healthcare private from their parents, thereby protecting their confidentiality.

Public Policy: Justice

When irreconcilable differences arise between what the practitioner determines to be in the minor patient's best interests and the patient's or parents' views, institutional policies and procedures provide for consultation with an ethics committee. It is crucial to understand that although these committees excel at conflict resolution, precise language, and fostering respectful dialogue, an ethics consultation will not yield a healthcare decision but only a healthcare recommendation. The responsibility for making treatment decisions remains with the parties involved and is not usurped by agreeing to an ethics consultation.

Societal laws have been enacted to safeguard vulnerable populations, including children, the elderly, and the intellectually handicapped, who depend on others for their healthcare and basic needs. Healthcare neglect of a minor falls under child neglect, governed by child abuse laws. Such neglect often occurs when caregivers either fail to seek necessary healthcare treatment or do not follow the healthcare advice given by the practitioner. In instances of child neglect, the legal policies of social justice to protect vulnerable populations hold greater moral significance than the parent's autonomy (informed consent) authority, prioritizing justice (be fair) over individual decision-making.

CONCLUDE

Practitioners are generally required to obtain a parent's consent before administering healthcare treatment to a minor. However, in emergencies when there is no time or opportunity to secure consent, the practitioner is legally, professionally, and ethically obligated to provide necessary treatment. If a minor patient wishes to keep information confidential from their parents, the practitioner should inquire why and encourage the minor to engage in open dialogue with their parents. The minor patient's best interests limit parental consent, and the practitioner must remain vigilant in maximizing those interests.

In summary, practitioners navigate unique challenges when treating minor patients, as they must deftly balance the rights and responsibilities of the patient, parents, and themselves. Although parental consent is typically required for healthcare treatment, exceptions exist for specific circumstances. Practitioners must always place the best interests of the minor patient at the forefront, offer suitable support, and foster open communication among all involved parties. This approach ensures optimal care and protection for minors within the healthcare system.

* * *

REVIEW QUESTIONS

1. With minor patients, there is a three-way relationship involving the patient, parents, and practitioner.
    A. True
    B. False

2. Minor patients capable of assent should be kept substantially informed of their diagnosis, prognosis, and treatment options according to their cognitive and emotional maturity.
   A. True
   B. False

3. Minor patients are a vulnerable population, meaning that neither the patient, parent(s), nor practitioner has the authority to deny standards of care if doing so will significantly increase the risk of harm to the patient.
   A. True
   B. False

4. The minor patient has no right to keep health information private from the patient's parents.
   A. True
   B. False

5. If a minor patient wishes to keep their treatment confidential, then the practitioner has a professional responsibility to inquire why the minor patient does not want their parents to be informed. The practitioner should encourage patient–parent discussion and offer to help facilitate an open dialogue with the patient's parents.
   A. True
   B. False

6. Publicly funded social services can bypass the need for insurance billing and therefore are much better options for a minor patient who wishes to keep aspects of their healthcare services private from parents.
   A. True
   B. False

7. Parents always have the authority to make treatment decisions for their children. This is an example of how the principle of autonomy always has more moral weight than the principles of beneficence (do good), nonmaleficence (do no harm), and justice (be fair).
   A. True
   B. False

Answers: 1A, 2A, 3A, 4B, 5A, 6A, 7B

* * *

CLINICAL VIGNETTES
1. Ms. Elizabeth Johnson is a 16-year-old high school student who has been referred to a pediatrician for evaluation of acute abdominal pain and vaginal bleeding. She reports that she is sexually active and has been using condoms inconsistently. She is concerned that her parents will learn about her sexual activity if she consents to the treatment. Her parents are present in the room and insist on being informed of all treatment decisions. The differential diagnosis includes pelvic inflammatory disease, ectopic pregnancy, or a urinary tract infection. Which of the following actions should the pediatrician take?
   A. The pediatrician should inform the parents of the healthcare condition, diagnosis, and treatment plan.
   B. The pediatrician should inform Ms. Johnson that her parents must be informed of the healthcare condition, diagnosis, and treatment plan.
   C. If the parents insist on being informed, then the pediatrician should inform the parents.
   D. The pediatrician should inform Ms. Johnson of available social resources for services that will not show up on the parents' insurance if she wishes to keep her healthcare

information confidential from her parents, and at the same time, the pediatrician should encourage Ms. Johnson to open a dialogue with her parents and offer to facilitate the discussion.

Explanation: Ms. Johnson is a minor patient, so her parents have decisional authority over her healthcare treatment. However, if she wishes to keep certain healthcare information confidential from her parents, the pediatrician must respect her autonomy and right to privacy. In this case, the pediatrician should inquire why Ms. Johnson does not want her parents to be informed and then inform her of available social resources for services that will not appear on the parent's insurance. At the same time, the pediatrician should encourage Ms. Johnson to open a dialogue with her parents and offer to facilitate a discussion with her parents. The pediatrician is responsible for ensuring that Ms. Johnson is informed about what information will be disclosed to her parents through their insurance company's healthcare billing. The other options disregard the importance of Ms. Johnson's autonomy and right to privacy, which are protected by ethical and legal principles. The pediatrician must encourage open communication between Ms. Johnson and her parents, but ultimately, the decision to inform the parents lies with Ms. Johnson. *Answer*: D

2. Ms. Destiny Roberts is a 16-year-old high school student recently diagnosed with an STI. She comes to her primary care provider for further information and treatment. She is mature and emotionally stable, but she requests that her healthcare information be kept confidential from her parents. The ethical question is, What should the practitioner do in this situation?
   A. The practitioner should inform Ms. Roberts' parents of her STI diagnosis and treatment despite her request for confidentiality.
   B. The practitioner should inform Ms. Roberts of the potential benefits of disclosing the STI diagnosis to her parents but ultimately respect her autonomy and keep her healthcare information confidential.
   C. The practitioner should inform Ms. Roberts' parents of the STI diagnosis but not disclose the treatment to find a middle ground between the patient's autonomy and parental rights.
   D. The practitioner should inform Ms. Roberts that her parents have the right to know about her STI diagnosis and treatment and that it is in her best interest to involve her parents in her healthcare.

   Explanation: The practitioner's professional and ethical duty is to respect Ms. Roberts' request for confidentiality. The practitioner must explain the potential benefits of informing her parents of the STI diagnosis. Still, ultimately, it is Ms. Roberts' autonomy to decide whether or not to inform her parents. Based on the principle of beneficence, the practitioner should inquire why Ms. Roberts does not want her parents informed and should offer to facilitate open dialogue between Ms. Roberts and her parents. It is important to remember that healthcare billing through insurance will disclose information to the parents, so the practitioner must educate Ms. Roberts on available social resources for services that will not appear on the parent's insurance. Disclosing the information to the parents without Ms. Roberts' consent would violate her privacy and autonomy. Although the parents have decisional authority, it is limited to the minor's best interests and overridden in healthcare neglect. *Answer*: B

3. Mr. Kurt Thomas, a 16-year-old high school student, is brought in by his parents for chest pain that he has been experiencing for the past 2 days. Mr. Thomas is a competitive athlete pushing himself hard to prepare for an upcoming competition. He is otherwise healthy and has no history of chest pain. The differential diagnosis includes musculoskeletal, cardiac, or anxiety-related chest pain. The practitioner orders an electrocardiogram, which reveals evidence of myocardial infarction—a heart attack. The practitioner informs Mr. Thomas and his parents of the standards of care: Immediate transfer to a specialized hospital for cardiac

catheterization. Mr. Thomas and his parents are hesitant to proceed with the treatment and suggest that he rest for a few days instead. What should the practitioner do in this situation, in which a minor patient is at risk of significant harm, but the patient and parents are hesitant to proceed with the standards of care?

　A. Respect the decision of Mr. Thomas and his parents, even if it may result in harm to the patient
　B. Attempt to persuade Mr. Thomas and his parents to follow the standards of care but ultimately respect their decision
　C. Seek ethics consultation to resolve the conflict between the practitioner's duty to provide a standard of care and the patient's and parent's hesitation
　D. Proceed with the standards of care, seeking court intervention if necessary, to protect the minor patient's best interests

Explanation: The minor patient is a vulnerable population, and the standards of care must be provided to protect the patient's best interests. If Mr. Thomas and his parents are hesitant to proceed with the standards of care, the practitioner must attempt to persuade them to follow the treatment plan. However, if the minor patient's and parent's reluctance puts the minor patient at risk of significant harm, then the practitioner must proceed with the standards of care, seeking court intervention if necessary to protect the patient's best interests. Seeking ethics consultation is an available option, but it is not the immediate action when the patient is at risk of significant harm in an emergency. The other options are incorrect because the practitioner has a duty to protect the patient's best interests; even if the patient or parents refuse the standards of care, the practitioner cannot simply respect the decision of the patient and parents if it may result in significant harm to the minor patient. There is not enough time to seek an ethics consultation that can only provide a recommendation.
Answer: D

4. Ms. Wendy Rodriguez is a 16-year-old high school student experiencing symptoms of depression, anxiety, and insomnia. Her parents have noticed a change in her behavior and have taken her to see a practitioner for treatment. Ms. Rodriguez has expressed that she does not feel comfortable disclosing all of her symptoms to her parents and wishes to keep certain aspects of her healthcare private. The practitioner informs Ms. Rodriguez that publicly funded social services can bypass the need for insurance billing and are therefore much better options for a minor patient who wishes to keep aspects of their healthcare services private from their parents. What is the ethical recommendation for the practitioner in this situation?
　A. The practitioner should inform Ms. Rodriguez's parents of all her symptoms and disclose all aspects of her healthcare to them, even if she objects.
　B. The practitioner should tell Ms. Rodriguez that keeping any aspect of her healthcare private from her parents is impossible and encourage her to discuss everything with them.
　C. The practitioner should inform Ms. Rodriguez about publicly funded social services that can keep her healthcare private from her parents and encourage her to seek treatment through those services.
　D. The practitioner should ignore Ms. Rodriguez's wish for privacy and treat her according to her parent's wishes.

Explanation: In this scenario, the practitioner has an ethical responsibility to respect Ms. Rodriguez's wish for privacy and inform her about publicly funded social services that can bypass the need for insurance billing and keep her healthcare private from her parents. The practitioner should not disclose Ms. Rodriguez's confidential and private information to her parents without her consent. In addition, the practitioner should not ignore Ms. Rodriguez's wish for privacy and treat her according to her parent's wishes. The practitioner must maximize Ms. Rodriguez's best interests and encourage her to seek treatment that will not appear on her parent's insurance. *Answer*: C

5. Ms. Darian Taylor, a 16-year-old high school student, presents to the clinic seeking advice on contraception. She is sexually active and would like to start taking birth control pills. She explains that she does not want her parents to know about her sexual activity or her desire for contraception. Her clinical examination is unremarkable, and she has no healthcare history that would preclude her from using birth control pills. What is the most appropriate course of action for the practitioner?
   A. Refuse to prescribe birth control pills without parental consent
   B. Prescribe birth control pills but inform the patient's parents
   C. Prescribe birth control pills and keep the patient's information confidential
   D. Refer the patient to another practitioner who will prescribe birth control pills without informing the patient's parents

Explanation: The practitioner is responsible for keeping the patient's information confidential if the patient requests it. In this case, the patient has expressed a desire for confidentiality and has demonstrated an understanding of the nature of the decision and its consequences. The practitioner should provide the patient with information about the available social resources for services that will not appear on the parent's insurance. The practitioner should also inform the patient about what information will be disclosed to the parents through their insurance company's healthcare billing and suggest publicly funded social services that bypass the need for insurance billing. Refusing to prescribe birth control pills or informing the patient's parents against the patient's wishes would violate the principle of patient autonomy and the professional principles of beneficence and nonmaleficence. Referring the patient to another practitioner who will prescribe birth control pills without informing the patient's parents is unnecessary because the practitioner can fulfill the patient's request while respecting her confidentiality. *Answer*: C

REFLECTION VIGNETTES

1. Practitioner Jane Smith, a pediatrician, meets with a minor patient and their parent for an annual checkup. During the private conversation with the patient, the minor reveals that they have been sexually active and think that they have a venereal disease of some type. The minor patient vehemently opposes having either parent informed of their sexual activity or lab tests.

   What is the ethical issue(s)?

   How should the ethical issue(s) be addressed?

2. Practitioner Maria Rodriguez, a pediatrician, is on duty in the emergency department when a minor patient is brought in and requires immediate surgery. Despite explaining the healthcare necessity of the surgery, both parents and the minor child refuse consent for the procedure.

   What is the ethical issue(s)?

   How should the ethical issue(s) be addressed?

* * *

WEB LINKS

1. *Code of Medical Ethics.* American Medical Association. https://code-medical-ethics.ama-assn.org
   The AMA *Code of Medical Ethics*, established by the American Medical Association, is a comprehensive guide for healthcare practitioners. It addresses issues and challenges, promotes adherence to standards of care, and is continuously updated to reflect contemporary practices and challenges.

2. "Principles of Clinical Ethics and Their Application to Practice." Basil Varkey. 2021. https://www.ncbi.nlm.nih.gov/pmc/articles/PMC7923912
   This source discusses the inherent and inseparable part of clinical medicine as the physician has an ethical obligation to benefit the patient, avoid or minimize harm, and respect the values and preferences of the patient.
3. "Patient Rights and Ethics." Jacob P. Olejarczyk and Michael Young. https://www.ncbi.nlm.nih.gov/books/NBK538279
   This source discusses the 1973 American Hospital Association patient bill of rights, the first bill of rights written specifically for patients.
4. "Involving Minors in Medical Decision Making: Understanding Ethical Issues in Assent and Refusal of Care by Minors." Abigail Lang and Erin Talati Paquette. 2018. https://pubmed.ncbi.nlm.nih.gov/30321891
   This source discusses the complexity of the clinician–patient relationship when caring for minors and how children become more capable of participating in the decision-making process as they age.
5. "Protecting Adolescent Patient Privacy: Four Key Questions." Richelle Marting. 2019. https://www.aafp.org/pubs/fpm/issues/2019/0100/p7.html
   This source discusses physicians' responsibility to engage minor patients in the decision-making process at a developmentally appropriate level.
6. "Treating Minors." University of Missouri School of Medicine, Center for Health Ethics. https://medicine.missouri.edu/centers-institutes-labs/health-ethics/faq/treating-minors
   This source discusses how healthcare providers are normally expected to keep patient information confidential and obtain informed consent from patients before treating them.
7. "Confidential Health Care for Minors." American Medical Association. https://code-medical-ethics.ama-assn.org/ethics-opinions/confidential-health-care-minors
   This source discusses that physicians who treat minors have an ethical duty to promote the developing autonomy of minor patients by involving children in making decisions about their healthcare to a degree commensurate with children's abilities.
8. "Minors' Rights in Medical Decision Making." Kathryn Hickey. 2007. https://pubmed.ncbi.nlm.nih.gov/17728582
   This source discusses how in the past, minors were not considered legally capable of making treatment decisions and were viewed as incompetent because of their age.

NURSES AND ALLIED HEALTH PROFESSIONALS

37

Life is short, and the art of healing long; the occasion fleeting; experience fallacious, and judgment difficult. The practitioner must be prepared to do what is right and help the patient, attendants, and externals work together.
—Hippocrates

* * *

MYSTERY STORY

DEADLY AUTHORITY: THE UNRAVELING OF A HEALTHCARE TYRANT

The hospital corridors of Unity Healer's Hospital hummed with the familiar cacophony of beeping monitors and hurried footsteps. But beneath the ordinary bustle lurked a sinister secret, one that would shatter the trust within the walls of the healthcare facility and expose a dark underbelly of abuse and manipulation.

Practitioner William Thatcher, a once-revered healthcare professional, was the primary care provider for Mr. Johnson, who was bravely fighting the aftermath of a heart attack. But Nurse Samantha Davis, an astute and observant member of Mr. Johnson's care team, noticed something amiss. The medication that Thatcher had been administering was not prescribed for Mr. Johnson. A chill ran down her spine as she pondered the implications, her mind racing with terrifying possibilities.

United by concern, the team, including pharmacists and various healthcare professionals, confronted Thatcher. He responded with arrogance and cold defiance, claiming his authority overrode their collective healthcare judgment. Despite their persistent objections and growing fears, Thatcher dismissed them, leaving behind a trail of doubt, confusion, and suspicion.

As the team grappled with Thatcher's actions, they were acutely aware that he had willfully disregarded two fundamental principles of healthcare ethics: beneficence (do good) and nonmaleficence (do no harm). Instead of seeking to do good for his patients and avoid harm, Thatcher's self-assured authority had led him down a dangerous path, one that would ultimately claim lives.

The next morning, the unthinkable happened. Mr. Johnson was found lifeless, a victim of the very medication that had been the subject of heated debate. The hospital plunged into a state of shock and disbelief, the horror of the event reverberating through every hallway.

As the investigation peeled back the layers of Thatcher's healthcare authoritarianism, a pattern of betrayal emerged. Thatcher had become a healthcare tyrant, manipulating his authority to control patient care and disregarding the voices of his colleagues. His reign had left not just Mr. Johnson but also many others in its devastating wake, a grim testament to the fragility of trust.

The road to justice was a grim one, leading to Thatcher's arrest and conviction for murder and malpractice. The entire ordeal served as a searing reminder of the fragile line between authority and arrogance, and the essential human values that must guide the healthcare profession.

In the shadow of tragedy, the hospital staff at Unity Healer's Hospital rallied together. They vowed never to let the dark specter of unilateral decision-making overshadow the core principles of patient care, including the ethics of beneficence and nonmaleficence.

> Their renewed commitment to collaboration, communication, and the relentless pursuit of their patients' best interests became a beacon of hope, forging a path toward healing and redemption.
>
> The story of Practitioner Thatcher and Mr. Johnson is a stark warning of the deadly consequences of unchecked power within the healthcare system. It underscores the importance of interprofessional respect, the humility to recognize collective wisdom, and the unbreakable bond of trust that must exist among those entrusted with the sacred duty of healing. It is a tale that lingers, a somber reminder of what can go awry when power is left unchecked and ethical principles are cast aside.

* * *

THINK

In the complex landscape of modern healthcare, physicians are often viewed as the primary bearers of legal, professional, and ethical responsibility for patient care. However, nurses and allied health professionals are crucial in delivering personal care, administering medication, and monitoring patients. Traditionally, physicians have been responsible for diagnosing, prognosis, and offering treatment options to patients. This relationship often involves intimate patient interactions, including divulging private histories and performing physical examinations. However, as healthcare has evolved, nurses and allied practitioners have increasingly become key contact points for patients. The following sections explore the collaborative approach required to optimize patient care and the vital principles that guide interprofessional healthcare delivery.

ASSESS

Patient: Autonomy

The practitioner is responsible for diagnosis, prognosis, and presenting treatment options to the patient. The patient, in turn, provides informed consent authorizing the practitioner to provide the healthcare treatment. When the patient provides informed consent, the patient also provides general consent for institutional participation in insurance and billing and interprofessional participation with patient management, including nurses, pharmacists, and a consortium of other interprofessional allied health professionals.

Practitioner: Beneficence and Nonmaleficence

Physicians, nurses, and other interprofessional health professionals all share the same professional prime directive of maximizing the patient's best interests in accordance with the patient's reasonable goals, values, and priorities. This prime directive exemplifies the shared professional principles of beneficence (do good) and nonmaleficence (do no harm) and is the cornerstone of what constitutes the health professionals to be allied.

In this context of the patient's best interests, everyone in this interprofessional team must do the following:

1. Communicate clearly and professionally with all interprofessional team members about patient management.
2. Understand that the professional principle of nonmaleficence (do no harm) is paramount in patient care. Although nobody but the primary practitioner has the legal, professional, and moral authority to modify or change patient management, if an interprofessional team member has nonmaleficence (do no harm) concerns, then those concerns should be immediately discussed with the practitioner and resolved.

3. Provide emergency care when there is an immediate and severe risk of harm or death for the patient as a matter of beneficence (do good). Interprofessional team members have the authority and obligation to provide the standards of care unless a specific patient documented order in the healthcare record or an available living will communicate otherwise.
4. Resolve disagreements among the interprofessional team to maximize the patient's best interests as determined by the patient's reasonable goals, values, and priorities.
5. Avoid discussing with the patient disagreements about patient management. Discussing disagreements undermines the patient–practitioner relationship, maligns the practitioner's reputation, and negatively impacts patient healing.

Physicians have a professional responsibility to support and promote interprofessional healthcare delivery and the team approach to optimize patient care and outcomes in a complex healthcare system.

Public Policy: Justice
The government and healthcare institutions have implemented laws and policies to ensure that all nurses and other allied health professionals are adequately educated, trained, and competent for performing their roles as interprofessional healthcare team members.

CONCLUDE
The practitioner's professional responsibility is to communicate respectfully with the interprofessional team and take responsibility for addressing and resolving any patient management disagreements, conflicts, and concerns. The practitioner is primarily responsible for ensuring patient management is performed competently and professionally.

In summary, the practitioner bears the primary responsibility for patient care, but nurses and allied health professionals are vital interprofessional team members. They all share the same goal of maximizing the patient's best interests while minimizing harm. To achieve this goal, clear communication and cooperation are necessary among all team members. Practitioners must address and resolve any disagreements or concerns within the team, ensuring that patient management is conducted competently and professionally. This collaborative approach ultimately benefits patients and contributes to better healthcare outcomes.

* * *

REVIEW QUESTIONS
1. Physicians do not have the primary legal, professional, and ethical responsibility for the patient's healthcare.
    A. True
    B. False

2. The interprofessional team equally shares the responsibility for determining patient management.
    A. True
    B. False

3. Nurses and allied practitioners do much of the personal contact, medication administration, and patient monitoring.
    A. True
    B. False

4. When the patient provides informed consent, the patient also provides general consent for institutional participation in insurance and billing and interprofessional participation with patient management.
   A. True
   B. False

5. Physicians are responsible for supporting and promoting interprofessional healthcare delivery and the team approach.
   A. True
   B. False

6. All team members must communicate clearly and professionally about patient management with all interprofessional team members.
   A. True
   B. False

Answers: 1B, 2B, 3A, 4A, 5A, 6A

* * *

CLINICAL VIGNETTES

1. Ms. Jacqueline Torres is a 65-year-old retired librarian who presents with a sudden onset of severe abdominal pain, nausea, and vomiting. On examination, there is tenderness in the epigastric region, and she has a low-grade fever. Clinical differential diagnoses include peptic ulcer disease, acute pancreatitis, or cholecystitis. The interprofessional team comprises a physician, a nurse, and a pharmacist. As an interprofessional team member, what is the ethical responsibility when there is a disagreement about patient management and the physician does not agree with the nurse's concerns?
   A. The nurse should document their concerns and escalate to the healthcare institution's patient advocacy department for resolution.
   B. The nurse should communicate directly with the patient and discuss their concerns about patient management.
   C. The nurse should comply with the physician's management plan and continue to monitor the patient for any adverse effects.
   D. The nurse should refuse to participate in the management plan until their concerns are addressed.

   Explanation: In the context of the patient's best interests, all interprofessional team members should communicate clearly and professionally with each other about patient management. In the scenario, if the nurse has concerns about the patient's management plan, they should immediately discuss those concerns with the physician and resolve them as part of the shared professional principle of nonmaleficence (do no harm). However, if there is a disagreement about patient management, the nurse should document their concerns and escalate them to the healthcare institution's patient advocacy department for resolution. The patient–practitioner relationship should be maintained, and discussing disagreements about patient management with the patient should be avoided. Refusing to participate in the management plan until their concerns are addressed may cause harm to the patient and is not in line with the shared professional principle of beneficence (do good). *Answer*: A

2. Mr. Boris Brown is a 50-year-old construction worker who presents with chest pain and shortness of breath. On examination, he is tachycardic, has crackles in both lung fields, and has low blood pressure. Clinical differential diagnoses include acute myocardial infarction, pulmonary embolism, or septic shock. The interprofessional team comprises a physician, a

nurse, and a respiratory therapist. As an interprofessional team member, what is the ethical responsibility when the respiratory therapist disagrees with the physician's treatment plan?
   A. The respiratory therapist should immediately modify the treatment plan as per their professional judgment.
   B. The respiratory therapist should document their concerns and escalate them to the healthcare institution's patient advocacy department for resolution.
   C. The respiratory therapist should comply with the physician's management plan and continue to monitor the patient for any adverse effects.
   D. The respiratory therapist should communicate directly with the patient and discuss their concerns about patient management.

   Explanation: In the context of the patient's best interests, all interprofessional team members should communicate clearly and professionally with each other about patient management. If an interprofessional team member has concerns about the patient's management plan, those concerns should be immediately discussed with the practitioner and resolved. However, if there is a disagreement about patient management, the team member should document their concerns and escalate them to the healthcare institution's patient advocacy department for resolution. It is inappropriate for the respiratory therapist to modify the treatment plan as per their professional judgment. The shared professional principle of nonmaleficence (do no harm) requires the respiratory therapist to comply with the physician's management plan and continue to monitor the patient for any adverse effects. Discussing disagreements about patient management with the patient undermines the patient–practitioner relationship and negatively impacts patient healing. *Answer*: B

3. Mr. Michael Davis is a 50-year-old electrician recently diagnosed with hypertension. He presents to the clinic for a follow-up appointment with the physician and reports experiencing side effects from the medication prescribed, including dizziness and fatigue. The interprofessional team includes a physician, a nurse, and a pharmacist. As an interprofessional team member, what is the ethical responsibility when a patient reports adverse effects from medication prescribed by the physician?
   A. The nurse should document the patient's concerns and report them to the physician, who will determine the appropriate management plan.
   B. The pharmacist should adjust the dosage of the medication without consulting the physician to alleviate the patient's symptoms.
   C. The nurse should tell the patient to stop taking the medication and suggest an alternative treatment plan.
   D. The pharmacist should prescribe a new medication to the patient without consulting the physician.

   Explanation: In the patient's best interests, interprofessional team members must communicate effectively and provide emergency care, resolve disagreements, and avoid discussing patient management disputes with the patient. When a patient reports adverse effects from medication prescribed by the physician, the nurse should document the patient's concerns and report them to the physician, who will determine the appropriate management plan. The shared professional principle of nonmaleficence (do no harm) requires all team members to take immediate action to address any concerns regarding patient safety. Adjusting the medication dosage or prescribing a new medication without consulting the physician is not in line with the interprofessional team's professional prime directive of maximizing the patient's best interests and avoiding harm. Telling the patient to stop taking the medication without consulting the physician can also cause harm to the patient and is not in line with the shared professional principle of beneficence (do good). Therefore, the ethical responsibility of the nurse is to document the patient's concerns and report them to the physician, who is primarily responsible for ensuring that patient management is performed competently and professionally. *Answer*: A

4. Mr. Edward Wilson is a 50-year-old plumber with a chief complaint of shortness of breath and chest pain. On examination, he is found to have tachycardia, elevated blood pressure, and reduced oxygen saturation. Clinical differential diagnoses include acute myocardial infarction, pulmonary embolism, or pneumothorax. The interprofessional team comprises a physician, a nurse, and a respiratory therapist. As an interprofessional team member, what is the ethical responsibility when there is a disagreement about patient management, and the respiratory therapist disagrees with the physician's management plan?
   A. The respiratory therapist should comply with the physician's management plan and continue to monitor the patient for any adverse effects.
   B. The respiratory therapist should document their concerns and escalate to the healthcare institution's patient advocacy department for resolution.
   C. The respiratory therapist should communicate directly with the patient and discuss their concerns about patient management.
   D. The respiratory therapist should refuse to participate in the management plan until their concerns are addressed.

   Explanation: In the context of the patient's best interests, all interprofessional team members should communicate clearly and professionally with each other about patient management. If the respiratory therapist has concerns about the patient's management plan, then they should immediately discuss those concerns with the physician and resolve them as part of the shared professional principle of nonmaleficence (do no harm). However, if there is a disagreement about patient management, then the respiratory therapist should document their concerns and escalate to the healthcare institution's patient advocacy department for resolution, in line with the principle of beneficence (do good) and ensuring the patient's safety. Although it is important to comply with the physician's management plan, escalating concerns is essential when the therapist genuinely believes the plan may harm the patient. Discussing disagreements about patient management with the patient should be avoided to maintain the patient–practitioner relationship and prevent harm to the patient. Refusing to participate in the management plan until their concerns are addressed may cause harm to the patient and is not in line with the shared professional principle of beneficence (do good). *Answer*: B

5. Mr. Quentin Quintero, a 68-year-old retired teacher, is brought to the emergency room by ambulance after experiencing sudden-onset chest pain and shortness of breath. On arrival, his blood pressure is 150/90 mmHg, his pulse is 110 beats per minute, and he has an oxygen saturation level of 88% on room air. The nurse on duty immediately recognizes the severity of Mr. Quintero's condition and begins administering supplemental oxygen, aspirin, and nitroglycerin. The physician arrives shortly thereafter and orders an electrocardiogram (ECG) and a series of blood tests to assess for cardiac damage. The results of the ECG and blood tests confirm the nurse's suspicions of a heart attack, and the patient is rushed to the cardiac catheterization laboratory for emergent intervention. What is the ethical obligation of the nurse in this scenario?
   A. The nurse should not take any action until the physician arrives to provide orders.
   B. The nurse should begin administering emergency care according to established protocols and policies.
   C. The nurse should contact the patient's family to inform them of his condition.
   D. The nurse should wait for the patient to provide informed consent before administering any treatment.

   Explanation: The nurse's primary ethical obligation in this scenario is to provide emergency care as quickly as possible to minimize harm to the patient. Interprofessional team members have the authority and obligation to provide the standards of care unless a specific patient documented order in the healthcare record or an available living will communicate otherwise. In this case, the nurse recognizes the severity of the patient's condition and immediately begins administering supplemental oxygen, aspirin, and nitroglycerin according to established emergency care protocols. Waiting for the physician to arrive before taking action could result in a delay in care that could increase the risk of harm to the patient. Contacting the patient's family or waiting for the patient to provide informed consent before administering treatment are not immediate priorities in an emergency situation in which prompt action is necessary to maximize the patient's best interests. *Answer*: B

REFLECTION VIGNETTES

1. Practitioner Benjamin Whitmore is a 45-year-old practitioner with years of experience in cardiology. During rounds with the interprofessional team, Practitioner Whitmore presents a treatment plan for a patient with severe heart disease that includes a new medication regimen. However, the team members, including the nurse and the pharmacist, express their disagreement with the plan due to potential interactions with other medications and the patient's healthcare history. They suggest an alternative approach to treatment.

   What is the ethical issue(s)?

   How should the ethical issue(s) be addressed?

2. Practitioner Victoria Harrington, a 35-year-old oncologist, has created a patient treatment plan for a cancer patient with a rare form of cancer. One of the nurse practitioners, who is part of the patient's interprofessional team, refuses to comply with the treatment plan, stating that the treatment is too aggressive and could cause more harm than good. Despite Practitioner Harrington's efforts to explain the rationale for the treatment plan, the nurse practitioner remains resistant and seeks the input of other team members.

   What is the ethical issue(s)?

   How should the ethical issue(s) be addressed?

* * *

WEB LINKS

1. *Code of Medical Ethics*. American Medical Association. https://code-medical-ethics.ama-assn.org
   The AMA *Code of Medical Ethics*, established by the American Medical Association, is a comprehensive guide for healthcare practitioners. It addresses issues and challenges, promotes adherence to standards of care, and is continuously updated to reflect contemporary practices and challenges.
2. "Patient Rights and Ethics." Jacob P. Olejarczyk and Michael Young. https://www.ncbi.nlm.nih.gov/books/NBK538279
   This source discusses how every interprofessional team member, including nurses and allied health providers, should observe patients' rights.
3. "Principles of Clinical Ethics and Their Application to Practice." Basil Varkey. 2021. https://www.ncbi.nlm.nih.gov/pmc/articles/PMC7923912
   This source presents an overview of ethics and clinical ethics, defining and explaining the four main ethical principles: beneficence, nonmaleficence, autonomy, and justice.
4. "The Nurse's Role in Ethics and Human Rights: Protecting and Promoting Individual Worth, Dignity, and Human Rights in Practice Settings." American Nurses Association, Center for Ethics and Human Rights. 2016. https://www.nursingworld.org/~4af078/globalassets/docs/ana/ethics/ethics-and-human-rights-protecting-and-promoting-final-formatted-20161130.pdf
   This source discusses the nurse's role in ethics and human rights, emphasizing the protection and promotion of individual worth, dignity, and human rights in practice settings.

5. "Patient Rights and Ethics." Jacob P. Olejarczyk and Michael Young. https://pubmed.ncbi.nlm.nih.gov/30855863

   This source discusses the development of patient bills of rights to empower people to take an active role in improving their health and to strengthen the relationships people have with their healthcare providers.

6. "Patient Rights and the Essential Role of Nurses." Hawaii Pacific University. 2023. https://online.hpu.edu/blog/patient-rights

   This source discusses the essential role of nurses in ensuring that patient rights are respected, highlighting the obligation established in nurses' codes of ethics.

7. "Nurses Bill of Rights." American Nurses Association. https://www.nursingworld.org/practice-policy/work-environment/health-safety/bill-of-rights

   This source discusses the nurses bill of rights, emphasizing the role of nurses as the largest group of healthcare professionals in the United States.

8. "Ethics in Health Care: Improving Patient Outcomes." Tulane University, School of Public Health and Tropical Medicine. 2023. https://publichealth.tulane.edu/blog/ethics-in-healthcare

   This source discusses the principle of beneficence in healthcare, referring to practitioners' responsibility to act in their patients' best interests.

ORGAN DONATIONS

38

The practitioner should not treat the disease but the patient who is suffering from it.
—Moses Maimonides

* * *

MYSTERY STORY

> **ORGAN DECEPTION: THE SHADOWS OF LIFESAVERS TRANSPLANT CLINIC**
>
> In the sterile halls of LifeSavers Transplant Clinic, Practitioner Sarah Peters was known as a savior, a highly respected surgeon with a golden touch. Her dedication to patients and unmatched organ transplant skill gave many a second chance at life. Fresh from a successful liver transplant surgery, Practitioner Peters was preparing to rest when her phone rang, bringing a chill to the room.
>
> "Practitioner Peters, this is Detective Samuel Wright from the police department. I need your expertise on a case that's taken a sinister turn."
>
> "What case, Detective Wright? I'm not involved in any investigations."
>
> "It's about an organ donor. We received an anonymous tip that a patient in your care received a liver from a prisoner executed last week. Can you confirm this?"
>
> A cold shock ran down Practitioner Peters' spine. She had no knowledge of any such donation. "I assure you, Detective, we follow all legal protocols. I will investigate this matter at once."
>
> "Thank you, Practitioner Peters," Detective Wright replied. "This could be bigger than we think."
>
> Days passed, and the transplant coordinator confirmed a liver donation from a prisoner, but all seemed proper and approved. Practitioner Peters felt a momentary relief but could not dismiss the shadow of doubt that loomed over her.
>
> Soon, whispers began. Organ donation records vanished. Nurses reported overhearing Practitioner Peters arguing about an organ donation on the phone. Despite her denials, Detective Wright's suspicion grew.
>
> Then, on a night as cold as the operating table, Practitioner Peters was found dead in her office, a lethal injection of potassium still in her veins.
>
> Detective Wright was haunted by the mystery. Why would someone want to kill a respected surgeon such as Practitioner Peters? What was lurking in the shadows of LifeSavers Transplant Clinic?
>
> As the investigation unraveled, a horrifying truth emerged. Practitioner Peters had been ensnared in a web of greed, drawn into a black market for nonrenewable organs. Financial desperation led her to accept a monstrous bargain, one that cost her integrity and life.
>
> When a patient threatened exposure, Practitioner Peters silenced him with a lethal dose of potassium. But Detective Wright's relentless pursuit left her trapped, leading her to inject herself, ending a life that had once saved so many.
>
> In the end, Detective Wright's dogged determination led to the capture of the criminal network behind the illegal organ trade. The grim case served as a dire warning of the importance of ethics and adherence to the rules, even in the face of temptation and despair. The shadows of LifeSavers Transplant Clinic still whisper the tale of Practitioner Peters, a story of deception and tragedy in the pursuit of life.

* * *

THINK

Giving consent for donating one's organs and tissue is a voluntary decision. Nobody can force a donation, nor can anyone purchase or sell nonrenewable vital organs and tissues. This prevents people from selling their nonrenewable organs for profit, which would increase social healthcare disparities.

However, it is permissible to pay for renewable tissues such as blood, sperm, and unfertilized ova, usually stated as reimbursement for the donor's costs and compensation for services rendered.

Plasma Donations

The U.S. Food and Drug Administration sets the guidelines based on body weight. In general, a person can earn between $20 and $50 per donation and can donate every 28 days, up to 13 times per year, for a total of $260 to $650 per year.

Sperm Donations

Men who produce specimens twice a week can make upwards of $1,500 a month for a total of $18,000 per year.

Eggs Donations

Women can make upwards of $5,000 to $35,000 per donation cycle. The American Society for Reproductive Medicine (ASRM) guidelines for egg donation recommend that females only donate up to six times in their lifetime because of risks to the egg donor and of inadvertent consanguinity:

1. Risks to egg donors
    a. Increased risk of infection
    b. Ovarian hyperstimulation syndrome
    c. Surgical or procedural risks associated with a donation (procedural and anesthesia)
2. Inadvertent consanguinity risk of procreation with a half-sibling: ASRM attempts to limit the possibility of consanguinity by allowing no more than 25 pregnancies per donor (sperm or egg) in a population of 800,000.

A fertilized ovum becomes an embryo once the cells start to divide. In the United States, paying for an embryo, such as a frozen embryo, is illegal, but an embryo can be donated. An open donation means the donor knows the recipient, and a closed donation means the donor does not know the recipient. Other options are to donate to scientific research, where the embryo will never become a child, or to keep the embryos frozen indefinitely. Legal problems can and do occur regarding what to do with the frozen embryos in cases of divorce, death, and inheritance. Referring to embryonic donations as embryonic adoption is controversial because adoption refers to a person, and an embryo is legally not a person. The term embryonic donation is not meant to disparage personal and religious viewpoints on this subject; rather, the use is a matter of legal definition.

ASSESS

Patient: Autonomy

An autonomous donor can provide informed consent for donating their nonrenewable organs. Under no circumstance can a person be compelled to donate an organ or tissue, regardless of whether or not the donation is of little consequence to the donor or high consequence for the potential recipient.

Once the informed consent decision to donate has been made, the donor may withdraw their decision at any time for any or no reason. For example, if a patient is histocompatible and decides to donate to a designated family recipient but changes their mind, then that is permissible. To minimize the social coercion and negative judgment, the practitioner may respond, if asked, that the individual was not a suitable candidate for donation. This statement aligns with the principle of respecting the donor's decision without disclosing private health information (PHI), which would violate the Health Insurance Portability and Accountability Act federal regulations. It is crucial to emphasize that this approach is rooted in ethical practice, intending to respect and protect the donor's autonomy (informed consent) and privacy and ensuring that the decision to donate remains a deeply personal and confidential matter without any coercive influences.

Minors may also donate a nonvital organ as long as they

- are mature minors;
- understand all the risks of harm;
- assent to the procedure; and
- their parents or guardians also consent to the donation.

Practitioner: Beneficence and Nonmaleficence

Patients who donate organs and tissues put themselves in harm's way for the benefit of others. The professional maxim of nonmaleficence (do no harm) is a prime reason why it is professionally, morally, and legally impermissible for a practitioner to request a patient to donate a vital organ. Practitioners must ensure that patients and society continue to trust that practitioners will never do anything to potentially harm a patient.

It is professionally, legally, and morally impermissible for a practitioner to care for both the donor and the recipient because that is considered a serious conflict of interest and could erode the patient–practitioner relationship.

Public Policy: Justice

Federal law mandates that all families be provided the option to donate when death is imminent. However, only a representative of an organ procurement organization is allowed to initiate an organ donation request.

There are two fundamental reasons why it is illegal for a practitioner to pursue the obtaining of consent for organ donation: conflict of interest and donor success:

Conflict of interest occurs when the practitioner is an advocate for both the donor and the recipient.
2. *Donor success* networks have been shown to have a much greater success rate in getting organ donation consent than when attempted by the patient's practitioner or healthcare team.

A prisoner can only donate if

- the decision to donate was made before their conviction;
- harvesting of organs occurs after the prisoner is confirmed dead;
- the procedure is not done in the death chamber; and
- the method of execution is not altered for harvesting.

It is believed that exchanging money for nonrenewable organs would result in the unjust distribution of benefits for the wealthy and be burdensome for the poor. Society must trust that practitioners will never have a conflict of interest between their professional obligation to maximize their patients' best interests and wanting their patients' organs.

CONCLUDE

By federal law, it is illegal to pay for an embryo. However, donor reimbursement and compensation for renewable tissues are permissible but vary dramatically from state to state. It is also professionally, morally, and legally impermissible for a practitioner to request from a patient a donation of a vital organ because (a) it is considered to be a conflict of interest to be an advocate of both the donor and the recipient at the same time and (b) donor networks have been shown to have a much greater success rate in obtaining organ donation consents than the patient's practitioner or healthcare team.

In summary, organ and tissue donation is a complex yet essential component of healthcare, and it is regulated by federal law and professional ethics. Compensation for renewable tissues is permissible, whereas the sale of nonrenewable organs and embryos is not. Practitioners must ensure they do not engage in conflicts of interest when dealing with organ donation and must respect patients' autonomy, allowing them to make informed decisions. By adhering to these guidelines and prioritizing patient autonomy, healthcare professionals can maintain trust and contribute to the success of organ and tissue transplantation.

* * *

REVIEW QUESTIONS

1. In the United States, paying for an embryo, such as a frozen embryo, is illegal, but an embryo may be donated.
    A. True
    B. False
2. An open donation means that the donor does not know the recipient, and a closed donation means that the donor knows the recipient.
    A. True
    B. False
3. Referring to embryonic donations as embryonic adoption is controversial because adoption refers to a person, and an embryo is legally not a person.
    A. True
    B. False
4. Under no circumstance can a person be compelled to donate an organ or tissue, regardless of whether or not the donation is of little consequence to the donor or high consequence for the potential recipient.
    A. True
    B. False
5. Once the informed consent decision to donate has been made, the donor may withdraw their decision at any time for any or no reason. To minimize the social coercion and to not disclose the patient's PHI, the practitioner should respond, if asked, that the individual was not a suitable candidate for donation.
    A. True
    B. False
6. It is professionally, legally, and morally impermissible to care for a donor and the recipient because that is considered a serious conflict of interest and would erode the patient–practitioner relationship.
    A. True
    B. False
7. Federal law mandates that all families be provided the option to donate when death is imminent.
    A. True
    B. False

Answers: 1A, 2B, 3A, 4A, 5A, 6A, 7A

* * *

CLINICAL VIGNETTES

1. Ms. Paloma Barton, a 28-year-old nurse, was in a car accident resulting in traumatic brain injury. She was declared brain dead, and her family was approached by the hospital staff for organ donation consent. Her family is distraught and undecided, but the healthcare staff keeps stressing the importance of organ donation and the potential lives that could be saved. Ms. Barton's family struggles with the decision and asks for more time. What ethical issue is raised by the healthcare staff's request for organ donation consent, and how should the situation be approached?
   A. Coercion to donate organs and tissues
   B. Conflicts of interest
   C. Use of prisoner organs for donation
   D. Social justice implications of organ donation

   Explanation: In this vignette, the healthcare staff presses Ms. Barton's family to consent to organ donation. This pressure can be viewed as coercion, which violates autonomy. The principle of autonomy means that the patient or the patient's family has the right to make their own decisions about their healthcare, including organ donation. Coercion undermines autonomy and may lead to negative outcomes, such as mistrust of healthcare professionals. The other options are not directly applicable to this case. Conflicts of interest occur when a healthcare professional's priorities conflict with those of the patient, which is not applicable in this scenario. The use of prisoner organs for donation is only allowed if certain conditions are met, such as the decision to donate before the prisoner's conviction. Last, although social justice is an important consideration in organ donation, it is not the primary ethical issue raised in this case. In approaching this situation, the healthcare staff should provide the family with all the necessary information about organ donation, answer their questions, and give them time to make an informed decision. The healthcare staff should also respect the family's decision, whether they consent to organ donation or not, and provide support during this difficult time. *Answer*: A

2. Ms. Noelle Hernandez, a 30-year-old sales clerk, was in a car accident resulting in severe head trauma. Despite aggressive healthcare management, she is declared brain dead. The organ procurement organization approached Ms. Hernandez's family to consider organ donation. The clinical differential diagnosis is brain death, and the ethical question is whether Ms. Hernandez's family can be compelled to donate her organs. Which of the following is correct?
   A. Yes, Ms. Hernandez's family can be compelled to donate her organs because organ donation can save lives and is in the best interest of society.
   B. No, Ms. Hernandez's family cannot be compelled to donate her organs because organ donation is a voluntary act, and no one can be forced to donate.
   C. Ms. Hernandez's family can be offered compensation for organ donation to help sway their decision.
   D. Ms. Hernandez's family should be informed that they can decide whether or not to donate, but their decision will influence the allocation of organs in the future.

   Explanation: Ms. Hernandez's family cannot be compelled to donate her organs because organ donation is voluntary, and no one can be forced to donate. Federal law mandates that all families be provided the option to donate when death is imminent. Only a representative of an organ procurement organization is allowed to initiate an organ donation request. It is professionally, morally, and legally impermissible for a practitioner to request from a patient a donation of a vital organ because (a) it is considered to be a conflict of interest to be an advocate of both the donor and the recipient at the same time and (b) donor networks have been shown to have a much greater success rate in obtaining organ donation consents than the patient's practitioner. Furthermore, compensation for organ donation is not allowed, and organ donation must be voluntary. *Answer*: B

3. Ms. Elizabeth Thompson is a 32-year-old retired nurse who presents to the hospital with symptoms of acute kidney failure. She has a history of hypertension and type 2 diabetes. Clinical differential diagnosis: End-stage renal disease due to hypertension and diabetes. Her healthcare team has informed Ms. Thompson that she requires a kidney transplant, and her brother, a histocompatible match, has agreed to donate one of his kidneys. However, during the preoperative evaluation, the transplant coordinator discovers that the surgeon who will be performing the transplant also cared for Ms. Thompson's brother in the past, and there is a potential conflict of interest. Given the potential conflict of interest, should the transplant coordinator allow the surgeon to perform the kidney transplant?
   A. Yes, the surgeon has the necessary expertise and experience to perform the transplant.
   B. No, the surgeon should not perform the transplant due to the conflict of interest.
   C. Yes, as long as the surgeon recuses themselves from the care of Ms. Thompson's brother during and after the transplant procedure.
   D. No, the transplant should be performed by a different surgeon.

   Explanation: The conflict of interest arises because the surgeon has a relationship with both the donor and the recipient, and it is professionally, legally, and morally impermissible to care for both parties. The surgeon's primary responsibility is to the patient, and it is essential to prioritize patient autonomy and the maxim of nonmaleficence. Allowing the surgeon to perform the transplant could potentially erode the patient–practitioner relationship and harm the patient's trust in their healthcare team. Therefore, it is not permissible for the surgeon to perform the transplant due to the conflict of interest. *Answer*: B

4. Mr. John Peters is a 30-year-old accountant who has expressed interest in donating one of his kidneys to his brother, who suffers from end-stage renal disease. Mr. Peters has undergone the necessary evaluations and has been cleared for donation. However, on the day of the surgery, Mr. Peters informs his healthcare team that he has changed his mind and no longer wishes to donate his kidney. What is the most appropriate course of action for the healthcare team to take in response to Mr. Peters' decision to withdraw his consent for organ donation?
   A. Respect Mr. Peters' decision and cancel the surgery
   B. Convince Mr. Peters to reconsider his decision and proceed with the surgery
   C. Delay the surgery to allow Mr. Peters more time to consider his decision
   D. Proceed with the surgery without Mr. Peters' consent

   Explanation: It is essential to remember that organ and tissue donation is voluntary, and individuals have the right to withdraw their consent at any time. The healthcare team must respect Mr. Peters' decision and cancel the surgery. The other options are wrong because convincing Mr. Peters to donate may be considered ethically problematic as it could be perceived as pressuring or coercing him to reconsider his decision to donate. It is never permissible to proceed with any healthcare procedure without the patient's informed consent. *Answer*: A

5. Ms. Gabrielle Harper, a 32-year-old kindergarten teacher, had recently expressed an interest in donating her kidney to her brother, who has end-stage renal disease. She had undergone extensive healthcare workups and had received clearance for the procedure. However, during a follow-up appointment with her practitioner, she expressed hesitation about going through with the donation. She cited concerns about the risks of the surgery, the recovery process, and her ability to return to work. Her practitioner informed her that she had the right to withdraw her consent at any time and that her decision would be respected. The practitioner also explained to Ms. Harper that the risks associated with kidney donation are relatively low and that the chances of complications are typically lower than for other surgeries. Which of the following statements is correct about Ms. Harper's decision?
   A. She can be compelled to go through with the donation if her brother's condition worsens.
   B. Her decision to withdraw her consent cannot be respected.

C. She can only withdraw her consent if she provides a reason for doing so.
D. Her decision to withdraw her consent must be respected.

Explanation: Ms. Harper has the right to withdraw her consent for organ donation at any time, for any reason, or no reason at all. Her decision must be respected, and she cannot be compelled to go through with the donation, even if her brother's condition worsens. Although her practitioner may discuss the risks associated with the procedure, the final decision to donate is hers. It is imperative that healthcare professionals prioritize patient autonomy and respect their decisions. Therefore, Ms. Harper's decision to withdraw her consent should not be taken lightly but must be respected. *Answer*: D

REFLECTION VIGNETTES

1. Practitioner Sarah Brown is a 50-year-old nephrologist who is approached by a parent who is histocompatible with their child, who needs a kidney transplant. All necessary tests are done, and the transplant is planned. However, just before the surgery, the parent informs Practitioner Brown that they have changed their mind and no longer wish to donate a kidney. The parent requests that the family not be informed, as they fear that revealing their change of heart would damage their relationship with their family.

   What is the ethical issue(s)?

   How should the ethical issue(s) be addressed?

2. Practitioner Jane Lee, a 40-year-old practitioner, finds herself in a challenging situation in the intensive care unit. Her patient, who is terminally ill, has viable organs that could potentially save the lives of three other patients.

   What is the ethical issue(s)?

   How should the ethical issue(s) be addressed?

* * *

WEB LINKS

1. *Code of Medical Ethics.* American Medical Association. https://code-medical-ethics.ama-assn.org
   The AMA *Code of Medical Ethics*, established by the American Medical Association, is a comprehensive guide for healthcare practitioners. It addresses issues and challenges, promotes adherence to standards of care, and is continuously updated to reflect contemporary practices and challenges.
2. "Financial Incentives for Organ Donation." U.S. Department of Health and Human Services. 1993. https://optn.transplant.hrsa.gov/resources/ethics/financial-incentives-for-organ-donation
   The report from the Ethics Committee's Payment Subcommittee critically explores the proposal of financial incentives for organ donation, considering ethical concerns, public opinion, and potential impacts on the current altruistic system, while questioning its effectiveness and ethicality.
3. "Avoiding Conflict in Third-Party Reproduction." American Society for Reproductive Medicine. https://www.reproductivefacts.org/news-and-publications/fact-sheets-and-infographics/avoiding-conflict-in-third-party-reproduction/?_t_id=4cU5VQvrUoaNAhiHTJOafA==&_t_uuid=l1fDd_qoSRWMFIlAYof3Bg&_t_q=Third-party+Reproduction:+Sperm,+Egg,+And+Embryo+Donation+and+Surrogacy.+American+Society+for+Reproductive+Medicine+(ASRM).&_t_tags=siteid:db69d13f-2074-446c-b7f0-d15628807d0c,language:en,andquerymatch&_t_hit.id=ASRM_M

odels_Pages_ContentPage/_d04bd6ae-312c-46b8-b5c2-d8e96d76bb2f_en&_t_hit.pos=3

This document highlights the need for the following to mitigate potential emotional, legal, and healthcare conflicts: clear contracts; informed consent; transparency; psychological counseling; experienced professionals in third-party reproduction involving donated eggs, sperm, or embryos; and gestational carriers.

4. "Responsibilities of Primary Practitioners in Organ Donation." S. W. Tolle, W. M. Bennett, D. H. Hickam, and J. A. Benson, Jr. 1987. https://citeseerx.ist.psu.edu/document?repid=rep1&type=pdf&doi=a7679e164782f445c090a4d8525a784001819c37

This document discusses the increasing demand for donor organs due to improved transplantation success, the role of primary physicians in identifying donors while addressing survivors' needs, the criteria for tissue and organ donation, ethical considerations, and how offering organ donation can comfort survivors and benefit those on waiting lists.

5. "Ethics of Organ Donation & Transplantation." American Medical Association. https://www.ama-assn.org/topics/ethics-organ-procurement-transplantation

This document discusses the ethics of organ donation and transplantation within the healthcare community.

6. "Organ Donation as a Collective Action Problem: Ethical Considerations and Implications for Practice." Keren Ladin. 2016. https://journalofethics.ama-assn.org/article/organ-donation-collective-action-problem-ethical-considerations-and-implications-practice/2016-02

This source presents findings from a national survey of organ donation attitudes and behaviors in the United States.

7. "Consent in Organ Transplantation: Putting Legal Obligations and Guidelines into Practice." Farrah Raza and James Neuberger. 2022. https://bmcmedethics.biomedcentral.com/articles/10.1186/s12910-022-00791-y

This source discusses the complexities of consent in healthcare practice, with a focus on organ transplantation.

8. "Philosophy of Organ Donation: Review of Ethical Facets." Aparna R. Dalal. 2015. https://www.ncbi.nlm.nih.gov/pmc/articles/PMC4478599

Reviews the ethical facets of organ donation, emphasizing the role of altruism.

9. "Reexamining the Flawed Legal Basis of the "Dead Donor Rule" as a Foundation for Organ Donation Policy." Scott J. Schweikart. 2020. https://journalofethics.ama-assn.org/article/reexamining-flawed-legal-basis-dead-donor-rule-foundation-organ-donation-policy/2020-12

This source discusses the "dead donor rule," a requirement in organ donation that stipulates a donor must first be dead before the retrieval of any vital organs.

10. "The Ethics of Organ Donation in Patients Who Lack the Capacity for Decision Making." Lauren Lee. 2018. https://pubmed.ncbi.nlm.nih.gov/30567762

This source discusses the ethics of organ donation in patients who lack the capacity for decision-making.

11. "The Living Organ Donor as Patient: An Ethics Framework." https://global.oup.com/academic/product/the-living-organ-donor-as-patient-9780197618202?cc=us&lang=en&

This source discusses the ethical considerations of living organ donation, treating the donor as a patient.

12. "Organ Donation, Patients' Rights, and Medical Responsibilities at the End of Life." Ana S. Iltis, Michael A. Rie, and Anji Wall. 2009. https://pubmed.ncbi.nlm.nih.gov/19050624

This source examines the revised Uniform Anatomical Gift Act and its implications for end-of-life healthcare in the United States.

13. "AMA *Code of Medical Ethics*' Opinions on Organ Transplantation." AMA Council on Ethical and Judicial Affairs. 2012. https://journalofethics.ama-assn.org/article/ama-code-medical-ethics-opinions-organ-transplantation/2012-03

This source discusses the American Medical Association's opinions on organ transplantation, focusing on the lives and quality of life of patients with end-stage organ failure.

14. "Does It Matter That Organ Donors Are Not Dead? Ethical and Policy Implications." M. Potts and D. W. Evans. 2005. https://jme.bmj.com/content/31/7/406
    This source discusses the "standard position" on organ donation, which states that the donor must be dead to remove vital organs.
15. "The COVID-19 Pandemic and Organ Donation and Transplantation: Ethical Issues." Ban Ibrahim, Rosanne Dawson, Jennifer A. Chandler, et al.2021. https://bmcmedethics.biomedcentral.com/articles/10.1186/s12910-021-00711-6
    This source discusses the impact of the COVID-19 pandemic on organ and tissue donation and transplantation, including ethical challenges related to infection risks and resource allocation.
16. "Ethical Analysis of Living Organ Donation." Benita J. Walton-Moss, Laura Taylor, and Marie T. Nolan, 2005. https://www.ncbi.nlm.nih.gov/pmc/articles/PMC8976442
    This source discusses the emergence of paired donation and other new living organ donor protocols.
17. "Human Rights and Biomedicine: Ethical and Legal Aspects of Organ Donation." Council of Europe. https://www.coe.int/en/web/bioethics/human-rights-biomedicine-ethical-legal-aspects-of-organ-donation
    This source discusses the need for synergies between the protection of human rights and the promotion of healthcare and scientific progress, particularly in the process of organ donation and transplantation.
18. "Ethical Issues in Organ Transplantation." George M. Abouna. 2003. https://pubmed.ncbi.nlm.nih.gov/12566971
    This source discusses clinical organ transplantation as one of the most gripping healthcare advances of the century, which requires the participation of fellow human beings and society.
19. "Organ Donation, Brain Death and the Family: Valid Informed Consent." Anita S. Iltis. 2015. https://www.cambridge.org/core/journals/journal-of-law-medicine-and-ethics/article/organ-donation-brain-death-and-the-family-valid-informed-consent/94988690C2C719327964CEC662019E45
    This source discusses the role of informed consent in organ donation, particularly in cases involving brain death.

PALLIATIVE CARE

39

The part of life we really live is small, for all the rest of existence is not life but merely time.
—Seneca

* * *

MYSTERY STORY

> **PALLIATIVE BETRAYAL: THE IMPORTANCE OF PALLIATIVE CARE AND MEDICATION MANAGEMENT IN HEALTHCARE**
>
> In the bustling corridors of Gentle Winds Compassion Center, Practitioner Victoria Harrington had carved out a sanctuary for her patients. With kindness in her eyes and compassion in her touch, she had become an emblem of hope for those grappling with life's most dire challenges, such as terminal illnesses like cancer.
>
> When 65-year-old Mr. Johnson, a retired mechanic who had never once succumbed to the temptations of a cigarette, staggered into her office, his face etched with pain and confusion, Victoria knew she had another battle on her hands. The diagnosis was as unforgiving as it was unexpected: stage 4 lung cancer.
>
> Victoria's heart ached for Mr. Johnson, but she was no stranger to the unrelenting march of disease. She reassured him and his family that although the cancer was incurable, it was not untreatable. With palliative care, she could ease his pain, soothe his symptoms, and even enhance his quality of life. The family's eyes, filled with fear and uncertainty, finally sparkled with a glimmer of hope.
>
> Weeks turned into months, and Victoria, ever diligent and devoted, tended to Mr. Johnson with unwavering dedication. Medications were prescribed, comfort was offered, and tears were wiped away. Yet, the shadow of something ominous loomed. Mr. Johnson's pain intensified, his vitality waned, and the once vibrant glimmer in his eyes dimmed.
>
> One cold, bleak night, the phone rang, shattering the silence in Victoria's home. It was Mr. Johnson's daughter, her voice trembling with grief. Her father had passed away, his last hours filled with torment and agony.
>
> Victoria's world crumbled. Had her care been in vain? Had she failed this gentle man who had trusted her with his life? The questions gnawed at her soul, and she embarked on a journey to uncover the truth.
>
> The revelations were shocking. Someone had tampered with his healthcare records, increasing the dosage of pain medications he was supposedly taking. A dark, insidious secret was festering within the very institution she had devoted her life to.
>
> The truth emerged, cold and brutal. One of the nurses on Mr. Johnson's care team had betrayed her oath, stealing his medication for sinister gains and leaving him to suffer.
>
> Victoria's trust in humanity wavered, but her resolve hardened. The incident became a catalyst for change. With fierce determination, she fought for stricter medication management controls, ensuring that no other patient would fall victim to such treachery.
>
> In the end, Mr. Johnson's case was not a defeat but a turning point. Through pain and betrayal, a lesson was learned, and a crusade was born. Victoria's fight for palliative care and medication management became a beacon for others—a stark reminder that compassion, vigilance, and integrity must guide the hands that heal. At the heart of this

> transformation stood the Gentle Winds Compassion Center, a sanctuary where patients' comfort and dignity are paramount. Victoria's story, marked by both sorrow and triumph, will resonate in the hearts of practitioners and patients alike, a testament to the enduring human spirit and the unbreakable bonds of trust and care.

* * *

THINK

The word "palliative" comes from the Latin root *palliare*, which means "to cloak." Palliative care includes an interprofessional healthcare team that focuses on the quality of life and the mitigation of suffering of the patient and family. Although palliative care grew out of the hospice movement, it is now much broader in its practice but still focuses on patients with serious illnesses. Unlike hospice, which requires the terminal patient to have less than 6 months to live, there are no specific qualifications for palliative care.

The World Health Organization defines palliative care as follows:

> Palliative care is an approach that improves the quality of life of patients (adults and children) and their families who are facing problems associated with life-threatening illness. It prevents and relieves suffering through the early identification, correct assessment, and treatment of pain and other problems, whether physical, psychosocial, or spiritual.

In 2006, the American Board of Medical Specialties recognized hospice and palliative care as a healthcare subspecialty. Central to hospice and palliative care is advanced care planning for providing a forum for patients to express their healthcare goals, values, and priorities; communicate their treatment preferences; and identify who will make treatment decisions on their behalf if they can no longer do so.

ASSESS

Patient: Autonomy

Palliative care respects, promotes, and supports patient end-of-life planning as a reflection of patient autonomy (informed consent). Healthcare's success in extending life through medication and technology has had the unintended consequence of diminishing the patient's autonomous control of where, how, and with whom the patient will die. As a subspecialty, palliative care directly addresses these concerns without violating or weakening the traditional patient–practitioner relationship.

In the past, there was no need for advanced care planning because people generally lived and died in the home environment. This changed when technologies started to fill hospital wards, and patients started to die on ventilators and other healthcare equipment, alone and isolated in intensive care units. Healthcare adaptation was essential for the principles of patient autonomy (informed consent) to continue to have decisional importance for those with life-threatening illnesses.

Palliative care filled this newly created niche, maximizing the patient's autonomy (informed consent) by incorporating advanced care planning and interprofessional collaboration with hospice teams. A living will determines what treatment options are permissible for the practitioner to provide to the patient under various circumstances. If the patient no longer has decisional capacity, then the durable power of attorney determines who will make treatment decisions on the patient's behalf.

Practitioner: Beneficence and Nonmaleficence

The healthcare profession is a patient-centered healing art. However, a problem arises when healing is no longer possible, and abandoning the patient in their time of highest need is unacceptable.

Palliative care is the healthcare profession's interprofessional attempt to address this issue by having the team focus on maximizing the patient's best interests in accordance with the patient's reasonable goals, values, and priorities. Although the practitioner's professional maxim of healing may, in some cases, no longer be possible, that is not equivalent to being part of an activity of harming, a violation of the professional principle of nonmaleficence (do no harm). Palliative care, therefore, is not a violation of nonmaleficence (do no harm), and the palliation of pain and suffering is in accordance with beneficence (do good).

Public Policy: Justice

Justice (be fair) is implementing public policy to address various healthcare disparities. Technologically laden hospitals have created unanticipated healthcare disparities. In the attempt to heal, there are times when the patient and the practitioner must accept that any further treatment is futile and that death is unavoidable and imminent. In such circumstances, the only just or fair thing to do is to palliate any accompanied pain and suffering without violating the patient–practitioner relationship, patient autonomy (informed consent), and the practitioner's duty of nonmaleficence (do no harm).

CONCLUDE

Palliative care is a recognized subspecialty available to anyone with a serious illness. In contrast, hospice care is only available for those with 6 months or less to live. Patients with serious illnesses find it desirable to maximize the quantity and quality of their lives, and palliative care provides those services without violating the patient–practitioner relationship.

In summary, palliative care, as a recognized subspecialty, plays a crucial role in the healthcare system by addressing the needs of patients with serious illnesses. Unlike hospice care, which is reserved for patients with less than 6 months to live, palliative care is accessible to anyone with a serious illness. It ensures that patients can maximize both the quantity and quality of their lives without violating the patient–practitioner relationship or compromising core healthcare principles. By prioritizing patient autonomy and providing comprehensive, compassionate care, palliative care serves as a valuable solution to the challenges of end-of-life care. (See also Chapter 27.)

* * *

REVIEW QUESTIONS

1. Palliative care requires the patient to live less than 6 months.
    A. True
    B. False

2. In 2006, the American Board of Medical Specialties recognized hospice and palliative care as a healthcare subspecialty.
    A. True
    B. False

3. Palliative care maximizes patients' autonomy (informed consent) by incorporating advanced care planning and interprofessional collaboration.
    A. True
    B. False

4. Palliative care violates nonmaleficence (do no harm) and should not be a recognized specialty.
   A. True
   B. False

5. In the attempt to heal, there are times when the patient and the practitioner need to accept that any further treatment is futile and death is unavoidable and imminent.
   A. True
   B. False

Answers: 1B, 2A, 3A, 4B, 5A

* * *

CLINICAL VIGNETTES

1. Mr. Joshua William Ortiz is a 65-year-old retired carpenter who presents to the palliative care clinic complaining of worsening shortness of breath and fatigue over the past few weeks. He was diagnosed with lung cancer 6 months ago and has undergone chemotherapy. On examination, he is found to have tachypnea, decreased breath sounds in the left lung base, and oxygen saturation of 88% on room air. His differential diagnosis includes pulmonary embolism, pneumonia, and progression of the underlying lung cancer. The ethical question in this scenario is whether to provide palliative care and manage the patient's symptoms or pursue further aggressive treatment to manage the underlying disease. Which of the following actions should be taken?
   A. Provide symptom management and palliative care
   B. Start antibiotics for a suspected pneumonia
   C. Order imaging studies to rule out pulmonary embolism
   D. Increase chemotherapy to manage the underlying lung cancer

   Explanation: Providing symptom management and palliative care is the best option. The core principles of palliative care include patient autonomy (informed consent), beneficence (do good), nonmaleficence (do no harm), and justice (be fair). In this case, the patient has a life-limiting illness, and the goal of care should shift from curative to palliative. Palliative care aims to improve the quality of life and reduce suffering in patients and families facing life-threatening illnesses. The palliative care team can work with the patient and family to manage symptoms, improve comfort, and provide emotional support. Starting antibiotics or increasing chemotherapy may not be effective in managing the patient's symptoms and may cause harm, violating the principle of nonmaleficence (do no harm). Ordering imaging studies may delay necessary symptom management and may not align with the patient's reasonable goals, values, and priorities. Therefore, providing symptom management and palliative care is the most appropriate action. *Answer*: A

2. Mr. James Lee is a 68-year-old retired accountant diagnosed with pancreatic cancer. He has undergone chemotherapy but has experienced severe side effects, including nausea, vomiting, and loss of appetite. He also suffers from significant pain in his abdomen, affecting his ability to perform activities of daily living. His differential diagnosis includes pancreatic cancer, chemotherapy-induced nausea and vomiting, and chemotherapy-induced peripheral neuropathy. What is the ethical question in this case?
   A. How to increase the dosage of chemotherapy to increase the chances of curing the cancer?
   B. How to alleviate Mr. Lee's pain and suffering caused by the side effects of chemotherapy?
   C. How to encourage Mr. Lee to continue chemotherapy despite the side effects?
   D. How to persuade Mr. Lee to consider hospice care?

Explanation: The core principles of palliative care include beneficence (do good) and nonmaleficence (do no harm). In this case, the ethical question is how to alleviate Mr. Lee's pain and suffering caused by the side effects of chemotherapy, which is in accordance with the principle of beneficence. Palliative care aims to improve the quality of life and reduce suffering in patients with serious illnesses, including those undergoing cancer treatment. The primary goal of palliative care is to address the patient's physical, psychosocial, and spiritual needs rather than attempting to cure the underlying disease. Therefore, increasing the dosage of chemotherapy or persuading Mr. Lee to consider hospice care does not align with the core principles of palliative care. In addition, encouraging Mr. Lee to continue chemotherapy despite the side effects may not be in his best interests and may violate the principle of nonmaleficence. *Answer*: B

3. Mr. Nathaniel Harrison, a 65-year-old retired teacher, is admitted to the hospital with a diagnosis of metastatic lung cancer. His primary complaint is severe pain in his chest and difficulty breathing. The patient's daughter, his primary caregiver, reports that Mr. Harrison has been increasingly fatigued and has lost his appetite. The clinical team suspects that he may be experiencing other symptoms, such as anxiety, depression, and spiritual distress. What is the ethical question that arises in this case?
    A. Should the patient be placed on life support?
    B. Should the patient be transferred to hospice care?
    C. Should the patient be given high doses of opioids to manage his pain?
    D. Should the patient's daughter make the treatment decisions for him?

    Explanation: In this case, the ethical question is how to manage Mr. Harrison's severe pain. The core principle of palliative care is to provide patient-centered care that maximizes the patient's best interests in accordance with the patient's reasonable goals, values, and priorities. The goal of palliative care is to prevent and relieve suffering, whether physical, psychosocial, or spiritual. The use of opioids is an appropriate and ethical approach to managing pain in palliative care because it falls under the principle of beneficence (do good). However, the use of opioids should also be carefully monitored and managed to avoid the risk of addiction or other negative side effects. The other options are inappropriate because placing the patient on life support or transferring him to hospice care may not be appropriate or necessary at this stage. Although the patient's daughter may be involved in the decision-making process, the ultimate decision should be made in accordance with the patient's autonomous wishes and preferences. *Answer*: C

4. Mr. James Brown is a 65-year-old retired construction worker diagnosed with advanced lung cancer. He is experiencing shortness of breath, pain in his chest and back, and fatigue. His clinical differential diagnosis includes lung cancer, metastatic disease, and pleural effusion. He is considering his treatment options and desires to maintain his quality of life and minimize his pain and suffering. What is the appropriate approach to Mr. Brown's care?
    A. Begin aggressive chemotherapy to try and cure the cancer
    B. Refer Mr. Brown to a hospice care team for end-of-life care
    C. Provide palliative care services to manage symptoms and maximize the quality of life
    D. Encourage Mr. Brown to continue with the current course of treatment and manage his pain with painkillers

    Explanation: Palliative care is an approach that improves the quality of life of patients facing life-threatening illnesses, focusing on reducing suffering through the early identification, correct assessment, and treatment of pain and other problems, whether physical, psychosocial, or spiritual. In Mr. Brown's case, palliative care is the most appropriate approach to care, given his desire to maintain his quality of life and minimize pain and suffering. Referring Mr. Brown to hospice care would only be appropriate if he had less than 6 months to live, which is not indicated in the clinical vignette. Aggressive chemotherapy would not be effective at this stage of the disease and would cause unnecessary harm, violating the principle of nonmaleficence (do no harm). Encouraging Mr. Brown to continue with his current treatment and manage his pain with painkillers would not address his desire to maintain his quality of life and minimize suffering, violating the principle of beneficence (do good). Therefore, providing palliative care services to manage symptoms and maximize quality of life is the most ethical approach to Mr. Brown's care. *Answer*: C

5. Mr. David Johnson is a 72-year-old retired accountant who has been battling pancreatic cancer for several years. Despite undergoing various treatments, Mr. Johnson's cancer has progressed, and he is now experiencing severe pain and discomfort, impacting his quality of life. His clinical symptoms include abdominal pain, loss of appetite, and nausea. The differential diagnosis includes pancreatic cancer-related pain, side effects of cancer treatments, and underlying comorbidities. What ethical question arises when providing palliative care to Mr. Johnson?
   A. Should the palliative care team continue administering aggressive treatments to prolong Mr. Johnson's life?
   B. Should the palliative care team prioritize pain management over other aspects of Mr. Johnson's care?
   C. Should the palliative care team involve Mr. Johnson's family in decision-making?
   D. Should the palliative care team encourage Mr. Johnson to pursue experimental treatments?

   Explanation: Palliative care focuses on providing patient-centered care that maximizes the patient's best interests in accordance with the patient's reasonable goals, values, and priorities. The core principles of palliative care include patient autonomy (informed consent), beneficence (do good), nonmaleficence (do no harm), and justice (be fair). Palliative care aims to prevent and relieve suffering, and pain management is a critical aspect of this care. It is essential to prioritize pain management over other aspects of care to improve the patient's quality of life and reduce their suffering. Although the other options may have ethical implications in certain situations, prioritizing pain management aligns with the core principles of palliative care. *Answer*: B

REFLECTION VIGNETTES

1. Ms. Lisa Thompson is a 60-year-old retired teacher experiencing severe pain and suffering from a serious illness. Despite her illness, Ms. Thompson does not qualify for hospice care and is seeking options to improve her quality and quantity of life.

   What is the ethical issue(s)?

   How should the ethical issue(s) be addressed?

2. Practitioner Sarah Johnson is a 55-year-old practitioner with years of experience in palliative care. She has a patient who is terminally ill and desires to die in the comfort of their home surrounded by loved ones. However, the patient's condition requires a significant amount of healthcare attention, making it difficult to leave the hospital.

   What is the ethical issue(s)?

   How should the ethical issue(s) be addressed?

* * *

WEB LINKS

1. *Code of Medical Ethics*. American Medical Association. https://code-medical-ethics.ama-assn.org
   The AMA *Code of Medical Ethics*, established by the American Medical Association, is a comprehensive guide for healthcare practitioners. It addresses issues and challenges,

promotes adherence to standards of care, and is continuously updated to reflect contemporary practices and challenges.
2. "What Is Palliative Care?" Get Palliative CARE. https://getpalliativecare.org/whatis
This article discusses how palliative care is a specialized healthcare approach for individuals with serious illnesses, focusing on relieving symptoms and stress to improve the quality of life for both the patient and their family, provided by a team of healthcare professionals who work in conjunction with other doctors and can be beneficial at any stage of the illness alongside curative treatment.
3. "Palliative Care." World Health Organization. https://www.who.int/news-room/fact-sheets/detail/palliative-care
The World Health Organization defines palliative care as an approach aimed at improving the quality of life for patients with life-threatening illnesses and their families. This approach faces challenges such as global disparities in access, especially in low- and middle-income countries, and the need for better integration into health systems, training for health professionals, and public awareness.
4. "Ethical Considerations at the End-of-Life Care." Melahat Akdeniz, Bülent Yardimci, and Ethem Kavukcu. 2021. https://journals.sagepub.com/doi/full/10.1177/20503121211000918
This article discusses the ethical difficulties that healthcare professionals face in decisions regarding end-of-life care.
5. "Patient Rights at the End of Life: The Ethics of Aid-in-Dying." Mary Atkinson Smith, Lisa Torres, and Terry C. Burton. 2020. https://pubmed.ncbi.nlm.nih.gov/32000206
This paper explores patients' varying needs and desires at the end of life.
6. "Ethical Issues in Palliative Care." Erik K. Fromme. https://www.uptodate.com/contents/ethical-issues-in-palliative-care
This source discusses the ethical issues that arise in palliative care due to concerns about the extent and type of care for those with limited life expectancy.
7. "Palliative Care: An Ethical Obligation." Stephanie C. Paulus. https://www.scu.edu/ethics/focus-areas/bioethics/resources/palliative-care-an-ethical-obligation
This article discusses the comprehensive management of patients' needs in palliative care while respecting their personal, cultural, and religious values.
8. "Ethics in Palliative Care." David E. Weissman and Art Derse. https://mypcnow.org/wp-content/uploads/2019/03/Lectures-Ethics.pdf
This document provides foundational knowledge about ethical decision-making in palliative care.
9. "Human-Rights, Ethical and Legal Issues in Palliative Care: A Guide for Health Care Providers." Namati. https://namati.org/resources/human-rights-ethical-and-legal-issues-in-palliative-care-a-guide-for-health-care-providers
This guide provides insights into common legal and human rights issues encountered in palliative care.
10. "Palliative Care as a Human Right: A Fact Sheet." Open Society Foundations. 2016. https://www.opensocietyfoundations.org/publications/palliative-care-human-right-fact-sheet
This fact sheet argues that palliative care is a basic human right and is fundamental to health and human dignity.
11. "Ethical and Legal Issues in Palliative Care." Mary S. McCabe and Nessa Coyle. 2014. https://pubmed.ncbi.nlm.nih.gov/25361881
This article provides foundational knowledge about approaches to ethical decision-making in palliative care.
12. "Real-World Ethics in Palliative Care: A Systematic Review of the Ethical Challenges Reported by Specialist Palliative Care Practitioners in Their Clinical Practice." Guy Schofield, Mariana Dittborn, Richard Huxtable, Emer Brangan, and Lucy Ellen Selman. 2021. https://journals.sagepub.com/doi/full/10.1177/0269216320974277
This systematic review discusses the real-world ethical challenges that palliative care practitioners face.

13. "Relational Autonomy in End-of-Life Care Ethics: A Contextualized Approach to Real-Life Complexities." Carlos Gómez-Vírseda, Yves de Maeseneer, and Chris Gastmans. 2020. https://bmcmedethics.biomedcentral.com/articles/10.1186/s12910-020-00495-1
This article argues for a more contextualized approach to autonomy in end-of-life care ethics.
14. "Palliative Care, Hospice Care, and Bioethics: A Natural Fit." Nessa Coyle. 2014. https://journals.lww.com/jhpn/Fulltext/2014/02000/Palliative_Care,_Hospice_Care,_and_Bioethics__A.3.aspx
This article discusses the role of clinical ethics in palliative and hospice care.
15. "Ethics in Palliative Care." Bidhu K. Mohanti. 2009. https://www.ncbi.nlm.nih.gov/pmc/articles/PMC2902121
This article provides guidelines and codes for physicians regarding their duty, responsibility, and conduct in palliative care.
16. "Your Rights and Responsibilities as a Patient." Northside Hospital. https://www.northside.com/docs/default-source/default-document-library/patients-rights-and-responsibilities.pdf
This document outlines patients' rights, including the right to receive information on patient rights, responsibilities, and ethics.

PATIENT-PRACTITIONER RELATIONSHIP

40

The art of healing has three factors: the disease, the patient, and the practitioner. The practitioner is the servant of the art. The patient must participate with the practitioner in combatting the disease.
—Hippocrates

* * *

MYSTERY STORY

ABANDONMENT AFFLICTION: THE CASE OF THE ABANDONED PATIENT

In the tight-knit community of Willowbrook, Practitioner Sarah Richards was not just a healthcare professional; she was a lifeline. With her warm smile and soothing presence, she had become a beacon of hope for those suffering from chronic ailments. At LoyalPractitioner Wellness Center, her dedication to patient care and unwavering commitment to her patients had earned her the trust and respect of the entire community. But even the most illustrious reputations can falter, and Practitioner Richards found herself in the eye of a storm when a lawsuit was filed against her for patient abandonment by Mr. Charles Miller, a long-term patient.

Mr. Miller, a gentleman in his late sixties, had been battling a relentless heart condition. For years, Practitioner Richards had been his compass, guiding him through the turbulent waters of his illness. They shared not just a doctor-patient relationship but also a bond of trust and mutual respect.

However, a single letter shattered this relationship. Mr. Miller opened his mailbox one fateful day to find a note from Practitioner Richards, cold and curt, severing their ties without explanation or guidance to another provider. He was left adrift, panic setting in as he struggled to find another healthcare provider.

Time was a luxury Mr. Miller did not have. His heart condition worsened, and the once steady rhythm of his life turned chaotic. With no practitioner to turn to and crucial healthcare records withheld, he found himself in the emergency room, teetering on the brink of a life-threatening crisis.

The courtroom became the battleground where the truth would unravel. Practitioner Richards' actions, or lack thereof, were laid bare. She had abandoned Mr. Miller, forsaking not only her ethical obligations but also the unspoken covenant between a healer and a patient. The term "patient abandonment" hung heavily in the air, a damning label that sent a chill through the healthcare community.

The judgment was unequivocal. Practitioner Richards had failed Mr. Miller, and the ramifications were not just legal but also moral. Her once unblemished reputation at LoyalPractitioner Wellness Center was now marred, and she faced the arduous task of rebuilding not only her practice but also her ethos.

The aftermath of the trial saw a transformation in Practitioner Richards. Chastened and humbled, she became a vigilant advocate for patient autonomy, trust, and transparent communication. She made sweeping changes to her practice, vowing never to let the shadow of abandonment darken her door again.

> The case of the abandoned patient resonates far beyond the confines of LoyalPractitioner Wellness Center. It stands as a stark and cautionary tale for healthcare providers everywhere, underscoring the sanctity of the patient-practitioner relationship. The threads of care, empathy, and responsibility weave the fabric of healing, and to sever one is to unravel the whole.
>
> In a world in which healthcare advances at breakneck speed, the Miller case reminds us that the heartbeat of healthcare lies not in technology but in the human connection. The path to healing is paved with trust, and to abandon a patient is to forsake the very soul of medicine.

* * *

THINK

The patient–practitioner relationship is the cornerstone for establishing a patient-centered relationship that maximizes the patient's best interests in accordance with the patient's reasonable goals, values, and priorities.

ASSESS

Patient: Autonomy

Autonomy (informed consent) is based on the liberal notion that the patient knows their own goals, values, and priorities better than anyone else. Therefore, if the practitioner can communicate clearly and effectively, then the patient will be the best person to determine what healthcare choices will maximize their best interests.

The patient–practitioner relationship is essential for patient autonomy. Trust in the practitioner's confidentiality and privacy is a necessary condition for the patient to disclose protected health information willingly. Trust in the practitioner's healthcare knowledge, skill, and competence is also essential for the patient to be able to believe that the practitioner's diagnosis, prognosis, and risk–benefit analysis for various treatment options are reliable.

Being able to communicate with reflective understanding, empathy, and compassion is necessary to effectively

- attain an accurate patient history, lifestyle, and description of the physical symptoms; and
- describe, using language appropriate for the patient's developmental age and education, the diagnosis, prognosis, treatment options, risks, and benefits and address any questions the patient may have.

Effective communication is central for the patient to be autonomous because it empowers the patient to provide informed consent, authorizing the practitioner to provide treatments that maximize the patient's best interests.

Practitioner: Beneficence and Nonmaleficence

Central to the professional responsibility of all practitioners are the moral principles of beneficence (do good) and nonmaleficence (do no harm), with the prime directive being the maximization of the patient's best interests in accordance with the patient's reasonable goals, values, and priorities. Everything the practitioner does as a professional is to attain the prime directive of the patient's best interests.

Public Policy: Justice

Federal and state laws have legislated that patient abandonment is a form of healthcare malpractice. *Patient abandonment* is a legal claim when a practitioner terminates the patient–practitioner relationship without reasonable notice to the patient.

Inadvertent abandonment occurs when a patient becomes "lost in the system" due to a computer glitch or some other type of mismanagement. Even though there was no practitioner intent to abandon, this argument has consistently failed in the courts.

Transfer without proper instructions is another cause for patient abandonment malpractice. The termination of the patient–practitioner relationship must not only be done with proper time for the patient to find an alternative practitioner but also be done in the proper manner in which all the healthcare records, notes, and information necessary for proper healthcare are transferred to the new practitioner. The practitioner must actively facilitate the patient by finding an alternate caregiver if needed.

The healthcare profession is built on patient and community trust. As a result, in general, any healthcare provider's actions that weaken that trust are considered to be professional misconduct, morally inappropriate, and possibly illegal.

CONCLUDE

Understanding the importance of the patient–practitioner relationship and the social contract that becomes established once such a relationship has been established is of vital importance. Patient abandonment is a form of healthcare malpractice. If a patient–practitioner relationship is terminated, then there must always be ample time for the patient to find alternative care.

In summary, the significance of the patient–practitioner relationship cannot be overstated because it forms the basis for effective healthcare decision-making. Practitioners must prioritize patient autonomy, trust, and communication to facilitate informed consent and act in the best interest of their patients. It is essential for healthcare providers to recognize the importance of this relationship and adhere to legal requirements, ensuring that patients are not abandoned or left without proper care. By fostering a strong patient–practitioner relationship, healthcare providers can create an environment that respects patient autonomy and maximizes their well-being.

* * *

REVIEW QUESTIONS

1. Autonomy (informed consent) is based on the liberal notion that the patient knows their own goals, values, and priorities better than anyone else.
    A. True
    B. False

2. Trust in the practitioner's confidentiality and privacy is necessary for the patient to disclose protected health information.
    A. True
    B. False

3. The prime directive of the patient–practitioner relationship is to maximize the patient's best interests in accordance with the patient's reasonable goals, values, and priorities.
    A. True
    B. False

4. Patient abandonment is a legal claim when a practitioner terminates the patient–practitioner relationship without reasonable notice to the patient.
    A. True
    B. False

5. Inadvertent abandonment occurs when a patient becomes "lost in the system" due to a computer glitch or some other type of mismanagement. This argument has always worked well in the courts.
   A. True
   B. False

6. Transfer without proper instructions is a failure to provide all the healthcare records, notes, and information necessary for proper healthcare and is cause for patient abandonment malpractice.
   A. True
   B. False

7. Healthcare providers' actions that weaken the public trust in the profession are generally types of professional misconduct.
   A. True
   B. False

Answers: 1A, 2A, 3A, 4A, 5B, 6A, 7A

* * *

CLINICAL VIGNETTES

1. Ms. Ruby Franklin is a 45-year-old teacher who comes to your clinic with complaints of chronic back pain. She has been experiencing the pain for several months, and it is becoming increasingly difficult to perform daily activities. She reports no previous history of back pain and has tried over-the-counter pain medication without relief. On physical examination, you find tenderness in the lower lumbar region and pain on extension. Your differential diagnosis includes lumbar strain, herniated disc, and spinal stenosis. What is the most appropriate course of action for the practitioner to take regarding Ms. Franklin's treatment plan?
   A. The practitioner should make the treatment decision without seeking Ms. Franklin's input, as the practitioner knows best.
   B. The practitioner should provide Ms. Franklin with the available treatment options and allow her to make an informed decision regarding her care.
   C. The practitioner should provide Ms. Franklin with one treatment option and recommend that she pursue that course of treatment.
   D. The practitioner should refuse to treat Ms. Franklin, as her back pain is likely chronic and difficult to treat.

   Explanation: The practitioner's most appropriate course of action is to provide Ms. Franklin with the available treatment options and allow her to make an informed decision regarding her care. Because autonomy is a fundamental aspect of the patient–practitioner relationship, the practitioner should empower Ms. Franklin to make informed treatment decisions based on effective communication. Making the treatment decision without Ms. Franklin's input is inappropriate because this would violate her autonomy and be considered paternalistic. Similarly, providing Ms. Franklin with only one treatment option or refusing to treat her would also be considered paternalistic and a violation of her autonomy. The practitioner's ethical responsibility is to prioritize the patient's best interests and allow the patient to make informed decisions that maximize those interests. *Answer*: B

2. Ms. Yasmin Williams, a 65-year-old retired teacher, comes to your clinic complaining of chest pain and shortness of breath. She has a past healthcare history of hypertension, hyperlipidemia, and type 2 diabetes. She reports experiencing these symptoms for the past 2 days, and you notice that she looks tired and pale. You conduct a thorough physical examination and order laboratory tests to determine the cause of her symptoms. The differential diagnosis

includes acute coronary syndrome, pulmonary embolism, or exacerbation of heart failure. How should you proceed if Ms. Williams refuses hospitalization or further testing, citing her desire to avoid a costly and invasive workup that might not change her overall prognosis?
- A. Respect her autonomy and provide her with supportive care and symptom management
- B. Try to persuade her to undergo further testing and hospitalization, emphasizing the potential benefits of early intervention
- C. Override her refusal and order emergency hospitalization to investigate her symptoms further
- D. Discharge her with instructions for follow-up care, but document her refusal of hospitalization and any further testing

Explanation: Ms. Williams has the right to make informed decisions regarding her healthcare. Although hospitalization and further testing may be necessary, the patient's autonomy must be respected, and the healthcare provider should provide supportive care and symptom management in accordance with the patient's reasonable goals, values, and priorities. Trying to persuade her to undergo further testing and hospitalization disrespects her autonomy. Overriding her refusal and ordering emergency hospitalization is inappropriate unless it is necessary to prevent immediate harm. Discharging her with instructions for follow-up care and documenting her refusal of hospitalization and further testing is not sufficient to ensure the patient's well-being and could potentially result in healthcare malpractice if proper instructions and healthcare records are not transferred if the patient–practitioner relationship is terminated. *Answer*: A

3. Ms. Rosemary Jones is a 55-year-old accountant who has been experiencing unexplained weight loss, abdominal pain, and fatigue. She comes to her primary care provider, Practitioner Edward Ferguson, concerned about her symptoms. Practitioner Ferguson performs a physical examination and orders several diagnostic tests, including blood work and imaging studies. The results of the tests show evidence of pancreatic cancer, and Practitioner Ferguson refers Ms. Jones to an oncologist for further treatment. The differential diagnosis for Ms. Jones' symptoms could include a range of conditions, including pancreatic cancer, inflammatory bowel disease, celiac disease, or a gastrointestinal infection. How can Practitioner Ferguson ensure that he is providing Ms. Jones with the information and support she needs to make informed treatment decisions about her treatment?
- A. Practitioner Ferguson tells Ms. Jones that she needs to undergo aggressive treatment immediately without providing detailed information about her diagnosis or treatment options.
- B. Practitioner Ferguson withholds information about Ms. Jones' diagnosis to protect her from the emotional impact of the news.
- C. Practitioner Ferguson explains Ms. Jones' diagnosis and treatment options in a way that is easy for her to understand and helps her make an informed decision.
- D. Practitioner Ferguson refers Ms. Jones to another practitioner without discussing her diagnosis or treatment options with her.

Explanation: Providing Ms. Jones with detailed information about her diagnosis and treatment options is essential to ensuring her autonomy in decision-making. Practitioner Ferguson should take the time to explain Ms. Jones' diagnosis and treatment options in a way that is easy for her to understand and help her make an informed decision. Withholding information or rushing into treatment without providing adequate information undermines the patient's autonomy and may lead to noncompliance or regret. Referring Ms. Jones to another practitioner without discussing her diagnosis or treatment options would also be a form of patient abandonment. Therefore, the most appropriate approach is allowing for open communication and informed decision-making while respecting the patient's autonomy. *Answer*: C

4. Ms. Samantha Johnson is a 62-year-old retired nurse who presents to her primary care practitioner with a 2-month history of persistent cough and shortness of breath. She has a history of smoking, and her vital signs are within normal limits. Her physical exam is

unremarkable except for wheezing in both lung fields. The differential diagnosis includes chronic obstructive pulmonary disease, asthma, bronchitis, and lung cancer. What ethical issue arises if the practitioner fails to provide proper instructions and healthcare records when terminating the patient–practitioner relationship?
A. Patient abandonment
B. Nonmaleficence
C. Autonomy
D. Beneficence

Explanation: Patient abandonment is a form of healthcare malpractice and can result in legal consequences. If a practitioner terminates the patient–practitioner relationship, it must be done with proper time for the patient to find an alternative practitioner, and all healthcare records, notes, and information necessary for proper healthcare must be transferred to the new practitioner. Failure to provide proper instructions and healthcare records when terminating the patient–practitioner relationship would result in patient abandonment and is considered a form of healthcare malpractice. Nonmaleficence refers to the principle of doing no harm, autonomy refers to the patient's right to make informed treatment decisions based on effective communication with the practitioner, and beneficence refers to the principle of doing good and maximizing the patient's best interests. Although these ethical principles are important in the patient–practitioner relationship, they are not directly related to the ethical issue in the clinical vignette. *Answer*: A

5. Ms. Diana Stone, a 45-year-old schoolteacher, presents to her primary care practitioner with complaints of difficulty sleeping, feeling irritable, and having a lack of energy for the past few weeks. Upon further questioning, Ms. Stone reports that she has been going through a stressful divorce and has been experiencing a lot of anxiety and depression. Her primary care practitioner conducts a thorough physical examination and runs some laboratory tests to rule out any underlying healthcare conditions that may be contributing to her symptoms. The practitioner considers a clinical differential diagnosis of anxiety disorder, major depressive disorder, and adjustment disorder. What steps should the practitioner take to ensure Ms. Stone's autonomy in the decision-making process regarding her treatment plan?
    A. The practitioner should prescribe a medication without discussing potential risks and benefits with Ms. Stone.
    B. The practitioner should recommend a specific treatment plan without giving Ms. Stone any options.
    C. The practitioner should discuss treatment options and risks and benefits with Ms. Stone but not consider her reasonable goals, values, and priorities.
    D. The practitioner should discuss treatment options, risks, and benefits and consider Ms. Stone's goals, values, and priorities to create a personalized treatment plan.

    Explanation: The correct answer is option D, the practitioner should discuss treatment options, risks, and benefits and consider Ms. Stone's goals, values, and priorities to create a personalized treatment plan, because it aligns with the ethical principle of patient autonomy. Patient autonomy refers to the patient's right to make informed decisions about their healthcare based on their reasonable goals, values, and priorities. In order to empower Ms. Stone to make an informed decision, the practitioner must engage in effective communication with reflective understanding; provide her with all available treatment options, potential risks, and benefits; and consider her reasonable goals, values, and priorities. It is important to prioritize Ms. Stone's best interests while respecting her autonomy, and this can only be achieved through shared decision-making. The other options violate the ethical principle of patient autonomy and do not prioritize Ms. Stone's best interests. *Answer*: D

REFLECTION VIGNETTES

1. Practitioner Maria Rodriguez is a 50-year-old family practitioner who has been seeing a patient, Mr. Howard Green, for several years. Despite providing treatment to Mr. Green for various health concerns, Practitioner Rodriguez has grown to dislike the patient due to his constant complaining and lack of compliance with treatment plans. Practitioner Rodriguez would like to terminate the patient–practitioner relationship, but Mr. Green refuses to find another practitioner and insists on continuing to see Practitioner Rodriguez.

   What is or the ethical issue(s)?

   How should the ethical issue(s) be addressed?

2. Practitioner Emily Wilson, a 45-year-old practitioner with years of experience in endocrinology, has been treating Mr. John Davis for diabetes for the past 5 years. Despite her best efforts to adjust his medications, Mr. Davis has yet to be honest with Practitioner Wilson about his diet and lifestyle, and his blood sugar levels need to be better controlled. He also frequently misses appointments and does not follow up on recommended testing. Practitioner Wilson is frustrated with Mr. Davis' lack of compliance and believes she cannot provide effective care for him.

   What is the ethical issue(s)?

   How should the ethical issue(s) be addressed?

* * *

WEB LINKS

1. *Code of Medical Ethics*. American Medical Association. https://code-medical-ethics.ama-assn.org
   The AMA *Code of Medical Ethics*, established by the American Medical Association, is a comprehensive guide for healthcare practitioners. It addresses issues and challenges, promotes adherence to standards of care, and is continuously updated to reflect contemporary practices and challenges.
2. "Patient–Practitioner Relationship." Wikipedia https://en.wikipedia.org/wiki/Doctor-patient_relationship
   This entry on the doctor–patient relationship explores its centrality in healthcare, emphasizing the importance of trust, respect, communication, informed consent, and shared decision-making in ensuring effective healthcare care and addressing the varied dynamics, including the impact of physician biases, patient behavior, and the evolving role of technology in telehealth.
3. "Principles of Clinical Ethics and Their Application to Practice." Basil Varkey. 2021. https://www.ncbi.nlm.nih.gov/pmc/articles/PMC7923912
   This article illustrates a model for patient care that integrates ethical aspects with clinical and technical expertise.
4. "Patient Rights and Ethics." Jacob P. Olejarczyk and Michael Young. https://www.ncbi.nlm.nih.gov/books/NBK538279
   This source discusses the right to informed consent in the United States, protected by legislation at both state and federal levels.
5. "Patient–Physician Relationships." American Medical Association. https://code-medical-ethics.ama-assn.org/ethics-opinions/patient-physician-relationships
   This source emphasizes the importance of trust in the patient–physician relationship and the ethical responsibility of physicians to place patients' welfare above their own.
6. "Communication Patterns in the Doctor–Patient Relationship: Evaluating Determinants Associated with Low Paternalism in Mexico." Eduardo Lazcano-Ponce, Angelica

Angeles-Llerenas, Rocío Rodríguez-Valentín, et al. 2020. https://bmcmedethics.biomedcentral.com/articles/10.1186/s12910-020-00566-3

This article discusses the principle of respect for autonomy in the doctor–patient relationship.

7. "Ethics of Patient–Physician Relationships." American Medical Association. https://www.ama-assn.org/topics/ethics-patient-physician-relationships

This source discusses how the practice of bioethics can strengthen doctor–patient relationships and improve healthcare decisions.

8. "Medical Ethics." American Medical Association. https://www.ama-assn.org/topics/medical-ethics

This source discusses the importance of bioethics in strengthening doctor–patient relationships.

9. "Patient Rights." American Medical Association. https://code-medical-ethics.ama-assn.org/ethics-opinions/patient-rights

This source discusses the importance of a collaborative effort between patient and physician in a mutually respectful alliance.

10. "Examining Consent Within the Patient–Doctor Relationship." Marwan A. Habiba. 2000. https://jme.bmj.com/content/26/3/183

This article discusses the notion of consent in the patient–doctor relationship and its implications for patient autonomy.

11. "What Does the Evolution from Informed Consent to Shared Decision Making Teach Us About Authority in Health Care?" James F. Childress and Marcia Day Childress, 2020. https://journalofethics.ama-assn.org/article/what-does-evolution-informed-consent-shared-decision-making-teach-us-about-authority-health-care/2020-05

This source discusses the evolution of informed consent in the patient–practitioner relationship.

PRACTITIONER DISAGREEMENTS

41

Given one well-trained practitioner of the highest type, they will do better work for a thousand people than ten specialists.
—William James Mayo

* * *

MYSTERY STORY

DISPUTED PRESCRIPTIONS: THE CLASH OF CONVICTIONS IN A HEALTHCARE CENTER

In the bustling corridors of Guided Wisdom Medical Center, where the dance of medicine is performed daily, there thrived Practitioner Anna Martinez. A seasoned attending physician, she was the embodiment of expertise and precision, a mentor to the young minds eager to tread the path of healing.

Into this disciplined world walked Practitioner Jack Lee, a new resident brimming with ideas, fresh from his school of medicine and healthcare, and eager to etch his mark. With his very first steps under Attending Martinez's tutelage, he brought with him not just enthusiasm but also a clash of convictions that would ripple through the very ethos of patient care.

The disagreement unfolded on a quiet afternoon, over the life of a patient unseen but central to this drama. Attending Martinez had penned down a prescription, confident in her decision. However, Resident Lee's eyes narrowed as he read it. A different drug danced in his mind, one he believed was a better fit for the patient's ailment.

Courage fueling him, Resident Lee challenged the decision of his superior. He was met with the immovable resolve of Attending Martinez, her years of experience crystallizing into a firm directive. She asked him to comply, her voice a blend of authority and assurance.

But Resident Lee's mind was a whirlpool of frustration and ethical quandary. Should he go rogue and prescribe what he believed was right, risking not just his career but also the fragile trust that binds a healthcare team? Or should he swallow his disagreement and follow the well-worn path of hierarchy?

With the patient's well-being weighing heavily on him, Resident Lee chose wisdom over rebellion. He turned to the pages of the hospital's policy, seeking guidance from the shadows of protocol. His concerns reached the ears of the department chair, the ultimate arbitrator, who, after hearing both sides, nodded in agreement with Attending Martinez.

The incident was more than just a disagreement; it was an education. Resident Lee emerged with lessons inked into his professional psyche. He learned to wear the dual hats of deference and dialogue, understanding that the path of healing was lined with respect for experience and adherence to healthcare standards.

Months later, when another patient's fate threw Attending Martinez and Resident Lee into a disagreement, the echoes of their past guided them toward collaboration. This time, words flowed, concerns were shared, and a consensus was reached.

Resident Lee's growth was not just in his healthcare acumen but also in the art of communication. He knew that his voice mattered, but so did the symphony of teamwork and adherence to a higher standard of care. Through the guidance of experienced mentors

> and the collaborative environment fostered at Guided Wisdom Medical Center, he honed the ability to express his convictions while respecting the wisdom of those with more experience.
>
> Disputed Prescriptions was not just a tale of two practitioners; it was a microcosm of the healthcare world—a world in which conviction clashes with experience; guidelines are both compass and anchor; and at the heart of every decision lies the unspoken pledge to heal, to care, to make whole again. It was within the walls of Guided Wisdom Medical Center that Resident Lee learned the value of harmonizing differing perspectives and embracing the collective expertise of the healthcare team, recognizing that the true power of healing lay in unity and shared purpose.

* * *

THINK

The attending practitioner has the ultimate responsibility for the patient's treatment management. Therefore, if the attending disagrees with a resident's patient management, the resident must comply with the attending's treatment management. The attending practitioner is legally liable for the treatment the attending practitioner provides and directs.

If the resident disagrees with the attending's treatment management, changing the patient's care is never permissible without the attending practitioner's approval. The resident is also legally liable for any patient treatment provided and is held to the same standards of care as the attending practitioner. If the resident believes the attending is in error, then the resident must bring it to the attending practitioner's attention, discuss the disagreement, and come to a mutual understanding and agreement. If there is an irreconcilable disagreement, then it is permissible to bring it to the attention of the attending practitioner's superior, such as the division head, department chair, or chief of services, who has the institutional authority to intervene in the patient management and quality of care issues.

The patient–practitioner relationship is between the attending practitioner and the patient, and the resident is part of the supportive interprofessional healthcare team under the attending practitioner's supervision.

ASSESS

Patient: Autonomy

Patient autonomy (informed consent) must be maintained at the highest level. This means that when a resident is introduced, the resident must be identified as a member of the interprofessional team that the attending practitioner is supervising. Disagreements that the resident may have about patient management should be brought to the attending practitioner's attention, who is legally responsible for the care and supervision of the patient. Failure to follow this standard procedure is considered an assault on the patient–practitioner relationship. It can undermine the patient's trust in the practitioner and cause a reduction in the patient's willingness to provide informed consent. In addition, this failure is considered an assault on patient autonomy because patient management is grounded in the informed consent between the attending practitioner and the patient.

Practitioner: Beneficence and Nonmaleficence

Nonmaleficence (do no harm) is a cardinal value for the healthcare profession. Nonmaleficence is generally the concern when a resident disagrees with the attending practitioner's patient management.

If the resident has such a disagreement, then

- the resident must follow the institutional policies and procedures for dealing with the disagreement;
- the resident must seek out the attending practitioner for deliberation about the disagreement; and
- the attending practitioner must carefully listen to and address any questions, concerns, and disagreements from the resident or others on the interprofessional treatment team.

The healthcare record is not the location to debate disagreements in patient management. Instead, it is for documentation of patient management. Objective documentation of one's impression and plan will adequately reflect other options within the differential diagnosis and serve as a contemporaneous record.

The resident is not to follow what is thought to be harmful patient orders blindly; rather, the resident is directed to immediately seek out consultation with the attending practitioner to

- verify that the orders of the attending practitioner are harmful;
- recognize that not following the attending practitioner's orders may put the patient in harm's way; and
- if the resident is correct, then the attending practitioner, not the resident, will document any treatment changes in the healthcare record.

If harm or error has occurred, then the resident needs to learn how to analyze and report the error(s) and cooperate with the attending practitioner to communicate the harm or error(s) to the patient. Beneficence (do good) promotes this open and honest communication for maximizing the patient's best interests as determined by the patient's reasonable goals, values, and priorities. Knowing one's role and responsibility and the standards of care when dealing with patient management disagreements is essential.

Public Policy: Justice

Society has implemented collective legal liability for institutions, attending practitioners, and resident practitioners to ensure that its citizens do not receive suboptimal healthcare while students are being mentored into the healthcare profession. Institutions, attending practitioners, and residents are all legally culpable for any patient treatment that does not meet the standards of care. These liability laws strengthen the healthcare profession and society's perception of the healthcare profession as a patient–practitioner relationship.

CONCLUDE

The attending practitioner has the ultimate responsibility for patient care and management. It is not professionally acceptable for the resident to discuss with the patient disagreements the resident might have with the attending practitioner. The attending practitioner has a patient–practitioner relationship, which must always be supported and respected. Following the proper process for dealing with disagreements is a professional obligation based on the standards of care.

In summary, the attending practitioner's responsibility for patient care and management is paramount, and it is crucial that residents respect and support the patient–practitioner relationship. When disagreements occur, following the proper process and adhering to standards of care is a professional obligation. By doing so, residents and attending practitioners can work together to provide optimal care and maintain the integrity of the healthcare profession.

* * *

REVIEW QUESTIONS

1. The attending practitioner has the ultimate responsibility for the treatment management of the patient.
   A. True
   B. False

2. If the attending disagrees with a resident's patient management, the resident must comply with the attending's treatment management.
   A. True
   B. False

3. If the resident disagrees with the attending's treatment management, changing the patient's care is never permissible without the attending practitioner's approval.
   A. True
   B. False

4. The healthcare record is an excellent location to debate disagreements in patient management.
   A. True
   B. False

5. If harm or error has occurred, then the resident needs to learn how to analyze and report the error(s) and cooperate with the attending practitioner to communicate the harm or error(s) to the patient.
   A. True
   B. False

6. Institutions, attending practitioners, and residents are all legally culpable for any patient treatment that does not meet standards of care.
   A. True
   B. False

Answers: 1A, 2A, 3A, 4B, 5A, 6A

* * *

CLINICAL VIGNETTES

1. Mr. Nathanael Morris, a 68-year-old retired accountant, is admitted to the hospital with chest pain and shortness of breath. The resident assesses the patient and orders a chest X-ray, electrocardiogram, and blood work. The attending practitioner reviews the results and disagrees with the resident's diagnosis and treatment plan, believing Mr. Morris is having a heart attack and requires urgent intervention. The resident disagrees, stating that the presentation is more consistent with pneumonia and that antibiotics should be initiated. The attending instructs the resident to transfer care of Mr. Morris to another resident and proceeds to intervene directly, ordering the necessary cardiac interventions. The resident is left feeling unsupported and frustrated. Which is the appropriate action for the resident to take in this situation?
   A. The resident should continue to follow his treatment plan and document his disagreement in the healthcare record.
   B. The resident should confront the attending in front of the patient and nursing staff to express his disagreement.

C. The resident should speak to the division head, department chair, or chief of services to address the disagreement.
D. The resident should go along with the attending's plan, recognizing that the attending has ultimate responsibility for patient care.

Explanation: The attending practitioner is responsible for patient care and treatment management, and the resident must comply with the attending's directives. When the resident disagrees with the attending's plan, the resident must bring it to the attending's attention and seek mutual understanding and agreement. If there is an irreconcilable disagreement, it is permissible to bring it to the attention of the attending's superior. It is inappropriate to confront the attending in front of the patient or nursing staff because this undermines the patient–practitioner relationship and can lead to a loss of trust from the patient. In addition, it is inappropriate to continue to follow a treatment plan that the attending has explicitly disagreed with. The resident has a duty to act in the patient's best interest and support the attending practitioner's ultimate responsibility for patient care. *Answer*: D

2. Ms. Zoe Reynolds is a 65-year-old retired librarian with a history of recurrent pneumonia, worsening cough, and shortness of breath. A chest X-ray shows bilateral infiltrates, and she is admitted to the hospital with a diagnosis of community-acquired pneumonia. During her hospital stay, a resident practitioner and the attending practitioner disagree on the choice of antibiotics to treat Ms. Reynolds's pneumonia. The resident suggests a broad-spectrum antibiotic, whereas the attending recommends a narrow-spectrum antibiotic. The resident believes the attending's recommendation is inadequate and is concerned that the patient's pneumonia may worsen. The resident decided to start the broad-spectrum antibiotic without the attending's approval. Which of the following actions is appropriate?
   A. The resident should continue to treat Ms. Reynolds with the broad-spectrum antibiotic without consulting the attending practitioner.
   B. The resident should follow the attending's recommended treatment.
   C. The resident should bring up the disagreement with the patient and let them choose the treatment they prefer.
   D. The resident should escalate the disagreement to the division head, department chair, or chief of services.

Explanation: The attending practitioner is responsible for the patient's care, and the resident must follow the attending's treatment management. Disagreements should be brought to the attending's attention, and the healthcare record is not for debating disagreements. In this case, the resident should follow the attending's recommended treatment. The patient–practitioner relationship is between the attending practitioner and the patient, and the resident is part of the supportive interprofessional healthcare team under the attending practitioner's supervision. The other options are incorrect because it is never permissible to change the patient's care without the attending practitioner's approval, bringing up the disagreement with the patient would be a breach of the patient–practitioner relationship and would not be in the patient's best interest, and escalating the disagreement to a higher authority should only be done if the disagreement cannot be resolved between the attending and resident. *Answer*: B

3. Mr. John Smith is a 50-year-old retired construction worker who presents with abdominal pain and diarrhea for the past week. Upon evaluation, the resident suspects the patient may have inflammatory bowel disease (IBD), but the attending practitioner disagrees and suggests that the patient's symptoms may be due to an infectious cause. The resident has reviewed the patient's healthcare history and completed a physical examination, but the attending wants additional tests to be conducted before making a diagnosis. The resident believes that waiting for additional tests may delay the patient's diagnosis and treatment, leading to worsening of the patient's symptoms. What should the resident do?

A. Change the patient's treatment plan based on their clinical suspicion of IBD
B. Document their disagreement with the attending's plan in the patient's healthcare record
C. Discuss their clinical suspicion with the patient and suggest a second opinion
D. Bring their disagreement to the attention of the attending and follow institutional policies and procedures

Explanation: The resident's responsibility is to follow the institutional policies and procedures for dealing with disagreements with the attending practitioner. It is not permissible to change the patient's treatment plan without the attending's approval because the attending practitioner and the resident are legally liable for the patient's care. It is inappropriate for the resident to discuss their clinical suspicion with the patient or suggest a second opinion because this undermines the patient–practitioner relationship and may reduce the patient's willingness to provide informed consent. The healthcare record is not the place for documenting disagreements, as it is for documenting patient management. Bringing the disagreement to the attention of the attending and following institutional policies and procedures is the best course of action because it promotes open and honest communication, benefits the patient's best interests, and strengthens the patient–practitioner relationship. *Answer*: D

4. Ms. Abigail Davidson is a 60-year-old retired teacher who presents to the emergency department with acute-onset chest pain. After a thorough workup, the attending practitioner diagnoses Ms. Davidson with a non-ST elevation myocardial infarction and recommends starting her on aspirin and heparin therapy. The resident disagrees with the attending practitioner's choice of anticoagulant and prefers to start Ms. Davidson on a direct oral anticoagulant (DOAC) instead. What action should the resident take?
   A. The resident should follow the attending practitioner's recommendation and start Ms. Davidson on aspirin and heparin therapy.
   B. The resident should start Ms. Davidson on the DOAC instead of aspirin and heparin therapy, as the resident believes this to be the better choice.
   C. The resident should consult with the institutional review board.
   D. The resident should discuss their disagreement with Ms. Davidson and let her choose which anticoagulant she would prefer to start.

   Explanation: The attending practitioner is responsible for the patient's care and legally liable for the treatment. Therefore, although the resident may disagree with the attending practitioner's treatment management, the resident must comply with the attending practitioner's recommendations. In addition, the healthcare record is not the location to debate disagreements in patient management, and objective documentation of one's impression and plan will adequately reflect other options within the differential diagnosis and serve as a contemporaneous record. Furthermore, maintaining patient autonomy (informed consent) at the highest level is essential, and it is inappropriate for the resident to discuss disagreements with the patient. Finally, the resident must prioritize the patient's best interests by following the attending practitioner's recommendation. *Answer*: A

5. Mr. Theodore Johnson, a 65-year-old retired mechanic, presents to the clinic with a 3-day history of worsening chest pain that is nonradiating and associated with shortness of breath. After conducting a thorough physical examination, the resident suspects a diagnosis of unstable angina and orders an electrocardiogram (ECG) and cardiac biomarkers. The attending practitioner reviews the ECG and cardiac biomarkers and disagrees with the resident's diagnosis, stating that Mr. Johnson's symptoms are more consistent with gastroesophageal reflux disease (GERD) and recommends antacid treatment. The resident disagrees with the attending's diagnosis, stating that Mr. Johnson's presentation is inconsistent with GERD, and insists on further cardiac workup. What should the resident do in this situation?

A. Follow the attending's recommended GERD treatment plan and document their disagreement in the healthcare record
B. Order additional cardiac workup without the attending's approval
C. Discuss the disagreement with the attending practitioner and come to a mutual understanding and agreement
D. Report the attending practitioner to the department chair for inappropriate patient management

Explanation: The attending practitioner is responsible for the patient's care and management, and the resident is part of the interprofessional healthcare team under the attending's supervision. If the resident disagrees with the attending's diagnosis, they should bring it to the attending's attention and reach a mutual understanding and agreement. In this case, the resident should discuss their concerns and present the reasoning behind their diagnosis to the attending. The healthcare record is not the location to debate disagreements in patient management, and objective documentation of one's impression and plan will adequately reflect other options within the differential diagnosis. Following the proper process for dealing with disagreements is a professional obligation based on the standards of care. *Answer*: C

REFLECTION VIGNETTES

1. Practitioner Natalie Davis is a 28-year-old first-year resident currently working in the internal medicine department. Resident Davis is treating a patient with severe gastrointestinal issues, and the attending practitioner has suggested a treatment plan that Resident Davis disagrees with. The attending has recommended surgery, but Resident Davis thinks a more conservative approach would be appropriate. Resident Davis is considering discussing the disagreement with the patient to determine if they can support a more conservative approach and then approach the attending together.

    What is the ethical issue(s)?

    How should the ethical issue(s) be addressed?

2. Practitioner Victoria Clark is a 28-year-old resident in her second year of training in pediatrics. During rounds, Resident Clark presents a treatment plan for a young patient with a rare genetic condition; the plan includes a specific medication regimen. However, the attending practitioner disagrees with the plan and suggests a different approach to treatment. Despite several discussions and attempts to compromise, Resident Clark and the attending cannot reach a consensus on the best course of action for the patient.

    What is the ethical issue(s)?

    How should the ethical issue(s) be addressed?

* * *

WEB LINKS

1. *Code of Medical Ethics*. American Medical Association. https://code-medical-ethics.ama-assn.org
The AMA *Code of Medical Ethics*, established by the American Medical Association, is a comprehensive guide for healthcare practitioners. It addresses issues and challenges, promotes adherence to standards of care, and is continuously updated to reflect contemporary practices and challenges.
2. "Professional Liability Issues in Graduate Medical Education." Allen Kachalia and David M. Studdert. 2004. https://jamanetwork.com/journals/jama/fullarticle/199356
This article examines the legal responsibilities of resident physicians, attending physicians, and healthcare education institutions in providing patient care, emphasizing that residents

are held to the same standard of care as attending physicians, with a focus on the importance of supervision, potential malpractice risks, and the evolving legal expectations in graduate healthcare education.

3. "Principles of Clinical Ethics and Their Application to Practice." Basil Varkey. 2021. https://www.ncbi.nlm.nih.gov/pmc/articles/PMC7923912
   This article discusses the ethical obligation of physicians to benefit the patient, avoid or minimize harm, and respect the values and preferences of the patient.

4. "Patient Rights and Ethics." Jacob P. Olejarczyk and Michael Young. https://www.ncbi.nlm.nih.gov/books/NBK538279
   This source discusses the right to informed consent in the United States, protected by legislation at both state and federal levels.

5. "Moral Distress and Nurse–Physician Relationships." Ann B. Hamric. 2010. https://journalofethics.ama-assn.org/article/moral-distress-and-nurse-physician-relationships/2010-01
   This article discusses the ethical obligation of nurses to advocate for patients when they believe patients' rights or best interests are in jeopardy.

6. "Ethical Conflicts in Patient-Centered Care." Sven Ove Hansson and Barbro Fröding. 2021. https://journals.sagepub.com/doi/full/10.1177/1477750920962356
   This source discusses the concept of patient-centered care and the ethical conflicts that can arise in its implementation.

7. "Patient Rights." American Medical Association. https://code-medical-ethics.ama-assn.org/ethics-opinions/patient-rights
   This source discusses the importance of a collaborative effort between patient and physician in a mutually respectful alliance.

8. "Ethics in Health Care: Improving Patient Outcomes." Tulane University School of Public Health and Tropical Medicine. 2023. https://publichealth.tulane.edu/blog/ethics-in-healthcare
   This article discusses the principle of beneficence in healthcare, which refers to practitioners' responsibility to act in their patients' best interests.

9. "Medical Ethics." American Medical Association. https://www.ama-assn.org/topics/medical-ethics
   This source discusses the importance of bioethics in the practice of clinical medicine.

10. "Patient Rights and Ethics." StatPearls. https://www.statpearls.com/articlelibrary/viewarticle/26766
    This article discusses patient rights as a subset of human rights and the concept of ethics as customary standards for how persons should treat others.

11. "Chapter 3 Study Guide Flashcards." Quizlet. https://quizlet.com/225607810/chapter-3-study-guide-flash-cards
    This source provides a study guide on various ethical concepts, including the rights of patients.

12. "Issues in Biomedical Ethics." J. R. Vevaina, L. M. Nora, and R. C. Bone. 1993. https://pubmed.ncbi.nlm.nih.gov/8243220
    This source discusses the ethical problems arising in the practice of healthcare and the pursuit of biomedical research.

13. "Clinical Ethics and Law." Lisa V. Brock and Anna Mastroianni. https://depts.washington.edu/bhdept/ethics-medicine/bioethics-topics/detail/56
    This source discusses ethical and legal parameters in cases of disagreement among surrogate decision-makers and with advance directive/end of life/futility.

PRECISION MEDICINE

42

Trials are medicines that our gracious and wise practitioners prescribe because patients need them, and the practitioner proportions their frequency and weight to what the case requires. Let us trust the practitioner's skill and thank them for their prescription.
—Isaac Newton

* * *

MYSTERY STORY

GENETIC ORPHANS: UNCOVERING THE UNSEEN BETRAYAL IN PRECISION MEDICINE

In the bustling research lab of Beacon Genetic Center, Practitioner Emily Hart and Researcher Robert Kane were a dynamic duo, working tirelessly to advance the field of precision medicine. They believed in the power of personalized treatment, guided by the genetic makeup of each individual. However, an unexpected discovery turned their world upside down.

One day, while analyzing patient data, Practitioner Hart noticed a troubling trend. Certain genetic groups seemed to be systematically excluded from receiving innovative healthcare treatments.

"Robert, look at this," Practitioner Hart said, pointing to the screen. "There's a pattern here, and I don't like it."

Researcher Kane leaned in, his brow furrowing. "You're right. What could be the reason?"

A deeper investigation led them to GenTech, a major pharmaceutical company. Rather than making universally safe drugs, GenTech was using genetic data to exclude individuals prone to adverse reactions, focusing on wealthier demographics, and leaving others orphaned.

"Can they do this?" Researcher Kane asked, shocked.

"I don't know if it's legal, but it's certainly not ethical," Practitioner Hart replied, her voice filled with resolve.

The two faced a moral dilemma: expose the practice and risk everything or stay silent and allow the injustice to continue. Their sense of duty won, and they embarked on a perilous journey to uncover the truth.

Working tirelessly, they assembled evidence, spoke with whistleblowers, and even confronted GenTech executives. The risks were high, but they were driven by their belief in justice.

Finally, with enough proof, they blew the whistle, igniting a global conversation on the ethics of precision medicine. Legal battles ensued, reputations were shattered, and the healthcare community was forced to reevaluate its practices.

Their victory, however, was bittersweet. They had revealed an uncomfortable truth: The world of precision medicine, once celebrated for its promise, was now tainted by profit-driven motives that left patients behind. Working within the bustling research lab of Beacon Genetic Center, Practitioner Hart and Researcher Kane uncovered how precision medicine has evolved into a bifurcated field. On one side, there is individualized medicine tailored to the wealthy, excluding the poor. On the other side, there is a universalized approach that leaves out groups without the appropriate genetic makeup, environmental factors, and lifestyle.

> Practitioner Hart and Researcher Kane's exposure of this dark secret had profound consequences, both for them and the industry they loved. Together, they continued to navigate the complex world of healthcare ethics, determined to ensure that no patient was left orphaned by the system meant to heal them. Their courageous stand had opened doors to questions about science, profit, and humanity, and their journey had only just begun.

* * *

THINK

Precision medicine is a personalized approach for predicting, diagnosing, and treating disease based on the individual's lifestyle, environment, and genes variability:

Whole-genomic sequencing is the process of determining the entire DNA sequence of a patient's genome to help guide therapeutic intervention.
Predictive genomic testing is done on asymptomatic patients to determine disease risk.
Diagnostic genomic testing is done on symptomatic patients to verify a genetically based diagnosis.
Pharmacogenomic testing is done on a diagnosed patient to determine appropriate medication prescriptions.

> National Institutes of Health Director Francis Collins wrote in a blog post,
>
> Today, much of our medical healthcare is "one-size-fits-all," not tailored to the specific needs of the individual patient.
> If we are to make the biomedical breakthroughs necessary to realize the full promise of precision medicine, researchers need a lot more data that takes into account individual differences in lifestyle, environment, and biology.

Critics argue that precision medicine, although heralded as a way to advance individualized treatments, may fall victim to economic pressures that push pharmaceutical companies toward universal solutions. Rather than creating customized drugs to enhance individual efficacy, companies might "orphan" these specialized treatments. They could focus on developing more universal drugs, leaving behind or "orphaning" groups of people with genetic predispositions to higher risks of adverse reactions. This approach maximizes overall corporate profit, potentially at the cost of patient well-being.

These economic pressures may further skew pharmaceutical companies to develop drugs primarily for wealthier countries and demographics, neglecting those less profitable due to financial or genetic factors. If these economic motivations hold sway, as they often do, healthcare disparities will likely persist and widen.

Should individualized precision medicine truly be realized without careful oversight and ethical considerations, the result may be a deepening of existing healthcare disparities within an already unjust system. Far from being a panacea, precision medicine could inadvertently create new inequalities and further entrench existing ones, complicating efforts to create a more equitable healthcare landscape.

ASSESS

Patient: Autonomy

Making autonomous informed consent with precision medicine is much more complicated than a typical individual informed consent. The patient needs to understand that genetically identifying disease risk and treatment is personal and familial because genetic indicators may also identify other family members' disease risks.

Precision medicine has many unintended consequences, including the following:

1. The patient may have the added burden of informing, or not informing, a relative of potentially fatal or debilitating disease risk.
2. The patient's family members may not want to know their disease risks, especially if there is no preventable cure or treatment.
3. Being informed of risks will significantly increase anxiety for the patient and family members.
4. Employers, insurance companies, and other social organizations can adversely use the risk of disease.
5. Family relationships can be significantly strained and altered.
6. Significant changes in childbearing decision-making could result.
7. Diets and other cultural activities could be affected.
8. As technology and understanding expand, other disease risks will become available, increasing the anxiety about what is to come.
9. High costs associated with personalized medicine regimes will increase healthcare disparities.
10. Law enforcement uses personal genetic information databases, known as big data, to solve crimes while compromising confidential and private information.
11. Most practitioners have yet to be educated and trained to present all the risks and benefits that a patient should know about precision medicine before providing informed consent.
12. Most practitioners must still be educated and trained to understand and interpret genomic results.
13. Confidentiality and privacy related to big data will be deeply compromised, diminishing the patient–practitioner relationship.

Epigenetic influences are nongenetic influences on gene expression, which for many disease risks are much greater an influence than genes by themselves. However, patients tend to falsely understand that health and disease are purely genetic rather than the tripartite combination of lifestyle, environment, and genes.

The added complexities associated with precision medicine are manageable and can be effectively addressed. One start toward this would be for the practitioner to encourage the patient to include family members in the informed consent decision-making process.

Practitioner: Beneficence and Nonmaleficence

There is much to consider when deciding whether or not genomic testing will be more beneficial than harmful. This is why direct-to-consumer genomic marketing may be more of a disservice than a benefit to the individual without expert assistance and counseling, violating the principle of nonmaleficence (do no harm). However, the big data information generated from direct-to-consumer marketing is being promoted because big data is essential for corporations and researchers to advance their understanding of the genetic code and to develop profitable products from such knowledge. The practitioner's professional responsibility is to always maximize the patient's best interests according to the patient's reasonable goals, values, and priorities—not to balance the positive contribution that big data has for corporate profit and research compared to the harms of the patient.

Precision medicine creates some serious concerns regarding the principles of nonmaleficence (do no harm) and beneficence (do good) when predicting disease susceptibility is done without any ability to prevent, treat, or cure and when the underestimation of the importance and impact that epigenetic influences of lifestyle and environment have in the development of a disease is perpetuated.

As direct-to-consumer marketing for genetic tests becomes increasingly popular, more practitioners are being asked by their patients to interpret the data and to provide counsel. However, most practitioners are not qualified to interpret genomic test results or to provide genomic counseling. Under such conditions, the practitioner should refer the patient to a qualified clinical geneticist or genetic counselor.

Public Policy: Justice

One of the roles of public policy and social legislation is to decrease healthcare disparities. Because there is a very high probability that precision medicine will exacerbate and increase healthcare disparities, it logically follows that precision medicine will need to come under public policy and social control, either indirectly by whether or not Medicaid or Medicare will cover the costs or directly by laws such as the 2009 Genetic Information Discrimination Act.

More public debate and reflection are necessary to establish how to minimize healthcare disparities while advancing genetic technologies to improve equitable healthcare outcomes.

CONCLUDE

Precision medicine and genetic testing increase the complexity of informed consent; the importance of encouraging the patient to involve family members in the informed consent process; and the concern that such technologies may further increase, rather than decrease, healthcare disparities.

In summary, as precision medicine becomes increasingly prevalent, addressing the complexities it introduces to informed consent and the potential for exacerbating healthcare disparities is crucial. Public debate and reflection are necessary to establish effective strategies for minimizing disparities while advancing genetic technologies to improve equitable healthcare outcomes. Practitioners should be prepared to involve family members in the informed consent process and refer patients to qualified clinical geneticists or genetic counselors when necessary to ensure the responsible implementation of precision medicine.

* * *

REVIEW QUESTIONS

1. Precision medicine is a personalized approach for predicting, diagnosing, and treating disease based on the individual's lifestyle, environment, and genes variability.
    A. True
    B. False

2. If individualized precision medicine were to be actualized, it could exacerbate healthcare disparities within a system already fraught with injustice.
    A. True
    B. False

3. Patients generally understand that their health and disease are primarily the result of lifestyle and environment and not just purely genetic.
    A. True
    B. False

4. Big data information generated from direct-to-consumer marketing is being promoted because big data is essential for corporations and researchers to advance their understanding of the genetic code and develop profitable products from such knowledge.
    A. True
    B. False

5. Precision medicine risks the perpetuation of underestimating the importance and impact of epigenetic influences of lifestyle and environment in developing a disease.
   A. True
   B. False

6. Because there is a high probability that precision medicine will exasperate and increase healthcare disparities, it logically follows that precision medicine must come under public policy and social control.
   A. True
   B. False

Answers: 1A, 2A, 3B, 4A, 5A, 6A

* * *

CLINICAL VIGNETTES

1. Mr. Chandler Kirkland, a 45-year-old investment banker, presents to his primary care practitioner with a history of chronic migraines and a family history of Alzheimer's disease. He expresses interest in pursuing genetic testing to determine his risk for developing Alzheimer's disease and to explore personalized treatment options for his migraines. The practitioner discusses the potential risks and benefits of genetic testing, including the potential impact on Mr. Kirkland's autonomy and the implications for his family members. The differential diagnosis for Mr. Kirkland includes other causes of chronic headaches, such as tension headaches or cluster headaches. An ethical question arises regarding whether Mr. Kirkland's desire for genetic testing is outweighed by potential negative consequences for his family members, including the psychological burden of knowing their own risk for developing Alzheimer's disease. Which of the following actions should the practitioner take?
   A. The practitioner should recommend genetic testing for Mr. Kirkland without involving his family members in the informed consent process.
   B. The practitioner should discourage Mr. Kirkland from pursuing genetic testing, citing potential negative consequences for him and his family members.
   C. The practitioner should present genetic testing as an option for Mr. Kirkland and encourage him to involve his family members in the informed consent process.
   D. The practitioner should recommend genetic testing for Mr. Kirkland, but only if he agrees not to disclose the results to his family members.

   Explanation: The ethical concerns associated with precision medicine and genetic testing highlight the importance of informed consent and the need for practitioners to balance the risks and benefits of such testing carefully. In the case of Mr. Kirkland, it is important to consider the potential impact of genetic testing on his autonomy and his family members. Encouraging Mr. Kirkland to involve his family members in the informed consent process can help minimize the psychological burden of knowing their own risk for developing Alzheimer's disease. By doing so, the practitioner can ensure that all parties understand the potential risks and benefits of genetic testing and can make informed decisions about their healthcare. It is important to note that withholding information from family members may compromise the patient's autonomy and cause long-term psychological harm. *Answer*: C

2. Ms. Annabel Garcia is a 45-year-old kindergarten teacher who is experiencing severe headaches and vision changes. Her clinical symptoms and history suggest the possibility of a brain tumor, but further testing is needed to make a definitive diagnosis. Her practitioner recommends whole-genomic sequencing to identify any genetic markers that may be associated with the disease, as well as diagnostic genomic testing to confirm the diagnosis. Ms. Garcia is hesitant to undergo genomic testing because she is concerned about the potential

implications for her family members and the complexity of informed consent. Which of the following would be the practitioner's best response?
- A. Respect Ms. Garcia's wishes and forgo genomic testing, even if it may compromise the accuracy of the diagnosis and potential treatment
- B. Strongly encourage Ms. Garcia to undergo genomic testing, emphasizing the potential benefits of an accurate diagnosis and personalized treatment while minimizing the concerns about informed consent and family implications
- C. Refer Ms. Garcia to a genetic counselor to discuss the potential risks and benefits of genomic testing and to facilitate the informed consent process
- D. Leave the decision up to Ms. Garcia, even if she does not have all the necessary information to make an informed decision

Explanation: In this case, referring Ms. Garcia to a genetic counselor would be the most ethical decision. Genetic counselors are experts in genomic testing and counseling. They can provide Ms. Garcia with comprehensive information regarding the potential risks and benefits of genomic testing and help her make an informed decision. This approach also recognizes the importance of Ms. Garcia's autonomy and ensures that she is well-informed before making a decision. Encouraging Ms. Garcia to undergo testing without fully addressing her concerns would violate her autonomy and could lead to unintended consequences. Respecting Ms. Garcia's wishes and not recommending testing could compromise the accuracy of the diagnosis and the potential for personalized treatment. Leaving the decision solely up to Ms. Garcia without providing her with adequate information and support may also lead to unintended consequences. *Answer*: C

3. Mr. Fernandez Dawson is a 55-year-old retired businessman experiencing chronic joint pain and inflammation for the past year. He has tried various over-the-counter pain medications and physical therapy, but nothing seems to work. His primary care practitioner recommends precision medicine and suggests whole-genomic sequencing and pharmacogenomic testing to determine appropriate medication prescriptions. Mr. Dawson hesitates to proceed with testing, citing concerns about the potential impact on his autonomy and family members. He is worried about informing them of potential disease risks and the added burden of increased anxiety for himself and his family. He is also concerned about confidentiality and privacy related to big data and the potential impact on his relationship with his practitioner. Which of the following actions should Mr. Dawson take?
    - A. Mr. Dawson should involve his family members in the informed consent process and, if there is a consensus, then proceed with testing.
    - B. Mr. Dawson should proceed with testing and inform his family members of potential disease risks once they are identified.
    - C. Mr. Dawson should not proceed with testing, and alternative treatment options should be explored.
    - D. Mr. Dawson should proceed with testing but should not involve his family members in the informed consent decision-making process.

    Explanation: Option A, Mr. Dawson should involve his family members in the informed consent process and, if there is a consensus, then proceed with testing, is considered correct because it directly addresses Mr. Dawson's concerns about autonomy, family involvement, and anxiety related to potential disease risks. By involving his family members in the informed consent process and seeking a consensus before testing, this approach respects the autonomy of all parties involved. It recognizes the familial implications of whole-genomic sequencing. The other options either overlook Mr. Dawson's specific concerns or contradict the ethical considerations he has raised, making option A the most ethically considerate response given the information provided. *Answer*: A

4. Ms. Evangeline Adams, a 45-year-old office manager, presents to her primary care practitioner with a history of breast cancer in her family. She has no symptoms but is concerned

about her risk of developing the disease. Her practitioner discusses the possibility of using predictive genomic testing to determine her disease risk. Ms. Adams is hesitant and expresses concern about the impact that such testing may have on her family members. Her practitioner reassures her that she can choose not to share the results with her family members, emphasizes the importance of informed consent, and encourages her to involve her family members in the decision-making process. Ms. Adams ultimately decided to undergo predictive genomic testing. What ethical concerns are associated with predictive genomic testing, and how can they be addressed?

- A. Ms. Adams should not undergo predictive genomic testing because of potential negative consequences for her family members.
- B. Ms. Adams should undergo predictive genomic testing without involving her family members in the decision-making process to protect their privacy.
- C. Ms. Adams should undergo predictive genomic testing after involving her family members in the decision-making process to ensure that they are aware of the potential risks and benefits.
- D. Ms. Adams should not undergo predictive genomic testing because of potential risks associated with healthcare disparities.

Explanation: The correct answer is to involve family members in decision-making, ensuring they know the potential risks and benefits of predictive genomic testing. Informed consent is crucial for precision medicine, and the complexity of informed consent in the case of predictive genomic testing requires the involvement of family members. Predictive genomic testing has many unintended consequences, including potential negative consequences for family members, which must be addressed. Employers, insurance companies, and other social organizations can adversely use the risk of disease. Family relationships can be significantly strained and altered, and significant changes in childbearing decision-making could result. Involving family members in the decision-making process can help mitigate these negative consequences and protect patient autonomy. The other options are incorrect because not undergoing predictive genomic testing solely due to the potential negative consequences for family members does not necessarily address the ethical concerns associated with predictive genomic testing, not involving family members in the decision-making process is disrespecting the importance of informed consent, and the potential risks associated with healthcare disparities are not a reason for Ms. Adams not to undergo predictive genomic testing. *Answer*: C

5. Mr. Shiloh Hamilton is a 52-year-old retired construction worker experiencing severe joint pain in his knees for the past 6 months. He has been seeing his primary care practitioner, managing his pain with over-the-counter painkillers. However, the pain has worsened, and Mr. Hamilton is starting to experience mobility issues. His practitioner decides to refer him for precision medicine testing to determine if his joint pain has a genetic basis. The clinical differential diagnosis includes osteoarthritis, rheumatoid arthritis, and other autoimmune diseases. What ethical question is raised by this clinical vignette?
    - A. How can we ensure that the patient fully understands the implications of precision medicine testing before consenting to it?
    - B. Should the practitioner proceed with precision medicine testing without first attempting to manage the patient's pain with nongenetic interventions?
    - C. Is it ethical to conduct precision medicine testing on retired patients who may not have the financial means to pay for the associated costs?
    - D. Does precision medicine testing have the potential to exacerbate healthcare disparities?

Explanation: The correct ethical question raised by the clinical vignette is option A, "How can we ensure that the patient fully understands the implications of precision medicine testing before consenting to it?" This question is central to Mr. Hamilton's practitioner considering precision medicine testing for his severe joint pain. Ensuring that the patient comprehends the implications, including potential risks and benefits, is the primary ethical concern in this context. Options B–D, although potentially valid in other scenarios, do not directly relate to the specific details described in the vignette. *Answer*: A

REFLECTION VIGNETTES

1. Practitioner Samantha Lee, a 35-year-old practitioner specializing in internal medicine, sees a patient named Ms. Rachel Nguyen for a wellness check. During the visit, Ms. Nguyen asked Practitioner Lee if getting her genome sequenced by a direct-to-consumer genetic testing company such as 23andMe or AncestryDNA could help inform her future healthcare treatment and medication choices.

    What is the ethical issue(s)?

    How should the ethical issue(s) be addressed?

2. Ms. Elizabeth Williams, a 42-year-old practicing lawyer, presented to the practitioner with complaints of chronic fatigue, malaise, and unexplained weight gain. Upon initial assessment, the practitioner noted a history of smoking and obesity, which may increase the risk of developing lifestyle-related diseases such as type 2 diabetes, hypertension, and coronary heart disease. However, the practitioner also considered other potential causes of her symptoms, including environmental factors such as lead poisoning or lung cancer, as well as genetic predisposition to diseases such as breast cancer or Huntington's disease. A thorough examination and diagnostic workup were initiated to determine the underlying cause of Ms. Williams' symptoms and to formulate an appropriate treatment plan.

    What is the ethical issue(s)?

    How should the ethical issue(s) be addressed?

* * *

WEB LINKS

1. *Code of Medical Ethics*. American Medical Association. https://code-medical-ethics.ama-assn.org
    The AMA *Code of Medical Ethics*, established by the American Medical Association, is a comprehensive guide for healthcare practitioners. It addresses issues and challenges, promotes adherence to standards of care, and is continuously updated to reflect contemporary practices and challenges.
2. "Patients' and Professionals' Views Related to Ethical Issues in Precision Medicine." Anke Erdmann, Christoph Rehmann-Sutter, and Claudia Bozzaro. 2021. https://bmcmedethics.biomedcentral.com/articles/10.1186/s12910-021-00682-8
    This article discusses the ethical issues raised by the development of precision medicine, including the accuracy of tests, informed consent, and data misuse.
3. "Ethics in Precision Health." *AMA Journal of Ethics*. https://journalofethics.ama-assn.org/issue/ethics-precision-health
    This source discusses the ethical issues raised by precision health, including privacy, informed consent, and social justice.
4. "Ethical Issues in Precision Medicine." Jon J. Jonsson and Vigdis Stefansdottir. 2019. https://journals.sagepub.com/doi/full/10.1177/0004563219870824
    This source discusses the ethical issues in precision medicine from the perspective of a healthcare geneticist/clinical laboratorian and a genetic counselor.
5. "Applying Bioethical Principles for Directing Investment in Precision Medicine." Alison Finall and Kerina Jones. 2020. https://journals.sagepub.com/doi/full/10.1177/1477750919897380
    This article discusses the bioethical principles that should guide investment in precision medicine.

6. "The Top 10 Most-Read Bioethics Articles in 2021." Kevin B. O'Reilly. 2021. https://www.ama-assn.org/delivering-care/ethics/top-10-most-read-medical-ethics-articles-2021
   This source lists the most-read bioethics articles in 2021, some of which may discuss precision medicine.
7. "Precision Medicine, AI, and the Future of Personalized Health Care." Kevin B. Johnson, Wei-Qi Wei, Dilhan Weeraratne, et al. 2021. https://www.ncbi.nlm.nih.gov/pmc/articles/PMC7877825
   This article discusses the role of artificial intelligence in precision medicine and its implications for personalized healthcare.
8. "Cases in Precision Medicine: Concerns About Privacy and Discrimination." Deborah Stiles and Paul S. Appelbaum. 2019. https://www.ncbi.nlm.nih.gov/pmc/articles/PMC6715527
   This source discusses the ethical concerns about privacy and discrimination in precision medicine.
9. "Federal Privacy Protections: Ethical Foundations and Sources of Confusion in Clinical Medicine." *AMA Journal of Ethics*. https://journalofethics.ama-assn.org/article/federal-privacy-protections-ethical-foundations-sources-confusion-clinical-medicine/2016-03
   This article discusses the federal privacy protections in place for patient information and the ethical foundations of these protections.
10. "Medical Ethics." American Medical Association. https://www.ama-assn.org/topics/medical-ethics
    This source provides a general overview of bioethics, which may include discussions of precision medicine.
11. "Precision Health: Improving Health for Each of Us and All of Us." Centers for Disease Control and Prevention. https://www.cdc.gov/genomics/about/precision_med.htm
    This source discusses the goal of precision health and how it can improve health outcomes for individuals and populations.

PREGNANT PATIENTS

43

By examining the patient's tongue, practitioners discover the diseases of the body and philosophers the diseases of the mind.
—Justin Martyr

* * *

MYSTERY STORY

ETHICAL DANCE: NAVIGATING AUTONOMY, HEALTHCARE, AND LAW IN A PREGNANT PATIENT'S JOURNEY

Practitioner Sarah Pennington's office at PatientChoice Medical Center was a sanctuary where many pregnant women found guidance and care. A respected obstetrician, she was known for her ability to balance the ethics of her profession with empathy for her patients. But today, she faced an ethical dilemma that challenged her understanding of autonomy and the well-being of both the patient and fetus.

Her patient, Ms. Maria Hensworth, a 28-year-old woman pregnant for the first time, was diagnosed with a condition that required immediate healthcare intervention. The recommended standard of care treatment, while posing a minimal risk to the fetus, was essential for Ms. Hensworth's health.

Ms. Hensworth, however, was an ardent believer in natural healing and holistic approaches. She refused the standard of care treatment, firmly standing by her values and priorities. Practitioner Pennington understood Ms. Hensworth's autonomy (informed consent) but was concerned about the risks involved.

Practitioner Pennington pulled out the guidelines from the American College of Obstetricians and Gynecologists (ACOG) and the legal framework of their state. She meticulously explained, "I cannot force you, Ms. Hensworth, but I must ensure that you comprehend the risks involved. Your decision today might affect not just your well-being but also the well-being of your fetus," her voice betraying a hint of concern.

Ms. Hensworth's eyes welled with tears as she grasped the gravity of her decision. "I trust my body. I believe it knows what to do. Can't we find another way?"

Practitioner Pennington recognized that the patient–practitioner relationship was with Ms. Hensworth, not with the fetus, ensuring that Ms. Hensworth was not being viewed as a "fetal container," and Practitioner Pennington spent hours exploring alternative treatments that aligned with Ms. Hensworth's beliefs. Although no perfect substitute was found, they agreed on a less invasive approach that slightly increased the risk but respected Ms. Hensworth's goals, values, and priorities.

Weeks turned into months, and Practitioner Pennington monitored Ms. Hensworth's pregnancy with vigilance, maintaining a delicate balance between her patient's autonomy (informed consent) and the principles of practitioner beneficence (do good) and nonmaleficence (do no harm).

In the end, both Ms. Hensworth and her newborn child were healthy, albeit after a few nerve-wracking moments that tested Practitioner Pennington's expertise and Ms. Hensworth's faith.

Practitioner Pennington's practice at PatientChoice Medical Center continued to thrive, her reputation as a physician who respected the uniqueness of each pregnant patient's journey growing. And in her office, a photo of Ms. Hensworth and her baby served as a

> gentle reminder that medicine is not merely about procedures and laws but about understanding, compassion, and the courage to embrace the complexity of human choices. The ethos of PatientChoice Medical Center, with its focus on responsible and patient-centered care, resonated strongly in her approach, creating a lasting impact on the lives she touched.

* * *

THINK

Most of the time, the pregnant patient's best interests—as determined by the patient's reasonable goals, values, and priorities—maximize both the patient's and the fetus' well-being. However, there are times when a pregnant patient may wish to refuse or withdraw from a standard of care treatment meant for maximizing the well-being of the pregnant patient, the well-being of the fetus, or the well-being of both the patient and fetus. State laws differ as to whether or not a practitioner should respect a pregnant patient's autonomous decision to accept, decline, or withdraw from a particular healthcare treatment when such decisions affect a fetus or, in some states, even a fertilized ovum, known as an embryo.

In contrast to some state laws, ACOG states,

> Pregnancy is not an exception to the principle that a decisionally capable patient has the right to refuse treatment, even treatment needed to maintain life. Therefore, a decisionally capable pregnant woman's decision to refuse recommended medical or surgical interventions should be respected.

A joint guidance document from ACOG and the American Academy of Pediatrics states, "Any fetal intervention has implications for the pregnant woman's health and necessarily her bodily integrity, and therefore cannot be performed without her explicit informed consent." ACOG) also states,

> When the pregnant woman and fetus are conceptualized as separate patients, the pregnant woman and her medical interests, health needs, and rights can become secondary to those of the fetus. At the extreme, construing the fetus as a patient sometimes can lead to the pregnant woman being seen as a "fetal container" rather than as an autonomous agent.

To prevent the pregnant patient from being construed as a "fetal container" and for the pregnant patient to retain their autonomous (informed consent) decisional authority, the patient–practitioner relationship is by many considered to be with the pregnant patient, not with the fetus within her womb. In other words, the fetus is considered a part of the pregnant patient's body, not an independent entity, and it is incapable of having a fetal–practitioner relationship independent of the pregnant patient–practitioner relationship. However, some state laws argue for protecting all persons and define personhood as being established at fetal viability, fetal heartbeat, or even as early as fertilization.

ASSESS

Patient: Autonomy

The practitioner must obtain the pregnant patient's autonomous (informed consent) authorization before providing any healthcare treatment. It is the pregnant patient's informed consent as to whether the patient will accept, refuse, or withdraw healthcare treatment based on the pregnant patient's reasonable goals, values, and priorities. Usually, the pregnant patient's interests align with the practitioner's interests in maximizing the best interests of the pregnant patient and the patient's developing fetus. However, circumstances can occur in which the patient's health and the developing fetus' health are incompatible, making it impossible to attain both patient and fetal health.

ACOG) states,

> Circumstances may arise during pregnancy in which the interests of the pregnant woman and those of the fetus diverge.... For example, if a woman with severe cardiopulmonary disease becomes pregnant, and her condition becomes life-threatening as a result, her obstetrician gynecologist may recommend terminating the pregnancy.

Under such conditions, the practitioner should inform the pregnant patient of the treatment options, including no treatment, and the risks and benefits of the treatment options, ensuring that the patient's decision is not being manipulated through the disclosure of only selective information and not being coerced by credible threats. The practitioner should allow the pregnant patient to make the informed consent decision to accept, decline, or withdraw from the standards of care treatment options.

Practitioner: Beneficence and Nonmaleficence

The prime directive of the practitioner is to maximize the patient's best interests in accordance with the pregnant patient's reasonable goals, values, and priorities. The patient's best interests are materialized by the professional principles of beneficence (do good) and nonmaleficence (do no harm). Because the pregnant patient should not be considered a mere "fetal container," ACOG argues that the professional patient–practitioner relationship is with the pregnant patient, not the fetus. Currently, several state laws challenge this position.

Regardless of differing state laws regarding the status of the ovum and fetus, once a fetus is born, it is agreed that the newborn child has all the protective rights of the state. The practitioner now has a universally recognized patient–practitioner relationship with the child and the mother. This patient–practitioner relationship means that the practitioner must maximize the newborn child's best interests based on the standards of care regardless of whether the parent(s) consents or refuses to consent to the child's healthcare. Certainly, the practitioner should pursue informed consent from the parent(s) before treating the child. However, if consent is not attainable and if not providing the standards of care will significantly increase the risk of harm to the newborn, then the healthcare professional is obligated professionally, legally, and morally to provide the standards of care.

Public Policy: Justice

A pregnant patient's legal right to autonomously consent, refuse, or withdraw from healthcare treatment options is regulated by federal and various inconsistent state laws.

However, once the fetus has emerged from the uterus, the now child is conferred with all the civil rights associated with being a citizen, and the state is federally mandated to protect the well-being of the child as a vulnerable citizen. Once born, the state has protective interests and mandates that the practitioner provides the child with the standards of care. Failure to provide the standards of care to the minor is tantamount to child neglect and healthcare malpractice.

Parental consent is not legally required to provide a minor pregnant patient with prenatal care because as a matter of justice (be fair), the state is mandated by federal law to protect its vulnerable population's civil rights and liberties (Table 43.1). It is not that the parent(s) of the minor pregnant patient does not have the right to choose appropriate healthcare for their dependents, as clearly they do; rather, it is that the parent(s) does not have the right to refuse the standards of care for their pregnant child. The moral weight of the principles of beneficence (do good), nonmaleficence (do no harm), and justice (be fair) under certain circumstances is greater than the weight of the parent's autonomy (informed consent) when healthcare professionalism and social justice together maximize the minor's best interests in contrast with the parent's refusal of informed consent.

Table 43.1 Circumstances in Which Minors Do Not Require Parental Consent

Healthcare	Emergency care
	Sexually transmitted infections
	Contraception
	Substance abuse (most states)
	Prenatal care
Emancipated minor	Homeless
	Parent
	Married
	Military
	Financially independent

CONCLUDE

The practitioner should respect the pregnant patient's autonomous authority to consent, refuse, or withdraw from any standards of care treatment options per state law. However, after a fetus is born and becomes a child with all the conferred social rights, then the practitioner has the obligation under the principles of professional beneficence (do good) and professional nonmaleficence (do no harm) to provide the standards of care to the child even if it requires the practitioner to override the parent's lack of consent. The principle of social justice (be fair) requires that the practitioner provide standards of care to maximize the child's best interests. Refusing standards of care would violate the professional principle of nonmaleficence and the social principle of justice and would legally be considered child neglect, a type of healthcare malpractice.

In summary, pregnant patients have the right to refuse or withdraw from healthcare treatment, but state laws vary regarding whether such decisions can affect the fetus. ACOG emphasizes respecting the pregnant patient's autonomy and obtaining explicit informed consent for fetal intervention. The practitioner must prioritize the patient's best interests, informed by principles of beneficence and nonmaleficence. State laws on the status of the fetus as a person vary, but once the fetus is born, it has the protective rights of the state, and the practitioner must prioritize the newborn's best interests. Based on the principle of justice and the practitioner's professional and legal obligations, the practitioner must provide standards of care to the child, even if it requires overriding parental consent.

* * *

REVIEW QUESTIONS

1. State laws differ regarding whether or not a practitioner should respect a pregnant patient's autonomous decision to accept, decline, or withdraw from a particular healthcare treatment when such a decision affects a fetus or, in some states, even a fertilized ova, known as an embryo.
    A. True
    B. False

2. The American College of Obstetricians and Gynecologists argues that the professional patient–practitioner relationship is only with the pregnant patient, not the fetus.
    A. True
    B. False

3. Once a fetus is born, the practitioner must maximize the child's best interests based on standards of care, regardless of whether or not the parent consents to the care.
    A. True
    B. False

4. In all states, pregnant patients retain their fundamental legal right to autonomously consent, refuse, and withdraw from treatment options with or without considering the fetus.
   A. True
   B. False

5. Once a fetus is born, the state has protective interests and mandates that the practitioner provide standards of care. Failure to do so is tantamount to child neglect and healthcare malpractice.
   A. True
   B. False

Answers: 1A, 2A, 3A, 4B, 5A

* * *

CLINICAL VIGNETTES

1. Ms. Elena Price is a 30-year-old woman who is currently 14 weeks pregnant. She works as a nurse in a local hospital. Ms. Price has a healthcare history of depression and is currently experiencing depression symptoms. She comes to the obstetrician for her prenatal care appointment. She wants to stop taking her antidepressant medication due to concerns about the potential effects on the developing fetus. The obstetrician informs her of the potential risks of untreated depression during pregnancy and discusses the available treatment options. Ms. Price insists on stopping the medication and refuses any alternative treatment options. The obstetrician is concerned about the risks to Ms. Price and the developing fetus. What is the ethical question in this scenario?
   A. Should the obstetrician override Ms. Price's refusal of treatment and continue her on the antidepressant medication for the well-being of both Ms. Price and the developing fetus?
   B. Should the obstetrician respect Ms. Price's refusal of treatment and provide alternative treatment options, even if there are potential risks to Ms. Price and the developing fetus?
   C. Should the obstetrician report Ms. Price's refusal of treatment to Child Protective Services to ensure the well-being of the developing fetus?
   D. Should the obstetrician refuse to provide prenatal care to Ms. Price if she refuses treatment options for her depression?

   Explanation: In this scenario, the ethical question is whether the obstetrician should respect Ms. Price's refusal of treatment and provide alternative treatment options, even if there are potential risks to both Ms. Price and the developing fetus. ACOG recognizes the pregnant patient's autonomy and the right to refuse or withdraw from healthcare treatment. The practitioner should respect the patient's decision as long as it is informed and the patient understands the potential risks and benefits of the treatment options. In this scenario, the obstetrician should provide alternative treatment options to Ms. Price and respect her decision to refuse medication for her depression. However, the obstetrician should inform Ms. Price of the potential risks of untreated depression during pregnancy to ensure that her decision is an informed decision. Reporting Ms. Price's refusal of treatment to Child Protective Services is not an appropriate course of action because Ms. Price's refusal of treatment does not constitute child neglect or abuse. Refusing prenatal care to Ms. Price is inappropriate, violates the principle of justice (be fair), and could be considered healthcare malpractice. *Answer*: B

2. Ms. Jasmine Walsh is a 25-year-old schoolteacher in her third trimester of pregnancy who presents to the emergency department with vaginal bleeding and lower abdominal pain. She reports a history of hypertension and preeclampsia in her first pregnancy, which resulted in

an emergency cesarean section. She is currently receiving antihypertensive medication for her current pregnancy. Clinical examination reveals an elevated blood pressure of 160/110 mmHg, tender abdomen, and a fetal heart rate of 140 beats per minute. The differential diagnosis includes placental abruption, uterine rupture, or preeclampsia with severe features. After thoroughly discussing the treatment options, including an emergency cesarean section and blood transfusion, and each treatment's potential risks and benefits, Ms. Walsh declined healthcare treatment. She requested to be discharged to return home. What ethical principle is most important in this scenario?

A. Patient autonomy (informed consent)
B. Practitioner beneficence (do good) and nonmaleficence (no harm)
C. Public policy justice (be fair)—state laws regarding the fetus
D. Parental consent for a minor pregnant patient

Explanation: In this scenario, the most important ethical principle to consider is the autonomy of the pregnant patient. ACOG states that a pregnant woman's decision to refuse treatment should be respected, and any fetal intervention must have explicit informed consent from the pregnant woman. Ms. Walsh has the right to make an informed decision about her healthcare treatment, and the practitioner must respect her decision even if it poses a risk to her or the fetus. The practitioner should prioritize the patient's best interests, informed by the principles of beneficence and nonmaleficence, but only after obtaining informed consent. State laws regarding the fetus and parental consent do not apply to Ms. Walsh's case because she is an adult capable of making decisions regarding healthcare treatment. The practitioner should respect Ms. Walsh's autonomous decision to decline healthcare treatment, as it is her right to make an informed decision about her own body. The practitioner should continue counseling and support for Ms. Walsh and encourage her to return to the emergency department if her condition worsens. *Answer*: A

3. Ms. Isabelle Green is a 35-year-old accountant 12 weeks pregnant with her second child. During her prenatal visit, she expressed concerns about possibly having a cesarean section during delivery due to her previous difficult recovery. The obstetrician explains that a vaginal birth after cesarean (VBAC) is possible but has some risks, including uterine rupture. Ms. Green, who is fully informed about the risks and benefits, decides to attempt a VBAC. The obstetrician advises her to reconsider and explains that there is a higher risk of complications. However, Ms. Green is firm in her decision. What ethical dilemma does this case present?

A. The obstetrician should respect Ms. Green's autonomy and schedule the VBAC.
B. The obstetrician should explain the risks and benefits of both options again and try to convince Ms. Green not to attempt a VBAC.
C. The obstetrician should refuse to allow Ms. Green to have a VBAC and perform a cesarean section.
D. The obstetrician should schedule the VBAC, but only if Ms. Green signs a waiver acknowledging the risks involved.

Explanation: The ethical dilemma in this case is whether to respect Ms. Green's autonomy and allow her to attempt a VBAC despite the obstetrician's concerns about the higher risk of complications. Respecting patient autonomy is a fundamental principle in bioethics. According to ACOG (2017), "Women have the right to make informed decisions about their healthcare, including the right to accept or refuse healthcare or surgical treatment." In this case, Ms. Green has been fully informed about the risks and benefits of both options and has made an informed decision to attempt a VBAC. However, the obstetrician also has a duty to provide the best possible care and to avoid harm to the patient. ACOG guidelines state, "Obstetrician–gynecologists should be prepared to offer VBAC, but it is reasonable to preferentially recommend delivery via repeat cesarean delivery in individual cases." In this case, the obstetrician has expressed concerns about the higher risk of complications associated with a VBAC. Ultimately, the obstetrician should respect Ms. Green's autonomy and allow her to attempt a VBAC as long as she has been fully informed about the risks

and benefits and has made an informed decision. The obstetrician should continue to provide Ms. Green with information and support throughout her pregnancy and delivery and be prepared to respond promptly if any complications arise. Therefore, the correct answer is that the obstetrician should respect Ms. Green's autonomy and schedule the VBAC.
Answer: A

4. Mr. Reed Johnson is a 30-year-old electrician who has just become a father. His wife gave birth to a healthy baby girl. However, the baby displays unusual symptoms a few hours after the delivery, including lethargy, poor feeding, and breathing difficulties. The differential diagnosis includes various conditions, such as sepsis, pneumonia, and congenital heart disease. What should the practitioner do if the parents do not consent to treat the newborn with standards of care?

   A. Prioritize the parent's best interests and refuse to provide healthcare treatment to the newborn
   B. Respect the parent's decision to refuse healthcare treatment for the newborn and provide no further intervention
   C. Provide the standards of care to the newborn, even if the parents refuse to provide informed consent
   D. Allow the parents to make decisions regarding the healthcare treatment of the newborn without providing any guidance or recommendations

   Explanation: After the baby is born, the practitioner's obligation is to provide the standards of care to the newborn, even if the parents refuse to provide informed consent. The principle of social justice requires the practitioner to provide standards of care to maximize the child's best interests. In cases in which parental consent cannot be obtained, and failure to provide standards of care will significantly increase the risk of harm to the newborn, the healthcare professional is obligated professionally, legally, and morally to provide the standards of care. Refusing standards of care would violate the professional principle of nonmaleficence (do no harm) and the social principle of justice (be fair) and would legally be considered child neglect, a type of healthcare malpractice. Therefore, the other options are incorrect because they do not prioritize the newborn's best interests and may lead to healthcare malpractice.
   Answer: C

5. Ms. Patrick Davis is a 24-year-old graphic designer who is currently 12 weeks pregnant. She presents to the obstetrics clinic complaining of severe morning sickness, weight loss, and dehydration. On physical examination, her blood pressure is 90/60 mmHg, and her heart rate is 110 beats per minute. Laboratory tests show electrolyte imbalances and elevated liver enzymes. The differential diagnosis includes hyperemesis gravidarum and other causes of severe nausea and vomiting during pregnancy. What ethical question arises in this scenario?
   A. Should the practitioner prioritize the fetus' best interests over Ms. Davis' best interests in deciding on the appropriate treatment for hyperemesis gravidarum?
   B. Should the practitioner obtain explicit informed consent from the fetus before initiating treatment for hyperemesis gravidarum?
   C. Should the practitioner override Ms. Davis' autonomous decision if she were to refuse treatment for hyperemesis gravidarum if such refusal is likely to seriously harm the fetus?
   D. Should the practitioner obtain explicit informed consent from Ms. Davis before initiating treatment for hyperemesis gravidarum?

   Explanation: The ethical question in this scenario concerns the practitioner's obligation to obtain explicit informed consent from Ms. Davis before initiating treatment for her hyperemesis gravidarum. ACOG emphasizes that the patient–practitioner relationship is with the pregnant woman and not with the fetus to ensure the woman's autonomy and prevent her from being viewed as a "fetal container." Therefore, it is important to obtain explicit informed consent from the pregnant patient before initiating any treatment. In this case, the practitioner should explain the risks and benefits of various treatment options and obtain Ms. Davis' consent before initiating any treatment for her hyperemesis gravidarum. This ensures that the patient's autonomy is respected and that the practitioner can provide appropriate healthcare treatment while prioritizing the best interests of both the patient and the fetus. The other options are incorrect because they prioritize the fetus over the patient's autonomy and rights, which does not align with ACOG's position. *Answer*: D

REFLECTION VIGNETTES

1. Practitioner Jessica Lewis is a 45-year-old with years of experience in obstetrics and gynecology. She has a patient, Mrs. Sarah Johnson, who is 24 weeks pregnant and has just received the news that the fetus has a urinary tract blockage. The treatment is to implant a fetal shunt to allow drainage into the amniotic cavity. Without the procedure, the fetus's bladder will enlarge, leading to underdeveloped organs and facial deformities. However, Mrs. Johnson declines the treatment options for the fetus, stating that she wants to have the baby as God intends.

   What is the ethical issue(s)?

   How should the ethical issue(s) be addressed?

2. Practitioner Emily Rodriguez is a 32-year-old pediatrician who has just delivered a newborn in dire need of immediate healthcare attention. The newborn requires a treatment that is considered standard healthcare for survival. However, the parents refuse to consent to the treatment, citing religious beliefs and a desire to have the baby as God intends.

   What is the ethical issue(s)?

   How should the ethical issue(s) be addressed?

* * *

WEB LINKS

1. *Code of Medical Ethics*. American Medical Association. https://code-medical-ethics.ama-assn.org
   The AMA *Code of Medical Ethics*, established by the American Medical Association, is a comprehensive guide for healthcare practitioners. It addresses issues and challenges, promotes adherence to standards of care, and is continuously updated to reflect contemporary practices and challenges.
2. "Ethics Experts and Fetal Patients: A Proposal for Modesty." Dagmar Schmitz and Angus Clarke. 2021. https://bmcmedethics.biomedcentral.com/articles/10.1186/s12910-021-00730-3
   This article discusses the dependent moral status of fetuses and the beneficence-based obligations of healthcare professionals.
3. "Mothers Matter: Ethics and Research During Pregnancy." Anne Drapkin Lyerly and Ruth R. Faden. 2013. https://journalofethics.ama-assn.org/article/mothers-matter-ethics-and-research-during-pregnancy/2013-09
   This source discusses ethical concerns about exposing pregnant women and fetuses to the risks of research.
4. "Should a Patient Who Is Pregnant and Brain Dead Receive Life Support, Despite Objection from Her Appointed Surrogate?" Daniel Sperling. 2020. https://journalofethics.ama-assn.org/article/should-patient-who-pregnant-and-brain-dead-receive-life-support-despite-objection-her-appointed/2020-12

This article discusses the importance of respecting the rights of the pregnant woman in ethical deliberations.

5. "Refusal of Medically Recommended Treatment During Pregnancy." American College of Obstetricians and Gynecologists. 2016. https://www.acog.org/clinical/clinical-guidance/committee-opinion/articles/2016/06/refusal-of-medically-recommended-treatment-during-pregnancy

   This source discusses the ethical conflicts that can arise when a pregnant patient refuses recommended healthcare treatment.

6. "Patient Rights and Ethics." Jacob P. Olejarczyk and Michael Young. https://www.ncbi.nlm.nih.gov/books/NBK538279

   This source discusses the first bill of rights written specifically for patients by the American Hospital Association in 1973.

7. "Ethical Dilemmas in the Care of Pregnant Women: Rethinking 'Maternal–Fetal' Conflicts." Françoise Baylis, Sanda Rodgers, and David Young. 2008. https://www.cambridge.org/core/books/cambridge-textbook-of-bioethics/ethical-dilemmas-in-the-care-of-pregnant-women-rethinking-maternalfetal-conflicts/53CFB3ED3A1B9130B954753CD01E86A3

   This source discusses the ethical dilemmas in the care of pregnant women.

8. "Ethical Decision Making in Obstetrics and Gynecology." American College of Obstetricians and Gynecologists. 2007. https://www.acog.org/clinical/clinical-guidance/committee-opinion/articles/2007/12/ethical-decision-making-in-obstetrics-and-gynecology

   This source discusses the four principles of ethical decision-making in obstetrics and gynecology: respect for patient autonomy (informed consent), beneficence (do good), nonmaleficence (do no harm), and justice (be fair).

9. "An Ethically Justified Framework for Clinical Investigation to Benefit Pregnant and Fetal Patients." Frank A. Chervenak and Laurence B. McCullough. 2011. https://pubmed.ncbi.nlm.nih.gov/21534151

   This article discusses the ethical challenges to clinical investigators, institutional review boards, funding agencies, and data safety and monitoring boards in research to improve the health of pregnant and fetal patients.

10. "Cancer During Pregnancy: How to Handle the Bioethical Dilemmas?" Diogo Alpuim Costa. José Guilherme Nobre, and Susana Baptista de Almeida. 2020. https://pubmed.ncbi.nlm.nih.gov/33425755

    This article discusses the bioethical dilemmas in handling cancer during pregnancy.

11. "American College of Obstetricians and Gynecologists." VBAC: Vaginal Birth After Cesarean Delivery. 2017. Retrieved from https://www.vbacfacts.com/wp-content/uploads/2018/05/ACOG-PB184-VBAC-2017.pdf

PRODUCT SALES

44

To the extent that they are a practitioner, the practitioner considers only the patient's best interests in what they prescribe, not the practitioner's good.
—Plato

* * *

MYSTERY STORY

COMPROMISED CARE: THE MYSTERIOUS CASE OF THE CONFLICTED PRACTITIONER AND A COMMUNITY'S RECKONING

The small town of Springfield was a cocoon of familiarity and trust. People knew their neighbors, and most of all, they trusted their healthcare providers at HeartOverHustle Healing Center. Practitioner Olivia King was a beacon of this trust, renowned for her compassionate care and dedication to her patients. But one chilly autumn morning, the town's faith would be shattered.

Practitioner King was found dead in her office, her body cold and lifeless. An overdose of a commonly used pain medication was determined to be the cause of death. A quiet murmur of disbelief spread through the community. How could someone so vigilant about patient care fall victim to such a tragedy?

Detective Richard Rodriguez, known for his thorough investigations, was assigned to the case. As he delved into the mystery, he discovered something that shook the very foundation of the Springfield citizens' trust in their healthcare providers. Practitioner King had been engaging in the sale of vitamins and cosmetics to her patients, a clear violation of the healthcare profession's ethics code.

What was once a puzzling death soon unraveled into a web of deceit and greed. Detective Rodriguez found that Practitioner King had been receiving kickbacks from manufacturers, tainting her unbiased care with financial conflicts of interest.

The community was stunned. Practitioner King's image transformed from a symbol of integrity to a lesson in compromise. The case prompted a town hall meeting, where Detective Rodriguez took the opportunity to educate the healthcare providers and the general public about the ethical implications of product sales within their profession.

He spoke passionately about the importance of patient-centered care, emphasizing that the pursuit of profit should never overshadow the patient's well-being. He quoted the great philosopher, Plato, reminding them that "the medical healthcare profession is not a business but a sacred trust between the practitioner and the patient."

The town's reaction was a mix of anger, confusion, and, ultimately, resolve. Practitioner King's colleagues at HeartOverHustle Healing Center and throughout the community began to reevaluate their own practices, ensuring that they were aligned with the ethical guidelines that protect the patient–practitioner relationship. Health seminars were organized at the center and other venues to educate both providers and the public further, creating a renewed focus on patient-centered care.

In the end, Practitioner King's legacy was both a cautionary tale and an inspiration. Her death served as a wake-up call, forcing the healthcare community to recommit to their true purpose. Although her actions were inexcusable, they sparked a movement toward transparency, trust, and integrity within Springfield's healthcare system.

> The case of the conflicted practitioner left a lasting impact, turning a dark chapter into an opportunity for growth and reformation within HeartOverHustle Healing Center and the broader community. The story of Practitioner Olivia King is a haunting reminder that even the most trusted individuals can falter and that constant vigilance, education, and adherence to ethical principles are essential in preserving the sanctity of patient care. Her story lives on, not as a symbol of betrayal but as a catalyst for change in a small town that learned the hard way that trust must always be earned, nurtured, and never taken for granted.

* * *

THINK

There is a long-standing capitalistic doctrine in American contract law known as *caveat emptor*, Latin for "let the buyer beware." This capitalistic contract is the antithesis of the healthcare patient-center maxim of "patient's best interests."

Product sales or endorsements by a practitioner are considered a conflict of interest and therefore regulated legally, professionally, and ethically. Practitioners engaging in product sales threaten the patient–practitioner relationship and how society perceives and interacts with the healthcare profession.

The healthcare profession is not a business meant for profit but, rather, a patient-centered profession for the patient's best interests.

ASSESS

Patient: Autonomy

Product sales to patients risk the erosion of the patient–practitioner relationship, eroding patient autonomy (informed consent). Patient informed consent is essential for authorizing the practitioner to treat the patient. For informed consent, the patient must trust the practitioner's competence and the information provided to be unbiased and appropriate. "Let the buyer beware" is incompatible with trust and unbiased treatment options.

The American College of Physicians (ACP) *Ethics Manual* states, "The sale of products from the physician's [practitioner's] office might also be considered a form of self-referral and might negatively affect the trust necessary to sustain the patient–physician [practitioner] relationship." This is not an assessment against capitalism per se; rather, it is an assessment that capitalism within the patient–practitioner relationship is unprofessional and morally incompatible.

Practitioner: Beneficence and Nonmaleficence

The practitioner's professional prime directive is the patient-centered focus: to maximize the patient's best interests as determined by the patient's reasonable goals, values, and priorities using the principles of beneficence (do good) and nonmaleficence (do no harm). Although product sales do not necessarily violate the principles of beneficence or nonmaleficence, they violate the patient-centered focus, thereby violating the trusting relationship.

Adam Smith's invisible hand argued that acting in one's self-interest would result in the greater good for all society. Even if this were to be economically accurate, this is in total contradiction with the practitioner's professional prime directive of altruistically pursuing the patient's best interests, not the practitioner's nor society's best interests. The practitioner's interests are vital and should be pursued, but within the context of the patient–practitioner social contract, which is not part of the agreement.

The American Medical Association (AMA) *Code of Medical Ethics* states,

> Physician [practitioner] sale of health-related products raises ethical concerns about a financial conflict of interest, risks placing undue pressure on the patient, and threatens

to erode patient trust, undermine the primary obligation of physicians [practitioners] to serve the interests of their patients before their own and demean the profession of medicine. (9.6.4 Sale of Health-Related Products)

If the practitioner makes products available that are necessary for providing the standards of care but are not readily available and without any financial gain, such as crutches, then doing so is legally, professionally, and morally accepted. Selling readily available items such as vitamins and cosmetics to patients is unacceptable.

The ACP *Ethics Manual* states, "Most products should not be sold in the office unless the products are specifically relevant to the patient's care, offer a clear benefit based on adequate clinical evidence, and meet an urgent need of the patient."

Public Policy: Justice
Numerous states have laws that prohibit the "promotion of the sale of drugs, devices, appliances, or goods provided for a patient in such a manner as to exploit the patient for the financial gain of the physician [practitioner]" (Illinois Law 225ILCS22). Federal and state governments are interested in protecting their citizens from unscrupulous practitioners who exploit their social standing and patient vulnerability.

The ACP *Ethics Manual* states, "The sale of products from the physician's [practitioner's] office might also be considered a form of self-referral."

The federal physician self-referral law (42 U.S.C. 1395) states,

> The Physician Self-Referral Law, commonly referred to as the Stark law, prohibits physicians [practitioners] from referring patients to receive "designated health services" payable by Medicare or Medicaid from entities with which the physician [practitioner] or an immediate family member has a financial relationship. (Office of Inspector General DHHS)

Without Medicare, Medicaid, and the permissibility of for-profit insurance companies, the healthcare profession would not exist in the lucrative position it currently enjoys. In exchange, society requires that practitioners only focus on ensuring that their profession is patient-centered, maximizing the patient's best interests according to the patient's reasonable goals, values, and priorities.

CONCLUDE
Combining healthcare practice with product sales for capitalistic gain is legally, professionally, and morally impermissible. Society has already given high social and economic rewards to the healthcare profession, and to break that altruistic patient-centered social contract for economic gain is unacceptable. Providing necessary but not readily available products for patients is accepted as long as the products comply with standards of care and it is communicated to the patient that the products are being sold at cost with no added remuneration.

In summary, the sale of products by practitioners is considered a conflict of interest and regulated legally, professionally, and ethically because it threatens the patient–practitioner relationship and public trust in the healthcare profession. Product sales risk the erosion of patient trust, essential for informed consent and the pursuit of the patient's best interests. The sale of necessary but not readily available products is acceptable, whereas selling readily available items such as vitamins and cosmetics is unacceptable. Society has already given high social and economic rewards to the healthcare profession, and practitioners must only focus on ensuring that their profession is patient-centered.

* * *

REVIEW QUESTIONS

1. Product sales or endorsements by a practitioner are considered a conflict of interest and, therefore, regulated legally, professionally, and ethically.
   A. True
   B. False

2. The sale of products from the practitioner's office for profit is not considered a form of self-referral.
   A. True
   B. False

3. The practitioner's professional prime directive is patient-centered: to maximize the patient's best interests as determined by the patient's reasonable goals, values, and priorities using the principles of beneficence (do good) and nonmaleficence (do no harm).
   A. True
   B. False

4. If the practitioner makes available products necessary for providing the standards of care, not readily available, and without any financial gain, such as crutches, that is acceptable.
   A. True
   B. False

5. Selling to patients readily available items such as vitamins and cosmetics is not prohibited.
   A. True
   B. False

6. The physician self-referral law, commonly called the Stark law, prohibits practitioners from referring patients to "designated health services" payable by Medicare or Medicaid from entities with which the practitioner or an immediate family member has a financial relationship.
   A. True
   B. False

Answers: 1A, 2B, 3A, 4A, 5A, 6A

* * *

CLINICAL VIGNETTES

1. Mr. Alexander Simpson is a 55-year-old retired engineer who visited his primary care practitioner for his annual checkup. During the consultation, the practitioner recommended a new multivitamin supplement and offered to sell it to Mr. Simpson from the office. Mr. Simpson is unsure whether he needs the supplement but feels obligated to buy it to maintain a good relationship with his doctor. He has no known allergies, and his physical exam and lab results are normal. What ethical concern is raised by the practitioner's offer to sell Mr. Simpson the multivitamin supplement?
   A. Autonomy
   B. Beneficence
   C. Nonmaleficence
   D. Justice

   Explanation: The practitioner's offer to sell the multivitamin supplement to Mr. Simpson raises concerns about patient autonomy. Autonomy refers to the patient's right to decide about their healthcare. By offering to sell the supplement, the practitioner creates a conflict of interest that may undermine Mr. Simpson's ability to make an informed decision about his healthcare needs. The ACP *Ethics Manual* states that the sale of products from a practitioner's office may negatively affect the trust necessary to sustain the patient–practitioner relationship and erode patient autonomy. The patient's trust in the practitioner's competence and the information provided is crucial for informed consent. The practitioner's offer to sell the supplement may make Mr. Simpson feel obligated to purchase it to maintain a good relationship with his doctor, which can undermine his autonomy. Although beneficence and nonmaleficence are important ethical principles in healthcare, they are not the main issue in this scenario. Beneficence refers to the professional duty to do good for the patient, and nonmaleficence refers to the professional duty to do no harm. Although product sales do not necessarily violate the principles of beneficence or nonmaleficence, they do violate the patient-centered focus, thereby violating the trusting relationship. Justice refers to the fair and equitable distribution of resources. Although product sales can raise issues of justice, the main ethical concern in this case is patient autonomy. *Answer*: A

2. Ms. Hannah Mitchell, a 55-year-old accountant, visits her primary care practitioner complaining of knee pain. Physical examination and imaging reveal signs consistent with osteoarthritis. The practitioner recommends physical therapy and provides Ms. Mitchell with an over-the-counter pain reliever prescription. The practitioner also offered to refer Ms. Mitchell to a physical therapist. What should the practitioner do?
   A. Refer Ms. Mitchell to the physical therapist with whom the practitioner has a financial relationship
   B. Refer Ms. Mitchell to a physical therapist without a financial relationship with the practitioner
   C. Advise Ms. Mitchell to search for a physical therapist on her own
   D. Provide physical therapy in the practitioner's office

   Explanation: The physician self-referral law, commonly referred to as the Stark law, prohibits practitioners from referring patients to receive designated health services payable by Medicare or Medicaid from entities with which the practitioner or an immediate family member has a financial relationship. Referring Ms. Mitchell to a physical therapist with whom the practitioner has a financial relationship violates the Stark law and could damage the trust between the practitioner and the patient. Referring Ms. Mitchell to a physical therapist without a financial relationship with the practitioner is the appropriate course of action because it avoids the appearance of a conflict of interest and ensures that the patient's best interests are being served. Advising Ms. Mitchell to search for a physical therapist on her own may not be practical or helpful, and providing physical therapy in the practitioner's office may not be feasible or appropriate for the patient's needs. *Answer*: B

3. Ms. Victoria Adams is a 62-year-old retired teacher who has been visiting her primary care provider, Practitioner Rachel Williams, for joint pain in her knees. During her last visit, Practitioner Williams suggested that Ms. Adams try a dietary supplement she sells at her office. Ms. Adams is hesitant but trusts Practitioner Williams and purchases the supplement. However, after taking the supplement for a few days, Ms. Adams started experiencing stomach pain and dizziness. She calls Practitioner Williams's office, and a nurse informs her that these symptoms are normal and that she should continue taking the supplement. Ms. Adams is concerned about her health and seeks a second opinion. What is the ethical question raised in this clinical vignette?
   A. Is it ethically acceptable for a practitioner to sell products to patients?
   B. Is it ethically acceptable for a nurse to provide healthcare advice without consulting the practitioner?
   C. Is it ethically acceptable for a retired person to seek healthcare from a primary care practitioner?

D. Is it ethically acceptable for a patient to seek a second opinion without informing the primary care practitioner?

Explanation: The ethical question raised in this clinical vignette is whether it is ethically acceptable for a practitioner to sell products to patients. The healthcare profession's patient-centered maxim of "patient's best interests" conflicts with the capitalistic doctrine of caveat emptor (let the buyer beware). The sale of products by practitioners is considered a conflict of interest. It is regulated legally, professionally, and ethically because it threatens the patient–practitioner relationship and public trust in the healthcare profession. Practitioner Williams' suggestion and sale of the supplement to Ms. Adams pose a conflict of interest and a violation of the patient-centered focus, thereby violating the trusting relationship. The ACP *Ethics Manual* and the AMA *Code of Medical Ethics* both highlight concerns about financial conflicts of interest, erosion of patient trust, and the ethical implications of product sales. *Answer*: A

4. Mrs. Elizabeth Hall, a 70-year-old retired schoolteacher, visits her primary care practitioner with a sprained ankle. Physical examination reveals significant swelling and tenderness. The practitioner recommends that Mrs. Hall use crutches and rest her ankle for a few days. The practitioner inquires whether Mrs. Hall has a pair of crutches at home, and Mrs. Hall replies that she does not. Which of the following would be the most ethical action for the practitioner?
    A. Sell a pair of crutches to Mrs. Hall
    B. Recommend that Mrs. Hall purchase a pair of crutches from a healthcare supply store
    C. Refuse to lend Mrs. Hall a pair of crutches
    D. Lend a pair of crutches to Mrs. Hall free of charge

Explanation: According to the ACP *Ethics Manual*, if the practitioner is making products available that are necessary for providing the standards of care, not readily available, and without any financial gain, such as crutches, that is acceptable. In this case, the practitioner is lending Mrs. Hall a pair of crutches free of charge, which is permissible and ethical. This is consistent with the principle of beneficence, because it promotes the patient's best interests, and the principle of nonmaleficence, because it does not harm the patient. The other options involve a financial gain for the practitioner, which could be perceived as a conflict of interest and erode patient trust in the practitioner. Refusing to lend would not promote Mrs. Hall's best interests and could harm her by delaying her recovery. *Answer*: D

5. Ms. Lily Taylor, a 65-year-old retired teacher, visits her primary care practitioner complaining of hip pain. Physical examination reveals signs consistent with osteoarthritis. The practitioner recommends physical therapy, weight loss, and using a cane to alleviate Ms. Taylor's symptoms. The practitioner also offers to provide Ms. Taylor with a cane, which the practitioner claims is of high quality and will better suit Ms. Taylor's needs than a store-bought cane. What should the practitioner do?
    A. Offer to sell the cane to Ms. Taylor, explaining its benefits and any financial benefits to the practitioner
    B. Refer Ms. Taylor to a healthcare supply store to purchase a cane
    C. Explain to Ms. Taylor that the practitioner cannot sell products for financial gain but can provide a cane if it is necessary for her care and if it is not readily available
    D. Provide Ms. Taylor with a cane at no cost as a courtesy of the practitioner's office

Explanation: It is acceptable for a practitioner to make products available that are necessary for providing standards of care if the products are not readily available and are sold without any financial gain, such as crutches. In this case, the practitioner can provide Ms. Taylor with a cane if necessary for her care and if it is not readily available. However, the practitioner cannot sell the cane for financial gain, which would be considered a conflict of interest. The practitioner should explain to Ms. Taylor that the cane is being provided for her care and is not sold for financial gain. Referring Ms. Taylor to a healthcare supply store may only be necessary if the cane is readily available and it is necessary for her care. Providing Ms. Lee with a cane at no cost may be a kind gesture, but it is not required and could raise concerns about undue influence or favoritism. *Answer*: C

REFLECTION VIGNETTES

1. Practitioner Leah Clark is a 40-year-old practitioner who recently started selling vitamins and cosmetics in her local practice. She noticed that most patients are compelled to buy her products because they trust that she will only sell "good" products to them. For the patients, purchasing the products is also a sign of support for Practitioner Clark's practice, and if they buy the products in front of her, they believe they will get extra special care.

   What is the ethical issue(s)?

   How should the ethical issue(s) be addressed?

2. Practitioner Alex Nguyen, a 40-year-old practitioner, has been operating in a rural community for the past 10 years. The community does not have access to a magnetic resonance imaging (MRI) machine, so Practitioner Nguyen decides to invest in an MRI machine to serve his patients better and generate some additional income for his practice.

   What is the ethical issue(s)?

   How should the ethical issue(s) be addressed?

* * *

WEB LINKS

1. *Code of Medical Ethics*. American Medical Association. https://code-medical-ethics.ama-assn.org
   The AMA *Code of Medical Ethics*, established by the American Medical Association, is a comprehensive guide for healthcare practitioners. It addresses issues and challenges, promotes adherence to standards of care, and is continuously updated to reflect contemporary practices and challenges.
2. "Fraud & Abuse Laws." Office of the Inspector General, U.S. Department of Health and Human Services. https://oig.hhs.gov/compliance/physician-education/fraud-abuse-laws
   The article outlines the five key federal fraud and abuse laws applicable to physicians—the False Claims Act, Anti-Kickback Statute, physician self-referral law (Stark law), exclusion authorities, and Civil Monetary Penalties Law—detailing their provisions, enforcement by government agencies, and the severe consequences for violations, including criminal penalties, civil fines, exclusion from federal health programs, and potential loss of healthcare licensure.
3. "Principles of Clinical Ethics and Their Application to Practice." Basil Varkey. 2021. https://www.ncbi.nlm.nih.gov/pmc/articles/PMC7923912
   This article discusses the inherent and inseparable part of clinical medicine as the physician has an ethical obligation to benefit the patient, avoid or minimize harm, and respect the values and preferences of the patient.
4. "Patient Rights and Ethics." Jacob P. Olejarczyk and Michael Young. https://www.ncbi.nlm.nih.gov/books/NBK538279
   This source discusses the right to informed consent protected by legislation at the state and federal levels in the United States.
5. "Considerations for Applying Bioethics Norms to a Biopharmaceutical Industry Setting." Luann E. Van Campen, Tatjana Poplazarova, Donald G. Therasse, and Michael Turik; on behalf of the Biopharmaceutical Bioethics Working Group. 2021. https://bmcmedethics.biomedcentral.com/articles/10.1186/s12910-021-00600-y

This paper defines biopharmaceutical bioethics as the application of bioethics norms to the research, development, supply, commercialization, and clinical use of biopharmaceutical healthcare products.

6. "AMA *Code of Medical Ethics*' Opinions on the Sale and Dispensing of Health-Related Products." AMA Council on Ethical and Judicial Affairs. 2010. https://journalofethics.ama-assn.org/article/ama-code-medical-ethics-opinions-sale-and-dispensing-health-related-products/2010-12

   This source discusses the guidelines that physicians must follow when providing or selling products to patients.

7. "Medical Ethics." American Medical Association. https://www.ama-assn.org/topics/medical-ethics

   This source provides a moral framework for the practice of clinical healthcare, promoting awareness of and adherence to bioethics.

8. "Patient Rights." American Medical Association. https://code-medical-ethics.ama-assn.org/ethics-opinions/patient-rights

   This source discusses the collaborative effort between patient and physician in a mutually respectful alliance.

9. "Ethics in Health Care: Improving Patient Outcomes." Tulane University School of Public Health and Tropical Medicine. 2023. https://publichealth.tulane.edu/blog/ethics-in-healthcare

   This source discusses the practitioners' responsibility to act in their patients' best interests, including improving their well-being and health, giving treatments to relieve pain, avoid injury, and promote health.

10. "Patient Rights and Ethics." StatPearls. https://www.statpearls.com/articlelibrary/viewarticle/26766

    This source discusses the common ethical principles, including autonomy of the patient, beneficence, nonmaleficence, justice, patient–provider fiduciary relationship, and inviolability of human life.

RACIAL CONCORDANCE

45

Congratulate the practitioner on their choice of calling, which offers a combination of intellectual and moral interests found in no other profession.
—Sir William Osler

* * *

MYSTERY STORY

CONCORDANCE CONUNDRUM: RACIAL CONCORDANCE AND THE ETHICS OF PATIENT AUTONOMY IN MEDICINE

It was a frenzied Wednesday afternoon at OneHumanity Hospital, and the emergency room was abuzz with activity. Mrs. Chantelle Jackson, a woman known in her community for her vibrant spirit, was rushed in with a severe case of pneumonia. The attending practitioner, Robert Johnson, quickly reviewed her chart, his eyes widening at a specific request: Mrs. Jackson had asked for a racially concordant practitioner.

This posed a problem for Practitioner Johnson, a Caucasian male. Mrs. Jackson's request was clear: She wanted someone who shared her African American phenotype and cultural identity.

The ethical conundrum weighed heavily on Practitioner Johnson's mind. He knew he had a professional obligation to provide the best care possible. But did Mrs. Jackson's request conflict with the healthcare ethics of nondiscrimination and fairness?

With a sigh, Practitioner Johnson picked up the phone and called his colleague, Practitioner Malik Jones, an African American practitioner known for his empathy and wisdom. Practitioner Jones, without hesitation, agreed to take on Mrs. Jackson's case.

Upon meeting Mrs. Jackson, Practitioner Jones could feel her initial apprehension. Her eyes, filled with fear and doubt, searched his face, seeking reassurance. As they spoke about her condition and the evidence-based treatment that lay ahead, her tension eased, replaced by a trust that seemed to transcend the clinical environment.

Later, Practitioner Johnson found himself reflecting on the situation. Mrs. Jackson's request for racial concordance was not merely a preference; it was a plea for understanding and connection. But where did one draw the line between autonomy and discrimination? Was there a risk of implicitly reinforcing biases and healthcare disparities?

Determined to open a dialogue, Practitioner Johnson organized a meeting with OneHumanity Hospital's healthcare staff. The room was filled with practitioners, biochemists, and pharmacists, all eager to dissect the ethical implications of racial concordance.

The conversation was both heated and enlightening. They discussed patients' constitutional right to freedom of association, their vulnerability, and their need for mutual trust. They explored the practitioner's role in maximizing patient interests without bias.

Practitioner Johnson's voice resonated as he concluded, "We must recognize our implicit biases and remember that our prime directive is to do good and do no harm. Treating patients equally, regardless of phenotype or culture, is not just our professional character; it's our moral compass."

The room fell into contemplative silence, a shared understanding dawning. The meeting marked a turning point, not just in policy but in perspective.

> As the OneHumanity Hospital healthcare staff left the room, their faces reflecting resolve, Practitioner Johnson knew they were moving in the right direction. They were one step closer to reducing healthcare disparities and embracing an approach that honored both autonomy (informed consent) and justice (be fair).
>
> They had begun to unravel the concordance conundrum.

* * *

THINK

Race is a complex social construct linked to observable characteristics or traits. Racial concordance occurs when the practitioner and patient share a racial or ethnic identity. Research indicates that some patients prefer this. However, this descriptive account does not necessarily mean racial concordance should be prescriptive or desirable.

From a descriptive standpoint, research has shown that racial concordance might lead to enhanced communication, increased trust, and improved patient satisfaction. These benefits are often attributed to shared cultural backgrounds and experiences, fostering better understanding and rapport. However, it is crucial to recognize that these findings do not apply universally, and individual preferences and experiences may vary significantly.

The transition from a descriptive to a prescriptive understanding of racial concordance is fraught with ethical, practical, and philosophical challenges, including the following:

- *Autonomy and choice*: Mandating racial concordance could infringe on the autonomy of both patients and providers, potentially restricting choices based on factors such as expertise, communication style, or proximity.
- *Equity and fairness*: Emphasizing racial concordance may inadvertently perpetuate racial stereotypes or create divisions within healthcare. Healthcare aims to offer equal treatment to all patients, and racial concordance must be balanced against this core principle.
- *Practicality*: Enforcing racial concordance may be logistically challenging, with potential privacy concerns and issues relating to the availability of providers from particular racial or ethnic backgrounds.
- *Focus on cultural competency*: Instead of prioritizing racial concordance, enhancing cultural competency training for healthcare providers might better address diverse cultural needs without relying on racial matching.
- *Individualized approach*: Acknowledging that some patients may prefer racial concordance, a nuanced approach allowing individualized considerations might be more appropriate, such as giving patients the option to request a racially concordant provider if desired.

Although some patients may prefer racial concordance, translating this into a prescriptive approach raises complex ethical and practical concerns. A more nuanced, individualized, and culturally sensitive approach might better respect the diversity and autonomy of both patients and healthcare providers.

ASSESS

Patient: Autonomy

Patients are vulnerable, and as vulnerable individuals, it is understandable that patients will autonomously choose to have a patient–practitioner relationship with those most similar to themselves regarding racial or ethnic identity and cultural background. These autonomous preferences are based on the essential elements of the patient–practitioner relationship: mutual trust, effective communication, and empathy.

Society and the healthcare profession recognize that patients have the constitutional right of freedom of association to make individualized judgments of whom they wish to establish the patient–practitioner relationship, even though evidence-based care is independent of racial or ethnic identity and culture.

Practitioner: Beneficence and Nonmaleficence

Patients' individual rights and liberties in selecting their practitioner are not reciprocal. Practitioners do not have the right or liberty to select patients based on racial or ethnic identity or culture. The prime directive of the practitioner is to maximize the patient's best interests in accordance with the patient's reasonable goals, values, and priorities using the moral principles of beneficence (do good) and nonmaleficence (do no harm).

Most often, the patient's reasonable goals, values, and priorities converge on the same end, no matter the patient's racial or ethnic identity or culture. This shared end is the patient's health and recovery from illness. Treatment for this end is determined by evidence-based care using standards of care, independent of racial or ethnic factors.

The American Medical Association (AMA) 1.1.2 Prospective Patients states, "Physicians must also uphold ethical responsibilities not to discriminate against a prospective patient based on race, gender, sexual orientation or gender identity, or other personal or social characteristics that are not clinically relevant to the individual's care."

As healthcare professionals, practitioners are obligated to help reduce healthcare disparities by treating all patients needing their services, regardless of racial or ethnic identity or cultural background. Adherence to evidence-based care and healthcare standards must always be the primary focus.

Public Policy: Justice

Discriminations that result in the violation of fundamental rights and liberties are socially unjust. Title VII of the Civil Rights Act of 1964 and other federal laws prohibit discrimination based on national origin, race, color, religion, sex, age, disability, and genetic information.

Healthcare disparities may reflect broader societal inequalities, potentially exacerbated by both conscious and unconscious biases. These biases can lead to violations of anti-discriminatory laws and contribute to inequalities in healthcare access and outcomes. Although racial concordance may address some patients' preferences and needs, careful consideration within the broader context of healthcare equity and social justice is crucial. Implementing racial concordance without this consideration may perpetuate existing disparities rather than alleviate them.

CONCLUDE

The practitioner must recognize the importance of treating patients equally, understanding that unjust discrimination is illegal, unprofessional, and unethical. Practitioners must be self-reflective, recognizing implicit biases and adhering to standards of care that mitigate those biases. The professional character of practitioners, focusing on evidence-based care and cultural competency, is vital to societal stability and consistency in healthcare.

In summary, racial concordance refers to a patient's preference for practitioners with shared racial or ethnic identity. Whereas patients have autonomy in these choices, practitioners must prioritize patients' interests, regardless of racial or cultural factors. Discrimination is unacceptable, and practitioners must adhere to healthcare standards, acknowledging and overcoming implicit biases. A more nuanced, individualized, and culturally sensitive approach may better serve the stability and consistency needed in healthcare, always considering broader social justice goals. (See Chapter 56.)

* * *

REVIEW QUESTIONS

1. Racial concordance occurs when the practitioner and patient share racial or ethnic identity.
   A. True
   B. False

2. Racial concordance in medicine occurs when the practitioner and patient have the same phenotype and cultural identity.
   A. True
   B. False

3. The transition from a descriptive to a prescriptive understanding of racial concordance is fraught with ethical, practical, and philosophical challenges.
   A. True
   B. False

4. Society and the healthcare profession recognize that patients have the constitutional right of freedom of association to make individualized judgments regarding with whom they wish to establish the patient–practitioner relationship, even though evidence-based care is independent of phenotype and culture.
   A. True
   B. False

5. Patients' rights and liberties in selecting their practitioner are reciprocal. As a profession, practitioners have the same rights and liberties as patients to select patients based on phenotype or culture.
   A. True
   B. False

6. The prime directive of the practitioner is to maximize the patient's best interests in accordance with the patient's reasonable goals, values, and priorities using the moral principles of beneficence (do good) and nonmaleficence (do no harm).
   A. True
   B. False

7. Most of the time, the patient's reasonable goals, values, and priorities will differ from those of the practitioner, which is why phenotype or cultures are so important for determining what end the patients want.
   A. True
   B. False

8. Discriminations that violate fundamental rights and liberties are, by definition, socially unjust (not fair). Title VII of the Civil Rights Act of 1964 and other federal laws prohibit discrimination based on national origin, race, color, religion, sex, age, disability, and genetic information.
   A. True
   B. False

Answers: 1A, 2A, 3A, 4A, 5B, 6A, 7B, 8A

* * *

CLINICAL VIGNETTES

1. Mr. Kenneth White, a 46-year-old Caucasian man who works as a banker, presents to Practitioner Mary Thompson, a 32-year-old African American physician, with chest pain

and fatigue symptoms. The differential diagnosis includes coronary artery disease, angina, and muscle strain. Mr. White expresses a preference for a Caucasian doctor. What is the best course of action for Practitioner Thompson?
A. Refuse to treat the patient
B. Immediately transfer the patient to a Caucasian doctor
C. Address the patient's concerns and continue treatment while respecting his autonomy
D. Ignore the patient's preference and proceed without discussion

Explanation: The correct answer is option C, address the patient's concerns and continue treatment while respecting his autonomy. In this situation, the ethical issue revolves around the patient's autonomy and preference for racial concordance. Practitioner Thompson must balance the patient's preference with the need to provide appropriate care. The best course of action is to address the patient's concerns, seek to understand his preferences, and continue treatment, respecting his autonomy while not reinforcing racial stereotypes or biases. The other options may neglect the patient's autonomy, violate anti-discrimination laws, or fail to respond thoughtfully to the patient's concerns. *Answer*: C

2. Ms. Maria Gonzalez, a 55-year-old Hispanic woman working as a teacher, presents to Practitioner James Lee, a 40-year-old Korean American physician, with chronic headaches and nausea symptoms. The differential diagnosis includes migraines, tension headaches, and brain tumors. Ms. Gonzalez requested a Hispanic doctor due to cultural beliefs. How should Practitioner Lee proceed?

A. Transfer the patient to a Hispanic doctor without discussion
B. Insist on treating the patient despite her request
C. Engage in a conversation about her concerns and offer an individualized approach
D. Reject her request and terminate the patient–practitioner relationship

Explanation: The correct answer is option C, engage in a conversation about her concerns and offer an individualized approach. The ethical issue here is balancing racial concordance with patient care. Practitioner Lee should communicate openly with Ms. Gonzalez about her concerns, providing a nuanced, individualized approach that respects her autonomy and cultural beliefs without compromising her care. The other options may dismiss the patient's autonomy or fail to consider the broader healthcare equity and cultural competency context. *Answer*: C

3. Mr. Robert Harris, a 67-year-old African American retired firefighter, presents to Practitioner Emily Chang, a 35-year-old Chinese American physician, with difficulty breathing and persistent coughing symptoms. The differential diagnosis includes chronic obstructive pulmonary disease, asthma, and lung cancer. Mr. Harris' family insists on an African American doctor. What is the most appropriate response from Practitioner Chang?
A. Respect the family's wishes and immediately transfer Mr. Harris to an African American doctor
B. Ignore the family's request and continue treatment
C. Engage the family in dialogue about their concerns, respecting both Mr. Harris' autonomy and the importance of evidence-based care
D. Report the family for discrimination

Explanation: The correct answer is option C, engage the family in dialogue about their concerns, respecting Mr. Harris' autonomy and the importance of evidence-based care. The ethical issue in this scenario is the family's preference for racial concordance. Practitioner Chang's best approach is to engage the family in dialogue about their concerns, ensuring a culturally sensitive approach while emphasizing the importance of evidence-based care. This respects Mr. Harris' autonomy and the need for equitable, nondiscriminatory treatment. The other options lack the nuance and individualized consideration required in this situation. *Answer*: C

4. Ms. Linda Johnson, a 72-year-old Caucasian female retired nurse, presents to Practitioner Samuel Adams, a 50-year-old African American physician, with joint pain and swelling symptoms. The differential diagnosis includes osteoarthritis, rheumatoid arthritis, and lupus. Ms. Johnson expresses discomfort with Practitioner Adams' race. How should Practitioner Adams address the situation?
   A. Dismiss Ms. Johnson's concerns and continue treatment
   B. Engage Ms. Johnson in a conversation about her discomfort, respecting her autonomy while providing appropriate care
   C. Transfer Ms. Johnson to a Caucasian doctor immediately
   D. Report Ms. Johnson for racial bias

   Explanation: The correct answer is option B, engage Ms. Johnson in a conversation about her discomfort, respecting her autonomy while providing appropriate care. Ms. Johnson's discomfort presents an ethical challenge, and Practitioner Adams should address it with empathy and respect for her autonomy. Engaging Ms. Johnson in a conversation about her discomfort allows for a nuanced approach that acknowledges her autonomy while addressing the broader principles of healthcare equity, nondiscrimination, and evidence-based care. The other options do not adequately balance these complex considerations. *Answer*: B

5. Mr. Michael Davis, a 38-year-old African American software engineer, presents to Practitioner Karen Williams, a 28-year-old Caucasian practitioner, with symptoms of anxiety and insomnia. The differential diagnosis includes generalized anxiety disorder, panic disorder, and major depressive disorder. Due to perceived cultural understanding, Mr. Davis' family requests a practitioner of the same racial background. How should Practitioner Williams proceed?
   A. Comply with the family's request and immediately transfer Mr. Davis to an African American doctor
   B. Dismiss the family's request and continue treatment without addressing their concerns
   C. Engage Mr. Davis and his family in a conversation, focusing on individualized care, cultural competency, and evidence-based treatment
   D. Accuse the family of being racially biased and refuse treatment

   Explanation: The correct answer is option C, engage Mr. Davis and his family in a conversation, focusing on individualized care, cultural competency, and evidence-based treatment. The ethical issue in this scenario pertains to racial concordance and cultural understanding. Practitioner Williams should engage Mr. Davis and his family in dialogue about their concerns, focusing on individualized care, cultural competency, and evidence-based treatment. This approach respects patient autonomy while not reinforcing racial stereotypes or biases. It also acknowledges the importance of cultural sensitivity without resorting to racial matching. The other options may violate principles of autonomy, healthcare equity, or professionalism, failing to respond thoughtfully to the family's concerns. *Answer*: C

REFLECTION VIGNETTES

1. Practitioner Maria Rodriguez is a 35-year-old practitioner with a specialization in family medicine. She opened her practice in a neighborhood with a large Latino population because she identifies as Latina and believes having shared ethnicity and cultural background with her patients will result in better outcomes. She believes that her patients will feel more comfortable with her and trust her more readily. In addition, she believes that deeply understanding her patients' cultural and linguistic nuances will allow her to provide more effective care.

   What is the ethical issue(s)?

   How should the ethical issue(s) be addressed?

2. Practitioner Emma Thompson is a 35-year-old practitioner who recently opened her own healthcare practice in a diverse urban community. As she begins interviewing candidates for

healthcare office staff and other interprofessional positions, Practitioner Thompson notices that her patients respond differently to staff members of different phenotypes and sexes. To ensure her patients are comfortable and well served, Practitioner Thompson hires based on positive biases, selecting candidates whose phenotypes and sexes match those of her patients.

What is the ethical issue(s)?

How should the ethical issue(s) be addressed?

* * *

WEB LINKS

1. *Code of Medical Ethics*. American Medical Association. https://code-medical-ethics.ama-assn.org
   The AMA *Code of Medical Ethics*, established by the American Medical Association, is a comprehensive guide for healthcare practitioners. It addresses issues and challenges, promotes adherence to standards of care, and is continuously updated to reflect contemporary practices and challenges.
2. "The Effects of Race and Racial Concordance on Patient–Physician Communication: A Systematic Review of the Literature." Megan Johnson Shen, Emily B. Peterson, Rosario Costas-Muñiz, et al. 2018. https://www.ncbi.nlm.nih.gov/pmc/articles/PMC5591056
   This article discusses the existence of racial disparities in healthcare in the United States, with a focus on disparities between Black and White patients.
3. "Should a Healthcare System Facilitate Racially Concordant Care for Black Patients?" Lauren A. Taylor, Osaze Udeagbala, Adam Biggs, Helen-Maria Lekas, and Keisha Ray. 2021. https://publications.aap.org/pediatrics/article/148/4/e2021051113/183293/Should-a-Healthcare-System-Facilitate-Racially
   This source discusses the urgency of reducing racial disparities in health outcomes, leading to a continual search for novel organization-level interventions that can yield substantive health gains for Black patients.
4. "Disentangling Evidence and Preference in Patient–Clinician Concordance Discussions." Leah Z. G. Rand and Zackary Berger. 2019. https://journalofethics.ama-assn.org/article/disentangling-evidence-and-preference-patient-clinician-concordance-discussions/2019-06
   This source discusses the concept of patient–clinician concordance, in which both patients and clinicians share similar values about the goals of healthcare and personal beliefs.
5. "Association of Racial/Ethnic and Gender Concordance Between Patients and Physicians with Patient Experience Ratings." Junko Takeshita, Shiyu Wang, Alison W. Loren, et al. 2020. https://jamanetwork.com/journals/jamanetworkopen/fullarticle/2772682
   This article discusses the benefits and drawbacks associated with patient–physician racial/ethnic or gender concordance.
6. "Racial, Ethnic, and Language Concordance Between Patients and Their Usual Health Care Providers." Dulce Gonzalez, Genevieve M. Kenney, Marla McDaniel, and Claire O'Brien. 2022. https://www.urban.org/sites/default/files/2022-03/racial-ethnic-and-language-concordance-between-patients-and-providers.pdf
   This source discusses the concept of racial, ethnic, and language concordance between patients and their healthcare providers.
7. "The Case for Racial Concordance Between Patients and Physicians." Jeremy Spevick. 2003. https://journalofethics.ama-assn.org/article/case-racial-concordance-between-patients-and-physicians/2003-06
   This source discusses the concept of racial concordance in the context of patient–physician relationship.
8. "Association Between Provider–Patient Racial Concordance and the Maternal Health Experience During Pregnancy." Adaora Okpa, Miatta Buxton, and Marie O'Neill. 2022. https://journals.sagepub.com/doi/full/10.1177/23743735221077522

This source discusses the impact of racial concordance in healthcare on patient–physician relationship factors such as trust, knowledge, regard, and loyalty.

9. "The Relationship of Ethnic, Racial, and Cultural Concordance to Physician–Patient Communication." Ann Neville Miller, Andrew Todd, Robert Toledo, and Venkata Naga Sreelalitapriya Duvuuri. 2023. https://pubmed.ncbi.nlm.nih.gov/35502565

This source discusses the concept of racial, ethnic, or cultural concordance between a healthcare provider and a patient as a dimension of the patient–physician relationship that could influence health outcomes for minoritized patients.

REFERRALS AND FEE SPLITTING

46

It may seem a strange principle to enunciate as the very first requirement in a hospital that it should do the sick no harm.
—Florence Nightingale

* * *

MYSTERY STORY

REFERRAL RUSE: A TANGLED WEB OF DECEPTION AND REDEMPTION IN HEALTHCARE PRACTICE

Practitioner Rachel King at HonestPath Health Center was a beacon of integrity in the healthcare community. Her dedication to patient-centered care and the unwavering belief that every healthcare decision should prioritize patient well-being had won her respect and success. Her reputation for excellence made her the go-to physician for families in the community, and her association with HonestPath Health Center underscored her commitment to truthfulness and ethical practice.

However, the arrival of an old healthcare school friend, Practitioner Michael Harris, changed everything. Practitioner Harris was now a charismatic specialist with an enticing proposition: He wanted to split fees for referrals to his booming specialty practice. Although Practitioner King knew that fee splitting was illegal and unethical, Practitioner Harris was persuasive. He painted a picture of mutual profit and argued it was "industry standard."

Despite a nagging voice of conscience, Practitioner King found herself drawn into the scheme. At first, the extra income was thrilling, and everything seemed fine. Practitioner Harris' practice flourished, and Practitioner King's doubts began to wane.

But then tragedy struck.

One of Practitioner King's patients, a loving father of three, died during a procedure at Practitioner Harris' practice. The family was inconsolable, and their grief soon turned to anger. They filed a lawsuit, accusing both Practitioner King and Practitioner Harris of negligence and malpractice.

The ensuing investigation unraveled a web of deception. The patient had been referred without any evidence-based reasoning. His healthcare records were mishandled, violating Health Insurance Portability and Accountability Act (HIPAA) laws—most damning of all was the discovery of the illegal fee-splitting agreement between Practitioner King and Practitioner Harris.

Practitioner King's world crumbled. Her reputation was in tatters, her career in jeopardy, and her principles betrayed. The guilt was unbearable, and she knew she had to make amends.

With a determined resolve, Practitioner King cooperated fully with the investigation. She reached out to the patient's family, offering not only her deepest condolences but also her commitment to justice. She began a personal crusade to educate her colleagues at HonestPath Health Center about the ethics of proper referrals and the insidious dangers of fee splitting.

Her redemption was a slow and painful journey, filled with public humiliation and self-doubt. But Practitioner King's unwavering commitment to her principles helped her rebuild trust, both in herself and in the community.

In the end, Practitioner King's story was not just about a lapse in judgment or a tale of greed. It was a testament to the power of integrity, the importance of patient trust, and the unbreakable bond between a practitioner and her calling.

> Her legacy became a lesson for all in healthcare: that the path to true success is paved with integrity, empathy, and an unwavering commitment to the well-being of those entrusted to your care.

* * *

THINK

The practitioner's prime directive is to maximize the patient's best interests as determined by the patient's reasonable goals, values, and priorities. This will frequently mean that the practitioner must refer the patient to another practitioner for healthcare services, such as diagnostic laboratory services, or to a specialist for a particular diagnosis or treatment.

It is imperative when referring that the

- patient understands the objective evidence-based reasons and the standards of care rationale for the referral;
- referrals are only to professionals who have the specific knowledge, skills, and healthcare licensing for which the patient is being referred;
- healthcare record private health information (PHI) necessary for the referral is transferred in keeping with HIPAA confidentiality and privacy laws;
- contractual patient healthcare relationships are considered;
- the patient–practitioner relationship is not dependent on accepting the recommended referral; and
- conflict of interests, such as self-referral and fee-splitting violations, are not committed.

Conflicts of interest: Decisions or actions that have the potential or appearance of influencing the practitioner's clinical judgment.

Self-referral: Referring a patient to any facility, practice, equipment, or products in which the practitioner has an economic interest. Federal Stark law prohibits self-referral arrangements for Medicare or Medicaid patients to designated health services (DHS), such as a clinical laboratory in which the practitioner has a financial interest.

Fee splitting: Occurs when a fee or commission is paid to a practitioner or by a practitioner for a referral with the express intention of ensuring referrals to the payee. Fee splitting is also known as *getting a cut* or *kickback* and is regulated by state and federal law.

Online group marketing websites such as Groupon and online telemedicine fee-for-service often use fee-splitting online technology for marketing and paying for services. Online financial structures have started to break down the universal moral analysis that all fee-splitting activities are unprofessional and unethical.

ASSESS

Patient: Autonomy

Autonomy, or self-rule, depends on the shared decision-making process between the patient and the practitioner, where the practitioner is trusted to provide accurate, evidence-based information and unbiased clinical judgments. This patient–practitioner relationship requires that the practitioner disclose all conflicts of interest that may impact the practitioner's clinical judgments. Without such conflict of interest avoidances and disclosures, the practitioner risks diminishing the patient's trust in the practitioner's clinical information and clinical judgments, along with the actual diminishing of the practitioner's unbiased clinical judgments necessary for a patient's informed consent.

If patient autonomy can only exist within a patient–practitioner relationship, and if the patient cannot trust the practitioner because of a conflict of interest, then it logically follows that the patient cannot be autonomous under those conditions. Conflicts of interest are, therefore, incompatible with patient autonomy. Because healthcare decision-making, such as referrals, is to be autonomously chosen by the patient, and because self-referral and fee splitting are types of conflicts of interest, it follows that self-referral and fee splitting are incompatible with autonomous healthcare decision-making.

Practitioner: Beneficence and Nonmaleficence

The practitioner's prime directive is the professional maxim of maximizing the patient's best interests, as determined by the patient's reasonable goals, values, and priorities. This prime directive is accomplished using the two professional principles of beneficence (do good) and nonmaleficence (do no harm).

Therefore, it is professionally necessary for the practitioner to avoid and disclose any conflict of interest that, by definition, increases the risk of compromising the practitioner's ability to provide accurate, evidence-based information and unbiased clinical judgments, increasing the patient's risk of harm, a violation of nonmaleficence (do no harm).

The practitioner has a professional obligation to avoid any real and apparent conflict of interest based on the principle of beneficence (do good). Because self-referral and fee splitting are types of conflict of interest, they also violate the principles of beneficence (do good) and nonmaleficence (do no harm).

Professionally, the patient–practitioner relationship should be patient-centered, not practitioner-centered. Mixing self-centered capitalistic motivation with altruistic patient-centered care is logically incompatible. Self-referring and fee splitting are incompatible with patient-centered healthcare and, therefore, a conflict of interest.

Public Policy: Justice

Justice (be fair) is instantiated through the process of legislation of public policy for

- protecting individual rights and liberties;
- implementing fair procedures for equal opportunities; and
- distributing social goods equitably throughout society.

In parallel with the public policy purposes, conflicts of interest are incompatible with the patient's rights and liberties for

- unbiased information and judgments from the practitioner;
- equal opportunity for a patient–practitioner relationship; and
- living in an equitable society.

Stark law is a federal law that prohibits practitioner self-referral with any entity that the practitioner or family members have a financial interest in for "designated health services" payable by Medicare or Medicaid. Patients need to understand that in the healthcare profession, public policy views the conflicts of interest arising from practitioner self-referrals and fee-splitting as inconsistent with the principle of justice (be fair).

Telemedicine fee-for-service and fee-splitting online technology for marketing and paying for services have begun to challenge some aspects of these professional economic boundaries. Revisions will need to be made with regard to what will be economically accepted in light of new online technologies that use a different model for marketing, providing services, and pay compared to traditional healthcare practice.

CONCLUDE

Public policy, the healthcare professions, and moral analysis all argue for the practitioner to have honest and transparent discussions with patients about any activity that could be perceived as a patient–practitioner conflict of interest and the importance of taking steps to mitigate such occurrences. Self-referrals and fee splitting are illegal, and internet group marketing practices can potentially distract the field of medicine from its patient-centered, patient–practitioner relationship, whose prime directive is to maximize the patient's best interests as determined by the patient's reasonable goals, values, and priorities.

In summary, it is essential for practitioners to maintain honest and transparent communication with their patients regarding any potential conflicts of interest, such as self-referral and fee splitting. These practices are incompatible with patient-centered care and violate legal and ethical principles. As online technologies and telemedicine continue to evolve, it is imperative for public policy and healthcare professions to address these new challenges to ensure that the prime directive of maximizing patients' best interests remains at the forefront of healthcare practice.

* * *

REVIEW QUESTIONS

1. It is imperative when referring that
   A. The patient understands the objective evidence-based reasons and the standards of care rationale for the referral.
   B. Referrals are only to professionals with the specific knowledge, skills, and healthcare licensing for which the patient is referred.
   C. Healthcare record PHI necessary for the referral is transferred in keeping with HIPAA confidentiality and privacy laws.
   D. Contractual patient healthcare relationships are considered.
   E. The patient–practitioner relationship is not dependent on accepting the recommended referral.
   F. Conflicts of interest, such as self-referral and fee-splitting violations, are not committed.
   G. All of the above.

2. *Conflicts of interest*: Decisions or actions that have the potential or appearance of influencing the practitioner's clinical judgment.
   A. True
   B. False

3. *Self-referral*: Referring a patient to any facility, practice, equipment, or products that the practitioner has an economic interest in.
   A. True
   B. False

4. Federal Stark law prohibits self-referral arrangements for Medicare or Medicaid patients to DHS, such as a clinical laboratory where the practitioner has a financial interest.
   A. True
   B. False

5. *Fee splitting*: When a fee or commission is paid to a practitioner or by a practitioner for a referral with the express intention of ensuring referrals to the payee.
   A. True
   B. False

6. If patient autonomy can only exist within a patient–practitioner relationship, and if the patient cannot trust the practitioner because of a conflict of interest, then it logically follows that the patient cannot be autonomous under those conditions.
   A. True
   B. False

7. Mixing self-centered capitalistic motivation with altruistic patient-centered care is the professional ideal of patient care.
   A. True
   B. False

8. Justice (be fair) is instantiated through the process of legislation of public policy for
   A. Protecting individual rights and liberties
   B. Implementing fair procedures for equal opportunities
   C. Distributing social goods equitably throughout society
   D. All of the above

9. New online technologies, such as telemedicine, use a different marketing model, providing services and pay, compared to traditional healthcare practice.
   A. True
   B. False

Answers: 1G, 2A, 3A, 4A, 5A, 6A, 7B, 8D, 9A

* * *

CLINICAL VIGNETTES

1. Ms. Jane Bailey is a 45-year-old accountant who presents to her primary care practitioner with chronic back pain. After a thorough physical examination, her practitioner referred her to a local physical therapy clinic for further evaluation and treatment. Upon completing her therapy, Ms. Bailey was surprised to receive an invoice indicating that her practitioner's office would split the fee for her treatment with the physical therapy clinic. Which umbrella issue was raised by the fee-splitting arrangement between Ms. Bailey's practitioner's office and the physical therapy clinic?
   A. Beneficence
   B. Nonmaleficence
   C. Autonomy
   D. Conflict of interest

   Explanation: Fee splitting, or the division of fees between healthcare providers in exchange for referrals, is considered a conflict of interest. Conflicts of interest are unethical because they violate the principles of beneficence and nonmaleficence, as they compromise the practitioner's ability to provide unbiased clinical judgments, potentially harming the patient. In addition, it is incompatible with patient autonomy because it diminishes the patient's trust in the practitioner's clinical information and judgment. Federal law prohibits self-referral arrangements for Medicare or Medicaid patients to DHS, such as a clinical laboratory in which the practitioner has a financial interest. Fee splitting is regulated by state and federal law and is incompatible with a just and equitable society. *Answer*: D

2. Ms. Olivia Garcia is a 45-year-old self-employed artist who presents to Practitioner Blake Walton's clinic with shoulder pain. Practitioner Walton performs a physical examination and diagnoses Ms. Garcia with a rotator cuff tear. Practitioner Walton then informs Ms. Garcia that he has a close friend who specializes in rotator cuff repairs and can refer her to this surgeon. Practitioner Walton said he would like to follow up with Ms. Garcia after the surgery. Ms. Garcia agrees to the referral and later receives a bill for the surgery that shows that Practitioner Walton has billed her insurance for a consultation on the day of the surgery. Is it ethical for Practitioner Walton to self-refer Ms. Garcia to his friend, a surgeon specializing in rotator cuff repairs, and bill for a consultation fee on the day of the surgery?
   A. Yes, as long as Practitioner Walton has disclosed the self-referral to Ms. Garcia and she has consented to the referral.
   B. Yes, as long as Practitioner Walton believes that his friend is the best surgeon for Ms. Garcia's particular issue.

C. No, self-referral violates the principles of beneficence and nonmaleficence and is incompatible with patient autonomy.
D. No, Practitioner Walton is violating federal law by engaging in self-referral arrangements.

Explanation: Self-referral violates the principles of beneficence and nonmaleficence because it can influence the practitioner's clinical judgment, harming the patient. In addition, self-referral is incompatible with patient autonomy because it compromises the patient's ability to make informed decisions about their healthcare. Therefore, it is unethical for Practitioner Walton to self-refer Ms. Garcia to his friend, a surgeon specializing in rotator cuff repairs. The fact that Practitioner Walton billed Ms. Garcia's insurance for a consultation on the day of the surgery further raises concerns about his intentions. The other options are incorrect because disclosing the self-referral to Ms. Garcia does not make the arrangement ethical, the belief that Practitioner Walton's friend is the best surgeon for Ms. Garcia is not sufficient justification for engaging in self-referral, and just because an action may not be illegal does not mean that the action is professional or ethical. Ms. Garcia may not be a Medicare or Medicaid patient and, therefore, does not fall under the federal Stark law, which prohibits self-referrals, but it probably should be state law. *Answer*: C

3. Mr. Vance Peck is a 45-year-old software engineer experiencing back pain for several weeks. He visits his primary care provider, Practitioner Berry Spencer, who believes Mr. Peck may need to see a specialist for further evaluation. Practitioner Spencer can refer Mr. Peck to a specialist within his own healthcare group, in which he has a financial interest, or to an independent specialist. What should practitioner Spencer do?
   A. Refer Mr. Peck to the specialist within his healthcare group
   B. Refer Mr. Peck to an independent specialist
   C. Discuss both options with Mr. Peck and let him decide
   D. Do not refer Mr. Peck to a specialist

   Explanation: Practitioner Spencer should refer Mr. Peck to an independent specialist to avoid any potential conflict of interest associated with self-referral and fee splitting. Self-referral, which involves referring a patient to any facility, practice, equipment, or products in which the practitioner has a financial interest, is prohibited under the federal Stark law for any Medicare or Medicaid patients to any DHS. Fee splitting, which occurs when a fee or commission is paid to a practitioner or by a practitioner for a referral with the express intention of ensuring referrals to the payee, is regulated by state and federal law. By referring Mr. Peck to an independent specialist, Practitioner Spencer can uphold his professional obligation to act in Mr. Peck's best interests and avoid any potential harm resulting from a conflict of interest. The option of referring to a specialist in his healthcare group would be a conflict of interest. Discussing the referral option with Mr. Peck would be unethical because even if Mr. Peck did choose Practitioner Spencer's healthcare group, that choice may not be fully autonomous. Last, it would be healthcare negligence to not refer Mr. Peck to a specialist, which could result in harm and healthcare malpractice. *Answer*: B

4. Mr. Christopher Rodriguez, a 45-year-old construction worker, presents to Practitioner Kline Davis with persistent lower back pain symptoms. Practitioner Davis suspects Mr. Rodriguez may have a spinal condition and recommends further testing, including magnetic resonance imaging (MRI). Practitioner Davis provides Mr. Rodriguez with a list of imaging centers in the area that provide these tests and explains that he has no financial interest in any of these centers. Mr. Rodriguez selects an imaging center from the list and obtains the necessary tests. However, after the tests, Mr. Rodriguez received a bill with a higher amount than expected, as the imaging center charged an additional fee for Practitioner Davis' referral of the patient to the center. The ethical question is, was Practitioner Davis' referral ethical and professional?
   A. Yes, as long as the imaging center provides a high-quality service.
   B. No, Practitioner Davis should have disclosed any financial interest in the imaging centers.
   C. No, as the referral may be viewed as a violation of nonmaleficence.
   D. No, as the referral violates the principle of patient autonomy.

Explanation: Practitioner Davis provided Mr. Rodriguez with a list of imaging centers for further testing and explicitly stated that he had no financial interest in any of these centers. However, the fact that the imaging center charged an additional fee for Practitioner Davis' referral indicates that there was indeed a financial interest, which was undisclosed. This lack of transparency violates ethical guidelines, as Practitioner Davis had a duty to disclose any financial relationships that could influence the referral. The correct response is not dependent on the quality of the service provided, the principle of nonmaleficence, or patient autonomy in this context. The key issue is the undisclosed financial interest, which is a conflict of interest and undermines the trust and professionalism essential to the patient–practitioner relationship. *Answer*: B

5. Ms. Hillary Lopez, a 35-year-old restaurant worker, presents to Practitioner Alexander Brown with chronic lower back pain symptoms. Practitioner Brown examines Ms. Lopez and recommends an MRI of the lower back to determine the cause of the pain. Practitioner Brown informs Ms. Lopez that he owns a partial interest in an imaging center that provides MRI services but recommends that Ms. Lopez obtain the MRI at a different facility without any financial interest. Practitioner Brown explains that it is important to avoid any potential conflict of interest and to ensure that the referral is based solely on healthcare necessity. Ms. Lopez obtained the MRI at a different imaging center and returned to Practitioner Brown for further treatment. Which of the following is the appropriate action for Practitioner Brown to take when referring Ms. Lopez for an MRI?
   A. Refer Ms. Lopez to the imaging center in which he has partial ownership
   B. Refer Ms. Lopez to a different imaging center in which he does not have partial ownership
   C. Ask Ms. Lopez to obtain the MRI at the imaging center in which he has partial ownership but offer her a discount
   D. Ask Ms. Lopez to obtain the MRI at the imaging center in which he has partial ownership and split the fee with her

Explanation: Practitioner Brown's recommendation for Ms. Lopez to obtain the MRI at a different imaging center without financial interest is appropriate. It is essential to avoid any potential conflict of interest and ensure that the referral is based solely on healthcare necessity. Fee splitting is considered unethical and is regulated by state and federal law. Practitioner Brown should not discount Ms. Lopez to obtain the MRI at the imaging center in which he has partial ownership because this can still be viewed as a form of self-referral. In addition, asking Ms. Lopez to obtain the MRI at the imaging center in which he has partial ownership and splitting the fee with her violates the principles of nonmaleficence and beneficence because it could potentially influence Practitioner Brown's clinical judgment. Therefore, to maintain ethical practice, it is important for Practitioner Brown to refer Ms. Lopez to an imaging center in which Practitioner Brown has no financial interest. *Answer*: B

REFLECTION VIGNETTES

1. Practitioner Jane Smith is a 52-year-old practitioner with a lucrative business referring patients needing cataract removal or corrective eye surgery to a local surgeon. In return, the surgeon provides the practitioner with a 10% cut of the surgical procedure.

   What is the ethical issue(s)?

   How should the ethical issue(s) be addressed?

2. Practitioner Sarah Robertson, a 45-year-old family practitioner in a rural community, has recently invested in a local clinical laboratory to provide convenient and quick lab results for her patients. In addition to the benefits for her patients, practitioner Robertson views this as an opportunity for extra income.

What is the ethical issue(s)?

How should the ethical issue(s) be addressed?

* * *

WEB LINKS

1. *Code of Medical Ethics*. American Medical Association. https://code-medical-ethics.ama-assn.org
   The AMA *Code of Medical Ethics*, established by the American Medical Association, is a comprehensive guide for healthcare practitioners. It addresses issues and challenges, promotes adherence to standards of care, and is continuously updated to reflect contemporary practices and challenges.
2. "Fee Splitting." American Medical Association. https://code-medical-ethics.ama-assn.org/ethics-opinions/fee-splitting
   This source discusses the ethical implications of fee splitting in healthcare, particularly in relation to patient referrals.
3. "AMA Principles of Medical Ethics: II Opinion 11.3.4 Fee Splitting." https://code-medical-ethics.ama-assn.org/sites/amacoedb/files/2022-08/11.3.4.pdf
   This document provides a comprehensive overview of the ethical concerns surrounding fee splitting in the healthcare profession.
4. "AMA *Code of Medical Ethics*' Opinions on Clinical Research." AMA Council on Ethical and Judicial Affairs. 2015. https://journalofethics.ama-assn.org/article/ama-code-medical-ethics-opinions-clinical-research/2015-12
   This article discusses the ethical implications of fee splitting in clinical research.
5. "AMA *Code of Medical Ethics*' Opinions on Ethical Referral." AMA Council on Ethical and Judicial Affairs. 2011. https://journalofethics.ama-assn.org/article/ama-code-medical-ethics-opinions-ethical-referral/2011-06
   This source discusses the ethical considerations of patient referrals in healthcare.
6. "Patient Rights and Ethics." Jacob P. Olejarczyk and Michael Young. https://www.ncbi.nlm.nih.gov/books/NBK538279
   This source provides a comprehensive overview of patient rights and ethics in healthcare, including the implications of fee splitting.
7. "Fee-Splitting/Kickbacks." American Academy of Emergency Medicine. https://www.aaem.org/resources/key-issues/fee-splitting
   This source discusses the ethical implications of fee splitting and kickbacks in emergency medicine.
8. "Policy Statement Self-Referral, Markups, Fee Splitting, and Related Practices." American Society for Clinical Pathology. https://www.ascp.org/content/docs/default-source/policy-statements/ascp-pdft-pp-self-ref-fee-split.pdf?sfvrsn=2
   This policy statement discusses the ethical implications of self-referral, markups, and fee splitting in healthcare.
9. "The Practice of Medical Referral: Ethical Concerns." E. B. Anyanwu, Abedi Harrison O., and Efe A. Ohohwakpor. 2015. 2https://www.researchgate.net/publication/273495709_The_Practice_of_Medical_Referral_Ethical_Concerns
   This source discusses the ethical concerns surrounding the practice of healthcare referral.
10. "The Ethics and Economics of Kickbacks and Fee Splitting." Mark V. Pauly. 1979. https://www.jstor.org/stable/3003336
    This source discusses the ethical and economic implications of kickbacks and fee splitting in healthcare.
11. "Referral Fee Laws for Doctors." LegalMatch. https://www.legalmatch.com/law-library/article/referral-fee-laws-for-doctors.html
    This source provides an overview of the legal implications of referral fees for practitioners.

REPORTABLE INFECTIONS AND ILLNESSES

47

It is not always in a practitioner's power to cure the sick; the disease is sometimes more potent than trained art.
—Ovid

* * *

MYSTERY STORY

CONTAGION CONFLICT: BALANCING PATIENT CONFIDENTIALITY AND PUBLIC HEALTH IN REPORTING INFECTIOUS DISEASES

It was a busy day at ContagionCare Community Hospital for Practitioner Alfred Jameson. He had seen more than 20 patients already and still had a few more to go. His last patient of the day, Mr. Roger Rodriguez, came in with a persistent cough and fever. Practitioner Jameson suspected it might be a case of tuberculosis and immediately ordered some tests.

A few days later, the test results came back, and Practitioner Jameson's suspicion was confirmed. Mr. Rodriguez had tuberculosis, a reportable disease that needed to be reported to the state health department. Practitioner Jameson was torn between his duty to report the disease and his obligation to maintain patient confidentiality.

He sat down with Mr. Rodriguez at ContagionCare Community Hospital and explained the situation, including the legal requirement to report the disease to the authorities. Practitioner Jameson also explained that he would do everything possible to protect Mr. Rodriguez's identity and keep his personal information confidential. Mr. Rodriguez agreed to the reporting, knowing that it was the right thing to do for public health.

However, a few days later, Practitioner Jameson received a call from the state health department, notifying him that there were two more cases of tuberculosis in the same area. The department asked Practitioner Jameson for any information he might have about the source of the disease. Practitioner Jameson realized that he needed to inform Mr. Rodriguez's close contacts of his diagnosis to prevent the spread of the disease.

He called Mr. Rodriguez and explained that he needed to notify his close contacts of the diagnosis, but he would keep Mr. Rodriguez's identity confidential. Mr. Rodriguez agreed, and Practitioner Jameson notified his close contacts and the state health department.

Unfortunately, the situation took a turn for the worse. One of Mr. Rodriguez's contacts, an elderly woman, had a severe reaction to the tuberculosis medication and ended up in the hospital. Her family blamed Practitioner Jameson for not warning them about the disease, and the case went to court.

Practitioner Jameson had followed all the necessary procedures and had done everything possible to protect patient confidentiality. He had even informed Mr. Rodriguez of the potential risks of notifying his contacts. Despite this, he was still being blamed for the woman's hospitalization.

The case was eventually dismissed, and Practitioner Jameson learned an important lesson about the importance of reporting reportable infections and illnesses. He realized that although it was important to protect patient confidentiality, it was equally essential to prevent the spread of communicable diseases and illnesses.

> From then on, Practitioner Jameson made sure to inform his patients at ContagionCare Community Hospital of their legal obligation to report reportable diseases and infections. He also worked closely with the state health department to ensure that the appropriate reporting procedures were followed to protect both public health and patient confidentiality.

* * *

THINK

In the United States, there are more than 70 reportable infections and illnesses. Reportable infections and illnesses are divided into four main groups:

1. Written: A report of the infection must be made in writing.
   Examples: gonorrhea and salmonellosis
2. Telephone: The provider must make a report by phone.
   Examples: rubeola (measles) and pertussis (whooping cough)
3. Number of cases
   Examples: chicken pox and influenza
4. Cancer registry: Cancer illness cases are reported to the state cancer registry.

ASSESS

Patient: Autonomy

Patients have a patient–practitioner relationship with the practitioner, which creates the environment necessary for the patient to provide informed consent. Confidentiality and privacy are central to this relationship, giving the patient the trust necessary for disclosing protected health information to the practitioner. This high regard for confidentiality and privacy maximizes the patient's best interests.

Reportable infections have a significant negative social impact. As a result, practitioners have a legal, professional, and ethical duty to mitigate those adverse effects without breaching the patient–practitioner relationship. This has been accomplished by including the categories of mandated reporting as part of the patient–practitioner social contract.

The practitioner's legal and professional requirement to report must be disclosed to the patient as soon as possible. The goal is for the patient to understand the practitioner's responsibility and for the patient to consent to report voluntarily. However, if the patient does not provide informed consent to report, then the practitioner is obligated by legal mandate, professional duty, and moral analysis to report the minimal amount of information required to the proper authorities. Mandatory reporting is not a breach of the patient–practitioner relationship but, rather, an expectation of that relationship.

Practitioner: Beneficence and Nonmaleficence

Practitioners have a professional patient-centered obligation to maximize the patient's best interests as determined by the patient's reasonable goals, values, and priorities using the professional principles of beneficence (do good) and nonmaleficence (do no harm).

If a patient has exposed others to a communicable infection, then the practitioner should counsel the patient to notify, offer to inform contacts, and notify the health department, all while ensuring patient confidentiality:

Counsel the patient: Encourage the patient to notify their contact(s) voluntarily.
   Beneficence (do good)
Offer to inform contact(s): If, after counseling, the patient is emotionally or otherwise still unwilling to notify their contact(s), then the practitioner should offer to inform the contact(s) for the patient without revealing the patient's identity.
   Beneficence (do good) and nonmaleficence (do no harm)

Notify the health department: If the patient still refuses, then the practitioner must notify the health department and provide the minimum necessary information to accomplish the intended purpose. The health department will then request a voluntary interview with the patient to construct a list of contacts to be notified. There is no legal penalty for patient noncompliance.
Nonmaleficence (do no harm)

Confidentiality: The patient must know that their identity as the source-patient will always be kept confidential.
Nonmaleficence (do no harm)

Public Policy: Justice

Public policies are legislated to protect individual rights and liberties within a general structure that promotes equitable opportunities for all citizens as a matter of justice (be fair). Not reporting these designated reportable infections and diseases violates individual rights and liberties and therefore requires legislation, as a matter of justice (be fair), to balance the unequal distribution of benefits and burdens.

It is unfair not to disclose exposure to those exposed to a communicable infection. Therefore, if a patient refuses to notify a contact, and the practitioner knows who their contact is, then although it is not mandatory to notify the contact, the practitioner has legal protection if they notify the contact. Regardless, the practitioner must inform the contact or ask the health department to notify the contact. Not following this process is a practitioner's failure of the public policy duty of justice (be fair), making the practitioner legally liable for harm that occurs because of the failure to report.

Failure to report a communicable infection or illness is a misdemeanor punishable with a fine between $50 and $1,000 and up to 90 days imprisonment.

CONCLUDE

The practitioner must balance the duty of promoting patient autonomy and confidentiality with the legal mandate to report reportable infections and illnesses to the appropriate agencies and warn those at risk of harm.

In summary, practitioners must balance upholding patient autonomy and confidentiality while adhering to their legal obligation to report infections and illnesses. By disclosing their responsibility to report and working with patients to ensure the appropriate parties are informed, practitioners can help minimize the negative social impact of reportable infections and illnesses, promoting public health and individual well-being.

* * *

REVIEW QUESTIONS

1. Reportable infections and illnesses the practitioner must report should be disclosed to the patient as soon as possible.
    A. True
    B. False

2. For reportable infections, the goal is for the patient to understand the practitioner's responsibility to report and for the patient to consent to report voluntarily.
    A. True
    B. False

3. If the patient does not provide informed consent for the practitioner to report a reportable infection or disease, then the practitioner, in accordance with the patient–practitioner relationship, must keep patient confidentiality and privacy and not report.
    A. True
    B. False

4. Mandatory reporting is a legal and professional requirement in the patient–practitioner relationship. Not reporting would be the practitioner's violation of the patient–practitioner relationship, a form of professional misconduct.
   A. True
   B. False

5. If a patient has exposed others to a communicable infection, then
   A. Counsel the patient to notify the contact(s).
   B. Offer to inform the contact(s) if the patient refuses.
   C. Notify the health department if the patient refuses to notify the contact(s) or have the practitioner notify the contact(s).
   D. Confidentiality of source-patient will be kept from contact(s).
   E. All of the above.

6. If a patient with a reportable infection refuses to notify a contact(s), then the practitioner does not have legal protection to inform the contact(s), and the practitioner must respect the patient's decision and not report to authorities.
   A. True
   B. False

7. A practitioner's failure to report a reportable infection is a violation of the public policy duty of justice (be fair), making the practitioner legally liable for harms that occur because of the failure to report.
   A. True
   B. False

Answers: 1A, 2A, 3B, 4A, 5E, 6B, 7A

* * *

CLINICAL VIGNETTES

1. Ms. Tiffany Rodriguez, a 35-year-old office worker, presents to the clinic with a persistent cough, fatigue, and shortness of breath. On examination, her oxygen saturation is low, and she is diagnosed with tuberculosis (TB). Which of the following is the most appropriate action for the practitioner?
   A. Do nothing because the patient has a right to privacy
   B. Notify the patient's employer because she works in an office
   C. Report the case of TB to the health department and inform the patient of the legal obligation to report
   D. Keep the diagnosis confidential and only notify the patient's close contacts

Explanation: The most appropriate action for the practitioner is to report the case of TB to the health department and inform the patient of the legal obligation to report. Practitioners have a legal, professional, and ethical duty to report reportable infections and diseases, including TB, to the appropriate agencies. In this case, Ms. Rodriguez's diagnosis of TB is a reportable infection, and failure to report it violates the law. The other options are incorrect because the practitioner has a duty to report the case of TB to the health department, and failure to do so may result in harm to others; the patient's employer does not need to be informed of the diagnosis unless required by law; and the practitioner has a legal and ethical obligation to inform the patient of the legal obligation to report, and the patient must be involved in the process of notifying her close contacts. By informing the patient of the legal obligation to report and involving her in the notification process, the practitioner balances the duty to promote patient autonomy and confidentiality with the legal mandate to report and warn those at risk of harm. *Answer*: C

2. Ms. Abigail Myers is a 50-year-old accountant who presents to the clinic with fever, malaise, and joint pain. On examination, she is diagnosed with Lyme disease, a reportable infection. However, Ms. Myers refuses to consent to the reporting of her illness. Which of the following is the most appropriate action for the practitioner?
   A. Respect the patient's decision and do not report the case of Lyme disease
   B. Attempt to persuade the patient to change her mind and provide consent for reporting
   C. Report the case of Lyme disease to the health department without informing the patient
   D. Report the case of Lyme disease to the health department while informing the patient of the legal obligation to report

   Explanation: The most appropriate action for the practitioner is to report the case of Lyme disease to the health department while informing the patient of the legal obligation to report. Although the goal is for the patient to provide informed consent to report voluntarily, the practitioner must report reportable infections and diseases to the appropriate agencies even if informed consent is not obtained. Failure to report is a violation of the law and may result in harm to others. The other options are incorrect because the practitioner has a legal and ethical duty to report reportable infections and diseases, and failure to do so may result in harm to others; the patient's decision to provide informed consent to report voluntarily must be respected, but the practitioner still has a legal obligation to report the case; and the practitioner must inform the patient of the legal obligation to report while trying to involve the patient in the notification process. *Answer*: D

3. Mr. Jerry Johnson is a 35-year-old accountant who presents to the clinic with a fever, chills, and a painful rash. He reports having unprotected sexual contact with a new partner a week ago. Clinical examination reveals a rash on his genitals, mouth sores, and swollen lymph nodes. Clinical differential diagnosis includes herpes simplex virus (HSV) infection, syphilis, or HIV infection. What is the practitioner's ethical responsibility in reporting communicable infections, and how can they balance their duty to promote patient autonomy and confidentiality with their legal obligation to report and warn those at risk of harm?
   A. Counsel the patient to notify their contacts
   B. Offer to inform the patient's contacts without revealing their identity
   C. Notify the health department
   D. All of the above

   Explanation: When a patient has exposed others to a communicable infection, the practitioner should counsel the patient to notify their contacts voluntarily, offer to inform the patient's contacts without revealing their identity, and notify the health department. The patient's confidentiality must be ensured, per the nonmaleficence principle. If the patient refuses to notify their contacts, the practitioner must notify the health department and provide the minimum necessary information to accomplish the intended purpose. The health department will then construct a list of contacts to be notified. Not reporting these reportable infections and illnesses violates individual rights and liberties and, therefore, requires legislation, as a matter of justice, to balance the unequal distribution of benefits and burdens. Therefore, all of the options are ethically responsible for the practitioner. *Answer*: D

4. Mr. James Peterson is a 35-year-old plumber who comes to your clinic complaining of fever, headache, and a rash on his arms and legs. He tells you he spent the past week camping in

a wooded area with friends. On examination, you notice a red rash that is flat and non-itchy on his extremities, torso, and palms. He also has a low-grade fever and enlarged lymph nodes in his neck. You suspect Mr. Peterson may have contracted Rocky Mountain spotted fever, but a differential diagnosis includes other tick-borne illnesses such as Lyme disease and ehrlichiosis. What is the ethical question regarding the reporting of reportable infections and illnesses that arises in this case?

A. Should the practitioner inform Mr. Peterson's employer about his illness to protect other workers on the job site?
B. Should the practitioner inform Mr. Peterson's camping companions about his illness to protect them from further exposure?
C. Should the practitioner report the case to the local health department to help track the spread of the disease in the community?
D. Should the practitioner report the case to the state cancer registry to ensure the patient receives the appropriate treatment?

Explanation: In this case, the ethical question is whether the practitioner should report the case to the local health department to help track the spread of the disease in the community. The practitioner has a legal and professional obligation to report reportable infections and illnesses to the proper authorities. The goal is for the patient to understand this responsibility and to consent to reporting voluntarily. If the patient does not provide informed consent to report, then the practitioner is obligated by legal mandate, professional duty, and moral analysis to report the minimal amount of information required to the proper authorities. In this case, Rocky Mountain spotted fever is a reportable infection, and reporting the case to the local health department would be necessary to help track the spread of the disease in the community and prevent further infections. Therefore, the other options are incorrect. *Answer*: C

5. Ms. Abigail Johnson is a 30-year-old chef who presents to the emergency department with severe abdominal pain and bloody diarrhea. She reports that her symptoms began 2 days ago and have gradually worsened. The patient is severely dehydrated upon physical examination, and laboratory tests show an elevated white blood cell count. The working diagnosis is *Escherichia coli* O157:H7 infection. What is the ethical obligation of the practitioner in reporting the *E. coli* O157:H7 infection to the proper authorities?

A. The practitioner must obtain informed consent from the patient before reporting to the proper authorities.
B. The practitioner should offer to inform the patient's contacts for the patient without revealing the patient's identity.
C. The practitioner may choose to notify the health department only if the patient's contacts are at immediate risk of harm.
D. The practitioner is obligated by legal mandate, professional duty, and moral analysis to report the minimal amount of information required to the proper authorities.

Explanation: The practitioner has a legal and professional obligation to report *E. coli* O157:H7 infection to the proper authorities. The practitioner must balance their duty to promote patient autonomy and confidentiality with their legal obligation to report and warn those at risk of harm. In cases in which informed consent is not obtained, the practitioner must report the minimum amount of information required to the proper authorities. The failure to report a communicable infection violates individual rights and liberties. It therefore requires legislation, as a matter of justice, to balance the unequal distribution of benefits and burdens. Although the practitioner should encourage the patient to notify their contacts voluntarily and offer to inform the patient's contacts without revealing the patient's identity, the practitioner must notify the health department if the patient refuses to do so. *Answer*: D

REFLECTION VIGNETTES

1. Practitioner Isabella Brown, a 35-year-old family medicine practitioner, sees a patient with symptoms of a communicable infection. After a thorough examination, Practitioner Brown diagnoses the patient with a sexually transmitted infection. During the patient's counseling session, Practitioner Brown emphasizes the importance of notifying their partners and contacts of their exposure to the infection to prevent further spread. However, the patient becomes emotionally distraught and refuses to inform their partners and contacts.

   What is the ethical issue(s)?

   How should the ethical issue(s) be addressed?

2. Practitioner Heidi Dunn, a 38-year-old general practitioner, meets with a patient who has clinical symptoms consistent with a reportable communicable infection. The patient requests to be treated but only under strict confidentiality and privacy, meaning that Practitioner Dunn should not disclose any information to the state or local health department, the patient's sexual partners, or any other contacts. The differential diagnosis suggests some possibilities, including bacterial vaginosis, chlamydia, gonorrhea, or syphilis.

   What is the ethical issue(s)?

   How should the ethical issue(s) be addressed?

* * *

WEB LINKS

1. *Code of Medical Ethics.* American Medical Association. https://code-medical-ethics.ama-assn.org
   The AMA *Code of Medical Ethics*, established by the American Medical Association, is a comprehensive guide for healthcare practitioners. It addresses issues and challenges, promotes adherence to standards of care, and is continuously updated to reflect contemporary practices and challenges.
2. "Reportable diseases." MedlinePlus. https://medlineplus.gov/ency/article/001929.htm
   This article provides an extensive list of diseases of public health significance in the United States, such as infectious diseases, certain chronic conditions, and environmental health issues. These diseases are required to be reported by healthcare providers to local, state, and national health agencies, including the Centers for Disease Control and Prevention, for effective tracking, research, and management of public health responses. The reporting process aids in identifying disease trends, controlling outbreaks, and shaping health policies and programs related to various aspects such as immunization, food safety, and disease prevention.
3. "Principles of Clinical Ethics and Their Application to Practice." Basil Varkey. 2021. https://www.ncbi.nlm.nih.gov/pmc/articles/PMC7923912
   This source discusses the inherent and inseparable part of clinical medicine ethics as the physician has an ethical obligation to benefit the patient, avoid or minimize harm, and respect the values and preferences of the patient.
4. "Patient Rights and Ethics." Jacob P. Olejarczyk and Michael Young. https://www.ncbi.nlm.nih.gov/books/NBK538279
   This source provides a comprehensive overview of patient rights and ethics in healthcare, including the implications of reportable infections and illnesses.
5. "Medical Ethics: Serious Reportable Communicable Diseases." J. Gall. 2016. https://ncbi.nlm.nih.gov/pmc/articles/PMC7148681
   This source discusses serious reportable or notifiable diseases in the context of the International Health Regulations 2005 and its historical development.

6. "Cardiac Pacemakers from the Patient's Perspective." Mark A. Wood and Kenneth A. Ellenbogen. 2002. https://doi.org/10.1161/01.cir.0000016183.07898.90
   This source discusses the patient's perspective on cardiac pacemakers, which can be related to the topic of reportable infections and illnesses.
7. "NASPGHAN Guidelines for Training in Pediatric Gastroenterology." Alan M. Leichtner, Lynette A. Gillis, Sandeep Gupta, et al. 2013. https://doi.org/10.1097/mpg.0b013e31827a78d6
   This source provides guidelines for pediatric gastroenterology training, including dealing with reportable infections and illnesses.
8. "Is Bioethics Doing Its Job?" M. G. Hansson and R. Chadwick. 2011. https://doi.org/10.1111/j.1365-2796.2011.02348.x
   This source discusses the role of bioethics in protecting patient interests, including in the context of reportable infections and illnesses.

RESEARCH AND CLINICAL EQUIPOISE

48

The good practitioner treats the disease; the great practitioner treats the patient who has the disease.
—William Osler

* * *

MYSTERY STORY

EQUIPOISE ENIGMA: THE CASE OF CLINICAL EQUIPOISE

Practitioner Hannah Adams, a pioneering healthcare researcher, was discovered dead in her laboratory at the Equipoise Medical Research Labs. She had been leading a high-profile randomized clinical trial comparing two rare cancer treatments, and the trial was on the brink of a groundbreaking conclusion.

Detective Olivia Winters, a sharp and experienced investigator, was handed the puzzling case. As she started her inquiries, she learned about the mysterious trial that had consumed 3 years of Practitioner Adams' life. Patients had been randomly divided into two groups, with one receiving standard treatment and the other an untested experimental therapy.

Detective Winters interviewed patients and their families, and an alarming pattern emerged. Some patients who had suffered adverse effects from the experimental treatment had kept silent, fearing reprisals. What was even more disturbing was that Practitioner Adams had monopolized the trial's results, refusing to share them with anyone—not even her own team or the institutional review board.

Delving deeper, Detective Winters discovered Practitioner Adams' internal conflict with the notion of clinical equipoise. Without any clear proof to support one treatment over the other, Practitioner Adams had become obsessed with her research, shielding the results and even neglecting to report adverse effects.

The detective's relentless probing unearthed even more disturbing facts. Practitioner Adams had flouted ethical protocols, including failure to obtain informed consent and nondisclosure of the experimental treatment's potential risks. She had turned a blind eye to the fundamental principles of beneficence (do good) and nonmaleficence (do no harm), putting her own ambition above her patients' welfare.

Detective Winters found that Practitioner Adams' selection of trial participants lacked fairness and transparency, raising serious questions about her adherence to the principle of justice. Her methods were arbitrary and self-serving, seemingly aimed at fulfilling her desire for fame and professional advancement rather than equitable scientific inquiry.

As the investigation reached its climax, it became apparent that Practitioner Adams had been more than a researcher who had lost her way. Her unethical behavior and blatant disregard for the sacred principles of bioethics had culminated in a tragic end. The culprit, one of her research assistants who had discovered the unethical practices, took justice into their own hands to protect future patients from harm.

The Equipoise Enigma became a chilling reminder of the importance of bioethical principles in human research. Respect for persons, beneficence, and justice—these are not merely theoretical constructs but essential safeguards that uphold the dignity and

> well-being of human subjects. The tragic events at Equipoise Labs underscored this fact with profound clarity.
>
> Detective Winters' exposure of Practitioner Adams' fatal pursuit served as a stark warning to the healthcare community. The pursuit of scientific truth must never overshadow patient autonomy, confidentiality, and well-being. Ethical adherence is not merely a professional obligation; it is a moral imperative that must guide every practitioner's actions. The lesson learned from the Equipoise Medical Research Labs resonated far and wide, a sobering testament to the critical importance of ethics in healthcare research.

* * *

THINK

Clinical equipoise exists when no treatment arm of a therapeutic randomized clinical trial (RCT) has empirical evidence or theoretical foundation of superior therapeutic efficacy over the other therapeutic intervention(s). This equivocation is called the null hypothesis. The therapeutic RCT must be stopped once one of the treatment arms has passed the threshold of evidence for superiority so that the superior treatment can be given to the other patients in the inferior treatment arm(s).

However, confusion exists about the threshold of evidence for superiority:

Is it when the primary investigator is convinced that the equipoise no longer exists?
Is it when the expert healthcare community is convinced that equipoise no longer exists?

Confusion also exists about whether the primary investigator should be kept ignorant about any clinical data reports during the RCT so that the clinical investigator can sustain the null hypothesis state of mind until

- a predetermined threshold of evidence for superiority has been met;
- a specified period of RCT) time has elapsed; or
- the expert healthcare community has come to a consensus on the treatment superiority of one treatment arm.

Ethicists argue that clinical equipoise is, at minimum, a disingenuous approach because

- the clinical null hypothesis rarely, if ever, exists;
- the threshold of empirical proof is objectively undefinable;
- a unified expert healthcare community does not exist;
- keeping the primary investigator ignorant about clinical data throughout the trial could (a) increase the subject risk to be higher than minimal or (b) increase the time before getting the superior therapeutic treatment; and
- equipoise is not a relevant condition for determining the permissibility or impermissibility of research on human subjects; rather, it is the purpose, benefits, and minimal risks to the subject that are the relevant conditions for therapeutic treatment.

Clinical equipoise and the null hypothesis are not part of

- the Belmont Report, which presents the three principles of bioethics for human subject protection;
- Common Rule 45CFR46, the federal law for human subject protections that legislates the implementation of the Belmont Report; and
- the institutional review board (IRB) criteria for human subject protection that regulate all research on human subjects in the United States.

The three Belmont principles for the protection of human subjects are as follows:

- *Respect for persons*: Autonomy (informed consent) authorization from the research subject
- *Beneficence*: Professional researcher obligations of beneficence (do good) and nonmaleficence (do no harm) to the research subject
- *Public policy:* Justice (fair human subject selection)

Clinical equipoise and the null hypothesis are not human protection criteria but, rather, methodological criteria that would be accepted by the IRB as long as the methodology does not negatively impact the following:

- Patient: Autonomy (informed consent)
- Practitioner: Beneficence (do good) and nonmaleficence (do no harm)
- Public policy: Justice (be fair)

ASSESS

Patient: Autonomy

Whether or not research on human subjects is therapeutic, research practitioners must always attain informed consent from the research patient for authorization. According to the Belmont Report and Common Rule 45CFR46, the research subject must be informed of the following:

1. Procedures to be performed
2. Purpose of the research
3. Risks and benefits of the procedures
4. Alternatives, including no treatment
5. Opportunity to ask questions and be free to withdraw from the research at any time

The therapeutic research practitioner has the professional obligation to provide only treatment options that will maximize the research patient's best interests, which are usually health and recovery. From those positive treatment options, the research patient has the legal right to autonomously provide an informed consent decision for authorizing the practitioner to provide therapeutic treatment.

Practitioner: Beneficence and Nonmaleficence

Researchers are responsible for not exposing human research subjects to any more than minimal risk—nonmaleficence (do no harm). Common Rule defines minimal risk as follows:

> Minimal risk means that the probability and magnitude of harm or discomfort anticipated in the research are not greater in and of themselves than those ordinarily encountered in daily life or during the performance of routine physical or psychological examinations or tests. 45CFR46.102(i)

This requirement might be impossible when treating clinical patients with greater than minimal risk options. Greater than minimal risk research requires ongoing monitoring by the principal investigator and the IRB, and it may require monitoring by an independent data and safety monitory board (DSMB).

The therapeutic research practitioner is obligated to promote and honor the patient-centered patient–practitioner relationship by maximizing the patient's best interests as determined by the patient's reasonable goals, values, and priorities.

Public Policy: Justice

Justice (be fair) mandates that all research subject selection must be made equitably, with extra legal protections for vulnerable populations.

CONCLUDE

Practitioners must have the basic knowledge of the four principles of bioethics that reflect the Belmont Report's mandate to protect human subjects. Clinical equipoise is a good discussion topic for evaluating critical thinking skills and recognizing that the practitioner's professional obligation is to minimize risk and maximize their patient's best interests.

In summary, although clinical equipoise serves as an interesting topic for evaluating critical thinking skills in healthcare research, it is essential for practitioners to prioritize the ethical principles outlined in the Belmont Report. By focusing on protecting human subjects and adhering to these principles, practitioners can work to minimize risk and maximize the best interests of their patients in both research and therapeutic settings.

* * *

REVIEW QUESTIONS

1. Clinical equipoise exists when the treatment arms of a therapeutic RCT have no empirical evidence or theoretical foundation for establishing superior therapeutic merit for the principal investigator or the expert healthcare community.
    A. True
    B. False

2. Clinical equipoise equivocation between treatment arms is called the null hypothesis.
    A. True
    B. False

3. The RCT must be stopped once one of the treatment arms has passed the threshold of evidence for patient treatment superiority.
    A. True
    B. False

4. Ethicists argue that clinical equipoise is disingenuous because
    A. Null hypothesis rarely exists.
    B. The threshold of empirical proof is undefinable.
    C. A unified expert community does not exist.
    D. Keeping primary investigators ignorant about clinical data could decrease human protection.
    E. All of the above.

5. Clinical equipoise and the null hypothesis are
    A. Not part of the Belmont Report, which presents the three principles of bioethics for human subject protection
    B. Not part of Common Rule 45CFR46, the federal law for human subject protections that legislates implementing the Belmont Report
    C. Not part of the IRB criteria for human subject protections
    D. All of the above

6. The three Belmont Report principles for the protection of human subjects are mandated
    A. When the null hypothesis no longer exists
    B. When the threshold of evidence for superiority has been met
    C. When the expert community is convinced that equipoise no longer exists
    D. When research is conducted on human subjects

7. Clinical equipoise and the null hypothesis are human protection criteria.
    A. True
    B. False

8. Greater than minimal risk research requires ongoing monitoring by the principal investigator and the IRB, and it may require monitoring by an independent DSMB.
   A. True
   B. False

Answers: 1A, 2A, 3A, 4E, 5D, 6D, 7B, 8A

* * *

CLINICAL VIGNETTES

1. Mr. John Williams, a 65-year-old retired electrician, presents to his primary care practitioner with a 1-month history of left-sided weakness and difficulty with speech. A computed tomography scan and magnetic resonance imaging revealed an acute ischemic stroke in the right middle cerebral artery distribution. A neurologist evaluates the patient and determines the patient is a candidate for a thrombectomy. The patient is randomized to one of two treatment arms: the standard treatment of thrombectomy alone or a thrombectomy with a novel adjunctive agent that may improve outcomes. The patient is informed of both treatment options but will not know which treatment they will receive. What ethical question is raised in this scenario?
   A. Is it ethical for the patient to be informed of both treatment options?
   B. Should the patient be excluded from the clinical trial and receive the standard treatment only?
   C. Is including a novel adjunctive agent in the thrombectomy procedure ethical?
   D. Is there clinical equipoise in this trial?

   Explanation: In this scenario, the ethical question is whether clinical equipoise is in the trial. Clinical equipoise exists when there is no empirical evidence or theoretical basis for the superiority of one treatment arm over the other. In this case, the patient is randomized to one of two treatment arms, and it needs to be clarified which arm is superior. Therefore, clinical equipoise exists, and including the patient in the trial is ethical. The other options are incorrect because it is ethical for the patient to be informed of both treatment options, the patient should not be excluded from the trial if clinical equipoise exists, and it is ethical to include a novel adjunctive agent in a clinical trial if the agent has undergone appropriate preclinical testing and initial human studies. *Answer*: D

2. Mr. David Mitchell is a 60-year-old retired factory worker who presents to the clinic with shortness of breath and fatigue symptoms. Upon further examination, he is found to have bilateral crackles in his lungs and lower extremity edema and is diagnosed with chronic heart failure. Mr. Mitchell is eligible for a therapeutic RCT comparing two treatment arms, an innovative medication versus standard treatment, both lacking empirical evidence or theoretical foundation for superior therapeutic efficacy over the other. As the primary investigator, which criteria would be most reliable for determining which RCT treatment arm is superior?
   A. When the primary investigator is convinced that equipoise no longer exists
   B. When the expert healthcare community is convinced that equipoise no longer exists
   C. When a predetermined threshold of evidence for superiority has been met
   D. When a specified period of an RCT time has elapsed

   Explanation: The threshold for stopping an RCT and providing the superior treatment to the other patients in the inferior treatment arm(s) is when a predetermined threshold of evidence for superiority has been met. This predetermined threshold is set before the trial begins and can include a certain level of statistical significance or a predetermined effect size. This threshold ensures that the trial is conducted ethically and that patients in the inferior treatment arm(s) are not unnecessarily subjected to a less effective treatment. The other options are wrong because relying on the primary investigator's conviction that equipoise no longer exists, or the expert healthcare community's conviction, is insufficient to stop the trial as it is based on subjective opinion rather than objective evidence. The specified period of RCT time elapsing is also insufficient to stop the trial because it could lead to patients being subjected to an inferior treatment for an unnecessarily long time. Furthermore, it is important to consider bioethics principles, including respect for persons, beneficence, and justice, in conducting RCTs and protecting human subjects. The primary investigator should prioritize the patient's best interests and minimize risks of harm while maintaining patient autonomy through informed consent. In addition, vulnerable populations should receive extra legal protections to ensure fair subject selection. *Answer*: C

3. Mr. Mark Brown is a 55-year-old retired firefighter who presents to the emergency department with complaints of chest pain, shortness of breath, and fatigue. His past healthcare history is significant for hypertension and hyperlipidemia. On physical exam, his blood pressure is 150/90 mmHg, heart rate is 90 beats per minute, and oxygen saturation is 92% on room air. Electrocardiogram shows ST segment depression in leads V4–V6. His clinical differential diagnosis includes acute coronary syndrome, aortic dissection, and pulmonary embolism. The attending practitioner proposes enrolling Mr. Brown in a therapeutic RCT comparing two anticoagulant agents for treating pulmonary embolism. What action should the attending practitioner take?
   A. The attending practitioner should enroll Mr. Brown in the therapeutic RCT without disclosing the potential risks and benefits.
   B. The attending practitioner should obtain informed consent from Mr. Brown before enrolling him in the therapeutic RCT.
   C. The attending practitioner should enroll Mr. Brown in the therapeutic RCT and inform him of the potential risks and benefits after the trial is over.
   D. The attending practitioner should not enroll Mr. Brown in the therapeutic RCT due to the potential risks.

   Explanation: The correct answer is that the attending practitioner must obtain informed consent from Mr. Brown before enrolling him in the therapeutic RCT. Informed consent requires that the research subject is informed of the purpose of the research; the procedures to be performed; the potential risks and benefits of the procedures; and alternative treatment options, including no treatment. Informed consent also includes the opportunity for the research subject to ask questions and withdraw from the research at any time. The other options are incorrect because not disclosing risks and benefits violates the principle of informed consent, waiting to inform Mr. Brown of the potential risks and benefits after the trial is over also violates the principle of beneficence, and simply not enrolling Mr. Brown in the trial does not allow Mr. Brown to make an informed decision about his treatment options. *Answer*: B

4. Mrs. Hillary Hernandez, a 68-year-old retired teacher, has been diagnosed with advanced-stage ovarian cancer. The oncologist suggests two treatment options: (a) chemotherapy and (b) surgery, followed by radiation therapy. Chemotherapy has a higher success rate but could result in significant side effects, whereas surgery followed by radiation therapy is less successful but has a lower risk of side effects. Mrs. Hernandez's daughter, who lives out of town, wants her mother to choose chemotherapy, believing that this offers the best chance of a cure. Mrs. Hernandez's son, who lives nearby, thinks that his mother should opt for surgery followed by radiation therapy, citing concerns about the impact of the chemotherapy on her quality of life. Mrs. Hernandez is struggling to decide which option to choose. Which decision-making approach is most appropriate in this case: Should Mrs. Hernandez make the decision on her own, with the guidance of her practitioner, or should she involve her adult children in the decision-making process?
   A. Mrs. Hernandez should make the decision on her own, with the guidance of her practitioner.

B. Mrs. Hernandez should involve her children in the decision-making process so they can decide what her goals, values, and priorities should be.
C. Mrs. Hernandez should let her children make the decision for her.
D. It is unclear which decision-making approach is most appropriate.

Explanation: In this case, Mrs. Hernandez is the patient, and therefore, it is her decision to make regarding her healthcare treatment. Although it is important to consider the opinions of family members, ultimately, the decision should be based on Mrs. Hernandez's own reasonable goals, values, and priorities. Mrs. Hernandez should have the autonomy to make the decision that is in her best interests. Her practitioner should provide her with all the relevant information regarding the risks and benefits of each treatment option and assist her in making an informed decision. Family members may be present to offer support and guidance, but the ultimate decision should be left to Mrs. Hernandez. *Answer*: A

5. Mr. John Smith is a 45-year-old man diagnosed with stage 2 colon cancer. He has undergone surgery to remove the tumor. He is now faced with deciding whether to participate in an RCT comparing two chemotherapy regimens. One regimen has been used for many years and is known to be effective, whereas the other is relatively new and has not yet been widely tested. Mr. Smith is torn between wanting to receive the best possible treatment and the uncertainty of participating in a trial in which he might receive the less effective treatment. The ethical question is, Should Mr. Smith be enrolled in an RCT if there is equipoise, even though one of the treatment arms has been effective in the past?
   A. No, Mr. Smith should not be enrolled in a clinical trial because it is unethical to withhold an effective treatment.
   B. Yes, Mr. Smith should be enrolled in the clinical trial because it is the only way to determine which treatment is superior.
   C. It is up to Mr. Smith to decide whether he wants to participate in the clinical trial or receive the standard treatment.
   D. Mr. Smith should be enrolled in the clinical trial only if he meets the specific inclusion criteria for the study.

Explanation: The principle of respect for persons, including autonomy, requires that Mr. Smith be fully informed of his treatment options and allowed to decide whether to participate in the clinical trial or receive the standard treatment. The principle of beneficence requires that the patient's best interests be considered, and the principle of nonmaleficence requires that the patient not be exposed to more than minimal risk. Clinical equipoise exists when there is no empirical evidence or theoretical basis for the superiority of one treatment arm over the other. In this case, there is equipoise between the two treatment arms, and it is up to Mr. Smith to decide which treatment he would like to receive. It is important to note that clinical equipoise is not a requirement for conducting a clinical trial but, rather, a methodological criterion that would be accepted by the IRB as long as the methodology does not negatively impact patient autonomy, practitioner beneficence and nonmaleficence, or public policy justice. *Answer*: C

REFLECTION VIGNETTES

1. Practitioner Scarlett Harris, a 50-year-old general practitioner, sees a patient in her office complaining of symptoms associated with a chronic disease that has been difficult to control. After discussing the patient's treatment options, Practitioner Harris informs the patient about an ongoing clinical trial for a new medication that may offer better symptom control. The patient agrees to participate in the clinical trial, believing this will be the best option for their treatment.

   What is the ethical issue(s)?

   How should the ethical issue(s) be addressed?

2. The patient–practitioner relationship exists between the clinical therapeutic research practitioner and the patient, whereas no relationship exists between a nontherapeutic researcher and the patient.

What is the ethical issue(s)?

How should the ethical issue(s) be addressed?

* * *

WEB LINKS

1. *Code of Medical Ethics*. American Medical Association. https://code-medical-ethics.ama-assn.org
   The AMA *Code of Medical Ethics*, established by the American Medical Association, is a comprehensive guide for healthcare practitioners. It addresses issues and challenges, promotes adherence to standards of care, and is continuously updated to reflect contemporary practices and challenges.
2. "Uses of Equipoise in Discussions of the Ethics of Randomized Controlled Trials of COVID-19 Therapies." Hayden P. Nix and Charles Weijer. 2021. https://bmcmedethics.biomedcentral.com/articles/10.1186/s12910-021-00712-5
   This article discusses the concept of equipoise in the context of randomized controlled trials.
3. "The Question of Clinical Equipoise and Patients' Best Interests." Spencer Phillips Hey and Robert D. Truog. 2015. https://journalofethics.ama-assn.org/article/question-clinical-equipoise-and-patients-best-interests/2015-12
   This article discusses the concept of clinical equipoise and its implications for patient care.
4. "After Equipoise: Continuing Research to Gain FDA Approval." Allison Kerianne Crockett. 2015. https://journalofethics.ama-assn.org/article/after-equipoise-continuing-research-gain-fda-approval/2015-09
   This article discusses the ethical considerations in continuing research after achieving clinical equipoise.
5. "Social Value, Clinical Equipoise, and Research in a Public Health Emergency." Alex John London. 2018. https://www.buffalo.edu/content/www/romanell/blog/social-value/_jcr_content/par/download/file.res/Social%20value%2C%20clinical%20equipoise%2C%20and%20research%20in%20a%20public%20health%20emergency-AJLondon.pdf
   This article discusses the social value of health-related research and the concept of clinical equipoise in public health emergencies.
6. "Community Equipoise and the Architecture of Clinical Research." Jason H. T. Karlawish and John Lantos. 1997. https://www.cambridge.org/core/journals/cambridge-quarterly-of-healthcare-ethics/article/abs/community-equipoise-and-the-architecture-of-clinical-research/97D0787B136F0B1E26C8FB8AA77C0D92
   This article discusses the concept of community equipoise and its role in the architecture of clinical research.
7. "Clinical Equipoise and the Incoherence of Research Ethics." Franklin G. Miller and Howard Brody. 2007. https://pubmed.ncbi.nlm.nih.gov/17454420
   This article discusses the concept of clinical equipoise and the challenges it presents to research ethics.
8. "Equipoise and the Ethics of Clinical Research." Benjamin Freedman. 1987. https://www.nejm.org/doi/full/10.1056/NEJM198707163170304
   This article discusses the ethical requirements of clinical trials, including the concept of equipoise.

RESEARCH ON HUMAN SUBJECTS

Every practitioner must be rich in knowledge, not only of that written in books; patients should be the book. Patients will never mislead the practitioner.
—Paracelsus

* * *

MYSTERY STORY

CONSENT CRISIS: THE UNRAVELING OF PRACTITIONER NOVAK'S UNETHICAL RESEARCH

Practitioner Orion Novak, once celebrated for his pioneering work, had just released a groundbreaking study on a promising new cancer treatment. The scientific community was abuzz, but whispers of something sinister lurked beneath the accolades.

When rumors emerged that Practitioner Novak had bypassed critical ethical guidelines, conducting research on human subjects without their informed consent, a shadow fell over his achievement. Student Amelia Hartley, driven by her curiosity and sense of justice, and her insightful mentor, Practitioner Oliver Heisenberg, felt a moral obligation to uncover the truth.

Equipped with knowledge from the Belmont Report and Common Rule (45 Code of Federal Regulation 46 [45CFR46]), they delved into the intricacies of research ethics, unraveling the principles of respect for persons (informed consent), beneficence (do good and do no harm), and justice (be fair). Armed with this understanding, they began to peel back the layers of Practitioner Novak's study.

Their initial inquiries revealed disturbing inconsistencies. Some participants were elderly, and their comprehension was hampered by the legal jargon in the consent forms. Other patients, weakened by severe illnesses, seemed incapable of making truly informed decisions.

The plot thickened as they discovered that Practitioner Novak had brazenly bypassed the institutional review board (IRB) approval, flouting federal law and the Common Rule. The absence of this essential oversight was a grave breach of trust.

Even more unsettling was Novak's choice of research participants. He had preyed on the vulnerable, targeting nursing homes and institutions and exploiting those who were most in need of care. This flagrant violation of the principle of justice left both Ms. Hartley and Practitioner Heisenberg aghast.

Their evidence was compelling, and they presented it to the institution's IRB. A rigorous investigation was launched, the ethical fabric of Novak's research unraveling with every passing day.

Finally, the shocking truth was laid bare: Practitioner Novak had indeed violated ethical guidelines, compromising the integrity of his study and the dignity of his subjects. The research community reeled from the betrayal, and Novak's study was denounced as both unethical and invalid.

The fallout was swift and devastating. Practitioner Novak lost his funding, his groundbreaking publication was retracted, and his reputation lay in tatters. The institution, determined to prevent any recurrence, introduced stringent training programs, strengthening its commitment to ethical compliance.

> The Consent Crisis, as it became known, sent shock waves through the healthcare community. It was a sobering testament to the unwavering importance of ethical adherence. For Ms. Hartley and Practitioner Heisenberg, it was a personal victory, a triumphant stand for the principles that must always guide research: respect for persons, beneficence, and justice. Their unyielding pursuit of the truth had laid bare a grave injustice, reminding all practitioners that in the quest for innovation, the human heart must never be forgotten.

* * *

THINK

Critical to research integrity are the three principles of biomedical and behavioral research:

- *Respect for persons*: Autonomy (informed consent) authorization from the research subject
- *Beneficence*: Professional researcher obligations of beneficence (do good) and nonmaleficence (do no harm) to the research subject
- *Public policy:* Justice (fair human subject selection)

These three principles were defined in the 1979 Belmont Report guidelines for research on human subjects. They were developed and formalized, with the protection of human research subjects being its central focus. The biomedical and behavior ethics principles are generalizable and are derivable from, consistent with, or at least not in contradiction with most worldviews.

The Belmont Report's three principles of respect for persons, beneficence, and justice are equivalent to the four principles that biomedical ethics and bioethics use: autonomy (informed consent), beneficence (do good), nonmaleficence (do no harm), and justice (be fair). Nonmaleficence is the first thing discussed in the Belmont Report under beneficence (do good). However, because nonmaleficence (do no harm) is a principle of inaction rather than action, it is not explicitly included in the Belmont Report listing of the three principles of action. Healthcare ethics includes nonmaleficence because "do no harm" is one of the core healthcare principles since even before the Hippocratic Oath:

> I will keep my patients from harm and injustice.

To implement and administer these biomedical and behavior principles for guiding all research involving human subjects, Congress legislated federal law 45CFR46, also known as the Common Rule. In addition to establishing additional protections for certain groups, such as prisoners and children, it also established IRBs. The IRB is a mandated diverse group of overseers that every institution conducting research on human subjects must have to ensure human subject protection compliance.

Congress legislated these IRBs in response to the individual, institutional, and systematic injustices accompanying research on human subjects. Unjust research practices arose because the biomedical and behavioral professions failed to self-regulate research to protect human subjects.

Action was taken following the public outcry that occurred when the American public found out that the Centers for Disease Control and Prevention, the American Medical Association, and the National Medical Association were fully supportive of the infamous syphilis study at Tuskegee conducted on 600 African American males from 1932 to 1972. The human research subject abuse at Tuskegee was the catalyst for the Belmont Report's exposition of the three biomedical and behavioral principles of ethics and the stimulus for Congress to pass federal law 45CFR46 (Common Rule) mandating moral compliance and the formation of IRBs for every institution in the country that conducts any biomedical and behavioral research on human subjects.

The penalty for IRB noncompliance is denial of researcher(s) publication(s) in academic journals without IRB approval and the loss of federal funding for the entire institution.

ASSESS

Human Subject: Autonomy

The Belmont Report and Common Rule 45CFR46 federally mandate that all human subject participants provide autonomous informed consent for research authorization. Research subject autonomy (informed consent) requires that the researcher explain to the human subject in clear and understandable language that is age, education, and mentally appropriate and fully understood. In addition, the following information must be documented in writing:

1. Purpose of the research
2. Procedures to be performed
3. Risks and benefits of the procedures
4. Alternatives, including no treatment
5. Opportunity to ask questions
6. Opportunity to withdraw from the research at any time

Autonomy is based on the liberal notion that the human subject is in the best position to determine what will maximize the subject's best interests according to their reasonable goals, values, and priorities.

Researcher: Beneficence and Nonmaleficence

Researchers using the principle of nonmaleficence (do no harm) are not to expose human subjects to risk greater than minimal if possible. Common Rule defines minimal risk as follows:

> Minimal risk means that the probability and magnitude of harm or discomfort anticipated in the research are not greater in and of themselves than those ordinarily encountered in daily life or during the performance of routine physical or psychological examinations or tests. 45CFR46.102(i)

Greater than minimal risk research requires ongoing monitoring by the principal investigator, the IRB, and may require monitoring by an independent data and safety monitory board (DSMB).

The researcher has the professional obligation of beneficence (do good) of conducting research on human subjects that maximizes benefits to the human subject. *Benefit* is defined by the research subject's reasonable goals, values, and priorities and standards of care.

Public Policy: Justice

When selecting human subject participants, researchers have a social obligation of justice (be fair), not to burden one social group for the benefit of another social group. There must be fair procedures and outcomes in the selection of research subjects, both individually and socially.

Researchers are prohibited by federal law (Common Rule 45CFR46) from selecting research participants based on ease of access, such as incarcerated prisoners, institutionalized mentally infirm, nursing home residents, and educational settings. The Belmont Report and federal law (Common Rule 45CFR46) have special protections for vulnerable populations such as prisoners, children, and the mentally infirm requiring that the IRB has increased protections and designated representatives for such groups when conducting the research approval assessment.

CONCLUDE

It is vital for practitioners to recognize the inherent failures and weaknesses that may arise when an organization attempts to self-regulate without independent, pluralistic, and diverse oversight, particularly involving representation by affected vulnerable populations. Self-regulation must be complemented by independent checks and balances. The National Commission for the Protection of Human Subjects established the Belmont Report's guiding principles for biomedical and behavioral ethics. These principles were subsequently legislated into law by the U.S. Department of Health, Education, and Welfare in 1979, becoming known as Common Rule 45CFR46.

In summary, the ethical principles outlined in the Belmont Report and enforced by Common Rule (45CFR46) are essential for maintaining the integrity and moral compliance of research involving human subjects. Understanding and adherence to these principles ensure that research is conducted ethically and that the rights and well-being of human subjects are protected. This framework underscores the importance of both self-regulation and independent oversight, which is crucial in preventing a repeat of past injustices and in ensuring the responsible conduct of future research.

* * *

REVIEW QUESTIONS

1. The three principles of biomedical and behavioral research are (a) respect for persons (autonomy [informed consent]), (b) beneficence (do good) and nonmaleficence (do no harm), and (c) justice (be fair).
   A. True
   B. False

2. The biomedical and behavioral research principles were developed and formalized to protect the principal investigator from professional, institutional, and public accountability.
   A. True
   B. False

3. Congress legislated federal law 45CFR46, also known as the Common Rule, implementing the Belmont Report principles and establishing IRBs.
   A. True
   B. False

4. In human subject research, the penalty for noncompliance with the Belmont Report's biomedical and behavioral principles of ethics includes
   A. Researcher(s) publication(s) in academic journals will be denied without IRB approval of the human subject research.
   B. Institutional IRB noncompliance will result in the loss of federal funding for the entire institution.
   C. Both A and B.

5. Research subject autonomy (informed consent) requires that the researcher explain to the human subject in clear and understandable language that is age, education, and mentally appropriate and fully understood. In addition, the following information must be documented in writing:
   A. Purpose of the research
   B. Procedures to be performed
   C. Risks of harm and benefits of the procedures
   D. Alternatives, including no treatment
   E. Opportunity to ask questions and to withdraw from the research at any time
   F. All of the above

6. Greater than minimal risk research requires ongoing monitoring by the principal investigator and the IRB, and it may require monitoring by an independent DSMB.
   A. True
   B. False

7. There must be fair procedures and outcomes in selecting research subjects, both individually and socially.
   A. True
   B. False

Answers: 1A, 2B, 3A, 4C, 5F, 6A, 7A

* * *

CLINICAL VIGNETTES

1. Mrs. Lucy Mitchell is a 78-year-old retired sales clerk who is admitted to the hospital with severe dehydration and confusion. She has been diagnosed with advanced dementia and cannot provide informed consent for a new study investigating the effectiveness of a new treatment for dementia. Her daughter, who has healthcare power of attorney, believes that the new treatment could be helpful for her mother and wants her to be enrolled in the study. The research team explains the study and its risks and benefits to the daughter, but Ms. Mitchell refuses to participate. What should the research team do?
   A. Enroll Ms. Mitchell in the study without her consent, as her daughter has healthcare power of attorney
   B. Enroll Ms. Mitchell in the study despite her refusal, as she lacks decisional capacity and her daughter believes it would be in her best interest
   C. Respect Ms. Mitchell's decision to decline participation in the study, even if she lacks decisional capacity
   D. Enroll Ms. Mitchell in the study only after obtaining consent from her daughter and consulting with the IRB

   Explanation: The Belmont Report and the Common Rule mandate that human research subjects have the right to make autonomous, informed decisions about their participation in research. However, when a patient lacks decisional capacity, a surrogate decision-maker may be consulted to make decisions on the patient's behalf. In this case, Ms. Mitchell lacks decisional capacity due to advanced dementia, and her daughter acts as her surrogate decision-maker. However, the principle of respect for persons requires that the research team respect the patient's autonomy even when a surrogate is involved. In this case, Ms. Mitchell has refused to participate in the study, and the research team should respect her decision even if her daughter believes it would be in Ms. Mitchell's best interest to participate. The research team must not enroll Ms. Mitchell in the study without her consent or overrule her decision. In addition, it is important to note that the surrogate's decision-making authority is not absolute and should always be guided by the patient's wishes, values, and best interests. In this case, the research team should respect Ms. Mitchell's refusal to participate in the study, as it aligns with her wishes and values, even though her daughter may disagree. If the study truly has clinical equipoise between the various treatment arms, then at this time, no added harm will befall Ms. Mitchell, regardless if she participates in the trial or not. *Answer*: C

2. Mr. Aaron Rogers is a 75-year-old retired businessman with early stage Alzheimer's disease. His daughter, who is his primary caregiver, is interested in enrolling him in a clinical trial for a new medication that may slow the progression of the disease. The medication has not

yet been approved by the U.S. Food and Drug Administration (FDA), and there is some risk of adverse effects. What should the research team do?
   A. The research team should enroll Mr. Rogers in the study without his consent, as it is in his best interests.
   B. The research team should only enroll Mr. Rogers if he has decisional capacity and can provide informed consent.
   C. The research team should enroll Mr. Rogers only if his daughter agrees to act as his surrogate decision-maker and provides informed consent on his behalf.
   D. The research team should not enroll Mr. Rogers in the study, as it may pose undue risk to his health and quality of life.

   Explanation: The principle of respect for persons requires that research subjects provide informed consent to participate in a study. In the case of Mr. Rogers, if he has decisional capacity, he should be fully informed about the nature and risks of the study and be given the opportunity to provide his informed consent. If he lacks decisional capacity, the research team should follow established procedures for identifying an appropriate surrogate decision-maker to provide informed consent on his behalf. In either case, the research team must prioritize Mr. Rogers' best interests and minimize the risks of the study. Enrolling Mr. Rogers in the study without his consent or solely based on his daughter's agreement is unethical because this would violate the principles of respect for persons and autonomy. The research team must ensure that Mr. Rogers' rights and interests are always protected. *Answer*: B

3. Mr. Tyler Johnson, a 75-year-old retired high school math teacher, is diagnosed with a rare type of cancer. His oncologist recommends a new treatment that is being tested in a clinical trial. The treatment has shown promising results in earlier phases of testing but is not yet approved by the FDA. Mr. Johnson is interested in participating in the trial, but his wife and children are concerned about the potential risks and want him to stick with conventional treatment options. What should the research team do?
   A. The research team should respect Mr. Johnson's decision to participate in the trial, even if his family disagrees.
   B. The research team should encourage Mr. Johnson to participate in the trial and disregard his family's concerns.
   C. The research team should attempt to persuade Mr. Johnson's family to support his decision to participate in the trial.
   D. The research team should exclude Mr. Johnson from the trial due to his family's concerns.

   Explanation: The principle of respect for persons (autonomy) requires that individuals have the right to make their own decisions regarding their healthcare, including whether to participate in research. The fact that Mr. Johnson has a rare type of cancer and the standard treatments have not been effective for him may make participating in the clinical trial a more appealing option. Although it is important to consider the opinions of his family, ultimately, the decision to participate in the trial should be made by Mr. Johnson. The Belmont Report states that "an agreement to participate in research constitutes a valid consent only if voluntarily given." Therefore, it is necessary for the research team to respect Mr. Johnson's autonomy and allow him to make his own decision about participation in the clinical trial, even if his family disagrees. *Answer*: A

4. Mr. Matthew Morris is a 70-year-old retired security officer with early stage prostate cancer. His practitioner has recommended that he participate in a randomized clinical trial comparing two standard treatments for prostate cancer. Mr. Morris is interested in participating but is concerned about the risks and benefits of the treatments. He is particularly worried about the possibility of adverse effects and the potential impact

on his quality of life. Which of the following is the most appropriate response by the research team?
- A. Convince Mr. Morris to participate in the study, highlighting the potential benefits and downplaying the risks
- B. Provide Mr. Morris with accurate and complete information about the study and the two treatment options, but let him make the final decision
- C. Tell Mr. Morris that he is unlikely to benefit from the study and encourage him to seek treatment outside of the trial
- D. Ask Mr. Morris' family members to make the decision for him, as they may have a better understanding of the risks and benefits

Explanation: The most appropriate response by the research team is to provide Mr. Morris with accurate and complete information about the study and the two treatment options but ultimately let him make the final decision. This is consistent with the principle of respect for persons, which requires that research participants are treated as autonomous individuals capable of making their own decisions. It is essential to inform Mr. Morris that the study is based on clinical equipoise, meaning that the two treatment options being compared are expected to be equally effective or beneficial. This would help him understand that there is no guarantee of benefit from being a research subject and that his participation is essential for advancing scientific knowledge and improving future treatments. The research team should also provide Mr. Morris with all of the information he needs to make an informed decision, including the potential risks and benefits of each treatment and the potential impact on his quality of life. This would ensure that he is fully informed and able to make the best decision for him. It is important to note that although the research team should provide accurate and complete information, the team should not use false or misleading information to convince Mr. Morris to participate. Doing so would be unethical and could compromise the integrity of the study. *Answer*: B

5. Mr. Brandon Thomas, a 75-year-old retired accountant, was recently diagnosed with stage 4 lung cancer. He has been started on chemotherapy, but his prognosis remains poor. His two children, a daughter and a son, are very involved in his care and attend all of his healthcare appointments. During one of his appointments, they learned about a clinical trial for a new cancer treatment and were eager to enroll their father in the study. Mr. Thomas was initially hesitant, but after much persuasion from his children, he agreed to participate in the study. After several weeks of participating in the clinical trial, Mr. Thomas began to experience significant side effects and discomfort. Despite the urging of his children to continue in the study, he decided that he wanted to discontinue his participation. However, his children strongly disagreed, insisting that he continue in the study because they believe it is the best option for their father. Which of the following is the most ethical course of action for the research team to take in this situation?
- A. The research team should continue to encourage Mr. Thomas to remain in the study because it is in his best interest.
- B. The research team should side with Mr. Thomas's children and encourage him to remain in the study.
- C. The research team should respect Mr. Thomas' decision to withdraw from the study, even if his children disagree.
- D. The research team should seek legal action against Mr. Thomas to ensure he remains in the study for the full duration.

Explanation: The most ethical course of action in this situation is for the research team to respect Mr. Thomas' decision to withdraw from the study, even if his children disagree. It is essential to obtain informed consent from the research subject, and if the subject decides to participate in the research study, it must be voluntary. Regardless of his decisional capacity, Mr. Thomas has the right to refuse to participate or to withdraw from the research at any time. Although his children's opinions may be important, the ultimate decision rests with Mr. Thomas and his right to autonomous informed consent. It is unethical to pressure or force Mr. Thomas to continue the study against his wishes, even if his children believe it is in his best interest. Seeking legal action against Mr. Thomas to force him to remain in the study would violate his autonomy and is not an ethical option. The research team must prioritize Mr. Thomas' well-being and respect his right to make decisions regarding his participation in the study. *Answer*: C

REFLECTION VIGNETTES

1. Ms. Isla Nelson, a 32-year-old teacher, has agreed to participate in a research study investigating the effectiveness of a new medication for treating anxiety. After undergoing a thorough informed consent process, including a detailed explanation of the study's procedures, possible risks and benefits, and an opportunity to ask questions, she agrees to participate. During the study, Ms. Nelson begins experiencing significant side effects from the medication, which cause her to feel extremely anxious and uncomfortable. Despite efforts to address her concerns and adjust her treatment, Ms. Nelson ultimately decides that she no longer wishes to participate in the study.

   What is the ethical issue(s)?

   How should the ethical issue(s) be addressed?

2. Ms. Chloe Lewis, a 45-year-old research practitioner, has conducted studies in a local nursing home for the past year. She has noticed that it is much easier to conduct research on institutionalized subjects because the subjects are all at the same location, she has efficient access to them, and she has better control over subject compliance. However, her study is now being criticized for not including a diverse enough sample population because the majority of her subjects are institutionalized.

   What is the ethical issue(s)?

   How should the ethical issue(s) be addressed?

* * *

WEB LINKS

1. *Code of Medical Ethics.* American Medical Association. https://code-medical-ethics.ama-assn.org
   The AMA *Code of Medical Ethics*, established by the American Medical Association, is a comprehensive guide for healthcare practitioners. It addresses issues and challenges, promotes adherence to standards of care, and is continuously updated to reflect contemporary practices and challenges.
2. "The Belmont Report." U.S. Department of Health and Human Services. https://www.hhs.gov/ohrp/regulations-and-policy/belmont-report/index.html
   This source discusses the Belmont Report, which identifies basic ethical principles and guidelines that address ethical issues arising from the conduct of research with human subjects.
3. "Principles of Clinical Ethics and Their Application to Practice." Basil Varkey. 2021. https://www.ncbi.nlm.nih.gov/pmc/articles/PMC7923912
   This paper discusses the ethical obligation of physicians to benefit the patient, avoid or minimize harm, and respect the values and preferences of the patient.
4. "Ethical, Legal, and Regulatory Framework for Human Subjects Research." National Library of Medicine. https://www.ncbi.nlm.nih.gov/books/NBK373547
   This source discusses the National Research Act of 1974, which created the National Commission for the Protection of Human Subjects of Biomedical and Behavioral Research.

5. "Revisiting the Ethics of Research on Human Subjects." Cynthia Tsay, 2015. https://journalofethics.ama-assn.org/article/revisiting-ethics-research-human-subjects/2015-12
This article reviews the ethics of clinical research on human subjects and its history.
6. "Ethical and Legal Issues in Research Involving Human Subjects: Do You Want a Piece of Me?" M. B. Kapp. 2006. https://www.ncbi.nlm.nih.gov/pmc/articles/PMC1860367
This paper discusses the ethical and legal issues that arise in biomedical research involving human participants.
7. "Why Human Subjects Research Protection Is Important." Michael G. White. 2020. https://www.ncbi.nlm.nih.gov/pmc/articles/PMC7122250
This paper reviews the history of human subjects participating in research, including examples of egregious events, and the ethical analyses that precipitated the evolution of the mandated protections afforded participants in research under current federal regulations.
8. "Federal Policy for the Protection of Human Subjects ('Common Rule)." U.S. Department of Health and Human Services. https://www.hhs.gov/ohrp/regulations-and-policy/regulations/common-rule/index.html
This source discusses the Federal Policy for the Protection of Human Subjects, also known as the Common Rule.
9. "What Is Ethics in Research & Why Is It Important?" David B. Resnik. 2020. https://www.niehs.nih.gov/research/resources/bioethics/whatis/index.cfm
This source discusses the importance of ethical standards in medicine and research.
10. "Federal Privacy Protections: Ethical Foundations, Sources of Confusion in Clinical Medicine, and Controversies in Biomedical Research." Mary Anderlik Majumder and Christi J. Guerrini, 2016. https://journalofethics.ama-assn.org/article/federal-privacy-protections-ethical-foundations-sources-confusion-clinical-medicine-and/2016-03
This source discusses the federal privacy protections and ethical foundations in clinical medicine.
11. "PDF International Ethical Guidelines for Health-Related Research Involving Humans." Council for International Organizations of Medical Science. 2016. https://www.who.int/docs/default-source/ethics/web-cioms-ethicalguidelines.pdf?sfvrsn=f62ee074_0
This source discusses the international ethical guidelines for health-related research involving human subjects.
12. "Ethics in Medical Research and Publication." Izet Masic, Ajla Hodzic, and Smaila Mulic. 2014. https://www.ncbi.nlm.nih.gov/pmc/articles/PMC4192767
This source discusses the principles of good scientific and good laboratory practice in healthcare research and publication.
13. "Ethics, Human Rights, and Clinical Research." Cheryl Erler and Cheryl Thompson. 2008. https://www.researchgate.net/publication/5397032_Ethics_Human_Rights_and_Clinical_Research
This source discusses ethics, human rights, and clinical research.
14. "Bioethics: Key Concepts and Research." Josephine Johnson and Elizabeth Dietz. 2018. https://daily.jstor.org/bioethics-key-concepts-research
This source discusses key concepts and research in bioethics.
15. "Global Health Ethics." World Health Organization. https://www.who.int/health-topics/ethics-and-health
This source discusses the importance of adhering to ethical principles in order to protect the dignity, rights, and welfare of research participants.
16. "Overview: Biomedical Ethics Research Program." Mayo Clinic. https://www.mayo.edu/research/centers-programs/biomedical-ethics-research-program/overview
This source discusses the Biomedical Ethics Research Program at Mayo Clinic, which assesses and addresses difficult questions about right and wrong in medicine and biomedical science.

SELF-TREATMENT AND FAMILY TREATMENT

50

A practitioner who treats oneself has a fool for a doctor and a fool for a patient.
—William Osler

* * *

MYSTERY STORY

> **TREATMENT TRAGEDY: THE LINE BETWEEN COMPASSION AND OBJECTIVITY AT ETHICAL BOUNDARIES HOSPITAL**
>
> The buzz of intellectual curiosity and moral inquiry filled the conference room at Ethical Boundaries Hospital. Students, residents, and attending practitioners gathered to engage in a dialogue about the ethical boundaries that guide their profession.
>
> Among the eager residents, Practitioner Brooke Mitchell, a bright and compassionate young doctor, stood up to share a harrowing personal experience that had left an indelible mark on her life and career.
>
> Practitioner Mitchell's cousin, someone she had grown up with and loved, had approached her one day with troubling symptoms. Her cousin's fear and desperation struck a chord with Mitchell, who, caught up in her desire to help, offered to examine her. Confident in her skills, Practitioner Mitchell believed that treating a family member was a compassionate act, demonstrating the trust and confidence she had in her abilities.
>
> Days turned into weeks, and Practitioner Mitchell's cousin's condition worsened. Despite prescribing medication, Practitioner Mitchell felt helpless as the symptoms persisted. Her frustration grew, clouded by the emotional connection she shared with her cousin. Then, tragedy struck. Her cousin collapsed and was rushed to the hospital, where the healthcare team fought a losing battle to save her life.
>
> Practitioner Mitchell was shattered. Her cousin's death, under her watch, was unbearable. As she poured over the healthcare records, guilt gnawed at her soul. She discovered missing information that her cousin had withheld, too embarrassed to share. Practitioner Mitchell realized she had also failed to perform a thorough and intimate examination, hindered by their familial relationship.
>
> The subsequent investigation revealed that Practitioner Mitchell's well-intentioned actions had violated the ethical and professional standards governing patient–practitioner relationships. Her breach of ethical boundaries had compromised her cousin's right to privacy, the right to receive adequate care, and the right to be able to provide autonomous informed consent.
>
> The case became a somber lesson for everyone at Ethical Boundaries Hospital, reinforcing the need for strict adherence to ethical and professional standards. It underscored the importance of recognizing the fine line between compassion and objectivity—a line that, once crossed, can have irreversible consequences.
>
> In the aftermath, Ethical Boundaries Hospital took the opportunity to educate its staff and reinforce policies that prevent self and family treatment. It also provided support to Practitioner Mitchell, helping her learn from the tragedy and become an advocate for ethical practice.
>
> The painful memory of Practitioner Mitchell's cousin serves as a constant reminder that the noble intent to heal must always be guided by ethical principles and professional distance. The tragedy was turned into a catalyst for growth, shaping a culture that values not just technical competence but also ethical integrity and wisdom.

* * *

THINK

Practitioners are not to treat themselves, family members, or friends because of the lack of professional objectivity that may influence their healthcare judgments, such as healthcare denial or attempting to self-treat ailments beyond the practitioner's professional expertise. When a practitioner treats a family member or friend, the "patient" may be uncomfortable and unwilling to openly disclose relevant private health information (PHI), which might be necessary for appropriate diagnosis and treatment options, because it would be too embarrassing to disclose, as well as a violation of the right to privacy. The practitioner may also be uncomfortable and unwilling to conduct a necessary intimate physical examination because doing so in that relationship would be inappropriate. Treating family members and friends can also have a negative social impact by inadvertently causing changes in role responsibilities and upending social hierarchies in the family structure or friendship. Practitioners are not to self-treat because of the lack of professional objectivity that may negatively influence their healthcare judgment. These are a few consequential negative reasons why practicing medicine on oneself, family members, and friends is against standards of care. This prohibition is based on the patient–practitioner relationship, which can be significantly compromised when treating oneself, family members, and friends.

The American Medical Association (AMA) *Code of Medical Ethics* Opinion 1.2.1 states,

> Treating oneself or a member of one's own family poses several challenges for physicians [practitioners], including concerns about professional objectivity, patient autonomy, and informed consent.... When the patient is an immediate family member, the physician's [practitioner's] personal feelings may unduly influence [their] professional medical judgment.... They may also be inclined to treat problems that are beyond their expertise or training.... In general, physicians [practitioners] should not treat themselves or members of their own families.

ASSESS

Patient: Autonomy

Patient autonomy requires the necessary conditions of confidentiality and privacy that are part of the patient–practitioner relationship. When a practitioner treats a family member or friend, confidentiality and privacy are at high risk of being violated. If the practitioner is socially sensitive with self-awareness, then the practitioner may avoid asking embarrassing and sensitive questions, and the practitioner will avoid performing intimate physical examinations. When treating family members and friends, the patient–practitioner relationship that is so essential for the practice of healthcare is threatened as either not existing or, if it does exist, becoming a conflict of interest for the practitioner between the role of the practitioner and the role of being a family member or friend.

If there is no patient–practitioner relationship, then it will be impossible for the patient to provide autonomous informed consent because the practitioner's diagnosis, prognosis, treatment options, and risk–benefit assessments will be compromised due to the lack of truthful patient communications, the inappropriateness of the practitioner to conduct intimate physical examinations, and compromised healthcare judgments due to emotional and other sociological factors.

Patients have a positive right to healthcare access, to qualified practitioners, and to form a patient–practitioner relationship required for the disclosure of PHI necessary for autonomous informed consent decisions. Because this does not happen when a practitioner treats oneself, family members, or friends, it is considered an ethical, professional, and social boundary that should not be breached.

Practitioner: Beneficence and Nonmaleficence

The practitioner's professional role is to maximize the patient's best interests as determined by the patient's reasonable goals, values, and priorities. With the practitioner treating oneself, family, or friends, it is easy for the practitioner to be self-deceived into thinking that they are in the best position to make such assessments because of the close personal relationship and knowledge of the patient's reasonable goals, values, and priorities. However, if the healthcare treatment of family members and friends results in a violation of patient autonomy because of a variety of conflicts of interest, then this can result in the violation of the professional obligation of beneficence (do good) and nonmaleficence (do no harm). Therefore, the healthcare profession has categorized self-treatment, family treatment, and friend treatment as unprofessional and a potential form of healthcare misconduct.

Public Policy: Justice

The treatment of self, family members, and friends violates the citizens' positive right to have access to a fair and impartial practitioner. The practitioner's treatment of self, family members, and friends violates that positive right. Patients have the civil right for access to a fair and impartial practitioner with whom patients can form a patient–practitioner relationship and share PHI, as regulated by the 1996 Health Insurance Portability and Accountability Act (HIPAA). This federal law aims to protect patients' privacy by setting guidelines for how healthcare information can be shared and who has access to it. It also includes provisions that make it easier for individuals to keep their health insurance when they change jobs, help the healthcare industry control administrative costs, and set standards for electronic billing and other processes. The Privacy Rule within HIPAA specifically addresses the saving, accessing, and sharing of patients' healthcare information and PHI.

When dealing with family members and friends, exceptions for treatment can be made in situations of a healthcare emergency, minor and commonplace conditions, and treatments independent of the necessity of having a patient–practitioner relationship. However, judgments regarding the exceptions to the standards of care boundaries must not become the accepted standard of care.

CONCLUDE

Making treatment decisions as a practitioner for oneself, family members, and friends violates patient autonomy because there is a conflict of interest between the role of the practitioner and the role of being a family member or friend, a violation of the professional obligation of beneficence (do good) and nonmaleficence (do no harm), as it is a violation of standards of care and the patient's civil rights for access to a fair and impartial practitioner with whom the patient can form a patient–practitioner relationship and share PHI. Exceptions to these practices for emergencies, minor and commonplace conditions, and situations independent of the necessity of having a patient–practitioner relationship should be uncommon and not lead to an unacceptable standard of care.

The AMA *Code of Medical Ethics* Opinion 1.2.1 states, "Physicians [practitioners] should not treat themselves or members of their own families."

In summary, practitioners should avoid treating themselves, family members, or friends to maintain the integrity of the patient–practitioner relationship and uphold the principles of patient autonomy, practitioner beneficence and nonmaleficence, and public policy justice. By adhering to this standard of care boundary, practitioners can protect the personal and social rights and liberties of their patients, family members, and friends. The AMA *Code of Medical Ethics* Opinion 1.2.1 reminds practitioners that they "should not treat themselves or members of their own families" to maintain the highest level of ethical and professional conduct. (For more information on professional boundaries, see Chapters 7, 51, and 53.)

* * *

REVIEW QUESTIONS

1. Practicing medicine on oneself, family members, and friends is not against standards of care and does not compromise the patient–practitioner relationship.
   A. True
   B. False

2. Patients have a positive right to healthcare access to qualified practitioners to form a patient–practitioner relationship necessary for autonomous informed consent decisions.
   A. True
   B. False

3. If the treatment of family members and friends results in a violation of patient autonomy because of a variety of conflicts of interest, then this can result in the violation of the professional obligation of beneficence (do good) and nonmaleficence (do no harm).
   A. True
   B. False

4. The treatment of self, family members, and friends violates the citizens' positive right to have access to a fair and impartial practitioner.
   A. True
   B. False

5. Exceptions to the treatment of self, family members, and friends can be made in situations of a healthcare emergency, minor and commonplace conditions, or independent of the necessity of having a patient–practitioner relationship.
   A. True
   B. False

Answers: 1B, 2A, 3A, 4A, 5A

* * *

CLINICAL VIGNETTES

1. Practitioner Ethan Parker, a 50-year-old cardiologist, has a sister who presents to his clinic complaining of chest pain and shortness of breath. Practitioner Parker faces an ethical dilemma regarding whether treating his sister as a patient is appropriate. What should Practitioner Parker do?
   A. Practitioner Parker should treat his sister as a patient.
   B. Practitioner Parker should decline to treat his sister and refer her to another cardiologist.
   C. Practitioner Parker should have his colleague assist in the treatment of his sister.
   D. Practitioner Parker should refer his sister to a different healthcare specialty.

   Explanation: Practitioner Parker should decline to treat his sister and refer her to another cardiologist. The healthcare practice of treating oneself, family members, or friends is discouraged due to a lack of professional objectivity and violation of patient autonomy. When a healthcare provider treats a family member or friend, the patient–practitioner relationship is compromised, causing issues with confidentiality, informed consent, and intimate examinations. The healthcare provider's personal feelings may unduly influence their professional healthcare judgment, leading to inadequate care. The ethical responsibility to prioritize the patient's best interests and provide impartial care falls on the healthcare provider. It is categorized as unprofessional and a potential form of healthcare misconduct to treat oneself, family members, or friends. Therefore, Practitioner Parker should decline to treat his sister and refer her to another cardiologist. *Answer*: B

2. Ms. Samantha Wilson, a 35-year-old pharmacist, presents to the pharmacy complaining of a persistent cough and sore throat. She informs her colleagues that she has diagnosed herself with a viral infection and has started taking an over-the-counter medication without a prescription. An ethical question arises regarding whether it is appropriate for healthcare providers to treat themselves. Which action should Practitioner Wilson take?
   A. Practitioner Wilson should continue to self-treat with over-the-counter medication.
   B. Practitioner Wilson should have her colleagues examine and treat the viral infection.
   C. Practitioner Wilson should see an impartial healthcare provider for an examination and treatment.
   D. Practitioner Wilson should visit a specialist for a second opinion.

   Explanation: Practitioner Wilson should see an impartial healthcare provider for an examination and treatment. The healthcare practice of treating oneself, family members, or friends is discouraged due to a lack of professional objectivity and violation of patient autonomy. When healthcare providers treat themselves, the lack of objectivity can lead to inadequate care, misdiagnosis, and inappropriate treatments. In addition, self-treatment can lead to personal feelings unduly influencing professional healthcare judgment, leading to inadequate care. It is categorized as unprofessional and a potential form of healthcare misconduct to treat oneself, family members, or friends. Therefore, Practitioner Wilson should see an impartial healthcare provider for an examination and treatment to ensure she receives the proper care and attention needed to address her symptoms. *Answer*: C

3. Practitioner Jessica Hall, a 30-year-old practitioner, is on a family vacation when her father suddenly experiences chest pain and shortness of breath. As a practitioner, she is faced with a dilemma regarding whether providing healthcare for her father is appropriate. What should Practitioner Hall do?
   A. Practitioner Hall should decline to provide healthcare for her father and refer him to another practitioner.
   B. Practitioner Hall should provide healthcare emergency care to stabilize her father's condition and then refer him to another practitioner for follow-up care.
   C. Practitioner Hall should not provide healthcare, especially if it is a healthcare emergency.
   D. Practitioner Hall should refer her father to an impartial practitioner for an examination and treatment.

   Explanation: Practitioner Hall should provide healthcare emergency care for her father until his condition stabilizes. Although it is generally discouraged for healthcare providers to treat their family members, exceptions can be made in situations of a healthcare emergency, minor and commonplace conditions, or situations independent of the necessity of having a patient–practitioner relationship. In this scenario, Practitioner Hall's father is experiencing chest pain and shortness of breath, which are symptoms of a potential healthcare emergency. As a practitioner, her professional duty is to provide healthcare to stabilize her father's condition until he can be transferred to an impartial practitioner for follow-up care. Providing healthcare in this situation is appropriate and justifiable under the exception of the general rule of not treating family members or oneself. Therefore, Practitioner Hall should provide healthcare for her father until his condition stabilizes. *Answer*: B

4. Practitioner David Clark, a 50-year-old psychiatrist, is at a family gathering when his cousin approaches him and requests a prescription for an antidepressant. The cousin explains that he has been feeling down and has read online that a specific antidepressant would be helpful for his symptoms. An ethical question arises regarding whether it is appropriate for healthcare providers to treat their family members and friends with prescription medications. What should Practitioner Clark do?
   A. Practitioner Clark should prescribe the antidepressant for his cousin.
   B. Practitioner Clark should refer his cousin to an impartial healthcare provider for an evaluation and prescription.

C. Practitioner Clark should discuss nonpharmacological treatment options with his cousin.
D. Practitioner Clark should explain the risks and benefits of the medication to his cousin and provide a prescription if the cousin agrees.

Explanation: Practitioner Clark should refer his cousin to an impartial healthcare provider for an evaluation and prescription. It is generally discouraged for healthcare providers to treat their family members or friends, as the lack of professional objectivity can lead to inadequate care, misdiagnosis, and inappropriate treatments. In addition, prescribing medication to family members or friends can lead to personal feelings unduly influencing professional healthcare judgment, leading to inadequate care. It is categorized as unprofessional and potential healthcare misconduct to treat oneself, family members, or friends with prescription medications. Moreover, prescribing medication to a family member or friend without a proper evaluation is against professional and legal regulations and may lead to disciplinary action or malpractice claims. Therefore, Practitioner Clark should refer his cousin to an impartial healthcare provider for an evaluation and prescription to ensure he receives the proper care and attention needed to address his symptoms. *Answer*: B

5. Ms. Maria Hernandez, a 35-year-old unemployed mother of two, visits the clinic with symptoms of depression, including sadness, loss of appetite, and difficulty sleeping. She is a close friend of the practitioner. The clinical differential diagnosis includes major depressive disorder, adjustment disorder, and dysthymia. As a practitioner, what is the ethical course of action in this situation?
    A. Treat Ms. Hernandez as a patient, but only after informing her of the ethical concerns with treating friends or family members
    B. Recommend that Ms. Hernandez see a different practitioner for her current symptoms
    C. Ask Ms. Hernandez to provide informed consent before treating her, but continue with the treatment
    D. Gracefully refuse to treat Ms. Hernandez and recommend she seek healthcare attention elsewhere, such as the Community Mental Health Center

Explanation: Treating friends or family members is generally discouraged due to the potential for compromised patient–practitioner relationships and violations of patient autonomy and healthcare obligations. In this case, the patient is a close friend of the practitioner, which could further complicate the situation. If the practitioner treats Ms. Hernandez, their relationship could be compromised, and the patient may feel uncomfortable sharing sensitive information necessary for proper diagnosis and treatment. In addition, the practitioner's personal feelings may unduly influence their healthcare judgment, leading to a violation of professional healthcare obligations of beneficence and nonmaleficence. Therefore, the ethical course of action is to gracefully refuse to treat Ms. Hernandez and recommend that she seek healthcare attention elsewhere, such as the Community Mental Health Center. *Answer*: D

REFLECTION VIGNETTES

1. Practitioner Karen Scott, a 35-year-old family medicine practitioner, encounters a family member seeking her healthcare expertise for a school physical and a urinary tract infection.

    What is the ethical issue(s)?

    How should the ethical issue(s) be addressed?

2. Ms. Lauren Davis, a 45-year-old practitioner, has been experiencing symptoms of a chronic disease for the past few months. Due to her concerns about confidentiality and privacy in the workplace and fear of being considered less competent, she has decided to self-treat her condition.

    What is the ethical issue(s)?

    How should the ethical issue(s) be addressed?

3. A seasoned faculty member at a well-respected school of medicine and healthcare was speaking with a junior colleague. In this conversation, the young practitioner said they "won't allow" their parents to be vaccinated because a well-known cardiologist said that there was an increased risk of myocarditis or pericarditis associated with being vaccinated. The young practitioner readily admitted that they and their three offspring were fully vaccinated with all the "usual" vaccines.

   What is the ethical issue(s)?

   How should the ethical issue(s) be addressed?

* * *

WEB LINKS

1. *Code of Medical Ethics*. American Medical Association. https://code-medical-ethics.ama-assn.org
   The AMA *Code of Medical Ethics*, established by the American Medical Association, is a comprehensive guide for healthcare practitioners. It addresses issues and challenges, promotes adherence to standards of care, and is continuously updated to reflect contemporary practices and challenges.
2. "Treating Self or Family." American Medical Association. https://code-medical-ethics.ama-assn.org/ethics-opinions/treating-self-or-family
   This resource discusses the fundamental role of informed consent in both ethics and law, emphasizing the patient's right to receive information and ask questions about recommended treatments.
3. "The Ethics of Treating Family Members." Colin Hutchison and Paul C. McConnell. 22019. https://pubmed.ncbi.nlm.nih.gov/30817390
   This article examines various ethical systems and how they relate to treating family members.
4. "Ethical Issues in Treating Self and Family Members." E. B. Anyanwu1, Harrison O. Abedi, and Efe A. Onohwakpor. pubs.sciepub.com/ajphr/2/3/6
   This article discusses the ethical issues that arise when physicians treat themselves or family members.
5. "The Ethics of Caring for Friends or Family." Kehinde Eniola. 2017. https://www.aafp.org/pubs/fpm/issues/2017/0700/p44.html
   This article discusses the ethical issues that can arise when physicians care for friends or family members.
6. "AMA *Code of Medical Ethics*' Opinion on Physicians Treating Family Members." AMA Council on Ethical and Judicial Affairs. 2012. https://journalofethics.ama-assn.org/article/ama-code-medical-ethics-opinion-physicians-treating-family-members/2012-05
   This article presents the American Medical Association's opinion on the ethical issues related to physicians treating family members.
7. "Ethics of Treating Family Members." Sandra Osswald. 2021. https://link.springer.com/chapter/10.1007/978-3-030-56861-0_19
   This article discusses the ethical principles involved in the care of family members, friends, and self.

SEXUAL BOUNDARIES

It is not enough to have a good mind; the main thing is to use it well.
—Rene Descartes

* * *

MYSTERY STORY

SEDUCTIVE MALPRACTICE: THE CASE OF THE UNETHICAL PRACTITIONER AT SACRED BOUNDARIES HOSPITAL

A bustling morning was in full swing at Sacred Boundaries Hospital, a reputable healthcare facility known for its stringent ethical standards. Among the skilled professionals was Practitioner Henry Baxter, a renowned cardiologist, lauded for his healthcare acumen but soon to be embroiled in a scandal that would shake the hospital's foundation.

The phone rang in Practitioner Baxter's office, and the voice on the other end was Mrs. Scarlett Rose, freshly discharged from Sacred Boundaries Hospital after a successful heart surgery performed by Practitioner Baxter. She called to thank the practitioner, her voice tinged with an admiration that transcended mere professional appreciation. Mrs. Rose had always found Practitioner Baxter charming, and during their interactions at Sacred Boundaries, she felt an inexplicable connection.

As their conversation unfolded, Practitioner Baxter sensed a flirtatious undertone in Mrs. Rose's words. The forbidden allure stirred something within him, and he found himself succumbing to temptation. Ignoring the nagging voice of reason, he asked Mrs. Rose out on a date.

Thrilled at first, Mrs. Rose accepted, but doubt soon clouded her excitement. She realized the inappropriateness of the situation and sought legal counsel. Learning about the ethical boundaries that govern the patient–practitioner relationship, she decided to report the incident to Sacred Boundaries Hospital's ethics committee.

The committee, rightfully alarmed, commenced an immediate investigation. Interviews with Mrs. Rose, Practitioner Baxter, and his colleagues at Sacred Boundaries revealed a disturbing pattern. Practitioner Baxter's history with patients was marred with unprofessional conduct, including instances of flirtation and inappropriate relationships that had caused emotional turmoil.

The committee's conclusion was swift and unequivocal: Practitioner Baxter had not only violated Sacred Boundaries Hospital's strict code of ethics but had also exploited his position, betraying the trust of vulnerable patients. The committee recommended termination of his employment and revocation of his healthcare license.

The downfall was immediate. Practitioner Baxter's once-thriving career crumbled, his reputation lay in ruins, and he was forced to leave the profession he once loved.

Sacred Boundaries Hospital, true to its commitment to ethical practice, used the incident as a learning opportunity, reinforcing the importance of professional boundaries and launching new training programs to prevent such incidents in the future.

The somber tale of Practitioner Baxter serves as a stark reminder of the sacred trust inherent in the patient–practitioner relationship. It illustrates the gravity of adhering to principles such as patient autonomy, beneficence, nonmaleficence, and justice, highlighting that the path to personal desire must never cross the unbreakable barriers of professional integrity and ethical conduct.

* * *

THINK

It is always unethical for a practitioner to have sexual relations with a current patient. For a psychiatrist, it is unethical to have any sexual relations with both present and former patients. However, it is unclear if any sexual contact with former patients is ethically acceptable by non-psychiatric practitioners.

The *American College of Physicians Ethics Manual* states,

> Issues of dependency, trust, transference, and inequalities of power lead to increased vulnerability on the part of the patient and require that a physician [practitioner] not engage in a sexual relationship with a patient. It is unethical for a physician [practitioner] to become sexually involved with a current patient even if the patient initiates or consents to the contact.

The American Medical Association (AMA) *Code of Medical Ethics* Opinion 9.1.1 states,

> Romantic or sexual interactions between physicians [practitioners] and patients that occur concurrently with the patient–practitioner relationship are unethical. Such interactions detract from the goals of the patient–physician [practitioner] relationship and may exploit the vulnerability of the patient, compromise the physician's [practitioner's] ability to make objective judgments about the patient's healthcare, and ultimately be detrimental to the patient's well-being.

Central to the art of healing is the patient–practitioner relationship related to the patient, practitioner, and public policy. This relationship can result in long-lasting, if not permanent, imprinting, making patients much more vulnerable to sexual exploitation, which is why it is unprofessional and unethical for a practitioner to have sexual relations with patients.

It is also unprofessional and unethical for a practitioner to have sexual relations with key third parties involved in the patient–practitioner relationship, such as parents, guardians, spouses, partners, or surrogates, when the interaction would exploit trust, knowledge, influence, or emotions derived from the patient–practitioner relationship that could result in compromised patient care.

ASSESS

Patient: Autonomy

Autonomy, with its resultant informed consent, can be deeply compromised due to the patient's emotional and physical vulnerabilities with the practitioner and the practitioner's emotionally compromised healthcare judgments. Therefore, maintaining strict sexual and emotional boundaries is essential for a legitimate patient–practitioner relationship and for attaining legitimate informed consent.

The practice of healthcare is supposed to be evidence-based, not emotionally based, because emotions and unconscious dependencies bring in so much decisional baggage that the practitioner's judgments are compromised and the patient's informed consent is compromised, showing the necessity of establishing clear and distinct patient–practitioner boundaries.

Practitioner: Beneficence and Nonmaleficence

The primary professional obligation is to maximize the patient's best interests in accordance with the patient's reasonable goals, values, and priorities using the principles of beneficence (do good) and nonmaleficence (do no harm).

The AMA) *Code of Medical Ethics* 9.1.1 states,

> Romantic or sexual interactions between physicians [practitioners] and patients... are unethical... detract from the goals of the patient–physician [practitioner] relationship

and may exploit the vulnerability of the patient, compromise the physician's [practitioner's] ability to make objective judgments ... and ... detrimental to the patient's well-being.

Sexual relations with a former patient or third party representative are also unprofessional, unethical, and sometimes illegal if the preceding patient–practitioner relationship, with its trust, protected health information (PHI), influence, and emotional attachments, is exploited in such a way that the previous patient, or involved third party, is vulnerable in the making of advances toward the practitioner or in responding to the practitioner's advances.

The AMA *Code of Medical Ethics* 9.1.1 states, "A physician [practitioner] who has reason to believe that nonsexual, nonclinical contact with a patient may be perceived as or may lead to romantic or sexual contact should avoid such contact."

These professional boundaries can also involve key third parties involved in the patient's care. The AMA) *Code of Medical Ethics* 9.1.2 states, "Physicians [practitioners] should avoid sexual or romantic relations with any individual whose decisions directly affect the health and welfare of the patient."

Public Policy: Justice

Justice (be fair) prohibits exploiting vulnerable citizens. The patient–practitioner relationship is a unique relationship with patients in a very vulnerable condition of needing to trust their practitioner to provide that which will maximize their best interests. The patient–practitioner relationship is not reciprocal because only the patient divulges PHI, which may include personal information about the patient's deepest desires and secrets. This places the patient in need of legal protection from practitioners who might consciously or unconsciously exploit their patients with unequal knowledge, vulnerability, dependency, and psychological transference.

CONCLUDE

It is imperative for the practitioner to keep the patient–practitioner relationship boundaries separate from any sexual, romantic, or familial association. These boundaries can be particularly difficult for practitioners who spend much of their time interacting mainly with patients. Professional, legal, and ethical boundaries that prohibit sexual and romantic relations with patients are essential for maximizing the patient's best interests.

In summary, maintaining professional and ethical boundaries in the patient–practitioner relationship is vital for preserving patient autonomy (informed consent), beneficence (do good), nonmaleficence (do no harm), and justice (be fair). Practitioners must avoid sexual and romantic relationships with patients or key third parties involved in the patient's care, as such relationships can compromise informed consent, exploit vulnerabilities, and ultimately be detrimental to the patient's well-being. By adhering to these boundaries, practitioners can uphold the highest standards of ethical and professional conduct. (See also Chapters 13, 22, 25, and 50.)

* * *

REVIEW QUESTIONS

1. Sometimes it is ethical for a practitioner to have sexual relations with a current patient.
    A. True
    B. False

2. The patient–practitioner relationship can result in long-lasting, if not permanent, imprinting, making patients much more vulnerable to sexual exploitation, which is why it is unprofessional and unethical for a practitioner to have sexual relations with patients.
    A. True
    B. False

3. It is unprofessional and unethical for a practitioner to have sexual relations with key third parties involved in the patient–practitioner relationship, such as parents, guardians, spouses, partners, or surrogates, when the interaction would exploit trust, knowledge, influence, or emotions derived from the patient–practitioner relationship resulting in compromised patient care.
   A. True
   B. False

4. Autonomy, with its resultant informed consent, can be deeply compromised due to the patient's emotional and physical vulnerabilities with the practitioner and the practitioner's emotionally compromised healthcare judgments.
   A. True
   B. False

5. The practice of healthcare is supposed to be evidence-based, not emotionally based, because emotions and unconscious dependencies bring in so much decisional baggage that both the practitioner's judgments and the patient's informed consent are compromised, showing the necessity of establishing clear and distinct patient–practitioner boundaries.
   A. True
   B. False

6. The patient–practitioner relationship is not reciprocal because only the patient divulges personal information about their deepest desires and secrets. This places the patient in need of legal protection from practitioners who might consciously or unconsciously exploit their patients with unequal knowledge, vulnerability, dependency, and psychological transference.
   A. True
   B. False

Answers: 1B, 2A, 3A, 4A, 5A, 6A

* * *

CLINICAL VIGNETTES

1. Ms. Sophia Roberts is a 38-year-old software engineer who has been seeing practitioner Emma Russell, a 45-year-old psychiatrist, treating depression and anxiety for the past 6 months. During one of their sessions, Ms. Roberts expresses her attraction to Practitioner Russell and asks her out on a date. Practitioner Russell knows that it is always unethical and illegal for a psychiatrist to engage in a romantic or sexual relationship with any patient, whether current or past, regardless of the circumstances. She also knows that abruptly terminating the therapeutic relationship could be considered patient abandonment, which is ethically and legally problematic. What should Practitioner Russell do?
   A. Practitioner Russell should explore her own feelings about Ms. Roberts and consider whether there is any transference or countertransference issue that might be affecting the therapeutic relationship.
   B. Practitioner Russell should discuss Ms. Roberts' feelings and explore any underlying issues that might be contributing to her attraction.
   C. Practitioner Russell should explain to Ms. Roberts that it is always unethical and illegal for a psychiatrist to engage in a romantic or sexual relationship with any patient, whether current or past, and that she is not able to accept the invitation for a date.
   D. Practitioner Russell should abruptly terminate the therapeutic relationship and provide Ms. Roberts with a referral to another psychiatrist.

   Explanation: Practitioner Russell should explain to Ms. Roberts that it is always unethical and illegal for a psychiatrist to engage in a romantic or sexual relationship with any patient, whether current or past, regardless of the circumstances. This response maintains clear boundaries and protects the patient's well-being. Practitioner Russell should also discuss Ms. Roberts' feelings and explore any underlying issues contributing to her attraction. Abruptly terminating the therapeutic relationship without providing adequate referrals or making arrangements for ongoing care could be considered patient abandonment and ethically and legally problematic. Exploring potential transference or countertransference issues can help Practitioner Russell address any underlying issues affecting the therapeutic relationship and ensure that the patient's needs are met safely and ethically. However, it is of first importance to establish in a non-accusatory fashion expected patient–practitioner boundaries. *Answer*: C

2. Mr. James Parker is a 28-year-old construction worker who has been seeing Practitioner Rachel Chen, a 35-year-old family practitioner, to treat chronic back pain. During one of their appointments, Mr. Parker makes a suggestive comment to Practitioner Chen and touches her inappropriately. Practitioner Chen is aware that it is always unethical and illegal for a practitioner to engage in a romantic or sexual relationship with a patient, and she is concerned about the safety and well-being of both Mr. Parker and herself. What should Practitioner Chen do?
    A. Practitioner Chen should explore her own feelings about Mr. Parker and consider whether there is any transference or countertransference issue that might be affecting the therapeutic relationship.
    B. Practitioner Chen should immediately end the appointment and refuse to see Mr. Parker again.
    C. Practitioner Chen should explain to Mr. Parker that his behavior is inappropriate and could be considered sexual harassment.
    D. Practitioner Chen should report Mr. Parker's behavior to the appropriate authorities and provide him with a referral to another practitioner.

    Explanation: Practitioner Chen should explain to Mr. Parker that his behavior is inappropriate and could be considered sexual harassment. This response maintains clear boundaries and protects the patient's well-being. Practitioner Chen should also explore potential transference or countertransference issues affecting the therapeutic relationship. Abruptly terminating the appointment without addressing the inappropriate behavior could be considered patient abandonment and ethically and legally problematic. Reporting Mr. Parker's behavior to the appropriate authorities and referring him to another practitioner may be necessary, depending on the severity of the situation. However, it is important to address the inappropriate behavior professionally and ethically and ensure that the patient's needs are met safely and respectfully. *Answer*: C

3. Ms. Jasmine Price is a 35-year-old restaurant owner who has been seeing Practitioner Alex Torres, a 42-year-old primary care practitioner, for her annual physical exams for the past 5 years. During one of their appointments, Practitioner Torres makes a comment about Ms. Price's appearance that makes her feel uncomfortable. Ms. Price is concerned that Practitioner Torres has overstepped professional boundaries and may have behaved inappropriately. What should Practitioner Torres do?
    A. Practitioner Torres should ignore Ms. Price's discomfort and continue with the physical exam.
    B. Practitioner Torres should apologize for the comment, assure Ms. Price that it will not happen again, and ask if there is anything he can do to make her feel more at ease.
    C. Practitioner Torres should try to make Ms. Price feel more comfortable by making additional comments or jokes.
    D. Practitioner Torres should acknowledge Ms. Price's discomfort.

    Explanation: Practitioner Torres should apologize for the comment, assure Ms. Price that it will not happen again, and ask if he can do anything to make her feel more at ease. This response shows that Practitioner Torres takes Ms. Price's feelings and concerns seriously and is willing to take steps to address them. Establishing and maintaining clear professional boundaries is essential to ethical healthcare practice, and Practitioner Torres should prioritize Ms. Price's well-being and comfort during the physical exam. Ignoring Ms. Price's discomfort or attempting to make light of the situation would be inappropriate and could further compromise professional boundaries and ethics. Acknowledging Ms. Price's discomfort is important, but once recognized, it is important to take responsibility and address the discomfort. *Answer*: B

4. Ms. Sarah Park is a 24-year-old graduate student seeking treatment for anxiety and depression from Practitioner Michael Kim, a 35-year-old clinical psychologist. During one of their therapy sessions, Ms. Park tells Practitioner Kim that she has feelings for him and asks him to engage in a romantic and sexual relationship with her. Practitioner Kim is surprised and unsure of how to respond. What should Practitioner Kim do?
   A. Practitioner Kim should accept Ms. Park's invitation and engage in a romantic and sexual relationship with her.
   B. Practitioner Kim should decline Ms. Park's invitation and continue their therapeutic relationship.
   C. Practitioner Kim should refer Ms. Park to another therapist and end their therapeutic relationship.
   D. Practitioner Kim should acknowledge Ms. Park's feelings and explain that it is inappropriate to explore a romantic and sexual relationship given the therapeutic relationship and work with her to maintain appropriate professional boundaries.

   Explanation: Practitioner Kim should acknowledge Ms. Park's feelings and explain that it is inappropriate to explore a romantic and sexual relationship given the therapeutic relationship. This response maintains clear boundaries and prioritizes Ms. Park's well-being and mental health. Practitioner Kim should explain to Ms. Park that the therapeutic relationship is a professional relationship and that it would be unethical to explore a romantic and sexual relationship with her. Accepting Ms. Park's invitation could create a conflict of interest and compromise the therapeutic process, and ending the therapeutic relationship would not be in Ms. Park's best interests. Referring Ms. Park to another therapist could be appropriate in some circumstances, but it is important to ensure that Ms. Park receives continuity of care and that her well-being and mental health needs are prioritized. It is important for Practitioner Kim to continue to work with Ms. Park to maintain appropriate professional boundaries and ensure that she feels safe and respected during her therapy sessions. Practitioner Kim should also consult with colleagues and ethical guidelines to ensure he follows the best practices and maintains ethical standards. *Answer*: D

5. Practitioner Meredith Hayes is a 45-year-old gynecologist who has seen Mrs. Fiona Martin, a 32-year-old patient, for annual checkups for the past 2 years. During a routine exam, Practitioner Hayes makes suggestive comments and behaves inappropriately toward Mrs. Martin, who is concerned that Practitioner Hayes may be overstepping professional boundaries. What should Mrs. Martin do?
   A. Mrs. Martin should ignore Practitioner Hayes' behavior and continue the checkup.
   B. Mrs. Martin should address Practitioner Hayes' behavior and tell her that it is making her uncomfortable.
   C. Mrs. Martin should report Practitioner Hayes to the appropriate authorities.
   D. Mrs. Martin should confront Practitioner Hayes and make suggestive comments in return.

   Explanation: Mrs. Martin should address Practitioner Hayes' behavior and tell her that it is making her uncomfortable. The practitioner is responsible for establishing and maintaining professional boundaries in the therapeutic relationship. Ignoring or reciprocating

Practitioner Hayes' behavior could compromise the therapeutic process and lead to potential ethical violations. Reporting Practitioner Hayes to the appropriate authorities may be necessary in some circumstances, but Mrs. Martin should first attempt to address the issue with her. Mrs. Martin should explain to Practitioner Hayes that the therapeutic relationship is a professional relationship and that it would be unethical to engage in any kind of romantic or sexual relationship. Mrs. Martin has the right to receive professional and ethical care during her appointments. Practitioner Hayes must maintain appropriate professional boundaries and ensure that Mrs. Martin feels safe and respected during her checkups. Practitioner Hayes should also consult with colleagues and ethical guidelines to ensure she follows the best practices and maintains ethical standards. *Answer*: B

REFLECTION VIGNETTES

1. Mr. Isaac Franklin, a 27-year-old third-year healthcare student, is being mentored into the practice of medicine at a local hospital. Ms. Chloe Eaton, a 24-year-old recent graduate, willingly provides informed consent for Mr. Franklin to observe and practice getting a patient's history and physical. Other activities included ensuring that Ms. Eaton was substantially informed of the risks and benefits of the treatment options and that all of her questions were answered under the supervision of the primary care practitioner. Mr. Franklin was diligent and responsible, spending extra time with Ms. Eaton. However, when Ms. Eaton and Mr. Franklin were together, they fell deeply in love.

    What is the ethical issue(s)?

    How should the ethical issue(s) be addressed?

2. Practitioner Lauren Simmons is a 34-year-old family practitioner at a busy healthcare center. One day, a patient named Mr. Brian Foster, a 28-year-old construction worker, comes to her clinic complaining of severe back pain that has been troubling him for several weeks. Practitioner Simmons diagnoses Mr. Foster with a herniated disc and prescribes him pain medication and physical therapy. Over the next few weeks, Mr. Foster visits the clinic for follow-up appointments. Practitioner Simmons works closely with him to monitor his progress and adjust his treatment plan. A couple of months later, Practitioner Simmons attends a friend's party, where she runs into Mr. Foster. They enjoy each other's company and spend most of the night talking and laughing together. As they leave the party, Mr. Foster asks if he can take Practitioner Simmons out on a date. The request takes her aback, and she realizes she has also developed feelings for Mr. Foster. As the days pass, Practitioner Simmons finds herself thinking about Mr. Foster increasingly more often. She wonders if she should pursue a relationship with him or distance herself to maintain professional boundaries.

    What is the ethical issue(s)?

    How should the ethical issue(s) be addressed?

* * *

WEB LINKS

1. *Code of Medical Ethics*. American Medical Association. https://code-medical-ethics.ama-assn.org
    The AMA *Code of Medical Ethics*, established by the American Medical Association, is a comprehensive guide for healthcare practitioners. It addresses issues and challenges, promotes adherence to standards of care, and is continuously updated to reflect contemporary practices and challenges.

2. "Patient–Physician Relationships." American Medical Association. https://code-medical-ethics.ama-assn.org/ethics-opinions/patient-physician-relationships
   This source discusses the ethical implications of romantic or sexual interactions between physicians and patients, emphasizing the importance of maintaining professional boundaries.
3. "Sexual and Romantic Boundary Violations." American Medical Association. https://edhub.ama-assn.org/code-of-medical-ethics/interactive/18014979
   This module explores why romantic and sexual interactions between physicians and patients undermine trust in the patient–physician relationship.
4. "Caring for the Trafficked Patient: Ethical Challenges and Recommendations for Health Care Professionals." Wendy L. Macias-Konstantopoulos. 2017. https://journalofethics.ama-assn.org/article/caring-trafficked-patient-ethical-challenges-and-recommendations-health-care-professionals/2017-01
   This article discusses the ethical challenges that healthcare professionals face when caring for trafficked patients.
5. "Engagement Without Entanglement: A Framework for Non-Sexual Patient–Physician Boundaries." Jacob M. Appel. 2023. https://jme.bmj.com/content/49/6/383
   This article discusses the importance of maintaining professional boundaries in the patient–physician relationship, focusing on avoiding sexual boundary transgressions.
6. "Quality Medical Care Should Be a Given, No Matter a Patient's Sexual or Gender Identity." Oregon Health & Science University. 2017. https://news.ohsu.edu/2017/06/26/quality-medical-care-should-be-a-given-no-matter-a-patient-s-sexual-or-gender-identity
   This source emphasizes the importance of providing quality healthcare regardless of a patient's sexual or gender identity.
7. "Georgia Law Change Would Require Transparency in Handling of Doctor Sex." Carrie Teegardin. 2021. https://www.ajc.com/news/investigations/legislation-to-require-transparency-reporting-of-doctor-sex-abuse/GEUZ6OOTKZHIRNQV63RJLCESNE
   This article discusses proposed legislation that would require transparency and new training related to professional boundaries and sexual misconduct in healthcare.
8. "Ethical and Professional Conduct of Health Care Providers." Indian Health Service. https://www.ihs.gov/IHM/pc/part-3/p3c23
   This source discusses healthcare providers' ethical and professional conduct, emphasizing the importance of respecting and maintaining appropriate provider–patient boundaries.

SEXUALLY TRANSMITTED INFECTIONS

52

A practitioner is an unfortunate person who is required to perform a miracle, namely to reconcile health with intemperance.
—Voltaire

* * *

MYSTERY STORY

INFECTION INTRIGUE: THE LOTUSBLOOM STI CLINIC DILEMMA

It was a brisk Monday morning at LotusBloom STI Clinic when a young woman named Ms. Isabella Walker walked in, her face flushed with anxiety. I watched from my intern's desk, filled with both excitement and nerves, as I was about to embark on my first real experience in the field of sexually transmitted infections (STIs). My attending practitioner, Practitioner Anthony Murphy, known far and wide for his skill, was to be my guide.

As Ms. Walker was led to the welcoming and discreetly designed examination room of LotusBloom, I noticed her eyes darting around, seemingly taking solace in the tranquil ambiance yet still appearing nervous. Practitioner Murphy, with his kind eyes and reassuring demeanor, calmed her fears, displaying an empathy I was eager to learn.

During the examination, Practitioner Murphy gently probed Ms. Walker's sexual history, uncovering a recent encounter that had left her worried. The testing was thorough, carried out with the utmost professionalism, and chlamydia was confirmed. Practitioner Murphy's face remained compassionate as he explained the infection, its treatment, and the road to recovery.

He then carefully emphasized the legal obligation to report the infection to the health department, reassuring Ms. Walker that LotusBloom STI Clinic upheld the strictest confidentiality. The intricate process of containing the spread involved contacting her sexual partners, something that clearly distressed her. With patience and wisdom, Practitioner Murphy guided her to recognize the greater societal responsibility, and she consented.

The day wore on, filled with learning and experience, until the waiting room's quiet was shattered by an irate man demanding answers about an STI report. His voice was thunderous, his anger palpable, and it became evident that he was one of Ms. Walker's recent partners, infuriated by what he perceived as an intrusion into his private life.

Practitioner Murphy, a pillar of calm amid the storm, explained the reasoning behind LotusBloom's procedure and the universal need to manage STIs responsibly. It was a master class in conflict resolution as the man's anger subsided and understanding dawned. He eventually agreed to treatment, a changed outlook in his eyes.

As the clinic's door closed behind us that evening, I pondered the intricate ethical landscape of managing STIs. Practitioner Murphy's tactful navigation between patient rights and community well-being was more than a lesson; it was an inspiration. LotusBloom STI Clinic's policy, founded on principles of care, trust, and responsibility, provided a framework for ethical practice, beautifully balancing patient autonomy, practitioner beneficence and nonmaleficence, and public policy justice.

The experiences of that day at LotusBloom would resonate with me throughout my healthcare career, a vivid reminder of the importance of empathy, ethics, and education in the delicate world of sexual health.

* * *

THINK

All sexually transmitted diseases (STDs) start as an STI. Once a patient is infected by a foreign bacteria, virus, or protozoan, the infection can develop into a disease. An STI can be asymptomatic, but an STD usually manifests itself with signs and symptoms.

Because the term venereal *disease* connotes undesirable symptoms, there has developed a social stigma associated with the term. Therefore, many practitioners now use the more accepted term, sexually transmitted *infection*.

STIs are transferred from one person to another usually through vaginal, anal, or oral sexual contact. Examples of bacterial infections are chlamydia, gonorrhea, and syphilis. Viral infections include hepatitis B, HIV, herpes simplex virus, and human papillomavirus (HPV). Examples of protozoal infections are trichomoniasis and parasites such as pubic lice.

Mandatory reporting laws are government surveillance laws that require the practitioner, institution, and laboratories to report STIs, including the patient's name and date of birth, to the health department. This allows the government to implement mitigation measures to curb infectious outbreaks.

Mandatory reporting laws do not breach the patient–practitioner relationship with its duty of patient confidentiality and privacy regarding the patient's protected health information (PHI). Rather, the categories of mandated reporting are part of the patient–practitioner social contract because that is what a reasonable person would consider just or fair public policy. The breach of the patient–practitioner relationship would be the practitioner not reporting a mandatory reporting incident.

ASSESS

Patient: Autonomy

The patient–practitioner relationship is central for patients to trust their caregiver with confidential PHI. Without such trust, informed consent would not be possible because the patient would fear providing honest disclosures and would not trust that the practitioner only had the patient's best interests in mind. However, like every personal relationship, the patient–practitioner relationship has specific boundaries and responsibilities. The practitioner is obligated to the patient and the community to respect mandatory reporting laws regarding STIs as part of the patient–practitioner relationship.

Although all contact reporting will be done anonymously to respect the patient's confidentiality and privacy, most patient partners will know immediately who their sexual partner(s) has been and who the patient is who reported them as a patient contact. As such, the notion of keeping "patient anonymity," although semantically true, is practically untrue.

Therefore, it is imperative that the practitioner disclose such reporting mandates to the patient as soon as possible and explain the purposes and justifications of the mandates and that practitioner reporting is not a breach of the patient–practitioner relationship but, rather, a part of the social boundaries of that relationship. The goal for the practitioner should be to help the patient understand the importance of mandated reporting and, if possible, obtain the patient's informed consent to report.

Practitioner: Beneficence and Nonmaleficence

The professional principles of beneficence (do good) and nonmaleficence (do no harm) are patient-centered principles for the maximization of the patient's best interests as determined by the patient's reasonable goals, values, and priorities. In treating STIs, it is imperative to treat not just the patient's infection but also the infection of the patient's sexual partner(s); otherwise, it will be impossible to maximize both the patient's best interest, which is to cure, and society's interest in halting the recycling of contagious infections.

Mandatory reporting, in addition to effective treatments, is a professional obligation of beneficence (do good) that maximizes patients' best interests by breaking the cycle of infection in accordance with the principle of nonmaleficence (do no harm).

Expedited partner therapy (EPT) occurs when the practitioner provides STI treatment to the patient's partner without seeing the partner in person. Some, but not all, states approve of EPT. The argument is that it would be futile to treat the patient without treating the patient's sexual partner, who would reinfect the patient if not treated. If the partner is unwilling or unable to come in for an appointment, then it would be better for the patient, partner, and the community to treat the partner from a distance rather than providing no treatment.

The American Medical Association (AMA) *Code of Medical Ethics* 8.9 states,

> Expedited partner therapy potentially abrogates the standard informed consent process, compromises continuity of care for patients' partners, encroaches upon the privacy of patients and their partners, increases the possibility of harm by a medical or allergic reaction, leaves other diseases or complications undiagnosed, and may violate state practice law.

If EPT is conducted by a practitioner in a state that allows it, then it is imperative that the practitioner try to mitigate each of those concerns by

- initiating a conversation with the partner by phone or by video conferencing;
- discussing all aspects of the informed consent protocol with the partner;
- providing educational materials;
- disclosing that the EPT will not effectively treat other infections, if present; and
- referring the partner to an appropriate healthcare provider.

As telemedicine is adopted by the healthcare profession, institutions, and patients as an accepted alternative for specific checkups and treatments, EPT, as implemented by telemedicine, will also become more adopted as a standard of care. EPT could be modified to include an online telemedicine visit by the partner before obtaining healthcare treatment. This effectively addresses the AMA 8.9 concerns about EPT.

Public Policy: Justice

Classical liberalism purports that maximizing each citizen's individual rights and liberties will maximize all citizens' individual good. Because the community comprises groups of individuals, it logically follows that the community's goodwill will be maximized. However, if the individual rights and liberties of others are harmed by the inappropriate exercise of individual rights and liberties, then mitigation measures must be mandated as a matter of justice. When STIs fit these social conditions, it becomes necessary for public health authorities to implement mandatory reporting laws to protect other individuals' rights and liberties that make up the community's good.

Ethically, there is no difference between an act of omission and an act of commission when the consequences of harm are equivalent. Justice (be fair) recognizes this, giving governments and social institutions authority to mandate the reporting of STIs and to inform the patient's partner(s) of communicable exposure risks as an act of omission by not having mandatory reporting laws, which would have the negative social consequence of an act of commission—a contagious infection epidemic.

Epidemiologists, the scientific experts on mitigating STIs, have incontrovertible empirical evidence that mandatory reporting of STIs is required to keep contagious infections from becoming a community epidemic. The role of government is to prevent and mitigate such contagious harms as a principle of upholding citizens' constitutional right and liberty to assemble and interact safely.

A practitioner's failure to report a communicable disease is a misdemeanor punishable by a fine of between $50 and $1,000 and up to 90 days imprisonment.

CONCLUDE

Mandatory reporting compliance for STIs is a necessary duty as a practitioner. The practitioner's goal should always be to obtain the patient's informed consent to report by helping the patient understand how mandatory reporting maximizes the patient's best interests. Although EPT is not yet accepted in all states, it behooves the practitioner to think through and assess the strengths and weaknesses of EPT and know what actions can be taken to mitigate the ethical and practical concerns.

In summary, mandatory reporting of STIs is essential for practitioners to protect public health and prevent the spread of infections. Practitioners must strive to obtain the patient's informed consent for reporting and help the patient understand the importance of these measures. Although EPT is a controversial practice and not accepted in all states, it is important for practitioners to consider the ethical implications and potential benefits of this approach. Ultimately, the healthcare community must continue to explore innovative and ethical solutions to manage and treat STIs effectively.

* * *

REVIEW QUESTIONS

1. Categories of mandated reporting are part of the patient–practitioner social contract, as that is what a reasonable person would consider just or fair public policy. The breach of the patient–practitioner relationship would be the practitioner not reporting a mandatory reporting incident.
   A. True
   B. False

2. The goal for the practitioner should be to help the patient understand the importance of mandated reporting and, if possible, obtain the patient's informed consent to report.
   A. True
   B. False

3. Mandatory reporting, with effective treatments, is a professional obligation of beneficence (do good) that maximizes patients' best interests by breaking the cycle of infection in accordance with the principle of nonmaleficence (do no harm).
   A. True
   B. False

4. Expedited partner therapy occurs when the practitioner prioritizes a face-to-face clinical appointment, expediting the patient's partner therapy.
   A. True
   B. False

5. When the consequences of harm are equivalent, there is ethically a significant difference between an act of omission and an act of commission.
   A. True
   B. False

Answers: 1A, 2A, 3A, 4B, 5B

* * *

CLINICAL VIGNETTES

1. Mr. Timothy Bryant is a 25-year-old graphic designer who comes to the clinic complaining of burning urination and a rash on his genitals. Upon further questioning, he admits to having unprotected sex with multiple partners in the past month. He denies any fever, chills, or swollen lymph nodes. The differential diagnosis includes gonorrhea, chlamydia, syphilis, and genital herpes. An ethical question arises about whether to inform his sexual partners of his condition. What should the practitioner do?
   A. The practitioner should inform Mr. Bryant's sexual partners without his consent to protect the community's interest.
   B. The practitioner should inform Mr. Bryant's sexual partners only with his consent, respecting both Mr. Bryant's autonomy and the community's interest.
   C. The practitioner should solely encourage Mr. Bryant to inform his sexual partners and provide resources to facilitate the process.
   D. The practitioner should report the STI as mandated by law and encourage Mr. Bryant to inform his sexual partners while also providing resources to facilitate the process.

   Explanation: The practitioner must balance respect for Mr. Bryant's autonomy and the community's interest while adhering to mandatory reporting laws. The patient–practitioner relationship is based on informed consent and confidentiality. However, mandatory reporting laws are not a breach of the patient–practitioner relationship because they are part of the social contract. The practitioner should explain to Mr. Bryant the purpose of mandated reporting and encourage him to inform his sexual partners. In this situation, the practitioner should report the STI as required by law and support Mr. Bryant in informing his partners. *Answer*: D

2. Ms. Charlotte Lewis is a 30-year-old accountant who comes to the clinic with a diagnosis of chlamydia. She reports having unprotected sex with her boyfriend, who lives in a different state. She is concerned about informing him of the diagnosis, as he cannot come to the clinic for testing and treatment due to work commitments. The practitioner discusses EPT as a potential option and explains the process to Ms. Lewis. What action does the practitioner take?
   A. The practitioner provides Ms. Lewis with a prescription for her boyfriend without any further discussion or consent.
   B. The practitioner advises Ms. Lewis to call her boyfriend and provide him with the prescription and instructions on how to take the medication.
   C. The practitioner arranges a telemedicine appointment for the boyfriend to consult with the practitioner and obtain a prescription.
   D. The practitioner encourages Ms. Lewis to have her boyfriend go to a local clinic or healthcare provider to obtain testing and treatment.

   Explanation: The AMA *Code of Medical Ethics* states that EPT may violate state practice law and compromise the continuity of care for patients' partners. Therefore, the practitioner should arrange a telemedicine appointment for the boyfriend to consult with the practitioner and obtain a prescription, which addresses informed consent concerns and continuity of care. The other options are incorrect because they do not involve informed consent or a consultation with the practitioner; assume the boyfriend will follow the instructions and may not address any questions or concerns he may have; or put the burden on Ms. Lewis and do not involve the practitioner in the process, and having the boyfriend go to a local clinic may not be feasible for the boyfriend due to work commitments. EPT is a potential option in cases in which the patients' partners are unable to visit the clinic or when partner notification is not sufficient or feasible. However, it is important for the practitioner to follow ethical and legal guidelines to ensure patient safety and privacy. Telemedicine is an emerging field that offers opportunities for providing EPT services while mitigating the ethical concerns associated with it. Overall, EPT can be a valuable tool in preventing the spread of STIs, and practitioners should consider it an option when appropriate. *Answer*: C

3. Ms. Abigail Wilson is a 26-year-old nurse who visits the clinic with a complaint of genital warts. She reports having unprotected sex with multiple partners during the past year. The practitioner performs an examination and confirms a diagnosis of HPV. As mandated by law, the practitioner explains the mandatory reporting laws and the importance of reporting the case to the health department. Ms. Wilson is concerned about the consequences of reporting, as she is afraid of losing her job and reputation as a nurse. What should the practitioner do?
    A. The practitioner should not report the case to the health department to protect Ms. Wilson's confidentiality.
    B. The practitioner should report the case to the health department but not disclose Ms. Wilson's name to protect her privacy.
    C. The practitioner should report the case to the health department and disclose Ms. Wilson's name as required by law.
    D. The practitioner should refer Ms. Wilson to another clinic that does not report cases to the health department.

    Explanation: The correct answer is option C because the practitioner must legally report the case to the health department. Mandatory reporting laws require the practitioner to report the case, including the patient's name and date of birth, to the health department. This allows the government to implement mitigation measures to curb infectious outbreaks. The other options are incorrect because they do not address the community's interest in preventing the spread of STIs, are in noncompliance with the mandatory reporting laws to provide the patient's name and date of birth, or do not specifically address the practitioner's obligation to report the case to the health department. Although respecting patient confidentiality and privacy is important, mandatory reporting laws exist to protect public health and prevent the spread of STIs. Practitioners must balance patient autonomy and confidentiality with their obligations under the law. It is important for practitioners to inform patients about mandatory reporting laws and the consequences of noncompliance. In summary, the practitioner should report the case to the health department, including Ms. Wilson's name, as the law requires. The practitioner should also work with Ms. Wilson to provide resources to deal with the consequences of mandatory reporting. *Answer*: C

4. Mr. Thomas Russell is a 35-year-old man who presents to the clinic with a complaint of painful urination and discharge from his penis. He reports having unprotected sex with a sex worker while on a business trip to a neighboring state 2 weeks ago. The practitioner examines him and diagnoses him with gonorrhea, an STI. As mandated by law, the practitioner explains the mandatory reporting laws and the importance of reporting the case to the health department. Mr. Russell expresses his concern about being reported to his employer, as he works for a large government agency and is required to obtain a security clearance. What should the practitioner do?
    A. The practitioner should not report the case to the health department to protect Mr. Russell's confidentiality.
    B. The practitioner should report the case to the health department but not disclose Mr. Russell's name to protect his privacy.
    C. The practitioner should report the case to the health department and disclose Mr. Russell's name as required by law.
    D. The practitioner should encourage Mr. Russell to self-report the case to the health department.

    Explanation: The practitioner must legally report the case to the health department, including Mr. Russell's name and other identifying information. Mandatory reporting laws require the practitioner to report the case, which is necessary to prevent the spread of the STI and protect public health. The other options are wrong because they do not address the community's interest in preventing the spread of STIs; violate the mandatory reporting laws requirement to disclose the patient's name; or put the responsibility on Mr. Russell to self-report, which may not be effective in preventing the spread of the STI. Although respecting patient confidentiality and privacy is important, mandatory reporting laws exist to protect public health and prevent the spread of STIs. Practitioners must balance patient autonomy and confidentiality with their obligations under the law. It is important for practitioners to inform patients about mandatory reporting laws and the consequences of noncompliance. In this case, Mr. Russell is concerned about the potential consequences of mandatory reporting, but the practitioner must comply with the law and report the case to the health department. The practitioner should explain the importance of reporting to Mr. Russell and provide resources to deal with the consequences of mandatory reporting. It is important for practitioners to be aware of the potential consequences of mandatory reporting, such as the impact on employment and security clearances, and to work with patients to mitigate these risks while complying with the law. *Answer*: C

5. Ms. Rachel Baker is a 28-year-old woman who presents to the clinic with a complaint of vaginal itching and a burning sensation during urination. She reports having unprotected sex with her new boyfriend 2 weeks ago. The practitioner examines her and diagnoses her with a chlamydia infection, an STI. As mandated by law, the practitioner explains the mandatory reporting laws and the importance of reporting the case to the health department. Ms. Baker expresses her concern about her immigration status, as she is currently undocumented and fears that reporting the case may lead to deportation. What should the practitioner do?
    A. The practitioner should not report the case to the health department to protect Ms. Baker's confidentiality.
    B. The practitioner should report the case to the health department and disclose Ms. Baker's name and date of birth as required by law.
    C. The practitioner should report the case to the health department but not disclose Ms. Baker's name to protect her privacy.
    D. The practitioner should encourage Ms. Baker to self-report the case to the health department.

   Explanation: Mandatory reporting laws for STIs require the practitioner to report the patient's name and date of birth to the health department. This information is necessary to identify and notify the patient's sexual partners and to prevent the spread of STIs. The other options are incorrect because they ignore the community's interest in preventing the spread of STIs; put Ms. Baker's privacy at risk; or put the responsibility on Ms. Baker to self-report, which may not be effective in preventing the spread of the STI. It is important for practitioners to balance patient autonomy and confidentiality with their obligations under the law. In this case, Ms. Baker's fear of deportation is a valid concern, and the practitioner should provide her with resources to deal with this issue. The practitioner should explain the importance of reporting the case to the health department and provide Ms. Baker with information about the consequences of noncompliance. Mandatory reporting laws exist to protect public health and prevent the spread of STIs. However, it is important for practitioners to be aware of the potential consequences of reporting, such as the impact on patients' privacy and immigration status, and to work with patients to mitigate these risks while complying with the law. In cases in which patients are undocumented, practitioners can still report the cases to the health department while taking steps to protect their privacy and minimize the risk of deportation. *Answer*: B

REFLECTION VIGNETTES
1. Ms. Daisy Burke is a 25-year-old receptionist who comes to the clinic complaining of genital itching and burning during urination. During her physical exam, the practitioner observes genital ulcers and bumps, leading to the diagnosis of genital herpes. The practitioner is mandated by law to report the STI to the state health department and to inform

Ms. Burke's sexual partner to ensure proper treatment and prevent further spread of the disease. However, Ms. Burke becomes agitated and anxious upon learning that her partner will be informed, even though the practitioner assures her that the information will be kept confidential and her identity will be protected. Ms. Burke expresses concerns about the negative impact on her personal and professional life if her partner finds out, and she worries about the stigma associated with having an STI.

What is the ethical issue(s)?

How should the ethical issue(s) be addressed?

2. Ms. Isabel Jensen is a 27-year-old nurse who presents to the clinic with symptoms suggestive of an STI. Following an examination, the practitioner diagnoses Ms. Jensen with chlamydia, and she provides informed consent for treatment. The practitioner advises Ms. Jensen's partner to seek healthcare attention for treatment, but the partner refuses due to various reasons. The patient requests EPT, a practice in which a practitioner provides a prescription for medication to a patient's sexual partner without examining them. As per the practice guidelines, the practitioner provides EPT to Ms. Jensen and instructs her on how to deliver the medication to her partner.

What is the ethical issue(s)?

How should the ethical issue(s) be addressed?

* * *

WEB LINKS

1. *Code of Medical Ethics.* American Medical Association. https://code-medical-ethics.ama-assn.org
   The AMA *Code of Medical Ethics*, established by the American Medical Association, is a comprehensive guide for healthcare practitioners. It addresses issues and challenges, promotes adherence to standards of care, and is continuously updated to reflect contemporary practices and challenges.
2. "STD vs STI: Common Types, Symptoms, and Treatment." State Urgent Care. https://starkvilleurgentcareclinic.com/std-vs-sti-common-types-symptoms-and-treatment
   The article differentiates between STDs and STIs by explaining that many STDs start as STIs with the invasion of bacteria or viruses, and it emphasizes the change in terminology to reduce stigma and encourage testing, given that STIs often show no initial symptoms.
3. "Sexually Transmitted Infections, Public Health, and Ethic." Mary Ott and John Santelli. 2019. https://www.oxfordhandbooks.com/view/10.1093/oxfordhb/9780190245191.001.0001/oxfordhb-9780190245191-e-36#ref_oxfordhb-9780190245191-e-36-note-21
   The source discusses the core public health function of controlling STIs, focusing on the ethical challenges of balancing individual liberties with population sexual health and addressing disparities in STI outcomes and access to information and treatment, suggesting a human rights approach to address underlying structural health determinants.
4. "Patient Rights and Ethics." Jacob P. Olejarczyk and Michael Young. https://www.ncbi.nlm.nih.gov/books/NBK538279
   This source discusses the rights of patients with STIs and the ethical considerations involved in their treatment.
5. "Ethical Decision Making in the Diagnosis and Management of STDs." *Journal of Ethics.* 2005. https://journalofethics.ama-assn.org/issue/ethical-decision-making-diagnosis-and-management-stds
   This article discusses the ethical controversies associated with the prevention, diagnosis, and management of STDs.

6. "Risks, Ethics & Sexually Transmitted Diseases." NYU Langone Health. https://med.nyu.edu/departments-institutes/population-health/divisions-sections-centers/medical-ethics/education/high-school-bioethics-project/learning-scenarios/risks-ethics-stds
   This learning scenario covers the basics of STDs, including types of infections, symptoms, and treatments.
7. "Ethical and Legal Considerations in STI Treatment for Adolescents." Quianta L. Moore. 2020. https://link.springer.com/chapter/10.1007/978-3-030-20491-4_4
   This source discusses the ethical and legal considerations in treating adolescents with STIs.
8. "The Stigma of Sexually Transmitted Infections." Amy S. D. Lee and Shameka L. Cody. 2020. https://pubmed.ncbi.nlm.nih.gov/32762851
   This article discusses the social stigma associated with STI testing and its impact on the utilization of prevention services.
9. "CDC Updates Guidance on Treatment of STIs." American Academy of Family Physicians. https://www.aafp.org/news/health-of-the-public/20210811stiguidlines.html
   This article discusses the CDC's updated guidelines on the treatment of STIs.
10. "Consent, Confidentiality and Curing Sexually Transmitted Infection: An Ethical Trilemma." K. P. Dunphy. 2011. https://pubmed.ncbi.nlm.nih.gov/21571977
    This article discusses the ethical dilemmas faced by health professionals when treating contacts of patients with STIs.
11. "Sexually Transmitted Infections (STIs)." World Health Organization. 2023. https://www.who.int/news-room/fact-sheets/detail/sexually-transmitted-infections-(stis)
    This fact sheet provides key facts about STIs worldwide.
12. "Biomedical Tools for STI Prevention and Management." National Center for Biotechnology Information. https://www.ncbi.nlm.nih.gov/books/NBK573166
    This source discusses the biomedical scientific advances that have provided numerous powerful tools to diagnose, prevent, and manage STIs.
13. "Sexually Transmitted Infections Treatment Guidelines, 2021." Centers for Disease Control and Prevention. https://www.cdc.gov/std/treatment-guidelines/default.htm
    This guideline provides current evidence-based prevention, diagnostic, and treatment recommendations for STIs.

SOCIAL MEDIA BOUNDARIES

53

Be kind. Every person is fighting a difficult battle.
—Plato

* * *

MYSTERY STORY

PRIVACY PERIL: THE MEDIAMANNERS HOSPITAL DILEMMA

In the shadowed corridors of MediaManners Hospital, a place known for its exemplary commitment to patient privacy, the staff was rocked by a chilling discovery. Practitioner Camilla Bello, a renowned and respected surgeon, was found lifeless in her office, her face etched with distress.

Detective Leonard McCoy, known for his analytical prowess, was summoned to unravel the perplexing mystery. The prestigious MediaManners Hospital had always prided itself on maintaining strict social media boundaries, and yet it was this very aspect that led to the unraveling of a twisted tale.

Through rigorous examination and careful interviews, Detective McCoy discovered that Practitioner Bello had been a hidden enthusiast of social media. Her passion was not just for medicine but also for sharing her experiences as a healer. Unbeknownst to her colleagues, she had created a blog in which she chronicled her surgical triumphs and challenges.

The blog, although filled with genuine insight and knowledge, was a Pandora's box. It was discovered that Practitioner Bello had unintentionally overstepped ethical boundaries, discussing patient cases and even sharing images without consent. This blatant breach of privacy laws was not just a violation of MediaManners Hospital's policies but also a legal infringement.

One particular patient, whose identity had been thinly veiled in one of Bello's posts, stumbled upon the blog. Enraged and feeling violated, he sought legal action. The pressure began to mount as Practitioner Bello realized the magnitude of her mistake.

But the nightmare was far from over. Detective McCoy unearthed an insidious layer to the story: blackmail. Someone had been holding Practitioner Bello hostage with threats of public exposure. Although she bravely refused to bow to the demands, the stress took its toll.

After weeks of relentless pressure, Practitioner Bello's strength finally waned. The exact details of her demise remained ambiguous, but the implications were clear. Her death was a tragic result of the clash between the virtual world and the sacred trust of the healthcare profession.

MediaManners Hospital was forced to confront an uncomfortable truth. New guidelines were implemented, and a renewed emphasis on digital ethics was infused into the hospital's culture. The incident served as a harrowing reminder to students and practitioners alike about the power and peril of social media.

As the investigation closed, the solemn words of Plato resonated throughout the halls of MediaManners Hospital: "Be kind. Every person is fighting a difficult battle." The story of Practitioner Bello's downfall was a lesson in empathy, caution, and the unbreakable bond of trust that must be maintained in the world of medicine. Her legacy became a beacon for others, a stern warning against the erosion of the very principles she had once held dear.

* * *

THINK

According to the Pew Research Center,[1] only 5% of the American population used social media in 2005; by 2011, that had risen to 50%. Currently, more than 72% of the population uses social media (Table 53.1). Young adults were the first to adopt social media, but usage by older adults has increased dramatically.

Table 53.1 **Social Media: U.S. Adults**

Demographic	%
Age (years)	
18–29	84
30–49	81
50–64	73
65+	45
Phenotype	
White	69
Black	77
Hispanic	80
Gender	
Male	66
Female	78
Income	
<$30,000	69
$30,000–$50,000	76
$50,000–$75,000	65
>$75,000	78
Education	
High school	64
Some college	76
College	77
Community	
Urban	76
Suburban	71
Rural	66

Social media networks, blogs, and video sites are an integral part of society and will only expand their influence. Students and practitioners can now have an online professional presence, communicating and disseminating health information instantly with millions of people.[2] However, care must be taken not to blur one's social online presence with one's professional online presence. This is particularly difficult for students who already have an online social presence and who are transitioning into a healthcare profession with significant professional, legal, and moral boundaries.

One practitioner boundary is the Health Insurance Portability and Accountability Act (HIPAA), a federal law enacted by Congress and signed into law in 1996, ensuring the confidentiality and privacy of patient's protected health information (PHI) from unauthorized disclosures with punishable fines of up to $250,000 and a jail term of up to 5 years.

A patient's right to confidentiality and privacy is so vital that the HIPAA Privacy Rule prohibits accessing a patient's records without prior authorization for any reason other than treatment, payment, and healthcare operations. This was exemplified in the high-profile Jussie Smollett case, in which at least 50 employees, including nurses, at Northwestern Memorial Hospital in Chicago were fired after being accused of improperly reviewing the actor's healthcare records.

---

1. Pew Research Center. https://www.pewresearch.org/internet/fact-sheet/social-media/?menuItem=d102d cb7-e8a1-42cd-a04e-ee442f81505a.
2. "Professionalism in the Use of Social Media." American Medical Association. https://www.ama-assn.org/delivering-care/ethics/professionalism-use-social-media.

Before a student or practitioner uses social media, they must determine precisely what their institution's social media policies are, as HIPAA is the minimum professional boundary set by federal law. Because of risk management issues, accreditation, and professional reputation, institutions usually put much stricter professional boundary standards in place than those enumerated in HIPAA.

If the student or practitioner wears the white coat for social media, then they become a spokesperson for all the institutions that the speaker is associated with, and institutions are very wary and strict about any actual or perceived risk of misperception. Without institutional authorization, the student or practitioner should never wear the white coat or associate themself with an institution when on social media. Personal and professional boundaries must be clearly and distinctly defined.

No patient identities, diagnoses, or treatment options should ever be discussed on social media because that would be a HIPAA violation, and the student or practitioner could be held professionally, legally, and institutionally liable. When a student uses social media, they must be aware of their school's social media policies and any other institutional policies they may be associated with while doing clerkships and sub-internships. Students and practitioners must set clear boundaries between their social and professional lives because failure to do so could have permanent career and legal consequences.

ASSESS

Patient: Autonomy

A practitioner might think that obtaining the patient's written informed consent before sharing the patient's information on social media would be a sufficient condition for doing so, but that is not correct. Although obtaining written informed consent from the patient would be a necessary condition, that may not be a sufficient condition for sharing on social media. If a practitioner's action violates professional, institutional, or public policy disclosure standards, then that would be a sufficient condition for the practitioner to face possible professional, institutional, and legal disciplinary actions.

Practitioner: Beneficence and Nonmaleficence

The healthcare profession's prime directive is the maximization of the patient's best interests in accordance with the patient's reasonable goals, values, and priorities. However, for the patient to share their PHI, the recognized prerequisite is for there to be a patient–practitioner relationship. This relationship will only be actualized if the patient and society trust that the healthcare profession will keep their PHI confidential and private. Any exceptions to confidentiality and privacy must be clearly stated and understood by the patient. Examples of confidentiality and privacy exceptions are when the practitioner is compelled by mandatory reporting laws for contagious infections, child and elder abuse, and gunshot wounds. Confidential and privacy boundaries must not be breached for social media purposes, as doing so would be a professional violation of beneficence (do good) and nonmaleficence (do no harm).

Public Policy: Justice

Social media has expanded the influence of an individual's words and thoughts. This can result in increased opportunities for both good and harmful social consequences. If harmful consequences violate individual civil rights and liberties, such as healthcare confidentiality and privacy, then the government has the authority to intervene to protect its citizens' rights based on the public policy principle of justice (be fair). HIPAA is one such federal protective law that has established the confidentiality and privacy of patients' PHI. Because social media can communicate and disseminate information to millions of people instantaneously, the magnitude and potential for a wide spectrum of influences and consequences are greater than at any other time in history.

Clear and distinct social media boundaries between one's social and professional lives must be understood, accepted, and implemented, and doing so will socially maximize the patient's best interests.

CONCLUDE

Social media has become a communication platform associated with many potential healthcare violations, ranging from the patient's moral and legal right to privacy to healthcare professionalism and institutional healthcare policies. Failure to set clear and distinct social media boundaries between one's social and professional activities can result in legal, professional, and moral consequences. It is important always to be cognizant of patient, professional, legal, and ethical social media boundaries and always to err on the side of caution. Being regularly informed about one's institutional social media policies is part of standards of care. Professional and institutional obligations of the healthcare provider usually outweigh and supersede any personal freedoms if and when they come into conflict.

In summary, the integration of social media into the healthcare profession demands clear and distinct boundaries to avoid potential healthcare violations and to uphold patients' rights to privacy and confidentiality. Students and practitioners must be aware of institutional social media policies and ensure that their online presence does not blur the line between their personal and professional lives. By prioritizing professional and institutional obligations and adhering to ethical social media boundaries, healthcare professionals can leverage the advantages of social media while minimizing potential legal, professional, and moral consequences.

* * *

REVIEW QUESTIONS

1. Students who use social media must be aware of their school's social media policies and any other institutional policies they may be associated with while doing clerkships and sub-internships, as HIPAA is the minimum professional boundary set by federal law.
   A. True
   B. False

2. If the student or practitioner wears the white coat for social media, then they become a spokesperson for all the institutions that the speaker is associated with, and institutions are very wary of potential consequences of this.
   A. True
   B. False

3. It is permissible for the student or practitioner to wear the white coat while on social media.
   A. True
   B. False

4. Students and practitioners must set clear boundaries between their social and professional lives, as failure to do so could have career-changing and legal consequences.
   A. True
   B. False

5. If a practitioner's action violates professional, institutional, or public policy disclosure standards, then that would be a sufficient condition for the practitioner to face professional, institutional, and legal disciplinary actions.
   A. True
   B. False

6. A patient who authorizes a practitioner to commit a crime or moral indiscretion would not absolve the practitioner from that crime or moral indiscretion as it relates to the patient.
   A. True
   B. False

7. Confidential and privacy boundaries must not be breached for social media purposes, as doing so would be a professional violation of beneficence (do good) and nonmaleficence (do no harm).
   A. True
   B. False

8. Clear and distinct social media boundaries between one's social and professional lives must be understood, accepted, and implemented, and doing so will socially maximize the patient's best interests.
   A. True
   B. False

Answers: 1A, 2A, 3B, 4A, 5A, 6B, 7A, 8A

* * *

CLINICAL VIGNETTES

1. Ms. Nora Jenkins, a 68-year-old retired teacher, presents with symptoms of chronic fatigue, joint pain, and swelling. A student observes the consultation and wishes to discuss the case with peers on social media. The student is unsure of the school's social media policies. What should the student do?
   A. Share the case without mentioning Ms. Jenkins's name
   B. Obtain written informed consent from Ms. Jenkins before discussing the case
   C. Consult the healthcare school's social media policies before taking any action
   D. Avoid discussing the case on social media

   Explanation: The student should consult their healthcare school's social media policies before taking action. This ensures that they follow the proper guidelines and avoid ethical breaches. Sharing the case without mentioning Ms. Jenkins' name still risks patient confidentiality, obtaining written informed consent may not be enough to protect privacy, and avoiding discussion altogether is unnecessary if the school has guidelines. *Answer*: C

2. A practitioner shares a blog post containing general health tips and advice on social media. In the comments, a patient asks for personalized healthcare advice. What should the practitioner do?
   A. Provide personalized healthcare advice in the comments
   B. Ask the patient to send a private message for personalized healthcare advice
   C. Encourage the patient to schedule an appointment for a proper consultation
   D. Ignore the patient's request

   Explanation: To maintain professional boundaries and provide appropriate care, the practitioner should encourage the patient to schedule an appointment for a proper consultation. Providing personalized healthcare advice in the comments or through a private message can blur professional boundaries and may not address the patient's specific needs. Ignoring the request is not helpful or professional. *Answer*: C

3. Practitioner Katherine Woods posts a picture on social media wearing a white coat with the hospital's logo visible. The hospital has strict policies against the unauthorized use of its logo. What should Practitioner Woods do?
   A. Keep the picture posted because it shows professionalism
   B. Remove the hospital's logo from the picture and repost it
   C. Remove the picture from social media
   D. Ask for the hospital's permission to use the logo in the picture

   Explanation: Practitioner Woods should remove the picture from social media because it may violate the hospital's policies. Keeping the picture posted or removing the logo and reposting the picture still disregards the hospital's rules, and asking for permission may be inappropriate if policies prohibit logo use. *Answer*: C

4. Practitioner Saomi Rivera is an attending practitioner who supervises several healthcare residents. She notices that one of the residents has been posting photos and updates about patients on social media, potentially violating patient privacy. What should Practitioner Rivera do?
   A. Report the resident to the hospital administration
   B. Ignore the situation, as the resident is responsible for their own actions
   C. Privately discuss the situation with the resident and provide guidance on appropriate social media use
   D. Publicly reprimand the resident on social media

   Explanation: As a supervising practitioner, Practitioner Rivera should privately discuss the situation with the resident and provide guidance on appropriate social media use. This helps the resident understand the importance of maintaining professional boundaries and patient privacy. Reporting the resident to the hospital administration may be too drastic as a first step, ignoring the situation is irresponsible, and publicly reprimanding the resident on social media is unprofessional. *Answer*: C

5. A practitioner wants to create a social media account to share general healthcare and healthcare advice and engage with the community. They are unsure how to handle friend requests from patients. What should the practitioner do?
   A. Accept all friend requests from patients
   B. Accept friend requests from patients but limit their access to personal information
   C. Politely decline friend requests from patients and direct them to a professional page or website
   D. Ignore friend requests from patients

   Explanation: The practitioner should politely decline friend requests from patients and direct them to a professional page or website. This allows patients to access the practitioner's healthcare and healthcare advice and updates without compromising professional boundaries and patient privacy. Accepting all friend requests or limiting access to personal information still risks blurring boundaries, and ignoring requests is unprofessional. *Answer*: C

* * *

REFLECTION VIGNETTES
1. Ms. Ivy Griffin is a first-year healthcare student who has amassed a large following on a social media site, where she shares her experiences and advice on how to navigate the journey into the healthcare profession. Ms. Griffin has been advising others for more than a decade on nutrition, exercise, sleep, studying techniques, exam preparation, and meeting the school of medicine and healthcare admissions requirements. Even after the white coat ceremony, Ms. Griffin continues offering advice on maintaining resilience and balance in

the school of medicine and healthcare while wearing her white coat. Her followers consist mainly of other students who are inspired by her insights and success in the field.

What is the ethical issue(s)?

How should the ethical issue(s) be addressed?

2. Ms. Cynthia Forrest, a 42-year-old intensive care unit (ICU) practitioner at a prominent school of medicine and healthcare, has been expressing frustration on the hospital's professional social media website about the increasing number of unvaccinated patients requiring extracorporeal membrane oxygenation (ECMO) machines in the ICU. She strongly advocates for prioritizing vaccinated patients for ECMO treatment because they have a higher success rate. Ms. Forest also suggests that unvaccinated patients should bear higher insurance rates and be responsible for the economic loss, suffering, and death caused by their lack of vaccination. She firmly believes that individualistic and selfish political ideologies should not supersede public health and safety measures and that vaccine hesitancy is not a matter of civil liberties but, rather, a manifestation of ignorance and recklessness.

What is the ethical issue(s)?

How should the ethical issue(s) be addressed?

* * *

WEB LINKS

1. *Code of Medical Ethics*. American Medical Association. https://code-medical-ethics.ama-assn.org
   The AMA *Code of Medical Ethics*, established by the American Medical Association, is a comprehensive guide for healthcare practitioners. It addresses issues and challenges, promotes adherence to standards of care, and is continuously updated to reflect contemporary practices and challenges.
2. "Social Media Fact Sheet." Pew Research Center. https://www.pewresearch.org/internet/fact-sheet/social-media/?menuItem=d102dcb7-e8a1-42cd-a04e-ee442f81505a
   This source presents comprehensive data on social media usage in the United States, revealing that approximately 72% of Americans use social media, with platforms such as YouTube and Facebook being the most popular. The usage varies by demographics such as age, gender, and education, and the data also show that a significant portion of users engage with these platforms daily for various purposes, including connection, information sharing, and entertainment.
3. "Professionalism in the Use of Social Media." American Medical Association. https://www.ama-assn.org/delivering-care/ethics/professionalism-use-social-media
   This source discusses the impact of the internet and social media on the healthcare profession, emphasizing the need for physicians to maintain patient privacy, separate personal and professional online content, monitor their online presence, and adhere to professional ethics. It highlights the responsibility of physicians to address unprofessional content by colleagues and be aware of the potential impact of their online actions on their reputation and public trust in the healthcare profession.
4. "Ethical Issues of Social Media Usage in Healthcare." K. Denecke, P. Bamidis, C. Bond, et al. 2015. https://www.ncbi.nlm.nih.gov/pmc/articles/PMC4587037
   This article discusses the ethical issues of social media usage in healthcare, focusing on the roles and responsibilities of healthcare professionals.
5. "Professional Guidelines for Social Media Use: A Starting Point." Terry Kind. 2015. https://journalofethics.ama-assn.org/article/professional-guidelines-social-media-use-starting-point/2015-05
   This article discusses the guidelines on online healthcare professionalism issued by the American College of Physicians and the Federation of State Medical Boards.

6. "Why Can't We Be Friends? A Case-Based Analysis of Ethical Issues with Social Media in Health Care." Kayhan Parsi and Nanette Elster. 2015. https://journalofethics.ama-assn.org/article/why-cant-we-be-friends-case-based-analysis-ethical-issues-social-media-health-care/2015-11
This article presents a case-based analysis of ethical issues with social media in healthcare.
7. "Ethics of Social Media Use for Doctors." British Medical Association. 2021. https://www.bma.org.uk/advice-and-support/ethics/personal-ethics/ethics-of-social-media-use
This article discusses the benefits and ethical considerations of social media use for practitioners.
8. "Patient Rights and Ethics." Jacob P. Olejarczyk and Michael Young. https://www.ncbi.nlm.nih.gov/books/NBK538279
This article provides a brief history of bioethics and human rights, including patient rights in modern medicine.
9. "Social Media Guidance for Physicians Taps Timeless Principles." American Medical Association. 2018. https://www.ama-assn.org/delivering-care/ethics/social-media-guidance-physicians-taps-timeless-principles
This article discusses the American Medical Association's guidance on social media use for physicians.
10. "Medical Ethics." American Medical Association. https://www.ama-assn.org/topics/medical-ethics
This article provides an overview of bioethics, including the ethical considerations of caring for patients.
11. "Social Media, E-Health, and Medical Ethics." Mélanie Terrasse, Moti Gorin, and Dominic Sisti. 2019. https://pubmed.ncbi.nlm.nih.gov/30790306
This article discusses the ethical issues at the intersection of social media and health, including the impact of social networking sites on the doctor–patient relationship.
12. "Bioethics and the Use of Social Media for Medical Crowdfunding." Brenda Zanele Kubheka. 2020. https://bmcmedethics.biomedcentral.com/articles/10.1186/s12910-020-00521-2
This article discusses the ethical considerations of using social media for healthcare crowdfunding.

STERILIZATION

54

The practitioner treats, but nature heals.
—Hippocrates

* * *

MYSTERY STORY

SABOTAGED STERILITY: A BATTLE BETWEEN PERSONAL BELIEFS AND PROFESSIONAL ETHICS

The hushed atmosphere of the sterile surgical room was broken only by the gentle rustle of gowns and the occasional clink of instruments. In a renowned city hospital, a skilled healthcare team was preparing for an important sterilization procedure. At Choice Haven Hospital, Practitioner Addison Perry was ready. A gynecologist with an unblemished record of hundreds of successful sterilizations, her dedication to patient autonomy and women's reproductive rights was well known and respected.

As final checks were made, Practitioner Perry's trained eye noticed something amiss. An essential instrument was missing. The situation quickly escalated into an urgent search, tension mounting among the healthcare staff. Time was of the essence, and the patient was already prepped and ready.

The nurse's search in the storage room was futile. There was no sign of the missing instrument. A disturbing realization settled over the room: The instrument had been deliberately hidden. Practitioner Perry's mind raced. Who would do this, and why?

The hospital administration was alerted, and an internal investigation was launched. The clues led to a shocking discovery. Charles Winchester, a young practitioner with strong religious convictions, was found to be behind the sabotage.

Practitioner Winchester had a history of being vocal against sterilization procedures, holding that they were a violation of his deeply held religious beliefs. He had even declined to assist Practitioner Perry in the past. The investigation, however, uncovered a deeper layer of conflict.

Under intense questioning, Practitioner Winchester finally broke down and confessed. He saw his act not as sabotage but, rather, as a desperate attempt to protect what he believed to be a patient's divine right to have children. He argued that he was, in fact, upholding a moral duty.

Choice Haven Hospital's response was swift but measured. Practitioner Winchester was suspended, but more important, an extensive dialogue was initiated. The incident exposed the delicate balance between personal beliefs and professional obligations.

In the following weeks, Choice Haven Hospital hosted seminars and workshops focusing on patient-centered care, informed consent, and the necessity of respecting patients' decisions regarding their own bodies. Practitioner Winchester, although misguided, was not cast aside; he was included in these discussions, his views challenged but also respected.

The incident served as a profound lesson for all healthcare providers. It emphasized the importance of sterilization as a constitutional right and the imperative to provide unbiased information, regardless of personal beliefs. Practitioners were reminded that their primary duty was to the patients, respecting their autonomy and their informed choices in line with standards of care.

In summary, the incident at Choice Haven Hospital was not just about a missing instrument. It was about the core principles that govern healthcare practice. The commitment to justice, reproductive rights, and the unswerving respect for patient autonomy were reaffirmed. It served as a stark reminder that personal convictions must never overshadow the sacred duty to provide compassionate, unbiased care.

* * *

THINK

All adults with decisional capacity have the individual autonomous right to request, have access to, and consent to contraception and sterilization, independent of the judgment of the spouse, partner, parents, or anyone else. The patient's autonomous right to choose to be sterilized as a form of contraception has been determined to be a fundamental constitutional right.

The 14th Amendment provides the due process clause from which several fundamental constitutional rights have been justified even though they are not explicitly listed in the Constitution, including the right to use contraception. The 14th Amendment, Section 1, states, "No state shall deprive any person of life, liberty, or property, without due process of law."

Sterilization as a form of birth control is the intentional treatment of a patient to make that individual unable to reproduce, and the methods used are surgical or nonsurgical. Typically, sterilization for female patients is by tubal ligation, and sterilization for male patients is by vasectomy. Long-acting reversible contraception (LARC) methods, such as intrauterine devices and implants, are available for female patients and are almost as effective as permanent sterilization.

Currently, sterilization is recognized as a constitutional right and considered to be both a negative right (obligation of others not to interfere) and a positive right (obligation of others to provide access) (Table 54.1).

Table 54.1 Civil Rights

Negative right	Obligation of others to not interfere
Positive right	Obligation of others to provide access

As a negative right, others have an obligation to not interfere with the patient's individual right to get sterilized. That means the patient's spouse, partner, parents, or anyone else does not have a legal right to interfere with the patient being sterilized.

As a positive right, others have an obligation to provide the patient with access to sterilization referrals, information, and services. The American College of Physicians recognizes the patient's positive right to sterilization when it explicitly states, "On abortion, sterilization, contraception, or other reproductive services... the physician [practitioner] has a duty to inform the patient about care options and alternatives or refer the patient for such information, so that the patient's rights are not constrained."

The prime directive of the healthcare profession is to maximize the patient's best interests as determined by the patient's reasonable goals, values, and priorities.

ASSESS

Patient: Autonomy

Patient autonomy is one of the four cardinal principles of bioethics. The patient has the right to provide informed consent, authorizing the practitioner to proceed with healthcare according to established standards of care. This autonomy includes the right to choose sterilization.

As with all healthcare procedures, the practitioner must ensure that the patient understands the procedure; acts voluntarily; and the decision is coherent and logically consistent with the patient's reasonable goals, values, and priorities. These conditions together comprise what is referred to as having decisional capacity.

For a patient to be considered autonomous in providing informed consent for sterilization, the following necessary conditions must be met:

- The patient understands what sterilization is, its purpose, how the procedure is performed, and the associated risks and benefits.

- The patient makes the choice for sterilization freely, without coercion (a credible threat) or manipulation (selective information guiding a particular decision).
- The patient can provide logical reasons for choosing sterilization that align with their reasonable goals, values, and priorities in authorizing the procedure.
- The patient actively chooses to authorize the practitioner to perform the procedure.

When these necessary conditions are met, the patient has fulfilled the sufficient conditions for being considered autonomous in providing informed consent to be sterilized.

Practitioner: Beneficence and Nonmaleficence

Practitioners have a professional obligation to exercise beneficence (doing good) by providing patient-centered care that aligns with the patient's best interests, as determined by the patient's reasonable goals, values, and priorities. Although these typically align with the practitioner's views, they may not always do so. The practitioner must respect the patient's autonomous informed consent for procedures that fall within the standards of care and adhere to the principle of nonmaleficence (doing no harm).

Sterilization falls within the standards of care. If a patient has the decisional capacity, then they have the right to authorize the practitioner to perform the procedure.

Should the practitioner decline or be unable to perform the requested procedure, they must, in a timely and nonjudgmental manner, provide unbiased, objective information and refer the patient to a qualified practitioner willing to carry out the requested sterilization procedure.

Public Policy: Justice

Reproductive justice through legal sterilization is as vital to public policy as the prohibition of reproductive injustices such as coercive and forcible sterilization. In the 1970s, during an era of paternalism in which the "doctor knows best," some obstetrician–gynecologists used a criterion for sterilization that involved multiplying a patient's age by the number of their children; sterilization was only permissible if the result was greater than 120. In addition, between 1909 and 1979, state and federally funded programs forcibly sterilized 60,000 females without consent.

To address these injustices, the U.S. Department of Health, Education, and Welfare developed protective Medicaid procedures that remained in effect until April 30, 2022. Medicaid regulations

- prohibit the sterilization of patients younger than age 21 years;
- require a 30-day waiting period after the practitioner counsels the patient about other nonpermanent and effective methods of birth control; and
- required that patients understand that they can refuse sterilization at any time and will not lose any health services or benefits provided by federal funds if they decline sterilization.

Unfortunately, these policies increased healthcare disparities. Tubal sterilization, a popular contraception method, is available to females with private insurance but not Medicaid beneficiaries unless arranged well in advance. This disadvantage restricts access to sterilization for Medicaid patients compared to those with private insurance.

The U.S. Department of Health and Human Services' "Consent for Sterilization" stipulates that at least 30 days must have passed between the date of the individual's signature on the consent form and the date the sterilization is performed.

Ethicists contend that allocating healthcare access based on payment source is unjust, and thus regulations requiring a 30-day wait are unfair, contributing to wealth-based healthcare disparities.

Ethicists also acknowledge the "dignity of risk," recognizing that regret is possible in autonomous decisions. The harm from restricting patient autonomy is often viewed as greater than the risk of regret. As a matter of justice (being fair), practitioners and society have professional, legal, and moral obligations to respect patients' rights to pursue or prevent pregnancy through sterilization.

CONCLUDE

Practitioners must recognize and honor the decision-making authority of patients who, demonstrating decisional capacity, have provided informed consent for sterilization. Although the evidence required for such a decision may be more substantial compared to less consequential choices, once decisional capacity and informed consent have been confirmed, those decisions must be respected.

In summary, sterilization is both a negative right (the obligation of others not to interfere) and a positive right (the obligation of others to provide access). The existing Medicaid regulations that mandate a 30-day wait before sterilization are viewed by ethicists as unjust, contributing to disparities in healthcare access based on wealth. Upholding the patient's decision-making authority when informed consent has been given to be sterilized promotes justice. It also highlights the importance of reproductive rights within public policy, ensuring a fair and equitable approach to this significant aspect of healthcare.

* * *

REVIEW QUESTIONS

1. All adult persons with decisional capacity have the individual autonomous right to request, access, and consent to contraception and sterilization, independent of the judgment of the spouse, partner, parents, or anyone else.
    A. True
    B. False

2. The patient's right to choose to be sterilized as a form of contraception is not a fundamental constitutional right.
    A. True
    B. False

3. As a positive right, others have an obligation to not interfere with the patient's individual right to get sterilized.
    A. True
    B. False

4. As a negative right, others have an obligation to provide the patient with access to sterilization referrals, information, and services.
    A. True
    B. False

5. If the practitioner declines or cannot perform the requested procedure, then in a timely manner, without judgment, the practitioner needs to respectfully provide unbiased, objective information and a referral to a qualified practitioner willing to provide the requested sterilization procedure.
    A. True
    B. False

6. Medicaid prohibits the sterilization of patients younger than age 21 years.
    A. True
    B. False

7. Before sterilization, Medicaid requires a 30-day waiting period after the practitioner counsels the patient about other nonpermanent and effective birth control methods.
   A. True
   B. False

8. Tubal sterilization after childbirth is one of the most popular forms of contraception and is readily available to females with private insurance but not for Medicaid beneficiaries unless the forms are filled out at least 30 days in advance, but not more than 3 months, before giving birth.
   A. True
   B. False

9. Ethicists argue that it is unjust to allocate healthcare access based solely on the source of payment and that, therefore, Medicaid regulations that require a 30-day wait are unjust and contribute to healthcare disparities based on wealth.
   A. True
   B. False

Answers: 1A, 2B, 3B, 4B, 5A, 6A, 7A, 8A, 9A

* * *

CLINICAL VIGNETTES

1. Ms. Grace Campbell is a 35-year-old accountant who is married and has two children. She is interested in permanent sterilization as a form of contraception and has consulted with her gynecologist, Practitioner Riley Morgan, about this option. She has decisional capacity and has provided informed consent for the sterilization procedure. Practitioner Morgan, however, declines to perform the procedure due to personal religious beliefs that do not support sterilization. Ms. Campbell asks for a referral to another qualified practitioner who can perform the procedure, but Practitioner Morgan refuses, stating that providing a referral would be going against her beliefs. What should Practitioner Morgan do?
   A. Practitioner Morgan is within her rights to decline performing the sterilization procedure and does not have an obligation to provide a referral.
   B. Practitioner Morgan must respect Ms. Campbell's autonomy and provide a referral to another qualified practitioner who can perform the sterilization procedure.
   C. Practitioner Morgan must perform the sterilization procedure because it is Ms. Campbell's right to access it.
   D. Practitioner Morgan must inform Ms. Campbell about the procedure but has the right to refuse to perform it and is not obligated to provide a referral.

   Explanation: Practitioner Morgan has a professional obligation to respect the patient's legal choices, even if it doing so goes against her personal beliefs. As a healthcare professional, Practitioner Morgan has an obligation to provide unbiased information and a referral to a qualified practitioner who can perform the sterilization procedure. This is supported by the American College of Physicians, which recognizes the patient's positive right to sterilization and states that practitioners have a duty to inform the patient about care options and alternatives or refer the patient for such information. Refusing to provide a referral would be a violation of the patient's autonomy and a failure to provide patient-centered care. The other options are incorrect because Practitioner Morgan has a professional obligation to provide a referral; the patient's right to access sterilization does not require Practitioner Morgan to perform the procedure if it goes against her personal beliefs; and Practitioner Morgan has an obligation to provide a referral to a qualified practitioner, not just inform the patient about the procedure. *Answer*: B

2. Mr. Frank Marlowe, a 35-year-old software engineer, presents to his primary care practitioner requesting a vasectomy. He is married and has two children, and he and his wife have decided that they do not want any more children. He understands the procedure, including the risks and benefits, and provides logical reasons for his decision. Which of the following best represents the ethical approach for the practitioner?
   A. Deny Mr. Marlowe's request due to his young age and current number of children
   B. Provide Mr. Marlowe with biased information to discourage him from the procedure
   C. Refer Mr. Marlowe to another practitioner willing to perform the procedure
   D. Provide Mr. Marlowe with unbiased information, and if he meets the criteria for decisional capacity, respect his autonomous right to provide informed consent for the procedure

   Explanation: The concept of autonomy is central to the ethical approach to sterilization. All adults with decisional capacity have the autonomous right to request, have access to, and consent to contraception and sterilization independently of the judgment of the spouse, partner, parents, or anyone else. Mr. Marlowe has provided logical reasons for his decision, and as long as he meets the necessary conditions for autonomous informed consent, his decision should be respected. The practitioner has a professional obligation to provide patient-centered care that maximizes the patient's best interests as determined by the patient's reasonable goals, values, and priorities. Referring Mr. Marlowe to another practitioner willing to perform the procedure is the appropriate ethical approach if the practitioner cannot perform the requested procedure. Providing biased information to discourage the procedure would violate the practitioner's professional obligation to respect the patient's autonomous informed consent. Denying Mr. Marlowe's request based on his age and current number of children would infringe on his individual right to get sterilized. *Answer*: D

3. Mrs. Maria Rivera, a 32-year-old elementary school teacher, and her husband, Mr. Jose Rivera, a 35-year-old engineer, present to their family practitioner requesting permanent sterilization. Mrs. Rivera wishes to undergo a tubal ligation, but Mr. Rivera is strongly against the procedure because he wants the option of having more children in the future. Mrs. Rivera has decisional capacity and meets the necessary conditions for autonomous informed consent because she understands the procedure, including the risks and benefits, and provides logical reasons for her decision. The practitioner provides counseling and education to the couple, but they are unable to reach a mutual decision. Which of the following best represents the ethical approach for the practitioner?
   A. Respect Mr. Rivera's wishes and deny Mrs. Rivera's request for sterilization
   B. Provide Mrs. Rivera with biased information to discourage the procedure
   C. Refer Mrs. Rivera to another practitioner willing to perform the procedure
   D. Respect Mrs. Rivera's autonomous right to provide informed consent for the procedure, and perform the sterilization as requested

   Explanation: All adult persons with decisional capacity have the autonomous right to request, access, and consent to contraception and sterilization, independent of the judgment of the spouse, partner, parents, or anyone else. Mrs. Rivera has decisional capacity and meets the necessary conditions for autonomous informed consent, so her decision to be sterilized must be respected. The practitioner provides counseling and education to the couple, but they cannot reach a mutual decision. Respect for patient autonomy is central to the ethical approach to sterilization, and spousal objections do not provide a legal or ethical basis for refusing a patient's request for sterilization. Providing biased information to discourage the procedure or denying Mrs. Rivera's request would violate the practitioner's professional obligation to respect the patient's autonomous informed consent. Therefore, the ethical approach in this scenario is for the practitioner to respect Mrs. Rivera's autonomous right to provide informed consent for the procedure and perform the sterilization as requested, as long as she has decisional capacity and meets the necessary conditions for autonomous informed consent. The practitioner should continue counseling and educating the couple to help them reach a mutual decision. However, ultimately, the decision to be sterilized is Mrs. Rivera's autonomous right. *Answer*: D

4. Mrs. Stella Hughes, a 25-year-old woman, has been married for 5 years and has two children. She works part-time at a grocery store and is covered by Medicaid. She wants to undergo a tubal ligation as a form of permanent sterilization. Mrs. Hughes has decisional capacity and meets the necessary conditions for autonomous informed consent because she understands the procedure, including the risks and benefits, and provides logical reasons for her decision. Which of the following is true regarding Medicaid's requirement for a 30-day waiting period before sterilization?
   A. Mrs. Hughes can undergo the procedure immediately because the waiting period does not apply to her.
   B. The practitioner can waive the waiting period if Mrs. Hughes is a low-income patient.
   C. The waiting period must be enforced, except for a healthcare emergency or if the patient has a healthcare condition that necessitates sterilization.
   D. Mrs. Hughes can shorten the waiting period to 15 days if she completes an expedited consent form.

   Explanation: Medicaid regulations require a 30-day waiting period after the practitioner counsels the patient of other nonpermanent and effective birth control methods before sterilization can be performed. The waiting period must be strictly enforced, regardless of the patient's circumstances. The waiting period is intended to give patients adequate time to consider their decision and explore alternative forms of contraception. Mrs. Hughes, being a Medicaid patient, is subject to this waiting period and must wait 30 days after receiving counseling on alternative forms of contraception before undergoing tubal ligation. The practitioner cannot waive the waiting period, even if Mrs. Hughes is a low-income patient. The waiting period can only be waived in the case of a healthcare emergency or if the patient has a healthcare condition that necessitates sterilization. In this scenario, the waiting period must be enforced, and Ms. Jackson cannot undergo the procedure until 30 days after counseling on alternative forms of contraception. The practitioner should provide counseling and education to Ms. Jackson on alternative forms of contraception during this waiting period to help her make an informed decision. *Answer*: C

5. Ms. Gabriella Taylor is a 27-year-old woman who recently gave birth to her third child. She works part-time at a retail store and is enrolled in Medicaid. During her postpartum visit, Ms. Taylor expressed to her obstetrician that she was interested in getting a tubal ligation as a form of permanent birth control. However, her obstetrician informs her that she cannot schedule the procedure until 30 days after the visit, as Medicaid regulations require. Ms. Taylor is frustrated because she knows that women with private insurance can get the procedure immediately after giving birth. Which of the following is correct?
   A. Ms. Taylor does not have decisional capacity
   B. Ms. Taylor has decisional capacity but should not be sterilized due to her age
   C. Ms. Taylor has decisional capacity and has the right to be sterilized
   D. Ms. Taylor does not have decisional capacity and should be sterilized for her own good

   Explanation: According to the ethical principles of patient autonomy, beneficence, and justice, Ms. Taylor has the right to be sterilized as a form of birth control as long as she has decisional capacity and provides informed consent. Medicaid regulations requiring a 30-day waiting period before sterilization after childbirth may be considered unjust and biased, as women with private insurance can receive the procedure immediately after giving birth. The obstetrician should provide Ms. Taylor with unbiased information on contraceptive options and refer her to a qualified practitioner if necessary. The 30-day waiting period required by Medicaid is an administrative requirement that does not affect Ms. Taylor's decisional capacity or her right to informed consent. *Answer*: C

REFLECTION VIGNETTES

1. Practitioner Samantha Green, a 35-year-old gynecologist, is approached by an adult patient with private insurance who is seeking permanent sterilization. The patient has decisional capacity and is fully informed of the procedure's permanency and the availability and effectiveness of LARC methods. Despite being informed of the LARC options, the patient declines and insists on the permanent sterilization procedure immediately.

    What is the ethical issue(s)?

    How should the ethical issue(s) be addressed?

2. Ms. Sarah Miller, a 30-year-old patient on Medicaid with decisional capacity, provides informed consent for getting sterilized. Despite being informed about the permanency of the procedure and being offered LARC methods that are equally safe and effective, Ms. Miller declines LARC options and immediately requests a permanent sterilization procedure.

    What is the ethical issue(s)?

    How should the ethical issue(s) be addressed?

3. Practitioner Sarah Park, a 34-year-old obstetrician–gynecologist, sees a patient in her office who is interested in permanent sterilization. The patient has decisional capacity and has expressed a clear desire to undergo sterilization after considering all options. Practitioner Park explains all the risks, benefits, and consequences of the procedure, as well as other LARC methods. However, the patient decides to proceed with sterilization. Soon after, the patient's spouse calls the office in distress, pleading with Practitioner Park not to perform the procedure because they wish to have biological children and believe that the patient may regret the decision.

    What is the ethical issue(s)?

    How should the ethical issue(s) be addressed?

* * *

WEB LINKS

1. *Code of Medical Ethics*. American Medical Association. https://code-medical-ethics.ama-assn.org
    The AMA *Code of Medical Ethics*, established by the American Medical Association, is a comprehensive guide for healthcare practitioners. It addresses issues and challenges, promotes adherence to standards of care, and is continuously updated to reflect contemporary practices and challenges.
2. "Consent for Sterilization." U.S. Department of Health and Human Services. https://opa.hhs.gov/sites/default/files/2020-07/consent-for-sterilization-english-updated.pdf
    This document outlines the informed consent process for sterilization, emphasizing that the decision is voluntary and will not affect eligibility for federal benefits, and it includes a stipulation that the operation will not be performed until at least 30 days after signing the consent form. This 30-day waiting period can create undue burdens for those who decide on sterilization late in their pregnancy or after childbirth, as they must wait for this period to pass before undergoing the procedure despite their immediate desire for sterilization.
3. "Sterilization of Women: Ethical Issues and Considerations." American College of Obstetricians and Gynecologists. https://www.acog.org/clinical/clinical-guidance/committee-opinion/articles/2017/04/sterilization-of-women-ethical-issues-and-considerations
    This source highlights the complexity of sterilization as a contraceptive method, emphasizing the need for an ethical approach that ensures access for willing women while protecting against coercive practices, within a reproductive justice framework acknowledging women's rights to pursue and prevent pregnancy.

4. "How Should a Physician Respond to Discovering Her Patient Has Been Forcibly Sterilized?" Kluchin, Rebecca. 2021. https://journalofethics.ama-assn.org/article/how-should-physician-respond-discovering-her-patient-has-been-forcibly-sterilized/2021-01
   This article discusses the ethical implications of forced sterilization.
5. "Sterilization of Women: Ethical Issues and Considerations." American College of Obstetricians and Gynecologists, April 2017. https://www.acog.org/clinical/clinical-guidance/committee-opinion/articles/2017/04/sterilization-of-women-ethical-issues-and-considerations
   This source provides an ethical approach to the provision of sterilization.
6. "Patient Rights and Ethics." Jacob P. Olejarczyk and Michael Young. https://www.ncbi.nlm.nih.gov/books/NBK538279
   This source discusses the rights of patients undergoing sterilization and the ethical considerations involved.
7. "Principles of Clinical Ethics and Their Application to Practice." Basil Varkey. 2021. https://www.ncbi.nlm.nih.gov/pmc/articles/PMC7923912
   This article discusses the ethical obligations of physicians in providing sterilization.
8. "Reproductive Rights and Access to Reproductive Services for Women with Disabilities." Anita Silvers, Leslie Francis, and Brittany Badesch. 2016. https://journalofethics.ama-assn.org/article/reproductive-rights-and-access-reproductive-services-women-disabilities/2016-04
   This article discusses the rights of women with disabilities to access sterilization services.
9. "The Top 10 Most-Read Medical Ethics Articles in 2021." Kevin B. O'Reilly. 2021. https://www.ama-assn.org/delivering-care/ethics/top-10-most-read-medical-ethics-articles-2021
   This source lists the most-read articles on bioethics in 2021, including topics related to sterilization.
10. "Principles of Bioethics." Thomas R. McCormick. https://depts.washington.edu/bhdept/ethics-medicine/bioethics-topics/articles/principles-bioethics
    This article provides an introduction to the use of ethical principles in healthcare ethics, including sterilization.
11. "The Right to Choose and Refuse Sterilization." United Nations Human Rights. https://www.ohchr.org/en/stories/2014/06/right-choose-and-refuse-sterilization
    This source discusses the right to choose and refuse sterilization based on international human rights law.
12. "Ethics of a Mandatory Waiting Period for Female Sterilization." Jessica Amalraj and Kavita Shah Arora. 2022. https://pubmed.ncbi.nlm.nih.gov/35993104
    This article discusses the ethics of a mandatory waiting period for female sterilization.
13. "Ethical Issues in Sterilization of Women." MDedge. 2017. https://www.mdedge.com/obgyn/clinical-edge/summary/clinical-guidelines/ethical-issues-sterilization-women
    This source reviews ethical issues related to the sterilization of women.

STRIKES—UNIONIZATION

55

The practitioner's highest calling, the only calling, is to make sick people healthy—to heal, as it is termed.
—Samuel Hahnemann

* * *

MYSTERY STORY

> **BARGAINING DILEMMA: THE POWER AND LIMITS OF COLLECTIVE BARGAINING IN HEALTHCARE**
>
> Practitioner Karen Chen had spent years as a loyal member of the local healthcare association, believing strongly in collective bargaining's potential to enhance both practitioners' livelihoods and patient care. However, when the association declared its intention to strike in 2 weeks—demanding higher pay, better benefits, and, puzzlingly, less patient contact—Chen's heart sank. These demands seemed to conflict with the principles of beneficence (do good) and nonmaleficence (do no harm) that she and her colleagues had pledged to uphold.
>
> The turmoil within Practitioner Chen was profound. She was torn between the loyalty she felt toward her association and her unwavering commitment to her patients. Knowing that she had to act, Practitioner Chen decided to engage her colleagues, hoping to persuade them to reevaluate the strike's principles.
>
> Walking into the meeting, Practitioner Chen found herself amidst a storm of impassioned debates and rising tempers. The room's energy was palpable, charged with conflicting emotions. Just as she was about to speak, a sudden cry broke through the cacophony. One of the practitioners collapsed, clutching their chest.
>
> Instincts and training took over. Practitioner Chen's hands flew into action, administering cardiopulmonary resuscitation as she barked out instructions to call 911. The room, once filled with discord, now moved in harmony, united by a common purpose.
>
> As the ambulance whisked the stricken practitioner away, the others found themselves huddled in a hospital waiting room, uncertainty and anxiety in the air. Practitioner Chen's mind whirled with the situation's profound irony. Here they were, embroiled in a battle for personal gains, yet it was their shared duty to their patients that had brought them together.
>
> The incident served as a catalyst for reflection and transformation. The association's leaders convened an emergency meeting, during which Practitioner Chen shared her insights and concerns. Her heartfelt words, combined with the dramatic events of the day, led to a profound shift in perspective.
>
> In the end, the strike was called off. The practitioners realigned their demands to focus on tangible improvements in patient care, such as better staffing and enhanced facilities. The episode had revealed to them that collective bargaining could indeed be a force for good, but it had to be wielded with wisdom and a true commitment to the Oath they had all taken.
>
> Practitioner Chen's courage and conviction had not only saved a life but also redirected the energies of her colleagues toward a path that honored the sacred patient-practitioner relationship. It was a lesson in leadership, compassion, and the delicate balance between self-interest and selfless devotion to a higher calling. Her dilemma had turned into a defining moment, illuminating the true power and limits of collective action in healthcare.

* * *

THINK

If practitioners are going to unionize and engage in collective bargaining pressures, then those actions are professionally justifiable if and only if they strengthen the patient–practitioner relationship, maximize the patient's best interests, and decrease healthcare disparities.

As the number of practitioners in private practice decreases and the number of practitioners in institutional salaried positions increases, there has been a renewed interest regarding whether or not a practitioner's collective actions employing unionization and strikes are compatible with the patient–practitioner relationship.

ASSESS

Patient: Autonomy

Patient autonomy is actualized through the process of informed consent, in which the patient authorizes the practitioner to provide the agreed-upon treatment option(s). The necessary conditions for a patient's informed consent are that the patient with decisional capacity has a patient–practitioner relationship; is sufficiently informed about their disease or injury, prognosis, accepted treatment options, and risks and benefits of the options; and has all their questions answered. After these conditions are met, the patient provides informed consent.

The extent to which unionizing and engaging in collective bargaining pressures would result in the improvement and better actualization of one or more of the necessary conditions for informed consent would be the degree to which the principle of patient autonomy would not be violated if there were to be practitioner unionization or collective bargaining. For example, practitioners using collective bargaining to acquire better foreign language translation technologies for improved communication with patients or adding services that would improve patient access to care would not violate the patient–practitioner relationship.

However, the extent to which unionization and strikes decrease the patient–practitioner relationship, change the patient-centered focus to practitioner-centered, and decrease the patient's ability to provide timely informed consent is the extent to which the principle of patient autonomy would be incompatible with practitioner unionization and strikes.

Practitioner: Beneficence and Nonmaleficence

Practitioners have the professional duty to be patient-centered and maximize the patient's best interests as determined by the patient's reasonable goals, values, and priorities. Practitioners accomplish this using the principles of beneficence (do good) and nonmaleficence (do no harm).

Therefore, it is considered unprofessional and unethical for practitioners to strike if it would put patients at increased risk of harm. The practitioner must always be patient-centered and respect the patient–practitioner relationship. The typical self-interested justifications for unionization and strikes, such as increased pay, increased benefits, and less patient contact, are not compatible with the professional altruistic character traits expected of practitioners.

However, if unionization and strikes can be organized or conducted in such a way that (a) there are no increases in the risk of harm to patients, (b) there is no violation of the patient–practitioner relationship, and (c) they increase the patient's best interests (e.g., better access to standardly accepted diagnostic technologies and treatment), then it can be argued that the principles of beneficence (do good) and nonmaleficence (do no harm) are not being violated.

To the extent to which practitioners shift their focus away from being patient-centered, away from maximizing the patient's best interests, is the degree to which the principles of beneficence (do good) and nonmaleficence (do no harm) are violated. The healthcare profession would be obligated to oppose practitioner unionization and strikes.

Public Policy: Justice

Public policy focuses on justice, the fair distribution of benefits and burdens related to patient care, and is generally referred to as decreasing healthcare disparities. To the extent that unionizing and engaging in collective bargaining pressures would increase patient fairness, such as lower co-pays, greater healthcare coverage, and decreasing healthcare disparities, would be the degree to which the public policy principle of justice (be fair) would not be violated.

However, the extent to which patient healthcare disparities increase because of unionization and strikes is the extent to which the principle of justice would be incompatible with practitioner unionization and strikes.

CONCLUDE

Unionization and strikes can only be justified if taking these actions is not motivated by self-interested justifications such as increased pay, increased benefits, and less patient contact; does not violate the patient–practitioner relationship; does not increase the risk of harm to patients; and prioritizes the patient's best interests while decreasing healthcare disparities. Practitioners must carefully consider these factors before engaging in collective actions to ensure they maintain their ethical obligations and commitment to patient-centered care.

In summary, whether practitioners should unionize and engage in collective bargaining is increasingly debated as private practices decrease and salaried positions increase. The answer is that such actions are only justified if they prioritize the patient's interests, strengthen the patient–practitioner relationship, and reduce healthcare disparities. The principles of patient autonomy, practitioner beneficence and nonmaleficence, and public policy's principle of justice must all be considered. Unionization and strikes can only be considered ethical if they do not violate the patient–practitioner relationship, do not increase the risk of harm to patients, and prioritize patient-centered care.

* * *

REVIEW QUESTIONS

1. If practitioners are going to unionize and engage in collective bargaining pressures, then those actions are professionally justifiable if and only if they strengthen the patient–practitioner relationship, maximize the patient's best interests, and decrease healthcare disparities.
   A. True
   B. False

2. The extent to which unionization and strikes decrease the patient–practitioner relationship, change the patient-centered focus to practitioner-centered, and decrease the patient's ability to provide timely informed consent is the extent to which the principle of patient autonomy would be incompatible with practitioner unionization and strikes.
   A. True
   B. False

3. Practitioners are no different from other citizens allowed to unionize and strike for self-interested reasons.
   A. True
   B. False

4. It is considered unprofessional and unethical for practitioners to strike if it would put patients at increased risk of harm. The practitioner must always be patient-centered and respect the patient–practitioner relationship.
   A. True
   B. False

5. The typical self-interested justifications for unionization and strikes, such as increased pay, increased benefits, and less patient contact, are not compatible with the professional altruistic character traits expected of practitioners.
   A. True
   B. False

6. To the extent that unionizing and engaging in collective bargaining pressures would increase patient fairness, such as lower co-pays, greater healthcare coverage, and decreasing healthcare disparities, would be the degree to which the public policy principle of justice (be fair) would not be violated.
   A. True
   B. False

7. The extent to which patient healthcare disparities increase because of unionization and strikes is the extent to which the principle of justice would be incompatible with practitioner unionization and strikes.
   A. True
   B. False

Answers: 1A, 2A, 3B, 4A, 5A, 6A, 7A

* * *

CLINICAL VIGNETTES
1. Ms. Samantha Walker is a 30-year-old nurse who works in a public hospital. She and her colleagues have been negotiating with the hospital administration for better pay, benefits, and working conditions for the past few months. However, despite their efforts, the negotiations have not been successful, and they have decided to go on strike. What is the primary ethical dilemma in this scenario?
   A. Balancing the right of healthcare professionals to strike with the potential impact on patient care
   B. Determining whether the hospital administration is obligated to meet the demands of the striking healthcare professionals
   C. Deciding if Ms. Walker's colleagues should support her in the strike
   D. Assessing the legal implications of the strike for the hospital and its employees

   Explanation: The primary ethical dilemma in this scenario is balancing the right of healthcare professionals to strike with the potential impact on patient care. This dilemma captures the core conflict faced by healthcare workers, such as Ms. Walker and her colleagues, in weighing their right to negotiate for better pay, benefits, and working conditions against their professional duty to prioritize patient well-being. Unionization and strikes can only be justified if they do not violate the patient–practitioner relationship, do not increase the risk of harm to patients, and prioritize patient-centered care. *Answer*: A

2. You are a first-year resident at a teaching hospital, where you are assigned to Ms. Jane Davis, a 58-year-old accountant who presents to the primary care practitioner with complaints of persistent fever, night sweats, weight loss, and fatigue. She has no significant healthcare history, and she does not smoke or drink. The differential diagnosis includes infectious and neoplastic diseases, such as tuberculosis, lymphoma, leukemia, and other malignancies. The practitioner is a member of a newly formed practitioner union negotiating with the hospital administration for better compensation and resources. The union has announced plans to strike in 2 days. You have heard that some more senior residents are considering participating in the strike. As a student, you are concerned about the impact of the strike on patient care and wonder whether the practitioner's participation in the strike is justifiable given Ms.

Davis' health status and the need for timely diagnosis and treatment. Should the practitioner participate in the strike?
A. The practitioner should participate in the strike because the union's demands are justified and necessary to improve patient care.
B. The practitioner should not participate in the strike because it would violate the patient–practitioner relationship and compromise Ms. Davis' timely diagnosis, treatment, and patient safety.
C. The practitioner should participate in the strike because it is necessary to decrease healthcare disparities and improve access to care.
D. The practitioner should participate in the strike because although current patients may suffer, it will result in the greater good for future patients.

Explanation: As a first-year resident, you have a duty to prioritize the patient's best interests and ensure that the patient receives timely and appropriate care. Ms. Davis' symptoms and differential diagnosis suggest a potentially serious illness that requires prompt diagnosis and treatment. The patient–practitioner relationship is essential to ensuring the patient's autonomy is respected and the patient's best interests are prioritized. The practitioner's participation in a strike would compromise the patient–practitioner relationship and violate the principle of patient autonomy. The principles of beneficence and nonmaleficence require that the practitioner act in the patient's best interests and avoid harm to the patient. Participating in a strike compromising Ms. Davis' timely diagnosis, treatment, and patient safety would violate these principles and put her at increased risk of harm. Therefore, the practitioner should not participate in the strike because it would violate the patient–practitioner relationship and compromise Ms. Davis' timely diagnosis and treatment. However, as a resident, you can work with the practitioner to find alternative ways to improve patient care while maintaining patient safety. Prioritizing the greater good for future patients over the well-being of current patients violates the principles of beneficence and nonmaleficence. *Answer*: B

3. Mr. Eric Johnson is a 40-year-old teacher who has been experiencing chest pain for the past few days. He goes to the local hospital, and the practitioner recommends an angiogram to rule out any heart-related issues. However, due to an ongoing strike by the hospital's practitioners, the procedure cannot be performed, and Mr. Johnson's condition worsens. What is the ethical question in this scenario?
A. Should healthcare professionals be allowed to strike?
B. Should the hospital management hire temporary staff to maintain patient care during the strike?
C. Should the government intervene to prevent healthcare professionals from striking?
D. Should healthcare professionals have to prove that striking is in the patient's best interests?

Explanation: The ethical question in this scenario is whether healthcare professionals should be allowed to strike. The issue of healthcare professionals' right to strike is a matter of debate. Practitioners must prioritize the patient's best interests, and any actions that violate the patient–practitioner relationship, decrease the patient's ability to provide timely informed consent, and increase healthcare disparities are incompatible with the principle of patient autonomy. Striking by healthcare professionals can significantly impact patients because it can affect patient care, increase the risk of harm to patients, and violate the patient–practitioner relationship. However, like other employees, healthcare professionals have the right to bargain collectively and engage in unionization and strikes. Therefore, the ethical question is whether or not the right to strike should be extended to healthcare professionals, given the potential harm to patients. The decision to allow healthcare professionals to strike should consider the balance between the right to strike and the potential harm to patients. The other options are not ethically justifiable because they do not address the fundamental ethical question. *Answer*: A

4. Ms. Jane Adams is a 60-year-old retired nurse diagnosed with cancer. After several rounds of chemotherapy, her condition worsens, and the practitioner recommends surgery to remove the tumor. However, due to a strike by the hospital's surgical team, the surgery was postponed, and Ms. Adams' condition continued to deteriorate. What is the ethical question in this scenario?
   A. Should the hospital have informed Ms. Adams of the potential strike risk and provided options for seeking surgical care at a different facility?
   B. Should the hospital administration take measures to prevent the surgical team from striking?
   C. Should the hospital hire temporary staff to perform the surgery during the strike?
   D. Should the practitioner recommend an alternative treatment plan for Ms. Adams during the strike?

   Explanation: It is considered unprofessional and unethical for practitioners to strike if it would put patients at increased risk of harm. In this scenario, the strike has put Ms. Adams' health and well-being at risk, violating the patient–practitioner relationship. Unionization and strikes can only be justified if they do not violate the patient–practitioner relationship, do not increase the risk of harm to patients, and prioritize patient-centered care. The correct answer is that the hospital administration should prevent the surgical team from striking to maintain the ethical obligations mentioned previously. The other options are not ethically justifiable. Informing Ms. Adam to get surgery elsewhere does address the violation of the patient–practitioner relationship that potentially puts patients at risk of harm. Hiring temporary staff to perform the surgery during the strike may still violate the patient–practitioner relationship and the principles of beneficence and nonmaleficence. Recommending an alternative treatment plan for Ms. Adams during the strike may not ensure the best possible care for her because the strike situation could impact the quality of available treatments and violate the principle of beneficence. Therefore, the main ethical question is whether or not the hospital administration, to ensure that the hospital maintains its ethical obligations and commitment to patient-centered care, should take measures to prevent the surgical team from striking. *Answer*: B

5. Ms. Sarah Thomas is a 35-year-old software engineer admitted to the hospital with a severe allergic reaction to medication. After several days of healthcare tests and examinations, the practitioner recommends a specific course of treatment that is being delayed due to a planned strike by the hospital's practitioners. What is the ethical question in this scenario?
   A. Should the practitioners be allowed to strike to express their grievances?
   B. Should the practitioners be prohibited from striking to ensure patients receive timely and appropriate care?
   C. Should the hospital be required to hire temporary staff to provide healthcare services during the strike?
   D. Should the practitioner continue with the treatment plan regardless of the strike?

   Explanation: Because practitioners are already allowed to strike, the ethical question in this scenario is whether, to ensure that patients receive timely and appropriate care, practitioners should be prohibited from striking. The issue of healthcare professionals striking is a subject of renewed interest in the healthcare industry. Healthcare professionals have the right to express grievances, such as fair compensation and better working conditions. However, they also have a professional duty to prioritize the patient's best interests. If the strike is motivated by self-interest, it may affect the patient–practitioner relationship and decrease patient autonomy, violating ethical principles. However, if the strike risks patients' health, it is considered unprofessional and unethical, violating the patient–practitioner relationship. Therefore, it is an ethical question to determine whether practitioners should have the right to strike or not. Prohibiting practitioners from striking is the most ethically justifiable option in this scenario because the strike has delayed Ms. Sarah Thomas' treatment, which may have serious consequences for her health. The hospital has a duty of care to ensure that its patients receive timely and appropriate treatment, and a strike that puts patients' health at risk is incompatible with this duty. The other options are not ethically justifiable because they may violate one or more ethical principles of bioethics. *Answer*: B

REFLECTION VIGNETTES

1. Practitioner Amanda Taylor, a 28-year-old recent school of medicine and healthcare graduate, matched at a prominent hospital. The hospital's resident practitioners, in solidarity, voted to unionize. The purpose of the union is to advocate for better working conditions, better pay, and other benefits that the resident practitioners believe are inadequate. Union membership is voluntary, and Practitioner Johnson is unsure if she should join the union.

   What is the ethical issue(s)?

   How should the ethical issue(s) be addressed?

2. Practitioner Kimberly Lee, a 30-year-old resident doctor employed at the regional hospital, is torn between her practitioner duties and her colleagues' decision to go on strike. The strike was called after collective bargaining for better salaries and more time off failed.

   What is the ethical issue(s)?

   How should the ethical issue(s) be addressed?

   * * *

WEB LINKS

1. *Code of Medical Ethics*. American Medical Association. https://code-medical-ethics.ama-assn.org
   The AMA *Code of Medical Ethics*, established by the American Medical Association, is a comprehensive guide for healthcare practitioners. It addresses issues and challenges, promotes adherence to standards of care, and is continuously updated to reflect contemporary practices and challenges.
2. "Global Medicine: Is It Ethical or Morally Justifiable for Doctors and Other Healthcare Workers to Go on Strike?." Sylvester C. Chima. 2013. https://bmcmedethics.biomedcentral.com/articles/10.1186/1472-6939-14-S1-S5
   This article discusses the motivations for healthcare worker strikes and the shift in modern healthcare practice.
3. "What Should Physicians Consider Prior to Unionizing?" Danielle Howard. 2020. https://journalofethics.ama-assn.org/article/what-should-physicians-consider-prior-unionizing/2020-03
   This source evaluates the ethics of physician unionization and the health system's corporate social responsibility.
4. "The Justification for Strike Action in Medical Healthcare: A Systematic Critical Interpretive Synthesis." Ryan Essex and Sharon Marie Weldon. 2022. https://journals.sagepub.com/doi/full/10.1177/09697330211022411
   This article discusses the unique qualities of strikes by healthcare workers and the ethical dilemmas they raise.
5. "Nursing Strikes: An Ethical Perspective on the US Healthcare Community." Paul Neiman. 2011. https://pubmed.ncbi.nlm.nih.gov/21646323
   This source discusses the ethical perspective of nursing strikes in the U.S. healthcare system.
6. "Patient Rights." American Medical Association. https://code-medical-ethics.ama-assn.org/ethics-opinions/patient-rights
   This source discusses patients' rights in the context of informed consent and privacy in healthcare.

7. "Moral Dilemma of Striking: A Medical Worker's Response to Job Duty, Public Health Protection and the Politicization of Strikes." Yao-Tai Li and Jenna Ng. 2022. https://journals.sagepub.com/doi/10.1177/0950017020981554
This article discusses the moral dilemma of striking for healthcare workers.
8. "Patient Data and Patient Rights: Swiss Healthcare Stakeholders' Ethical Awareness Regarding Large Patient Data Sets—A Qualitative Study." Corine Mouton Dorey, Holger Baumann, and Nikola Biller-Andorno. 2018. https://bmcmedethics.biomedcentral.com/articles/10.1186/s12910-018-0261-x
This source discusses the ethical issues of collecting and processing patient data.
9. "Ethics in Health Care: Improving Patient Outcomes." Tulane University School of Public Health and Tropical Medicine. 2023. https://publichealth.tulane.edu/blog/ethics-in-healthcare
This article discusses the importance of adhering to ethics in healthcare to improve patient outcomes.
10. "Doctor and Healthcare Workers Strike: Are They Ethical or Morally Justifiable: Another View." Sylvester C. Chima. 2020. https://pubmed.ncbi.nlm.nih.gov/31904696
This source reviews legal and ethical issues surrounding recent doctor and healthcare worker strikes.
11. "Is It Ethical for Health Workers to Strike? Issues from the 2001 QECH General Hospital Strike." Joseph Mfutso-Bengu and Adamson S. Muula. 2002. https://www.ncbi.nlm.nih.gov/pmc/articles/PMC3346014
This article discusses the ethical issues surrounding health worker strikes.
12. "The Ethics of Uninsured Participants Accessing Healthcare in Biomedical Research: A Literature Review." Hae Lin Cho, Marion Danis, and Christine Grady. 2018. https://www.ncbi.nlm.nih.gov/pmc/articles/PMC6133717
This source discusses the ethical issues related to uninsured participants accessing healthcare in biomedical research.
13. "Ethics of the Health Strike." Gonzalo Herranz. https://en.unav.edu/web/humanities-and-medical-ethics-unit/bioethics-material/etica-de-la-huelga-sanitaria
This article discusses the ethics of the health strike.
14. "Medical Ethics." American Medical Association. https://www.ama-assn.org/topics/medical-ethics
This source provides resources on patient data privacy and the ethics of patient–physician relationships.
15. "The Justification for Strike Action in Medical Healthcare: A Systematic Critical Interpretive Synthesis." Ryan Essex and Sharon Marie Weldon. 2022. https://pubmed.ncbi.nlm.nih.gov/35411830
This article discusses the ethical tension created by strike action in healthcare, particularly in relation to potential patient harm.

STRUCTURAL INJUSTICE

56

A fool contributes nothing worth hearing and takes offense at everything.
—Aristotle

* * *

MYSTERY STORY

> **INJUSTICE UNVEILED: CHAMPIONING EQUALITY AT JUSTICE CARE HOSPITAL**
>
> Fresh from her residency and filled with anticipation and passion for healing, Practitioner Emily Scott landed a promising role at a reputable city hospital. However, her excitement quickly turned to concern as she witnessed a pattern emerging. Certain patients, identifiable by their ethnic and racial backgrounds, seemed to receive unequal care, reflecting inequalities she had studied in education, wealth, and political representation. Structural injustice was present, and it was affecting patient outcomes.
>
> Appalled, Practitioner Scott delved into the complex web of practices that led to these healthcare disparities. She recognized that this inequality stemmed from biases deeply ingrained in the hospital's procedures, subtly favoring certain groups over others.
>
> Feeling a strong ethical duty to act, Practitioner Scott began to raise awareness, holding workshops that emphasized the unity of humanity, promoting the understanding that all people are part of a singular social group, and fighting against the divisive notion of "us" versus "them." Resistance from management was swift, but Practitioner Scott was resolute, recognizing that the patient–practitioner relationship was at stake.
>
> Her activism found a new home when she accepted a position at Justice Care Hospital, a pioneer in combating structural injustice in healthcare. Here, Practitioner Scott's mission flourished. She created programs that stressed the principles of patient autonomy (informed consent), practitioner beneficence (do good) and nonmaleficence (do no harm), and public policy justice (be fair) in patient care.
>
> Practitioner Scott's work at Justice Care Hospital began to radiate through the healthcare community, sparking a movement toward greater transparency and equality. She tackled not only the immediate issue of inequality in treatment but also the underlying structural problems that led to inequality in education, opportunity, and access to healthcare.
>
> Eventually, even her former employers, recognizing the moral imperative of her mission, sought guidance from Justice Care Hospital to reform their practices. Practitioner Scott's determination and empathy had not only changed the culture within Justice Care Hospital but also become a beacon of hope and change in the broader healthcare community.
>
> Her legacy stood as a testament to the importance of acknowledging and combating structural injustice, not only in healthcare but also in all aspects of life. Practitioner Scott's story reinforced the belief that quality healthcare is a right, not a privilege, echoing the principles of a democratic constitutional republic in which fairness and structural alignment are paramount.
>
> Practitioner Scott's story served as an enduring symbol of Justice Care Hospital's commitment to care, fairness, and empathy, reflecting the broader societal need to recognize the unity of humanity in combating structural injustice. Her work continues to inspire others to stand against healthcare disparities and work toward a more just and compassionate world.

* * *

THINK

Structural injustices are social, political, and institutional practices that result in different and sometimes unjust effects on individuals because they are members of some identifiable social group.[1] Examples of structural injustice are

- inequality of education;
- inequality of opportunity;
- inequality of wealth;
- inequality of political representation; and
- inequality of access to healthcare.

Structural injustice is strongly associated with group identification. When a person's group identity, such as politics, ethnicity, religion, or socially constructed race determinations, becomes a significant part of a person's identity, then the result can be social injustice if social benefits go to the tribal "us(es)" at the expense or burdens of the tribal "them(s)."[2]

Society and the healthcare professions must continually focus on how all persons are part of the unified social group of one humanity if there is going to be any chance of eliminating or even mitigating structural injustice with its resultant healthcare disparities.[3]

Structural injustice begins before initiating the patient–practitioner relationship, such as with education; employment; housing; diet; exercise; sleep; tobacco, alcohol, coffee, and drug use; clothing; relationships; sexuality; leisure; transportation; social entitlements; and health insurance.

When first establishing the patient–practitioner relationship, the patient is expected to provide a complete history and allow a complete physical. The practitioner is then expected to objectively interpret this protected health information for prevalence rate risk determinations and the possible identification of disease and afflictions. Structural injustice affects both the patient and the practitioner in the patient–practitioner relationship. Because of this, some have argued for racial concordance, in which patients and practitioners choose to have patient–practitioner relationships primarily with those who have the same phenotypic and ethnic similarities as themselves. Some think that racial concordance will promote trust, understanding, and empathy, resulting in better healthcare outcomes. Others think that racial concordance results in unacceptable segregation and increased healthcare disparities, as discussed in Chapter 45. Concerning racial concordance, it is essential to remember that there is no ethically reciprocal option for the practitioner to choose racially concordant patients. Individual rights and liberties that patients have in selecting their practitioner are not reciprocal. Practitioners do not have the right or liberty to select patients based on racial or ethnic identity or culture.

ASSESS

Patient: Autonomy

For the patient to be able to provide autonomous informed consent, it requires,

- the patient's trust in the unbiased knowledge and skills of the practitioner to provide an accurate diagnosis, prognosis, and treatment plan; and
- that the practitioner trusts that the patient is disclosing openly and honestly all necessary information for the practitioner to make a proper diagnosis, prognosis, and treatment plan.

---

1. Powers M, Faden R. *Structural Injustice: Power, Advantage and Human Rights*. Oxford University Press; 2019: 85.
2. Greene J. *Moral Tribes*. Penguin; 2013.
3. Sapolsky R. *Behave: The Biology of Humans at Our Best and Worst*. Penguin; 2017.

Recognizing these two necessary conditions clarifies why informed consent is considered a joint decision-making process between the patient and the practitioner. Informed consent output can only be as good as the informed consent input, and the degree to which structural injustice affects not only the input but also the output shows how profound a problem structural injustice influences can be when trying to attain autonomous informed consent.

Practitioner: Beneficence and Nonmaleficence

The practitioner's prime directive is to maximize the patient's best interests as determined by the patient's reasonable goals, values, and priorities. Ideally, the patient's best interests are determined by the autonomous informed consent of the patient. Practically, as healthcare becomes more complex and technologically complicated, the practitioner's expert knowledge and experience are indispensable for attaining the patient's best interests. This prime directive is attained when there is an alignment with the two professional principles of beneficence (do good) and nonmaleficence (do no harm). Implicit is that when fulfilling either of the professional principles, it implies, or results in, the fulfilling of the other principle.

Structural injustice is core to the collapse of the patient–practitioner relationship; the practitioner's failure to determine what the patient's reasonable goals, values, and priorities are; and the diminished ability to maximize the patient's best interests using the principles of beneficence (do good) and nonmaleficence (do no harm).

Public Policy: Justice

Structurally, the United States is not a democracy per se; rather, the United States is a democratic constitutional republic. Pure democracies tend to become the tyranny of the majority. Constitutions provide minority protections from the tyranny of the majority. A republic elects qualified representatives to champion the interests of its constituents. Together, this becomes a democratic constitutional republic that is more just (fair) than any one characteristic by itself. In other words, structural checks and balances are needed to keep each branch of government in structural alignment.

The executive, legislative, and judicial branches each have their powers and responsibilities. Unfortunately, these structural checks and balances can be corrupted with concerted political effort. When one branch of government is exercised without the checks and balances of the other branches of government, this is known as political structural injustice.

The four principles of biomedical ethics, like a democratic constitutional republic, are necessary structural checks and balances on each other. Emphasizing one moral principle at the expense of the others can result in unacceptable moral consequences, resulting in structural injustice with its resultant healthcare disparities.

CONCLUDE

When communicating with patients, practitioners must always stay neutral concerning politics, religion, gender identity, and other socially divisive issues. All patients need to feel that they are in a safe and neutral location. Structural injustice in healthcare occurs whenever there is unequal and unfair distribution of healthcare benefits and burdens. Because equality of healthcare access and treatment is ideal, practitioners must always promote moving closer to that end.

In summary, structural healthcare injustice refers to social, political, and institutional practices that result in unequal effects on individuals due to their membership in a certain social group, resulting in significant effects on healthcare disparities. Society and healthcare practitioners must recognize the unity of humanity in combating structural injustice to eliminate or mitigate healthcare disparities. The patient–practitioner relationship is impacted by structural injustice, which can affect informed consent and the principles of beneficence and nonmaleficence in healthcare practice. A democratic constitutional republic is essential in promoting fairness and structural alignment in government, and the four principles of biomedical ethics are necessary structural checks and balances on each other. Practitioners should stay neutral on socially divisive issues and instead promote moving closer to equality of healthcare access and treatment. (See Chapter 45.)

* * *

REVIEW QUESTIONS

1. Structural injustices are social, political, and institutional practices that result in different and sometimes unjust effects on individuals because they are members of some identifiable social group.
   A. True
   B. False

2. When a person's group identity, such as politics, ethnicity, religion, or socially constructed race determinations, becomes a significant part of a person's identity, then the result can be social injustice if social benefits go to the tribal "us(es)" at the expense or burdens of the tribal "them(s)."
   A. True
   B. False

3. Some think that racial concordance will promote trust, understanding, and empathy, resulting in better healthcare outcomes. However, nobody thinks that racial concordance results in unacceptable segregation and increased healthcare disparities.
   A. True
   B. False

4. Informed consent output can only be as good as the informed consent input, and to the degree to which structural racism affects not only the input but also the output shows how profound a problem structural injustice influences can be when trying to attain an autonomous informed consent.
   A. True
   B. False

5. Structural injustice is core to the collapse of the patient–practitioner relationship; the practitioner's failure to determine what the patient's reasonable goals, values, and priorities are; and the diminished ability to maximize the patient's best interests through the use of the principles of beneficence (do good) and nonmaleficence (do no harm).
   A. True
   B. False

6. When one branch of government is exercised without the checks and balances of the other branches of government, this is known as political structural injustice. This has a direct parallel in balancing the principles of bioethics.
   A. True
   B. False

Answers: 1A, 2A, 3B, 4A, 5A, 6A

* * *

CLINICAL VIGNETTES

1. Mr. Isaac Wong is a 56-year-old retired construction worker who presents to the emergency department with shortness of breath and chest pain. Living in a socioeconomically

disadvantaged neighborhood, he has struggled with access to regular healthcare. He reports that he has been feeling progressively more fatigued over the past few weeks and has been experiencing difficulty breathing even while at rest. His healthcare history is significant for hypertension, hyperlipidemia, and type 2 diabetes, for which he has been unable to afford medications. On examination, he is found to have a heart rate of 120 beats per minute, blood pressure of 150/90 mmHg, and an oxygen saturation of 88% on room air. His physical exam is notable for bilateral crackles on lung auscultation. A chest X-ray reveals bilateral pulmonary edema. The clinical differential diagnosis includes acute coronary syndrome, exacerbation of heart failure, and pneumonia. What ethical consideration should be taken into account in the management of Mr. Wong's care?
A. The need to address structural injustice in healthcare to ensure equitable access to healthcare
B. The importance of maintaining racial concordance in the patient–practitioner relationship
C. The patient's autonomy in selecting a specific medication brand for treatment
D. The practitioner's obligation to provide a private hospital room as a standard of care

Explanation: In this scenario, the correct answer is option A, the need to address structural injustice in healthcare to ensure equitable access to healthcare, because it specifically addresses the underlying issue of structural injustice that has affected Mr. Wong's ability to access and afford necessary healthcare. The scenario emphasizes his socioeconomic background and inability to afford medications, highlighting the broader context of structural inequality in healthcare access. The other options are framed as distractors that do not directly relate to the central theme of structural injustice in the provided scenario. *Answer*: A

2. Mr. Marcus Washington, a 40-year-old African American man, works as a school teacher in a low-income community. He presents with uncontrolled hypertension, despite being on multiple antihypertensive medications. On further evaluation, he reveals that he struggles to afford healthy food and cannot afford to join a gym or hire a personal trainer due to financial constraints. He also reports that he has limited access to fresh produce in his neighborhood, and the only grocery store in the area sells mostly processed foods. He requests that his practitioner prescribe him cheaper medications, but the practitioner is concerned about the risk of adverse outcomes due to the potential side effects and drug interactions. The ethical question is, What is the most appropriate course of action in this situation?
A. Prescribe cheaper medications to accommodate the patient's financial situation
B. Refer the patient to a social worker to explore financial assistance for healthy lifestyle interventions
C. Offer the patient resources for finding healthy, affordable food options and exercise programs
D. Discuss the importance of the current medications and their potential benefits in managing his hypertension

Explanation: In this scenario, the most appropriate course of action would be to refer the patient to a social worker to explore financial assistance for healthy lifestyle interventions. Structural injustices, such as wealth inequality and access to healthy food options, can result in healthcare disparities and negatively impact patients' health outcomes. In this case, Mr. Smith is experiencing these structural injustices contributing to his uncontrolled hypertension. Prescribing cheaper medications may not address the root cause of his health issues and may potentially result in adverse outcomes due to drug interactions and side effects. Similarly, offering resources for finding healthy food options and exercise programs may not be practical for patients experiencing financial constraints. Although discussing the current medications' importance and potential benefits is important, it may not be sufficient to address the patient's overall health needs. Referring the patient to a social worker can help address the structural injustices contributing to his health issues and facilitate access to healthy lifestyle interventions, ultimately improving his health outcomes. *Answer*: B

3. Ms. Angela Cook, a 72-year-old woman living in a rural area, has been suffering from chronic kidney disease. Due to the lack of healthcare facilities in her area and limited transportation, she has struggled to attend regular dialysis sessions. Her condition has worsened, leading to an emergency hospital admission. What ethical consideration should be taken into account in the management of Ms. Johnson's care?
    A. Ensuring continuity of care across different healthcare facilities
    B. The need to address structural injustice in healthcare to ensure equitable access to healthcare in rural areas
    C. The practitioner's obligation to follow evidence-based guidelines for kidney disease
    D. Ensuring the patient's understanding of alternative treatment options

    Explanation: Ms. Cook's case emphasizes the need to address structural injustice leading to unequal access in rural areas. Therefore, the best answer is option B, the need to address structural injustice in healthcare to ensure equitable access to healthcare in rural areas. The other options are all legitimate, ethical considerations, but they do not highlight the specific issue of structural injustice affecting her care. *Answer*: B

4. Mr. Douglas Bennett, a 45-year-old factory worker, has faced challenges in accessing healthcare due to language barriers and lack of insurance. He has recently been diagnosed with depression but cannot afford therapy or medication. What ethical consideration should be taken into account in the management of Mr. Bennett's care?
    A. Ensuring informed consent for treatment despite language barriers
    B. The practitioner's obligation to consider generic medications to reduce costs
    C. The need to address structural injustice in healthcare to ensure equitable access for non-English speakers and uninsured patients
    D. The importance of regular follow-up for depression treatment

    Explanation: Mr. Bennet's case focuses on the structural injustices of healthcare due to language barriers and lack of insurance. Therefore, the best answer is option C, The need to address structural injustice in healthcare to ensure equitable access for non-English speakers and uninsured patients. Although the other options are relevant to his care, they do not address the underlying structural barriers to care. *Answer*: C

5. Ms. Lakshmi Patel, a 50-year-old immigrant, has struggled with managing her diabetes due to language barriers and unfamiliarity with the healthcare system. Her condition has deteriorated, and she now requires hospitalization. What ethical consideration should be taken into account in the management of Ms. Patel's care?
    A. Ensuring culturally sensitive care that respects the patient's background
    B. The need to address structural injustice in healthcare to ensure equitable access for immigrants and those with language barriers
    C. The importance of patient education on managing diabetes
    D. Collaboration with community resources to support ongoing diabetes care

    Explanation: Ms. Patel's case focuses on the structural barriers affecting Ms. Patel, including language barriers and unfamiliarity with the healthcare system. Therefore, the best answer is option B, the need to address structural injustice in healthcare to ensure equitable access for immigrants and those with language barriers. The other options are relevant but do not directly tackle the core issue of structural inequality affecting her care. *Answer*: B

REFLECTION VIGNETTES

1. Practitioner Jasmine Kim, a 35-year-old practitioner, meets with a new patient, Mr. Abdulrahman Al-Faisal, a 45-year-old male of Middle Eastern descent. Practitioner Kim and Mr. Al-Faisal have different phenotypes and cultural identities.

   What is the ethical issue(s)?

   How should the ethical issue(s) be addressed?

2. Ms. Melissa Davis, a 32-year-old journalist, presents to the practitioner for a general checkup. She reports feeling healthy and having no specific healthcare concerns. During the consultation, Ms. Davis expresses her passion for sports and politics and attempts to engage the practitioner in a discussion about current events.

   What is the ethical issue(s)?

   How should the ethical issue(s) be addressed?

* * *

WEB LINKS

1. *Code of Medical Ethics*. American Medical Association. https://code-medical-ethics.ama-assn.org
   The AMA *Code of Medical Ethics*, established by the American Medical Association, is a comprehensive guide for healthcare practitioners. It addresses issues and challenges, promotes adherence to standards of care, and is continuously updated to reflect contemporary practices and challenges.
2. "Structural Justice Ethics in Health Care." Wendy Dunne DiChristina. 2021. https://journals.library.columbia.edu/index.php/bioethics/article/view/8404
   This source discusses the effects of social determinants of health and racism in clinical care.
3. "Powers, M., & Faden, R. R. (2019). *Structural Injustice: Power, Advantage, and Human Rights*. Oxford University Press." Johns Hopkins Berman Institute of Bioethics. https://bioethics.jhu.edu/people/profile/ruth-faden
   This source discusses structural injustice, power, advantage, and human rights in healthcare.
4. "Justice: A Key Consideration in Health Policy and Systems Research Ethics." Bridget Pratt, Verina Wild, Edwine Barasa, et al. 2020. https://www.ncbi.nlm.nih.gov/pmc/articles/PMC7245410
   This source calls for interpreting the ethical principle of justice in a more expansive way for health policy and systems research relative to biomedical research.
5. "Segregation in Health Care." American Medical Association. 2023. https://journalofethics.ama-assn.org/issue/segregation-health-care
   This source discusses the delivery of healthcare according to patients' race and insurance status.
6. "'The Health Equity Curse': Ethical Tensions in Promoting Health Equity." Bernie Pauly, Tina Revai, Lenora Marcellus, Wanda Martin, Kathy Easton, and Marjorie MacDonald. 2021. https://bmcpublichealth.biomedcentral.com/articles/10.1186/s12889-021-11594-y
   This source discusses the ethical concerns related to the promotion of health equity in public health.
7. "Structural Justice Ethics in Health Care ." Wendy DiChristina. 2021. https://doaj.org/article/46017a112c44465bbd208354beda2369
   This source discusses the effects of social determinants of health, racism in clinical care, and necessary advocacy with the local community.

8. "Empathy and Structural Injustice in the Assessment of Patient Noncompliance." Yolonda Wilson. 2021. https://onlinelibrary.wiley.com/doi/10.1111/bioe.12996
   This source discusses the importance of empathy in healthcare environments in the context of structural injustice.
9. "Patient Rights and Ethics." Jacob P. Olejarczyk and Michael Young. https://www.ncbi.nlm.nih.gov/books/NBK538279
   This source discusses the establishment of prospective patient rights.
10. "Solidarity and the Problem of Structural Injustice in Medical Healthcare." Carol C. Gould. 2018. https://pubmed.ncbi.nlm.nih.gov/30044895
    This source discusses structural problems with the U.S. insurance system and the solidarity movements addressing its deficiencies.

STUDENT PATIENT CARE

57

The young practitioner starts life with 20 drugs for each disease, and the old practitioner ends life with one drug for 20 diseases.
—William Osler

* * *

MYSTERY STORY

> **IDENTITY CONFUSION: A LESSON IN HONESTY AT PATIENTTRUST HEALTHCARE CENTER**
>
> The corridors of PatientTrust Medical Center echoed with the age-old philosophy of *docere*, the Latin term meaning "to teach," the etymological origin of the title of Doctor. Amid this legacy, Practitioner Rachel Wilson, known for her dedication to teaching the next generation of practitioners, found herself guiding her student, Timothy Campbell.
>
> As they entered an exam room, an elderly patient waited, her eyes widening with curiosity at Tim's youth. Eager to initiate the teaching moment, Practitioner Wilson confidently said, "I'm Practitioner Wilson," then turned to Tim, mistakenly adding, "and this is Practitioner Campbell."
>
> The patient's confusion was evident, and Practitioner Wilson's heart sank. The principle of clear communication within the patient–practitioner relationship had been breached. Tim's face blushed, but he held back, unsure of how to address the oversight.
>
> Clearing her throat, Practitioner Wilson swiftly corrected, "I apologize for the oversight. Tim here is a healthcare student, currently in his third year, not a licensed practitioner. He'll be observing and assisting under my supervision today."
>
> Yet, the misstep cast a shadow of doubt. Although Practitioner Wilson strived to educate her patients about their health, she also held the moral responsibility to mentor students such as Tim. But this had to be within the bounds of trust and clarity. The patient's involvement in this educational process was vital but required her informed and willing participation.
>
> Post-appointment, Tim approached Practitioner Wilson, the weight of the day's lesson heavy on his mind. "I think we should address this more proactively in the future," he suggested.
>
> Practitioner Wilson nodded in agreement: "We must. It's vital for patients to know who is treating them and who is learning clearly. These principles aren't unique to PatientTrust Medical Center; they're fundamental across the healthcare profession, reflecting the universal value of transparency and teaching mentorship."
>
> The following day, they jointly met the patient, their regret evident. Together, they emphasized the importance of the patient–practitioner relationship, the role of practitioners in teaching students, the role of students in the learning process, and the healthcare profession's commitment to upholding the highest ethical standards.
>
> In the aftermath, PatientTrust Medical Center implemented clear protocols emphasizing full disclosure and informed consent regarding student involvement. This incident not only shaped Tim's understanding of professional responsibility but also reiterated for Practitioner Wilson the core values of healthcare education. The relationship between patient and practitioner was not just clinical; it was a bond of trust and transparency. The event underscored the imperative of teaching, supervision, mentorship, and adherence to ethical standards in nurturing students, preserving patient trust, and upholding the integrity of the profession.

* * *

THINK

Practitioners have a professional and moral obligation to teach, mentor, and induct students into the art of healing. The practitioner's title of Doctor comes from Latin *docere*, which means to teach. Practitioners are obligated to teach their patients their diagnosis, prognosis, and treatment options, and they are obliged to teach and mentor the next generation of practitioners with the knowledge and skills necessary for good patient care.

Good patient care is dependent on the patient–practitioner relationship. This means that teaching the art of healing to students can only occur within the context of the patient–practitioner relationship. Therefore, it must be clear to the patient who the student is, what year of the program they are in, their level of expertise, and who is supervising them. Any student status obfuscation violates the patient–practitioner relationship and violates legal, professional, and ethical mandates. The patient's willingness to participate in the education of students is imperative and can only occur within the patient–practitioner relationship.

ASSESS

Patient: Autonomy

Central to all medicine is the history and physical, which can only occur within the patient–practitioner relationship in which the patient trusts that the practitioner will keep all protected health information confidential and private. Most patients are more than willing to help aspiring students and residents learn the art of healing so that they can become licensed practitioners.

The patient must understand and trust that any patient engagements will only occur after full disclosure and the patient has given informed consent for allowing each engagement. This is especially true for student involvement when the patient is temporarily incapacitated.

Active patient participation starts with the patient providing informed consent giving the student or resident the authority to conduct a patient history and physical (H&P) under the supervision of the attending practitioner. However, the attending practitioner must personally review the patient's H&P and not just cosign the work of a student who is not a licensed practitioner.

Practitioner: Beneficence and Nonmaleficence

Professionally, practitioners ensure that the patient knows that not all doctors are licensed physicians who can practice medicine without supervision. Without specific clarification, it is never professionally, legally, or ethically permissible to introduce a nonlicensed practitioner as a "Doctor" to a patient, as that is fraud. If a student has a PhD, MD, Doctorate in Pharmacology, or any number of other doctorate degrees but is not licensed, and if the term "doctor" is used when introducing the student or resident to the patient, the introduction must be accompanied with a clarification that the student or resident is not licensed and is under the supervision of the attending practitioner.

A practitioner's professional obligation is to provide patient-centered care in which the patient's best interests are maximized in accordance with the patient's reasonable goals, values, and priorities. This is accomplished by practitioners only providing standards of care recognized by the healthcare profession as beneficent (good) and nonmaleficent (does no harm) to the patient.

Professionally, legally, and ethically, it is impermissible for any student or resident to practice unsupervised medicine without being a licensed practitioner. Therefore, the attending practitioner must exercise due diligence to supervise all patient–student and patient–resident interactions. If a student makes a healthcare judgment or acts independently of healthcare supervision, then the student is practicing medicine without a license. If true, then the student and the attending practitioner can be held liable for any patient injuries. Practicing medicine without a license is a form of healthcare quackery, healthcare misconduct, and, if harm befalls the patient, healthcare malpractice as well.

Public Policy: Justice

Both the public at large and patients generally give high regard to students. Students should embrace and own the limited amount of time they have to enjoy the privilege of being a student who wears the short white coat. Society, institutions, professions, and patients all know the importance of students being mentored into the healthcare profession, and as a result, a significant amount of respect and honor is conferred to the student.

However, under no condition is it ever permissible for an unlicensed student to independently write an order, prescribe and dispense drugs, or make independent healthcare judgments. In a healthcare emergency, staff and other interprofessional practitioners need to follow established healthcare protocols and the attending practitioner's orders, but never a student's advice or healthcare judgment.

All medically licensed nurses and other interprofessional practitioners have more legal authority, responsibility, and accountability than a student, no matter how much experience the student has and how close the student is to becoming a licensed practitioner. Under the law, someone is either licensed or not licensed; there is no in-between. It behooves the student to take advantage of the opportunity to learn and serve but also to be careful never to cross the line of making healthcare judgments that could result in the unacceptable practice or even perception of practicing medicine without a license.

CONCLUDE

Students and unlicensed practitioners must be correctly represented when being introduced to the patient, and the patient must provide informed consent authorization for each procedure a student conducts. General consent is not sufficient consent for student authorization. It is imperative that the student does not practice medicine without a license and that all patient–student interactions are under the supervision of the attending practitioner. This approach helps maintain the public trust and upholds the privilege of being a part of the healthcare profession.

In summary, practitioners have a professional and moral obligation to teach, mentor, and induct students into the art of healing, but the involvement of students in patient care must be with full disclosure and the patient's informed consent. Students must always be under the supervision of the attending practitioner, and they must not make independent healthcare judgments, prescribe drugs, or write orders. The attending practitioner must exercise due diligence to supervise all student– and resident–patient interactions, and students must not practice medicine without a license. All medically licensed nurses and other interprofessionals have more legal authority, responsibility, and accountability than a student, no matter how much experience the student has and how close the student is to becoming a licensed practitioner.

* * *

REVIEW QUESTIONS

1. The practitioner's title of Doctor comes from Latin *docere*, which means to practice medicine.
    A. True
    B. False

2. It must be clear to the patient who the student is, what year of the program they are in, their level of expertise, and who is supervising them.
    A. True
    B. False

3. Without specific clarification, it is never professionally, legally, or ethically permissible to introduce a nonlicensed practitioner as "Doctor" to a patient because that is fraud.
    A. True

B. False

4. If a student makes a healthcare judgment or acts independently of healthcare supervision, then the student is practicing medicine without a license.
   A. True
   B. False

5. In a healthcare emergency, staff and other healthcare practitioners need to follow established protocols and the attending's orders, but never the advice or healthcare judgment of a student.
   A. True
   B. False

6. The student should take advantage of the opportunity to learn and serve but also be careful never to cross the line of making healthcare judgments that could result in unacceptable practice or even the perception of practicing medicine without a license.
   A. True
   B. False

Answers: 1B, 2A, 3A, 4A, 5A, 6A

* * *

CLINICAL VIGNETTES

1. Mr. Timothy Nguyen is a 57-year-old retired engineer who presents to the clinic with a 6-month history of intermittent epigastric pain, bloating, and weight loss. He reports that the pain is usually relieved with antacids, but it recurs every few days. He denies any vomiting, hematemesis, or melena. His healthcare history is significant for hypertension and hyperlipidemia. A clinical differential diagnosis of gastroesophageal reflux disease, peptic ulcer disease, and gastric cancer is considered. A student is assigned to accompany the attending practitioner during the patient's visit. The attending practitioner introduces the student as Practitioner David Kim. The student proceeds to take an H&P under the supervision of the attending practitioner.
   A. This scenario violates the ethical principles of patient autonomy, beneficence, and nonmaleficence because the student's introduction as "Practitioner David Kim" is a form of misrepresentation and deception.
   B. This scenario upholds the ethical principles of patient autonomy, beneficence, and nonmaleficence because the patient has given informed consent for the student to participate in the H&P.
   C. This scenario violates the ethical principles of patient autonomy, beneficence, and nonmaleficence because the student is allowed to practice unsupervised medicine without the attending practitioner's review.
   D. This scenario violates the ethical principle of justice because the student is not given the opportunity to learn and serve under the attending practitioner's supervision.

   Explanation: The introduction of the student as "Practitioner David Kim" violates the ethical principles of patient autonomy, beneficence, and nonmaleficence. Introducing a nonlicensed physician as a "Practitioner" to the patient is a form of fraud and misrepresentation. It is ethically, legally, and professionally impermissible. The correct ethical approach is to introduce the student as a student, clearly stating the student's level of expertise, the year of the program they are in, and that the student is not a licensed practitioner. The patient must also provide informed consent for each student interaction, and the attending practitioner must exercise due diligence to supervise all patient–student interactions. The other options are incorrect because the introduction of the student as "Practitioner David Kim" is a fraud, the student did not practice unsupervised healthcare, the student was given the opportunity to learn and serve under the attending practitioner's supervision, and none of these options addressed the main ethical issue of misrepresentation and deception. *Answer*: A

2. Mr. Joseph Campbell is a 45-year-old accountant who presents to the clinic with a 2-week history of cough, shortness of breath, and fever. He reports that the symptoms started as a mild cough but have progressed to become more severe. He denies any chest pain or palpitations. His healthcare history is significant for asthma. A clinical differential diagnosis of pneumonia, bronchitis, and acute exacerbation of asthma is considered. A student is assigned to accompany the attending practitioner during the patient's visit. The attending practitioner introduces the student as a second-year student, and the student proceeds to get permission from the patient to perform an H&P under the direct supervision of the attending practitioner.
    A. This scenario upholds the ethical principles of patient autonomy, beneficence, and nonmaleficence because the patient has been informed of the student's status, and the student is under the direct supervision of the attending practitioner.
    B. This scenario violates the ethical principle of patient autonomy because the patient is not informed of the student's status.
    C. This scenario violates the ethical principle of justice because the student is not given the opportunity to learn and serve under the attending practitioner's supervision.
    D. This scenario violates the ethical principle of nonmaleficence because the student is allowed to prescribe medication without the attending practitioner's review.

    Explanation: This scenario upholds the ethical principles of patient autonomy, beneficence, and nonmaleficence. The patient has been informed of the student's status, and the student is under the direct supervision of the attending practitioner. The attending practitioner must diligently supervise all patient–student interactions. The correct ethical approach is to introduce the student as a student, not a licensed practitioner, and obtain the patient's informed consent for each student interaction. The other options are incorrect because the patient was informed of the student's status, the student did not prescribe medication without the attending practitioner's review, and the student was given the opportunity to learn and serve under the attending practitioner's supervision. *Answer*: A

3. Ms. Gabriella Brown is a 25-year-old teacher who presents to the emergency department with a sudden onset of severe chest pain and shortness of breath. She reports that the pain started while she was sitting at her desk and has been constant since then. She denies any history of heart disease or similar episodes. A clinical differential diagnosis of acute coronary syndrome, pulmonary embolism, and aortic dissection is considered. A student is assigned to assist the attending practitioner in the evaluation of the patient. The attending practitioner introduces the student as a third-year student, and the student proceeds to take a focused H&P under the direct supervision of the attending practitioner. During the examination, the student observes that the patient is pale, diaphoretic, and in severe distress. The student suggests that the patient be given oxygen and started on anticoagulation therapy immediately. The attending practitioner reminds the student that in a healthcare emergency, staff and other healthcare practitioners need to follow established protocols and the attending's orders, but never the advice or healthcare judgment of a student. The attending practitioner then proceeds to give orders for supplemental oxygen and a diagnostic workup, including an electrocardiogram and a chest X-ray.
    A. This scenario upholds the ethical principles of patient autonomy, beneficence, and nonmaleficence because the student is under the direct supervision of the attending practitioner, and the attending's orders are followed in a healthcare emergency.
    B. This scenario violates the ethical principle of patient autonomy because the patient is not informed of the student's status.

C. This scenario violates the ethical principle of justice because the student is not given the opportunity to learn and serve under the attending practitioner's supervision.
D. This scenario violates the ethical principle of nonmaleficence because the student is allowed to prescribe medication without the attending practitioner's review.

Explanation: This scenario upholds the ethical principles of patient autonomy, beneficence, and nonmaleficence. The student is under the direct supervision of the attending practitioner, and the attending's orders are followed in a healthcare emergency. In a healthcare emergency, staff and other healthcare practitioners must follow established protocols and the attending's orders, but never a student's advice or healthcare judgment. The correct ethical approach is prioritizing the patient's well-being and following established protocols and orders in a healthcare emergency. The other options are incorrect because the student is under the direct supervision of the attending practitioner, the patient is not informed of the student's status, the student is not given the opportunity to learn and serve under the attending practitioner's supervision, and the student is not allowed to prescribe medication without the attending practitioner's review. *Answer*: A

4. Mr. John Wilson is a 42-year-old truck driver who presents to the urgent care clinic with a 1-day history of severe abdominal pain and diarrhea. He reports that the pain is located in the left lower quadrant and is associated with bloating and nausea. He also reports passing watery stools with mucus and blood. His healthcare history is significant for hypertension and obesity. A clinical differential diagnosis of diverticulitis, inflammatory bowel disease, and infectious colitis is considered. A student is assigned to assist the attending practitioner in the patient's evaluation. The attending practitioner introduces the student as a third-year student, and the student proceeds to take a focused H&P under the direct supervision of the attending practitioner. During the evaluation, the student suggests obtaining a stool sample for microbiological analysis to identify the possible infectious agent causing the patient's diarrhea. The attending practitioner agrees with the student's suggestion and orders the test. The results come back positive for *Escherichia coli* O157:H7, a type of bacteria known to cause foodborne illness.
   A. This scenario upholds the ethical principle of nonmaleficence because the attending practitioner and the student work together to identify the possible infectious agent causing the patient's diarrhea and provide appropriate treatment.
   B. This scenario violates the ethical principle of patient autonomy because the patient is not informed of the student's status and their involvement in the evaluation.
   C. This scenario violates the ethical principle of justice because the student is not given the opportunity to learn and serve under the attending practitioner's supervision.
   D. This scenario violates the ethical principle of beneficence because the student is allowed to perform unsupervised healthcare decision-making without the attending practitioner's review.

Explanation: This scenario upholds the ethical principle of nonmaleficence because the attending practitioner and the student work together to identify the possible infectious agent causing the patient's diarrhea and provide appropriate treatment. The attending practitioner must diligently supervise all patient–student interactions. The correct ethical approach is introducing the student as a student, not a licensed practitioner, and obtaining the patient's informed consent for each student interaction. The other options are incorrect because the attending practitioner and the student work together to provide appropriate care, the patient's autonomy is not violated because they are informed of the student's involvement in the evaluation, the student is learning and serving under the attending practitioner's supervision, and the student is not allowed to perform unsupervised healthcare decision-making. *Answer*: A

5. Mr. Dylan Bailey is a 40-year-old accountant who presents to the clinic with a 2-day history of dyspnea, cough, and fatigue. He reports that the cough is productive of yellowish sputum

and that he has experienced chills and fever. He denies any chest pain, palpitations, or syncope. His healthcare history is significant for hypertension and diabetes mellitus. A clinical differential diagnosis of community-acquired pneumonia, bronchitis, and asthma is considered. A student is assigned to assist the attending practitioner in the patient's evaluation. The attending practitioner introduces the student as a fourth-year student and leaves the clinic to attend to an emergency in the hospital. The student proceeds to take an H&P, order a chest X-ray, and prescribe antibiotics for community-acquired pneumonia without consulting the attending practitioner. The patient returns to the clinic 2 days later with worsening symptoms, and a repeat chest X-ray shows no improvement in the pneumonia. The attending practitioner becomes aware of the situation and re-evaluates the patient, orders additional tests, and changes the antibiotics.

A. This scenario violates practitioner beneficence and nonmaleficence, and public policy justice, because the student was engaging in unsupervised decision-making and practicing medicine without a license.

B. This scenario upholds the ethical principle of patient autonomy because the patient was informed of the student's status and gave consent for their involvement in the evaluation.

C. This scenario did not violate the ethical principle of justice even though the student did not have the opportunity to learn and serve under the attending practitioner's supervision.

D. This scenario violates the ethical principle of beneficence because the student was not given the opportunity to perform unsupervised healthcare decision-making.

Explanation: This scenario violates practitioner beneficence and nonmaleficence, and public policy justice, because the student engaged in unsupervised decision-making and practiced medicine without a license. Healthcare student standard operating procedures dictate that students must not make independent healthcare judgments or act independently of healthcare supervision. The appropriate ethical approach involves introducing the student as a student and obtaining the patient's informed consent for each student interaction. The attending practitioner is responsible for exercising due diligence to supervise all patient–student interactions. The other answers are wrong because although the patient was informed of the student's status and consented to their involvement in the evaluation, the student still practiced medicine without a license, violating ethical principles and public policy justice. Informed consent does not justify the student's unsupervised decision-making. The scenario does violate the ethical principle of justice because the attending practitioner should have ensured that the student had the opportunity to learn and serve under their supervision, adhering to the proper ethical guidelines, and the student illegally practiced medicine without a license. Last, the issue is not whether the student was given the opportunity to perform unsupervised healthcare decision-making. The main concern is that the student practiced medicine without a license and proper supervision, violating practitioner beneficence, nonmaleficence, and public policy justice. *Answer*: A

REFLECTION VIGNETTES

1. Practitioner Emily Wilson, a 32-year-old attending practitioner, supervises a student at her clinic. The student, David Thomas, has recently graduated from healthcare school but has yet to pass Step 3 of the healthcare licensing exam. Practitioner Wilson asks David to introduce himself to a patient who has come in for a routine checkup.

    What is the ethical issue(s)?

    How should the ethical issue(s) be addressed?

2. Ms. Sarah Johnson, a 25-year-old woman, is a student in an obstetrician–gynecologist rotation. The attending practitioner instructs her to perform a pelvic examination on an

unconscious, sedated patient undergoing surgery. The attending practitioner tells her that the patient will be relaxed and that this would be an excellent opportunity for her to practice. When Ms. Johnson raises concerns about obtaining the patient's informed consent for the examination, the attending tells her that the patient's general consent to receive treatment at a teaching hospital covers the student's practice.

What is the ethical issue(s)?

How should the ethical issue(s) be addressed?

* * *

WEB LINKS

1. *Code of Medical Ethics.* American Medical Association. https://code-medical-ethics.ama-assn.org
   The AMA *Code of Medical Ethics*, established by the American Medical Association, is a comprehensive guide for healthcare practitioners. It addresses issues and challenges, promotes adherence to standards of care, and is continuously updated to reflect contemporary practices and challenges.
2. "Medical Student Involvement in Patient Care." American Medical Association. https://code-medical-ethics.ama-assn.org/ethics-opinions/medical-student-involvement-patient-care-0
   This source discusses the benefits of having students participate in patient care and the obligation of physicians to ensure that patients are aware of this involvement.
3. "Medical Student Involvement in Patient Care." AMA Council on Ethical and Judicial Affairs. 2001. https://journalofethics.ama-assn.org/article/medical-student-involvement-patient-care-report-council-ethical-and-judicial-affairs/2001-03
   This source discusses the benefits and challenges of student involvement in patient care.
4. "Cases in Medical Ethics: Student-Led Discussions." Chris Cirone. 2005. https://www.scu.edu/ethics/focus-areas/bioethics/resources/cases-in-medical-ethics-student-led-discussions
   This source presents case studies addressing patient autonomy and rights.
5. "The Care of Patients and Student's Rights." Regional West Medical Center, School of Radiolic Technology. https://www.rwhs.org/sites/default/files/722_8_20_40_the_care_of_patients_and_students_rights_2017.pdf
   This source discusses the expectations for students in the program to provide quality healthcare to all patients at all times.
6. "Building Connections Between Biomedical Sciences and Ethics for Medical Students." Oluwaseun Olaiya, Travis Hyatt, Alwyn Mathew, Shawn Staudaher, Zachary Bachman, and Yuan Zhao. 2022. https://bmcmededuc.biomedcentral.com/articles/10.1186/s12909-022-03865-y
   This source discusses the importance of integrating ethics into biomedical science education.

SURROGATE DECISION-MAKING

58

Excellence is never an accident. It always results from high intention, sincere effort, and intelligent execution; it represents the wise choice of many alternatives—choice, not chance, determines destiny.
—Aristotle

* * *

MYSTERY STORY

SURROGATE CHOICES: SURROGATE DECISION-MAKING IN END-OF-LIFE CARE

Practitioner Laura Carter, a seasoned neurosurgeon at DecisionCare Medical Center, knew well that healthcare's prime directive was to maximize the patient's best interests, guided by their reasonable goals, values, and priorities. A notion steeped in the liberal philosophy of John Locke, this principle emphasized individual rights and liberties and placed a firm responsibility on those entrusted with surrogate decision-making.

Her latest patient, Mr. Edward Phillips, a retired accountant diagnosed with a malignant brain tumor, had executed a durable power of attorney for healthcare (DPOA), appointing his daughter, Lisa, as his healthcare decision-maker. The initial diagnosis and consultation were held at DecisionCare Medical Center, where Practitioner Carter worked closely with patients and their surrogates.

They met several times, discussing treatment options, always keenly aware that without a living will, the DPOA was essential. The meetings were filled with progress until one day when Lisa's demeanor changed.

"I want what's best for my father, but I'm uncertain about his wishes," Lisa confessed, overwhelmed.

Practitioner Carter empathized, guiding her through the uncertainty. "We must align with your father's reasonable goals, values, and priorities," she counseled, her voice calm and reassuring, scheduling another meeting at DecisionCare Medical Center.

But time ran out. Mr. Phillips' condition worsened suddenly, and Lisa, as the appointed DPOA, found herself in the unimaginable position of making end-of-life decisions for her father. This weight of responsibility was shared with Practitioner Carter, who stood steadfast at her side.

At the hospital, Practitioner Carter comforted Lisa, "We have to consider your father's healthcare history, personal values, and any conversations about his wishes."

Together, they made decisions not based on Lisa's personal views but, rather, as if she were her father, striving to understand his values, goals, and priorities. They faced the agony of his deteriorating condition and the weight of their responsibility, guided by their unwavering commitment to Mr. Phillips' autonomy.

When Mr. Phillips' heart rate dropped, a code blue was initiated. Despite all efforts, he passed away. A somber debriefing followed, where Practitioner Carter stressed the importance of surrogate decision-making, with its prime directive being maximizing patient interests in accordance with the patient's reasonable goals, values, and priorities.

Practitioner Carter's and Lisa's experiences served as a vivid reminder of the delicate balance between law, ethics, and the human heart in healthcare practice.

* * *

THINK

Healthcare's prime directive is to maximize the patient's best interests as determined by the patient's reasonable goals, values, and priorities. Since the Age of Enlightenment, liberalism, as espoused by John Locke, has bolstered the position that patients not only have individual rights and liberties of self-determination but also are in the best position for determining what will maximize their own best interests.

But situations can occur when the patient no longer has decisional capacity and has no written living will to document and instruct what treatment options should be chosen or pursued. Without specific patient instructions in the healthcare records and without a living will, the next level of authority is the DPOA. The DPOA is a legal document that delineates who will be the patient's surrogate for making treatment decisions on their behalf if they lose decisional capacity.

If there is no DPOA, then decisional authority will be sought typically in the following order:

1. Guardian
2. Spouse
3. Adult offspring
4. Either parent
5. Any adult sibling
6. Any adult grandchild
7. Close friend
8. Guardian of the estate
9. Temporary custodian

The order of the list is based on the assumption that those higher up on the list will be in a better position to represent the patient's best interests. The goal of the surrogate is not to make decisions based on what the surrogate themself would do or based on the surrogate's goals, values, and priorities; rather, the surrogate must make decisions as if the surrogate were the patient using the patient's reasonable goals, values, and priorities.

ASSESS

Patient: Autonomy

If a patient lacks decisional capacity and has no documented decisional wishes in the healthcare record and no living will, then a surrogate will be determined by a DPOA or, if that does not exist, through the use of a state-determined prioritized list. The goal is for the surrogate to make an informed consent decision as close as possible to what the patient would have made if they still had decisional capacity.

This means that just like the patient, the surrogate must be substantially informed about the diagnosis, prognosis, treatment options, risks and benefits and have the opportunity to have all their questions answered. Thus, practitioners should encourage the patient to allow the surrogate to be part of the patient–practitioner relationship before the patient loses their decisional capacity so that the surrogate can gain a better understanding regarding how the patient would like their treatment decisions made.

Practitioner: Beneficence and Nonmaleficence

The practitioner's professional responsibility is the same as the surrogate's responsibility concerning maximizing the patient's best interests in accordance with the patient's reasonable goals, values, and priorities. The practitioner's and surrogate's obligations are also to use the moral principles of beneficence (do good) and nonmaleficence (do no harm) when making decisions on the patient's behalf.

During this collaborative decision-making process, if the practitioner and surrogate have unresolvable differences and disagree on what the patient would have chosen, then an ethics consultation would be the next step. If that does not resolve the issue, then a court order may be necessary. The extent to which the practitioner must pursue other decisional mandates for the protection of the patient is context-determined.

If there is no evidence of the adult patient's reasonable goals, values, and priorities, then the practitioner and patient's surrogate are obliged to default to the standards of care. If two or more surrogates, at the same level of decisional authority, cannot agree on what healthcare decision should be made for the patient, then that also is when an ethics consult could be of benefit. Although ethics consults are only recommending bodies, and not decision-makers, ethics consults will provide an outside perspective, clarify surrogate misunderstandings, and help the surrogates focus on their decision-making rationale.

Minor patients have never been legally competent; therefore, there are no reasonable goals, values, and priorities of the patient to appeal to for decision-making. With minors, the practitioner, surrogate, and the state have the interest to protect and maximize the minor patient's best interests, as determined by standards of care. If standards of care are violated, then the practitioner has a professional, legal, and ethical obligation to intervene through the use of an ethics consultation, institutional reporting, and, if necessary, a court order for the protection of the minor patient.

Public Policy: Justice

The Self-Determination Act of 1990 was put in place to ensure that patients are aware of their right to make informed decisions, with practitioners regularly inquiring as to whether or not a patient has executed an advance directive and, if executed, ensuring that the directive has been documented in the patient's healthcare record. As much as public policy has supported the liberal notion of patient informed consent for the authorization of a practitioner to provide treatment, public policy has also supported surrogate decision-making.

CONCLUDE

The practitioner must inform the surrogate of all relevant information, as if they were the patient, and ensure, as much as possible, that the healthcare decision of the surrogate is what the patient would have chosen for themself, not what the surrogate would choose if they were to be in that situation. Surrogate decision-making is vital in cases in which patients lack decisional capacity, and both practitioners and surrogates must work together to uphold the patient's autonomy and best interests.

In summary, when a patient lacks decision-making capacity, a DPOA is used to determine a surrogate who will make treatment decisions on the patient's behalf based on the patient's reasonable goals, values, and priorities. Surrogates must be informed about the diagnosis, prognosis, treatment options, risks, and benefits. The practitioner has a professional responsibility to maximize the patient's best interests and use the moral principles of beneficence and nonmaleficence. If differences cannot be resolved, an ethics consultation is the next step, with a court order being necessary in some cases. The Self-Determination Act of 1990 supports both patient informed consent and surrogate decision-making.

* * *

REVIEW QUESTIONS
1. Without specific patient instructions in the healthcare records and without a living will, the next level of authority is the DPOA.
    A. True
    B. False

2. The goal of the surrogate is to make decisions as if the surrogate were the patient using the patient's reasonable goals, values, and priorities.
   A. True
   B. False

3. Just like the patient, the surrogate must be substantially informed about the diagnosis, prognosis, treatment options, risks, and benefits and have the opportunity to have all their questions answered.
   A. True
   B. False

4. The practitioner's and surrogate's responsibility is to maximize the patient's best interests in accordance with the patient's reasonable goals, values, and priorities.
   A. True
   B. False

5. If two or more surrogates, at the same level of decisional authority, cannot agree on what healthcare decision should be made for the patient, then an ethics consult can be beneficial.
   A. True
   B. False

6. Ethics consults are decision-making bodies requiring compliance from the practitioner and surrogate.
   A. True
   B. False

Answers: 1A, 2A, 3A, 4A, 5A, 6B

* * *

CLINICAL VIGNETTES

1. Ms. Lily Reed is a 72-year-old retired teacher who was admitted to the hospital with severe heart failure. Despite initial interventions, her condition has worsened, and she is now experiencing multi-organ failure. She has been intubated and placed on a ventilator, and her healthcare team has determined that she lacks decisional capacity. Ms. Reed has no living will or advanced directive, but her daughter has been designated as her DPOA. The healthcare team has explained to the daughter that her mother's prognosis is poor and that she is unlikely to survive without life-sustaining measures. The daughter, however, is deeply religious and believes that it is not her place to end her mother's life. The ethical question is whether to follow the daughter's wishes or transition Ms. Reed to comfort measures.
   A. Follow the daughter's wishes and continue all life-sustaining measures
   B. Transition Ms. Reed to comfort measures, as recommended by the healthcare team
   C. Consult with an ethics committee to determine the best course of action
   D. Seek a court order to make the decision for the patient

   Explanation: In the absence of specific patient instructions in the healthcare record and a living will, the DPOA is the next level of authority in decision-making. The surrogate must make decisions based on the patient's reasonable goals, values, and priorities and must be informed about the diagnosis, prognosis, treatment options, risks, and benefits. In this case, the daughter has been designated as the surrogate and has the legal authority to make decisions on behalf of her mother. If differences cannot be resolved, an ethics consultation is the next step. The ethics committee can help identify ethical issues, clarify misunderstandings, and provide guidance on relevant policies and laws. However, the decision-making authority remains with the surrogate(s) unless there are unresolvable differences between the surrogate(s) and the practitioner, in which case a court order may be necessary. In this case, the daughter's deeply held religious beliefs conflict with the healthcare team's recommendations to transition Ms. Reed to comfort measures. Although it may be appropriate for the healthcare team to continue to educate the daughter about the risks and benefits of continuing aggressive measures, ultimately the decision lies with the surrogate, and following the daughter's wishes to continue all life-sustaining measures would be the correct course of action. *Answer*: A

2. Ms. Ana Hill is a 60-year-old retired nurse who was admitted to the hospital after experiencing chest pain, shortness of breath, and fatigue. Initial assessments suggest that Ms. Hill might have congestive heart failure or a pulmonary embolism. She has two adult children, both of whom are listed as having equal decisional authority as her healthcare surrogates. However, Ms. Hill's children cannot agree on what healthcare decision to make regarding their mother's care. One believes that their mother would want to receive aggressive treatment to prolong her life, whereas the other thinks that their mother would not want to go through invasive interventions if her condition worsens. An ethics consultation is called to help Ms. Hill's children reach a consensus on the best course of action for their mother's care.
   A. The ethics consult will make the decision for the children.
   B. One child's decision will be made because they are listed first.
   C. A court order will be needed to make a decision.
   D. The ethics consult will provide a recommendation.

   Explanation: An ethics consultation can be beneficial when surrogates with equal decisional authority cannot agree on what healthcare decision to make for a patient. The role of the ethics consult is not to make the decision for the surrogates but, rather, to provide recommendations and help the surrogates come to a consensus on the best course of action for the patient. The other options are wrong because the ethics consult does not make the decision for the surrogates but, rather, is only a recommending body; both surrogates are listed as having equal decisional authority; and a court order is typically only necessary in extreme cases in which there is a significant risk of harm to the patient. *Answer*: D

3. Ms. Victoria Wyatt is a 68-year-old retired schoolteacher who is admitted to the hospital with acute respiratory distress syndrome. She is intubated and placed on a ventilator. Ms. Wyatt has no advance directives or surrogate decision-maker identified. The healthcare team has exhausted all treatment options and the patient is not responding to therapy. The healthcare team believes that it is time to consider end-of-life care. The ethical question in this case is who should make the decision for Ms. Wyatt, given that she lacks decisional capacity, has no advance directives or identified surrogate decision-maker, and the healthcare team is recommending end-of-life care.
   A. The healthcare team should make the decision to withdraw care.
   B. The hospital ethics committee should make the decision.
   C. The state-appointed surrogate decision-maker should make the decision.
   D. The chaplain should make the decision.

   Explanation: In the absence of an identified surrogate decision-maker or advance directives, the state-appointed surrogate decision-maker should make the decision. The surrogate decision-maker is responsible for making decisions based on the patient's reasonable goals, values, and priorities. If the surrogate cannot be identified or located, the healthcare team may petition the court to appoint a surrogate. It is inappropriate for the healthcare team to make the decision to withdraw care without a surrogate decision-maker or court order. Although the hospital ethics committee may provide guidance and consultation, it is not responsible for making any decisions. In the end, the ethical principle of patient autonomy is upheld by ensuring that the appointed surrogate decision-maker is making decisions in accordance with the patient's reasonable goals, values, and priorities or, if those are not known, by focusing on the patient's best interests in accordance with healthcare standards of care. *Answer*: C

4. Ms. Julia Allen is a 70-year-old retired teacher who has been diagnosed with advanced lung cancer. She has been hospitalized for severe respiratory distress and cannot make her own treatment decisions. She has no DPOA and no known family members. As such, the healthcare team has been determined to have the legal decision-making authority, and it is considering the use of an artificial intelligence (AI) surrogate decision-making program for determining her healthcare. What is the most pressing ethical question in this scenario?
   A. Is it ethical to use AI surrogate decision-making in this situation?
   B. Who has the legal authority to make treatment decisions for Ms. Allen?
   C. Should the healthcare team continue to provide treatment for Ms. Allen despite the poor prognosis?
   D. Is it ethical to withhold healthcare treatment from a patient with no known family members?

Explanation: The ethical question in this scenario is whether it is ethical to use AI surrogate decision-making to determine Ms. Allen's healthcare. Although the use of AI technology can potentially assist in making treatment decisions, it raises ethical concerns related to patient autonomy, beneficence, and nonmaleficence. AI may not be able to fully consider the patient's individual circumstances and individual goals, values, and priorities, which could potentially result in decisions that are not in the patient's best interests. In addition, the use of AI technology in surrogate decision-making raises questions about who is ultimately responsible for the decision and the potential for legal and ethical conflicts. Therefore, it is important for practitioners to consider the limitations and ethical implications of using AI in surrogate decision-making and ensure that any decisions made align with the patient's reasonable goals, values, and priorities. For example, who would be responsible if an AI system made a decision that resulted in harm to the patient? Would the responsibility lie with the developer of the AI system, the healthcare provider who used it, or the surrogate decision-maker who relied on the AI system? In addition, how can it be ensured that the AI system is taking into account the patient's reasonable goals, values, and priorities in the same way a human surrogate decision-maker would? Furthermore, the use of AI in healthcare decision-making also raises concerns about the potential for bias and discrimination in the decision-making process. For instance, if an AI system is trained on data that are biased against certain patient groups, it may produce decisions that perpetuate that bias. Overall, although the use of AI in surrogate decision-making is still in its early stages, it is important for healthcare providers and policymakers to carefully consider the ethical and legal implications of this technology in order to ensure that patient autonomy and best interests are upheld. The practitioner must inform the surrogate of all relevant information, as if they were the patient, and ensure, as much as possible, that the healthcare decision of the surrogate is what the patient would have chosen for themself, not what the surrogate would choose if they were to be in that situation. *Answer*: A

5. Ms. Rachel Taylor is a 70-year-old retired teacher diagnosed with advanced lung cancer. She has been hospitalized for severe respiratory distress and is not able to make her own treatment decisions. She has no DPOA and no known family members in the country. The healthcare team is considering the use of telemedicine to connect with a surrogate decision-maker who is located in a different country. What is the ethical question in this scenario?
   A. Is it ethical to use telemedicine to connect with a surrogate decision-maker in a different country?
   B. Who has the legal authority to make treatment decisions for Ms. Taylor?
   C. Should the healthcare team continue to provide treatment for Ms. Taylor despite the poor prognosis?
   D. Is it ethical to withhold healthcare treatment from a patient with no known family members?

Explanation: The ethical question in this scenario is whether it is ethical to use telemedicine to connect with a surrogate decision-maker in a different country. Although telemedicine can assist in making treatment decisions, it raises ethical concerns about patient autonomy, beneficence, and nonmaleficence. The surrogate decision-maker may not be familiar with the healthcare system or cultural norms of the country where the patient is receiving care, which could potentially result in decisions that are not in the patient's best interests. In addition, the use of telemedicine in surrogate decision-making raises questions about who is ultimately responsible for the decision and the potential for legal and ethical conflicts. Therefore, it is important for practitioners to consider the limitations and ethical implications of using telemedicine in surrogate decision-making and to ensure that any decisions made are in line with the patient's reasonable goals, values, and priorities. For example, who would be responsible if the surrogate made a decision that harmed the patient? Would the responsibility lie with the healthcare provider who facilitated the telemedicine connection or the surrogate decision-maker who relied on the telemedicine system? Furthermore, the use of telemedicine in healthcare decision-making also raises concerns about the quality and security of communication. For instance, if the telemedicine connection is disrupted or insecure, it may affect the communication quality and the accuracy of the decision-making process. Overall, although telemedicine in surrogate decision-making is becoming more common, it is important for healthcare providers and policymakers to carefully consider this technology's ethical and legal implications to ensure that patient autonomy and best interests are upheld. The practitioner must inform the surrogate of all relevant information, as if they were the patient, and ensure, as much as possible, that the healthcare decision of the surrogate is what the patient would have chosen for themselves, not what the surrogate would choose if they were to be in that situation. *Answer*: A

REFLECTION VIGNETTES

1. Practitioner Olivia Lee, a 45-year-old attending practitioner, is caring for a patient in the intensive care unit (ICU) who lacks decisional capacity. The patient's surrogate has been making treatment decisions, but Practitioner Lee believes that the surrogate is making decisions based on their own goals, values, and priorities rather than those of the patient.

    What is the ethical issue(s)?

    How should the ethical issue(s) be addressed?

2. Practitioner Nora Sanders, a 38-year-old family practitioner, sees a 65-year-old retired teacher named Scarlett Hall in her office for a routine checkup. During the history and physical exam, Practitioner Sanders discovers that Ms. Hall still needs to fill out an advance directive with a living will or DPOA.

    What is the ethical issue(s)?

    How should the ethical issue(s) be addressed?

3. Practitioner Emma Parker, a 45-year-old ICU attending practitioner, is faced with a complex case involving a patient on a ventilator in the ICU for severe pneumonia. The patient lacks decisional capacity, and no oral or written preferences, living will, or DPOA are on record. The patient's condition is deteriorating, and the patient's older offspring wants to withdraw life support, but the patient's younger offspring disagrees. The differential diagnosis includes various life-support interventions, such as mechanical ventilation, extracorporeal membrane oxygenation0, and dialysis.

    What is the ethical issue(s)?

    How should the ethical issue(s) be addressed?

* * *

WEB LINKS

1. *Code of Medical Ethics.* American Medical Association. https://code-medical-ethics.ama-assn.org

   The AMA *Code of Medical Ethics*, established by the American Medical Association, is a comprehensive guide for healthcare practitioners. It addresses issues and challenges, promotes adherence to standards of care, and is continuously updated to reflect contemporary practices and challenges.

2. "Who Decides When a Patient Can't? Statutes on Alternate Decision Makers." Erin S. DeMartino, David M. Dudzinski, Cavan K. Doyle, et al. 2017. https://www.ncbi.nlm.nih.gov/pmc/articles/PMC5527273

   This source discusses the shift from paternalism toward a more patient-centered approach in healthcare, especially for patients who cannot participate in decision-making.

3. "AMA *Code of Medical Ethics*' Opinions on Patient Decision-Making Capacity and Competence and Surrogate Decision Making." Danielle Hahn Chaet. 2017. https://journalofethics.ama-assn.org/article/ama-code-medical-ethics-opinions-patient-decision-making-capacity-and-competence-and-surrogate/2017-07

   This source discusses the role of surrogate decision-makers in making decisions based on the patient's preferences and values.

4. "Surrogate Decision-Making." John C. Moskop. 2016. https://www.cambridge.org/core/books/ethics-and-health-care/surrogate-decisionmaking/AB0A6FC3324E75550B43A40FE048D522

   This source discusses the role of surrogate decision-making in healthcare.

5. "Decisions for Adult Patients Who Lack Capacity." American Medical Association. https://code-medical-ethics.ama-assn.org/ethics-opinions/decisions-adult-patients-who-lack-capacity

   This source discusses the ethical responsibility of physicians when a patient lacks decision-making capacity.

6. "The Ethics of Surrogate Decision Making." Ben A. Rich. 2002. https://www.ncbi.nlm.nih.gov/pmc/articles/PMC1071685

   This source discusses the importance of surrogate decision-making being consistent with the patient's values.

7. "Clinical Ethics and Law." Lisa V. Brock and Anna Mastroianni. https://depts.washington.edu/bhdept/ethics-medicine/bioethics-topics/detail/56

   This source discusses the ethical and legal parameters in surrogate decision-making.

8. "Who Makes Decisions for Incapacitated Patients Who Have No Surrogate or Advance Directive?" Scott J. Schweikart. 2019. https://journalofethics.ama-assn.org/article/who-makes-decisions-incapacitated-patients-who-have-no-surrogate-or-advance-directive/2019-07

   This source discusses the challenges of making decisions for incapacitated patients who do not have a designated surrogate.

9. "Ethical Aspects of Surrogate Decision Making." U.S. Department of Veterans Affairs.

   This source discusses the role of the surrogate as a partner in the process of shared decision-making.

10. "Patient Confidentiality and the Surrogate's Right to Know." Lynn A. Jansen and Lainie Friedman Ross. 2021. https://www.cambridge.org/core/journals/journal-of-law-medicine-and-ethics/article/patient-confidentiality-and-the-surrogates-right-to-know/9C21F5B60B7454EC273B1D149BFE88F5

    This source discusses the complexities of navigating surrogate decision-makers through a difficult course of treatment decisions.

11. "Competence, Capacity, and Surrogate Decision-Making." Robert D. Orr. 2004. https://www.cbhd.org/cbhd-resources/competence-capacity-and-surrogate-decision-making

    This source discusses respecting patients' autonomy and promoting their well-being in surrogate decision-making.

12. "Surrogate Decision Making: Reconciling Ethical Theory and Clinical Practice." Jeffrey T. Berger, Evan G. DeRenzo, and Jack Schwartz. 2008. https://pubmed.ncbi.nlm.nih.gov/18591637

    This source discusses the ethical considerations in surrogate decision-making for adult patients without decision-making abilities.

13. "Advance Directives and Surrogate Decision Making." University of Missouri School of Medicine, Center for Health Ethics. https://medicine.missouri.edu/centers-institutes-labs/health-ethics/faq/advance-directives

    This source discusses the importance of designating a surrogate for decision-making in the event of incapacitation.

14. "Ethical Conflicts in Surrogate Decision Making." Leah Conant and Piroska Kopar. 2022. https://link.springer.com/chapter/10.1007/978-3-030-84625-1_39

    This source discusses the application of the landmark principles of bioethics—i.e., autonomy, nonmaleficence, beneficence, and justice—in surrogate decision-making.

15. "The Ethics of Choosing a Surrogate Decision Maker When Equal-Priority Surrogates Disagree." Matthew Shea. https://muse.jhu.edu/article/800079

    This source discusses the ethical considerations governing who the surrogate should be and what criteria should guide the surrogate's decisions.

16. "Standards for Surrogate Decision-Making." Center for Bioethics, University of Minnesota. 2023. https://bioethics.umn.edu/events/standards-surrogate-decision-making

    This source discusses the process of surrogate decision-making and the importance of making decisions that the patient would most likely agree with.

TELEMEDICINE

59

A person will not address a practitioner as a nobody nor a magistrate as an everyday individual. The practitioner has the power of skill, and the magistrate has the power of position.
—Saint Basil

* * *

MYSTERY STORY

DIGITAL DECEPTION: THE TELEMEDICINE MYSTERY

In the digital age of medicine, the rapid expansion of telemedicine had opened doors for not only care but also complexity. There were synchronous appointments allowing real-time patient–practitioner interactions, asynchronous interactions for non-emergency communications, and remote monitoring of vital signs.

Practitioner Emma Ross, a pioneering telemedicine practitioner, prided herself on staying at the forefront of technology, following stringent privacy and security protocols, and establishing clear professional boundaries with her patients.

But all of these principles were about to be tested.

Late one evening, during a synchronous appointment, Practitioner Ross found herself in the middle of a concerning scenario. She was reviewing data with a patient when she noticed a sequence of unusual vital sign readings from a remote monitoring device connected to Mr. Frank Harris, a patient under her care but not scheduled for consultation that night. The vital signs were alarming.

Following protocol, Practitioner Ross immediately contacted Mr. Harris but received no response. She then notified emergency services, who rushed to Mr. Harris' location, only to find his house empty. The readings were coming from a different location entirely.

The mystery deepened as Practitioner Ross, in consultation with Detective Harry Caldwell, discovered that the readings were being artificially manipulated. Someone had hacked into the telemedicine system, playing a dangerous game with patient data.

As Practitioner Ross and Detective Caldwell dug into the case, they uncovered a twisted web of deceit, including evidence of state-based Medicaid fund dispersion across borders, which was highly illegal. Furthermore, they stumbled upon an underground network that was exploiting telemedicine technology to prescribe Schedule II drugs unlawfully.

They managed to trace the hackers, shut down the illegal network, and bring those responsible to justice.

This incident was not just an attack on one practitioner; it was an assault on the principles that held the telemedicine community together: autonomy, beneficence, nonmaleficence, and justice.

The incident sparked nationwide discussions on telemedicine regulations, patient–practitioner relationships, and the need for national healthcare licensure versus state-based licensure. It also paved the way for stricter enforcement of telemedicine parity laws and Health Insurance Portability and Accountability Act (HIPAA) regulations.

The mystery served as a stark reminder that as telemedicine continues to revolutionize healthcare delivery, it must be navigated with the utmost care, ethics, and responsibility, lest it become a playground for those with malicious intent.

* * *

THINK

Telemedicine is a technology that has resulted in changes in both the delivery and distribution of healthcare, providing increased patient convenience and expanded patient access for underserved populations.

Telemedicine is synchronous, asynchronous, and monitorial:

Synchronous: Patient–practitioner interactions occur live in real time. Synchronous communications are appropriate for personal and sensitive information disclosures.
Asynchronous: Patient–practitioner interactions occur non–real time, meaning not at the same time. Asynchronous communications are appropriate for patient educational materials and many non-emergency communications.
Monitoring: Patient's vital signs and other data are tracked from a distance. Information can be automatically graphed and periodically checked by the practitioner for red flags, at which point the practitioner will contact the patient. Other types of indicators can immediately alert the practitioner of an emergency.

Recent advances in computers, mobile phones, watches, wearables, audio–video, and security have opened the doors to expanding opportunities for telemedicine for care and monitoring that have historically only been available in the hospital or office-based setting.

ASSESS

Patient: Autonomy

Although not required by all states, the practitioner should always attain a written informed consent from the patient authorizing the use of telemedicine. The informed consent should include a simple explanation of how telemedicine works, services available, scheduling, confidentiality, privacy, prescribing policies (e.g., no Schedule II drugs, only Schedule III–V drugs), coordinating with other care providers, contingency plans for emergencies, security protocols, risk of patient information breach, professional boundaries, and billing (fees, costs, federal Medicare, state Medicaid, and private insurance reimbursements).

The practitioner may also assure the patient of their education, licensure, and skills by directing the patient to the Federation of State Medical Boards of the United States (FSMB) website, where the patient can select DocInfo.org and type in the name of their practitioner. The link will provide the practitioner's school of education, board certifications, states with active licenses, and any legal actions that have been taken against the practitioner. Such knowledge about the practitioner will enhance the patient–practitioner relationship.

Practitioner: Beneficence and Nonmaleficence

Telemedicine does not change the practitioner's prime directive of maximizing the patient's best interests as determined by the patient's reasonable goals, values, and priorities, using the biomedical principles of beneficence (do good) and nonmaleficence (do no harm). The benefits of telemedicine patient–practitioner interactions must always outweigh the increased risk of harm caused by online patient–practitioner interactions.

Because telemedicine allows for frequent patient–practitioner exchanges, it is imperative for the practitioner to establish clear and definable professional boundaries early on to avoid any perceived, implied, or actual patient–practitioner impropriety. Any practitioner conflict of interest must be disclosed to the patient.

Public Policy: Justice

Telemedicine is regulated primarily by state legislation. Most states have passed telemedicine parity laws that mandate private payers to reimburse telemedicine visits at the same rate as comparable in-person visits. Telemedicine, an electronic online service, is not geographically bounded by physical borders, unlike healthcare licensure, which is geographically bounded. Practitioners

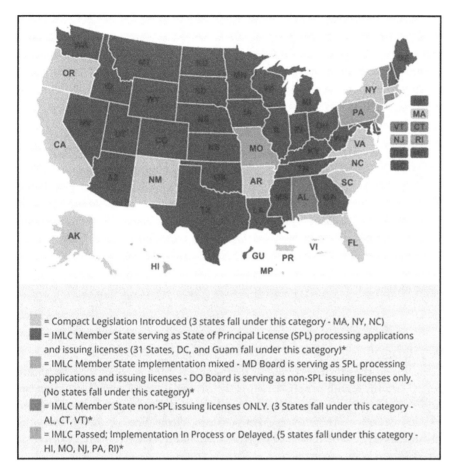

FIGURE 59.1 U.S. state participation in the Interstate Medical Licensure Compact (IMLC). Reproduced with permission from the Interstate Medical Licensure Compact Commission (IMLCC). http://www.imlcc.org.

must be licensed to practice medicine in each state that they practice. Crossing state borders through telemedicine technology that results in practicing medicine where the practitioner is not licensed is illegal, unprofessional, and unethical.

As of 2024, 39 states, the District of Columbia, and the Territory of Guam were part of the Interstate Medical Licensure Compact (IMLC), which makes licensure more straightforward if one has a primary license in one of the participating states. The IMLC uses a uniform standard for licensure, allowing practitioners to treat patients who reside in participating states (Figure 59.1, Table 59.1).[1]

Table 59.1 Interstate Healthcare Licensure Compact

Average number of licenses obtained:	3
Approval rate	90%
Average wait time for license	19 days
Licenses available within 1 week	51%

As telemedicine becomes more nationalized, there will be a greater need to consider whether or not healthcare licensure should also be national rather than state licensure. The FSMB and the Nation Board of Medical Examiners have already collaborated and established the Step exams as the national standard every state uses as the minimum standard for healthcare licensure eligibility.

1. CompHealth blog, "Interstate Medical Licensure Compact States List and Guide for 2023." https://www.imlcc.org.

Problems with nationalizing healthcare licensure for telemedicine include the vast differences in state laws, for which the practitioner must be held accountable. State healthcare licensure exams ensure that practitioners know the state's healthcare laws. Other issues arise with dispersing state Medicaid funds to practitioners residing in other states. Medicaid is managed by states and is based on income. Medicare is managed by the federal government and is based on age. Legislation and federal actions have sought to adapt healthcare policies, particularly in the realm of telehealth, to address the evolving public health crisis.

On March 6, 2020, the "Coronavirus Preparedness and Response Supplemental Appropriations Act" became law. In addition to providing healthcare agencies with more money to address the COVID-19 outbreak, it also loosens existing telehealth restrictions in order to enhance patient access to care services. On March 17, 2020, the Centers for Medicare and Medicaid Services (CMS) announced that they had expanded the waiver for telehealth in several areas, including the care of new patients for the diagnosis and treatment of COVID-19 and other conditions unrelated to the public health emergency. However, it does not open up telehealth billing to new practitioners.[2]

CONCLUDE

For telemedicine, the patient–practitioner relationship is still foundational for healthcare, and the federal mandate for full compliance with HIPAA regulations for confidentiality and privacy of patient's private health information is enforced. Telemedicine practitioners need to know the state healthcare laws of each state where they practice medicine. Telemedicine has revolutionized healthcare delivery, and its continued expansion and integration will further enhance patient care and access to healthcare services.

In summary, telemedicine is a technology that has revolutionized healthcare delivery and provides increased convenience and expanded access to underserved populations. There are three main types of telemedicine: synchronous, asynchronous, and monitoring. Telemedicine is primarily regulated by state legislation, with some states having parity laws mandating reimbursement for telemedicine visits at the same rate as in-person visits. As telemedicine becomes more widespread, there is a growing need to consider whether healthcare licensure should be national rather than state-based. This new technology augments the importance of patient autonomy, practitioner beneficence and nonmaleficence, and the need for establishing clear professional boundaries.

* * *

REVIEW QUESTIONS

1. Telemedicine can be synchronous, asynchronous, or monitorial.
    A. True
    B. False

2. Recent advances in computers, mobile phones, watches, wearables, audio–video, and security have opened the doors to expanding opportunities for telemedicine for care and monitoring that have historically only been available in the hospital setting.
    A. True
    B. False

---

2. Pharmacists Advancing Healthcare. April 9, 2020. https://www.ashp.org/advocacy-and-issues/key-issues/other-issues/additional-advocacy-efforts/issue-brief-covid-19?loginreturnUrl=SSOCheckOnly.

3. Telemedicine does not change the practitioner's prime directive of maximizing the patient's best interests as determined by the patient's reasonable goals, values, and priorities, using the biomedical principles of beneficence (do good) and nonmaleficence (do no harm).
   A. True
   B. False

4. Because telemedicine allows for frequent patient–practitioner exchanges, it is imperative for the practitioner to establish clear and definable professional boundaries early on to avoid any perceived, implied, or actual patient–practitioner impropriety.
   A. True
   B. False

5. Telemedicine allows the practitioner to have patients and practice medicine in states where the practitioner is not licensed.
   A. True
   B. False

6. The IMLC uses a uniform standard for licensure, allowing practitioners to treat patients who reside in participating states.
   A. True
   B. False

Answers: 1A, 2A, 3A, 4A, 5B, 6A

* * *

CLINICAL VIGNETTES

1. Ms. Amanda Jackson is a 45-year-old working mother who lives in a rural area. She has been experiencing symptoms of anxiety and depression for several months. She has been searching for a local mental health provider who can see her in person, but due to her work schedule and the limited availability of mental health professionals in her area, the search has been difficult. She has heard about telemedicine and is interested in receiving mental health services through this method. Which of the following is a potential benefit of telemedicine for Ms. Jackson's situation?
   A. Access to specialized in-person care
   B. Ability to avoid sharing personal information with a practitioner
   C. Convenience of being able to receive care from home
   D. Higher likelihood of receiving a prescription for medication

   Explanation: Telemedicine can provide greater convenience for patients with difficulty accessing in-person care due to location, work schedule, or other factors. By using telemedicine, Ms. Jackson could receive mental health services from the comfort of her own home, which can save time and reduce travel costs. Although telemedicine can also provide access to specialized care, in this case, the option of avoiding sharing personal information with a practitioner is not a benefit because it may hinder the ability to diagnose and treat Ms. Jackson's conditions. Last, the likelihood of receiving a prescription for medication is not necessarily higher in telemedicine compared to in-person visits, as that would depend on the practitioner's assessment and clinical judgment. *Answer*: C

2. Ms. Rachel Cooper is a 45-year-old woman with a history of hypertension and diabetes. She lives in a rural area and has difficulty accessing healthcare due to her busy work schedule and limited transportation options. Her primary care practitioner recommends a telemedicine consultation to monitor her blood pressure and blood glucose levels. Ms. Cooper is

provided with a wearable device that measures her vital signs and is connected to her smartphone. During the consultation, the practitioner notes that Ms. Cooper's blood pressure is elevated and her blood glucose levels are high. The differential diagnosis includes poorly controlled hypertension and diabetes. The practitioner orders medication adjustments and advises Ms. Cooper to follow a healthy diet and exercise regimen. Which ethical principles are exemplified in this scenario?

A. The practitioner's duty to obtain informed consent from Ms. Cooper for the use of telemedicine
B. The practitioner's obligation to ensure patient privacy and security protocols during the telemedicine consultation
C. The practitioner's responsibility to establish clear and definable professional boundaries to avoid any perceived, implied, or actual patient–practitioner impropriety
D. The practitioner's obligation to follow the principles of beneficence and nonmaleficence during the telemedicine consultation

Explanation: The correct answer is option D, the practitioner's obligation to follow the principles of beneficence and nonmaleficence during the telemedicine consultation. Telemedicine does not change the practitioner's prime directive of maximizing the patient's best interests as determined by the patient's reasonable goals, values, and priorities, using the biomedical principles of professional beneficence (do good) and professional nonmaleficence (do no harm). The practitioner must ensure that the benefits of telemedicine patient–practitioner interactions always outweigh the increased risk of harm caused by online interactions. In this scenario, the practitioner has a duty to make medication adjustments and provide advice on diet and exercise to improve Ms. Cooper's health. The other options are incorrect because informed consent, patient privacy and security protocols, and the practitioner's responsibility to establish clear and definable professional boundaries are not the central focus of this scenario. *Answer*: D

3. Mr. Matthew Walker is a 68-year-old retired carpenter who lives in rural Mississippi. He has been experiencing worsening shortness of breath and chest pain for the past week. He has a past healthcare history of hypertension, hyperlipidemia, and coronary artery disease. Mr. Walker is unable to travel to a healthcare facility due to the distance from his home to the nearest hospital. He contacts his primary care provider, Practitioner Keven Taylor, who practices in Tennessee. Practitioner Taylor is licensed in both Tennessee and Mississippi and is part of the IMLC. Practitioner Taylor decides to conduct a synchronous telemedicine visit with Mr. Walker to assess his condition and determine the best course of action. What ethical considerations must Practitioner Taylor consider when conducting a telemedicine visit with Mr. Walker?

A. Practitioner Taylor should not obtain informed consent from Mr. Walker, as this will delay treatment.
B. Practitioner Taylor does not need to establish clear professional boundaries with Mr. Walker, as they have a pre-existing patient–practitioner relationship.
C. Practitioner Taylor should ensure that Mr. Walker's privacy and security are protected during the telemedicine visit.
D. Practitioner Taylor should not be licensed in both Tennessee and Mississippi, as this violates state healthcare licensure laws.

Explanation: Practitioner Taylor must protect Mr. Walker's privacy and security during the telemedicine visit. This includes obtaining informed consent from Mr. Walker, explaining how telemedicine works, and establishing clear professional boundaries. Practitioner Taylor must also ensure that he is licensed to practice medicine in the state where the patient is located. Being licensed in Tennessee and Mississippi as part of the IMLC allows Practitioner Taylor to conduct telemedicine visits with patients in either state. The best option presented is that privacy and security protocols are crucial in telemedicine, and ensuring patient privacy and security is an ethical obligation for practitioners. The other options are incorrect because obtaining informed consent is a necessary ethical obligation for practitioners; establishing professional boundaries is essential for all patient–practitioner relationships, regardless of pre-existing relationships; and being licensed in both Tennessee and Mississippi is allowed as part of the IMLC. *Answer*: C

4. Ms. Olivia Grant is a 65-year-old retired teacher living in a rural area of North Dakota. She has a history of hypertension, hyperlipidemia, and type 2 diabetes mellitus. Recently, she has been experiencing shortness of breath, chest pain, and palpitations. She contacts her primary care practitioner, who is licensed in North Dakota, but due to a variety of reasons, the practitioner is unable to see her in person. Instead, the practitioner recommends a telemedicine consultation with a specialist who is licensed in Minnesota, which is a participating state in the IMLC. Which ethical principle is best upheld by the practitioner recommending a telemedicine consultation with a specialist licensed in Minnesota?
   A. Autonomy
   B. Beneficence
   C. Nonmaleficence
   D. Justice

   Explanation: The ethical principle of beneficence, which requires practitioners to act in their patient's best interests, maximize the benefits, and minimize the harms, is best upheld by the practitioner recommending a telemedicine consultation with a specialist licensed in Minnesota. By doing so, the practitioner ensures that the patient receives the appropriate care and treatment for her symptoms despite not seeing her in person. This also promotes access to healthcare for patients living in rural areas. The other options are incorrect because autonomy, nonmaleficence, and justice, although important in telemedicine consultations, are not the central focus of this scenario of the practitioner's decision to recommend a telemedicine consultation with a specialist in another state. *Answer*: B

5. Ms. Ella Robinson is a 62-year-old retired teacher who lives in a rural area with limited access to healthcare services. She reports to her primary care practitioner, through a telemedicine visit, that she has been experiencing shortness of breath and chest pain. Her practitioner suggests a differential diagnosis of either congestive heart failure exacerbation or pulmonary embolism (PE). The practitioner orders a chest X-ray and a D-dimer test for Ms. Robinson. The results indicate that the chest X-ray is negative, whereas the D-dimer test is positive, raising the possibility of a PE. The practitioner recommends that Ms. Robinson be referred to the nearest hospital for further testing, including a computed tomography scan of her chest, to confirm or exclude the diagnosis of PE. Ms. Robinson refuses to go to the hospital, stating that she is afraid of contracting COVID-19. The ethical question is whether the practitioner should override Ms. Robinson's refusal and insist that she go to the hospital for further testing.
   A. The practitioner should insist that Ms. Robinson go to the hospital for further testing because PE is a serious condition that requires immediate treatment.
   B. The practitioner should respect Ms. Robinson's autonomy and offer alternative options, such as arranging for home oxygen therapy and monitoring.
   C. The practitioner should consult with a bioethics committee before making any decision.
   D. The practitioner should consult with Ms. Robinson's family members and obtain their opinion before making any decision.

   Explanation: In telemedicine, practitioners must ensure patient autonomy and obtain informed consent while adhering to the principles of beneficence and nonmaleficence and establishing professional boundaries. In this scenario, Ms. Robinson's refusal to go to the hospital must be respected, and alternative options must be offered. The practitioner may suggest arranging for home oxygen therapy and monitoring, with periodic telemedicine follow-ups, to assess Ms. Robinson's condition and provide the necessary care. The practitioner may also discuss the potential risks and benefits of alternative options and provide Ms. Robinson with the necessary information to make an informed decision. In addition, the practitioner must ensure that Ms. Robinson's decision is voluntary, without coercion, and that her refusal does not result in harm or neglect. The practitioner may consider seeking a second opinion from a specialist or consulting with an ethics committee if necessary. However, consulting with Ms. Robinson's family members and obtaining their opinion may violate her confidentiality and privacy rights and compromise her autonomy. The practitioner must respect Ms. Robinson's decision and provide the necessary care through telemedicine while ensuring her safety and well-being. *Answer*: B

REFLECTION VIGNETTES

1. Practitioner Sarah Barnes, a 38-year-old telemedicine practitioner, receives a call from a patient, Mr. Joseph Nelson, who resides in an IMLC state where Practitioner Barnes is not currently licensed to practice medicine. Mr. Nelson is seeking healthcare for his chronic back pain and is interested in receiving telemedicine services from Practitioner Barnes. Practitioner Barnes holds licenses to practice medicine in multiple IMLC states but not in Mr. Nelson's state.

   What is the ethical issue(s)?

   How should the ethical issue(s) be addressed?

2. Practitioner Julia Miller, a 35-year-old board-certified family medicine practitioner, has been hired to lead the new telemedicine clinic proposed to meet the healthcare needs of the rural community after the local hospital closed due to financial challenges.

   What is the ethical issue(s)?

   How should the ethical issue(s) be addressed?

* * *

WEB LINKS

1. *Code of Medical Ethics*. American Medical Association. https://code-medical-ethics.ama-assn.org
   The AMA *Code of Medical Ethics*, established by the American Medical Association, is a comprehensive guide for healthcare practitioners. It addresses issues and challenges, promotes adherence to standards of care, and is continuously updated to reflect contemporary practices and challenges.
2. "Telemedicine Practice: Review of the Current Ethical and Legal Challenges." Giulio Nittari, Ravjyot Khuman, Simone Baldoni, et al. 2020. https://pubmed.ncbi.nlm.nih.gov/32049608
   This article reviews the current ethical and legal challenges in telemedicine practice.
3. "Application of Ethics for Providing Telemedicine Services and Information Technology." Mostafa Langarizadeh, Fatemeh Moghbeli, and Ali Aliabadi. 2017. https://www.ncbi.nlm.nih.gov/pmc/articles/PMC5723167
   This article discusses the application of ethics in providing telemedicine services and the use of information technology.
4. "Ethical Practice in Telehealth and Telemedicine." Danielle Chaet, Ron Clearfield, James E. Sabin, et al. 2017. https://pubmed.ncbi.nlm.nih.gov/28653233
   This article summarizes the report of the American Medical Association's Council on Ethical and Judicial Affairs on ethical practice in telehealth and telemedicine.

5. "Initial Drafting of Telemedicine's Code of Ethics Through a Delphi Method." Hamid Shafizadeh, Bagher Larijani, Rita Mojtahedzadeh, Ehsan Shamsi Gooshki, and Saharnaz Nedjat. 2021. https://www.ncbi.nlm.nih.gov/pmc/articles/PMC9078934
This article discusses the initial drafting of a code of ethics for telemedicine using the Delphi method.
6. "Promise and Peril: Defining Ethical Telehealth Practice from the Patient and Clinician Experience." Amanda Jane Keenan, George Tsourtos, and Jennifer Tieman. 2022. https://www.ncbi.nlm.nih.gov/pmc/articles/PMC8744182
This article discusses the promise and potential pitfalls of telehealth practice from the perspective of both patients and practitioners.
7. "Revisiting Health Information Technology Ethical, Legal, and Social Issues and Evaluation: Telehealth/Telemedicine and COVID-19." Bonnie Kaplan. 2020. https://www.ncbi.nlm.nih.gov/pmc/articles/PMC7831568
This article revisits the ethical, legal, and social issues related to health information technology, including telemedicine.
8. "Regulatory, Legal, and Ethical Considerations of Telemedicine." Barry G. Fields. 2020. https://pubmed.ncbi.nlm.nih.gov/32762973
This article discusses the regulatory, legal, and ethical considerations of telemedicine.
9. "Telemedicine's Potential Ethical Pitfalls." Shivan J. Mehta. 2014. https://journalofethics.ama-assn.org/article/telemedicines-potential-ethical-pitfalls/2014-12
This article discusses the potential ethical pitfalls of telemedicine.
10. "Ethical and Legal Challenges of Telemedicine in the Era of COVID-19." Renata Solimini, Francesco Paolo Busardò, Filippo Gibelli, Ascanio Sirignano, and Giovanna Ricci. 2021. https://pubmed.ncbi.nlm.nih.gov/34946259
This article discusses the ethical and legal challenges of telemedicine in the era of COVID-19.
11. "Ethical Challenges of Telemedicine and Telehealth." Bonnie Kaplan and Sergio Litewka. 2008. https://pubmed.ncbi.nlm.nih.gov/18724880
This article discusses the ethical challenges of telemedicine and telehealth.
12. "The Legal and Ethical Aspects of Telemedicine. 1: Confidentiality and the Patient's Rights of Access." B. Stanberry. 1997. https://pubmed.ncbi.nlm.nih.gov/9614731
This article reviews the principle of confidentiality and the rights of access by patients to their healthcare records in the context of telemedicine.
13. "Telemedicine, Ethics, and Law in Times of COVID-19: A Look Towards the Future." C. Gil Membrado, V. Barrios, J. Cosín-Sales, and J. M. Gámez. 2021. https://www.ncbi.nlm.nih.gov/pmc/articles/PMC8133381
This article discusses telemedicine, ethics, and law in times of COVID-19 and looks toward the future.

TERMINAL AND PALLIATIVE SEDATION

Death is a friend that alone can bring the peace that treasures cannot purchase and remove the pain that practitioners cannot cure.
—Mortimer Collins

* * *

MYSTERY STORY

SEDATION SECRETS: TERMINAL/PALLIATIVE SEDATION IN END-OF-LIFE CARE

It was a pivotal morning at TranquilRest Hospice, renowned for its groundbreaking palliative care unit. Practitioner Seren Caldwell, a sage in palliative care, scanned through her patients' charts when her seasoned eye caught something amiss. Mr. Anthony Strickland, a venerable patient battling a relentless disease, was manifesting a disturbing rise in distress and pain, even under potent pain medication.

TranquilRest Hospice had been Mr. Strickland's sanctuary for the past year, but now his condition was spiraling into a realm that demanded urgent intervention. Knowing that Mr. Strickland was navigating the ominous final stages of his illness, Practitioner Caldwell feared his symptoms were morphing into a refractory state, a phase defying conventional healthcare control.

With the gravity of the situation sinking in, Practitioner Caldwell convened an urgent meeting with the hospital's ethics committee. Mr. Strickland's case was sensitive and required the seamless fusion of healthcare expertise, ethical considerations, and legal boundaries. The discussion gravitated toward terminal sedation, a path neither tread lightly nor chosen frequently.

The ethics committee, after painstaking review and soul-searching consultations with Mr. Strickland's family, agreed that terminal sedation was a door that must be opened, given his declining condition. Mr. Strickland's daughter, the cornerstone of his support system, was brought into the fold.

Sitting under the warm glow of TranquilRest Hospice's family consultation room, Practitioner Caldwell, with a voice imbued with care and conviction, walked Mr. Strickland's daughter through the landscape of terminal sedation. She painted a picture of a life freed from the chains of suffering, but also one devoid of consciousness and interaction. It was an irrevocable step toward peace, a journey that Mr. Strickland would take alone.

With tears glinting and resolve solidifying, Mr. Strickland's daughter nodded her assent. Her father's torment was too painful to witness, and TranquilRest Hospice had provided a path to compassionate reprieve.

Under Caldwell's watchful guidance, the palliative care team commenced the delicate process of terminal sedation. With each passing hour, Mr. Strickland's pain ebbed away, replaced by a serene tranquility that mirrored the philosophy of TranquilRest Hospice.

Days merged into one another, and Mr. Strickland's life song approached its final note. Surrounded by the love of his family, he passed away with a gentle smile on his face, a testament to the humane care he received.

> Practitioner Caldwell immortalized Mr. Strickland's journey into an enlightening lesson for her protégés and the healthcare community at TranquilRest Hospice. The story became a beacon, emphasizing the essence of informed consent, ethical responsibility, and the thin line separating palliative sedation from euthanasia.
>
> Through Strickland's poignant story, TranquilRest Hospice continued to embody its name, offering a haven for those in the twilight of life, teaching invaluable lessons on compassion and healthcare ethics, and nurturing hope in the shadows of despair.

* * *

THINK

Terminal sedation, also known as palliative sedation, is morally accepted by the American Medical Association (AMA), American College of Physicians (ACP), Veterans Administration (VA), and many other professional associations. It is agreed that terminal sedation is not a form of "slow" euthanasia.

Terminal sedation is the use of medication to decrease a patient's consciousness to relieve refractory symptoms, making the final stages of dying more tolerable. The goal is to increase the quality and quantity of life by decreasing suffering and not to hasten death. Ninety-six percent of terminal sedation cases are for patients who have less than 7 days to live, and 40% of that group have less than 24 hours to live.[1]

In comparison to assisted suicide (Chapter 9), the doctrine of double effect (Chapter 18), and euthanasia (Chapter 21), terminal sedation is a relatively new concept, first appearing in healthcare literature in 1991.[2] The word "terminal" refers to end-of-life care, not that the sedation is terminal in its effect. However, because of such misunderstandings, there is an evolving consensus that the term "palliative sedation therapy" or simply "palliative sedation" should be the new preferred terminology.

ASSESS

Patient: Autonomy

Patient autonomy requires that the patient or surrogate provide informed consent authorizing the practitioner to provide palliative sedation. This means that the practitioner has instructed the patient or the patient's surrogate about the diagnosis, prognosis, treatment options, risk of harm, and benefits of sedation; answered their questions; and is assured that the patient or surrogate is freely making their decision without coercion or manipulation.

If a patient decides for intermittent sedation to unconsciousness, then a designated duration of sedation will be established, after which the patient, or surrogate, will be able to reassess autonomously what to do next.

If a patient decides for continuous sedation to unconsciousness (CSU), then the designated duration will be until death, meaning that the patient will have made an autonomous decision never to be autonomous again. The decision for CSU is permanent and irrevocable.

However, in reality, by the time the patient has reached the stage in which the symptoms are medically refractory and intolerable by any other means other than by sedation, most patients will already be in a state of diminished decisional capacity and will be at most days if not just hours away from death. Under these conditions, the surrogate will need to make a healthcare decision in accordance with the patient's reasonable goals, values, and priorities. If death is imminent, then the possibility of shortening the patient's quantity of life will at most be minimal,

---

1. Claessens P, Menten J, Schotsmans P, Broeckaert B. Palliative sedation: A review of the research literature. *J Pain Symptom Manage*. 2008;36(3):310–333. doi:10.1016/j.jpainsymman.2007.10.004.
2. Enk RE. Drug-induced terminal sedation for symptom control. *Am J Hosp Palliat Care*. 1991;8(5):3–5.

making it much easier for the surrogate to focus on the patient's quality of life rather than quantity of life.

The patient or surrogate must understand that when deciding for CSU, the patient will no longer be able to drink or eat. Without medically administered nutrition and hydration (MANH), the patient will die within 2 weeks. Ninety-six percent of patients who have CSU die within 7 days. Therefore, most patients and surrogates refuse MANH. If the patient dies in less than 2 weeks, then the disease is the proximate cause of death. If the patient dies after 2 weeks, then the absence of MANH is the proximate cause of death.

Autonomy, or self-rule, is essential for defining the human condition. The consequence of CSU is the ending of the ability to ever again experience one's goals, values, and priorities; interact with family, friends, and others; and, most important, have lucidity of mind. Once CSU has commenced, for the patient there is no mental or sociological difference between CSU and death. Patients therefore need to be clear to their practitioner and surrogate(s) the extent to which they are willing to sacrifice their ability to interact with family and friends and sacrifice lucidity of mind for the relief of pain and suffering. If the patient has already lost their decisional capacity, lucidity of mind, and the ability to interact with family and friends, then such considerations become irrelevant.

Practitioner: Beneficence and Nonmaleficence

The practitioner's professional responsibility and the surrogate's role for the patient are to maximize the patient's best interests in accordance with the patient's reasonable goals, values, and priorities, using the principles of beneficence (do good) and nonmaleficence (do no harm). The best way to determine the patient's best interests is by the patient providing informed consent, a living will, or a surrogate through the use of a durable power of attorney. Ideally, a living will had been created while the patient was in good health and able to express their wishes. If the patient's reasonable goals, values, and priorities are not known or available, then the practitioner and surrogate must promote the patient's best interests in accordance with standards of care.

If practitioners have a moral obligation to reduce suffering, then there is a moral obligation to provide proportionate sedation if other treatment methods are failing. With CSU, because 96% of CSU patients die within 7 days, and 40% of that group die within 24 hours, it logically follows that CSU is not a type of slow euthanasia. Every healthcare profession considers euthanasia incompatible with the healthcare profession's mission, purpose, and vision. CSU for imminently terminal patients is within the standards of care for the AMA, ACP, VA, and a host of other healthcare associations.

Public Policy: Justice

Terminal or palliative sedation is legal in every country in the world and every state in the United States. Any moral opposition to palliative sedation is personal, not religious, institutional, or governmental. Those who oppose palliative sedation usually do so because of the mistaken "fuzzy," "gray," and "conflated" boundary between terminal sedation and slow euthanasia or aid-in-dying.

CONCLUDE

Terminal sedation is the use of medication to decrease a patient's consciousness to relieve refractory symptoms, making the final stages of dying more tolerable. Terminal sedation is not a form of slow euthanasia because most patients will be at most days if not just hours away from death.

In summary, terminal sedation is a valuable healthcare practice that can make the final stages of dying more tolerable for patients experiencing refractory symptoms. This approach is not considered a form of slow euthanasia because it aims to improve the quality of life without hastening death. With informed consent and a focus on the principles of beneficence and nonmaleficence, terminal sedation can be an ethical and compassionate option for patients nearing the end of life.

* * *

REVIEW QUESTIONS

1. Terminal sedation, also known as palliative sedation, is morally accepted by the AMA, ACP, VA, and a host of other professional associations.
   A. True
   B. False

2. Terminal sedation is a form of "slow" euthanasia.
   A. True
   B. False

3. Terminal sedation is the use of medication to decrease a patient's consciousness to relieve refractory symptoms, making the final stages of life more tolerable.
   A. True
   B. False

4. The word "terminal" in "terminal sedation" refers to end-of-life care, not that the sedation is terminal in its effect. Because of such misunderstandings, there is an evolving consensus for the term "palliative sedation therapy" or simply "palliative sedation" as the new preferred terminology.
   A. True
   B. False

5. If a patient decides on CSU, then the designated duration will be until death, meaning that the patient will have made an autonomous decision never to be autonomous again.
   A. True
   B. False

6. Ninety-six percent of patients who have CSU die within 7 days. Therefore, most patients and surrogates request to have MANH.
   A. True
   B. False

7. Because 96% of CSU patients die within 7 days, and 40% of that group die within 24 hours, it logically follows that CSU is a type of slow euthanasia.
   A. True
   B. False

8. Moral opposition to palliative sedation is generally religious, institutional, or governmental.
   A. True
   B. False

Answers: 1A, 2B, 3A, 4A, 5A, 6B, 7B, 8B

* * *

CLINICAL VIGNETTES

1. Mr. Robert Edwards is a 75-year-old retired engineer who was diagnosed with pancreatic cancer that has metastasized to his liver. Despite receiving chemotherapy, Mr. Edwards has developed severe abdominal pain and nausea that has become intolerable. He is unable to eat or drink and is losing weight rapidly. His healthcare team has tried various medications

to alleviate his symptoms, but they have been unsuccessful. Mr. Edwards and the healthcare team are considering terminal sedation to relieve his refractory pain and suffering. What is the ethical question in this case?
- A. Should Mr. Edwards be given MANH during terminal sedation?
- B. Should Mr. Edwards's family members be involved in the decision-making process for terminal sedation?
- C. Should Mr. Edwards be given intermittent CSU?
- D. Should Mr. Edwards be given terminal sedation to unconsciousness for relieving his pain and suffering?

Explanation: The ethical question in this case is whether Mr. Edwards should be given terminal sedation to unconsciousness to relieve his pain and suffering. Terminal sedation is a healthcare practice to relieve refractory symptoms in the final stages of life. It is considered morally acceptable by many professional associations and aims to reduce suffering and improve the quality of life without hastening death. In this case, Mr. Edwards is experiencing severe pain and nausea that is not responding to other treatments, and terminal sedation is considered an option to alleviate his suffering. The other options are incorrect because most patients who undergo CSU will refuse MANH and will die within 2 weeks without it, because the decision to undergo terminal sedation is a personal one that the patient or surrogate should make with informed consent from the practitioner, and because the decision for CSU is permanent and irrevocable. Mr. Edwards' healthcare team should consider using terminal sedation to relieve his pain and suffering, aligning with the principles of beneficence and nonmaleficence. The decision to undergo terminal sedation should be made with informed consent from Mr. Edwards or his surrogate, and the healthcare team should ensure that it is acting in his best interests in accordance with his reasonable goals, values, and priorities. *Answer*: D

2. Mr. Roger Torres is a 75-year-old retired construction worker who has been battling pancreatic cancer for the past year. He is currently under hospice care and experiencing excruciating pain, nausea, and difficulty breathing despite aggressive symptom management. His healthcare team recommends CSU to provide him with relief. Mr. Torres' daughter, who is his healthcare surrogate, is struggling to make the decision because she fears that the CSU will hasten his death. What ethical question arises in this scenario?
- A. Is it ethical to withhold CSU if it is the only way to relieve refractory symptoms in a terminally ill patient?
- B. Is it ethical for a surrogate to make the decision for CSU without the patient's input?
- C. Is it ethical to offer CSU if it may hasten the patient's death?
- D. Is it ethical for healthcare providers to offer CSU when there is a risk of addiction?

Explanation: The ethical question in this scenario is whether it is ethical to offer CSU if it may hasten the patient's death. CSU is a healthcare practice aimed at relieving refractory symptoms in terminally ill patients and improving their quality of life by reducing suffering. However, there is a concern that it may hasten death, and this raises ethical issues around the balance between providing comfort to the patient and ensuring that the treatment does not cause harm. The other options are incorrect because CSU is considered necessary in some cases to relieve refractory symptoms in terminally ill patients; because in cases in which the patient has lost decision-making capacity, the surrogate is responsible for making decisions that align with the patient's reasonable goals, values, and priorities; and because there is no evidence that patients receiving palliative sedation are at risk of addiction. *Answer*: C

3. Ms. Isabella Hunter is a 68-year-old woman with advanced pancreatic cancer that has metastasized to her liver. She has been receiving hospice care for several weeks and her symptoms, including pain, nausea, and shortness of breath, have been managed with medications.

However, in the past few days, her pain has become refractory, even with the highest doses of opioids. Ms. Hunter is feeling very distressed and requests to be sedated until she passes away. What is the most appropriate action for the healthcare team to take?

- A. Respect Ms. Hunter's autonomy and proceed with CSU as per her request
- B. Suggest alternative pain management options, such as a nerve block or non-opioid analgesics, before considering sedation
- C. Consult with a palliative care specialist before proceeding with sedation
- D. Administer sedation intermittently rather than continuously to maintain some level of consciousness

Explanation: In this scenario, Ms. Hunter is experiencing refractory pain, defined as pain that is not adequately controlled despite optimal analgesics and other interventions. Palliative sedation, or CSU, aims to relieve her refractory symptoms and improve her quality of life by reducing her suffering rather than hastening her death. Ms. Hunter has the right to make treatment decisions and has requested sedation, so respecting her autonomy is crucial. Informed consent must be obtained, meaning the healthcare team must provide information about the risks and benefits of the procedure, as well as alternative options. Once Ms. Hunter has provided informed consent, the healthcare team can proceed with CSU to relieve her refractory symptoms. The other options are not correct because since Ms. Hunter's pain is refractory, alternative options have already been attempted and proven to be ineffective; because consulting with a palliative care specialist, which may be helpful, is not necessary before proceeding with sedation; and because administering sedation intermittently may not adequately relieve Ms. Hunter's refractory symptoms and may not respect her autonomy in choosing to be sedated continuously. Overall, respecting Ms. Hunter's autonomy and providing informed consent, with what is considered standards of care, before proceeding with CSU is the most appropriate action for the healthcare team in this scenario. *Answer*: A

4. Mr. Robert Martin is a 72-year-old retired electrician who has been recently diagnosed with stage IV pancreatic cancer. He has been experiencing severe abdominal pain, nausea, and vomiting, despite multiple rounds of chemotherapy. He is now bedridden and has lost significant weight. His healthcare team has recommended terminal sedation to relieve his symptoms, but Mr. Martin is hesitant to undergo the procedure because he fears losing control over his decision-making. What ethical principle is involved in this scenario?

    A. Autonomy
    B. Beneficence
    C. Nonmaleficence
    D. Justice

    Explanation: In this scenario, the ethical principle involved is autonomy. Patient autonomy requires the patient or surrogate to provide informed consent authorizing the practitioner to provide palliative sedation. However, Mr. Martin hesitates to undergo the procedure because he fears losing control over his decision-making. If Mr. Martin were to make an autonomous decision for CSU, it would be an autonomous decision never to be autonomous again. It is essential for the practitioner to ensure that Mr. Martin is making an informed decision without coercion or manipulation. The decision to undergo CSU is permanent and irrevocable. The patient or surrogate must understand that when deciding for CSU, the patient will no longer be able to drink or eat, resulting in death within 2 weeks without MANH. Therefore, patients must be clear to their practitioner and surrogate(s) the extent to which they are willing to sacrifice their ability to interact with family and friends and sacrifice lucidity of mind to relieve pain and suffering. *Answer*: A

5. Mrs. Ryan Cooper is an 80-year-old woman with advanced cancer who is being treated in the hospital for severe pain and other symptoms. She is no longer able to communicate

effectively due to her illness, and her family members have been involved in her care decisions. The healthcare team has informed the family that Mrs. Cooper's pain and suffering can be effectively managed with CSU. The family is hesitant to agree to CSU because they believe that it is a form of euthanasia and goes against their religious beliefs. They also worry that Mrs. Cooper may not be able to communicate her wishes if she were to regain consciousness while under sedation. What is the best way to proceed?

A. The family should refuse CSU and explore other treatment options.
B. The healthcare team should respect the family's wishes and not recommend CSU.
C. The healthcare team should explain the ethical and healthcare justifications for CSU and work with the family to reach a shared decision.
D. The healthcare team should override the family's objections and proceed with CSU.

Explanation: From the perspective of the patient's family, there may be religious or cultural beliefs that conflict with the use of CSU; however, the healthcare team has a responsibility to provide information about the benefits and risks of available treatment options, including CSU, to enable the family to make an informed decision. The healthcare team should explain that CSU is not a form of euthanasia and that the goal is to manage the patient's symptoms and improve quality of life. The team should also reassure the family that if Mrs. Cooper regained consciousness, she would be free to reassess her treatment options. Ultimately, the healthcare team and the family should work together to reach a shared decision that is in Mrs. Cooper's best interests and respects her autonomy. *Answer*: C

REFLECTION VIGNETTES

1. Practitioner Olivia Ross, a 65-year-old internist, is seeing a 72-year-old patient named Mr. Dennis Perez, who has end-stage cancer. During his visit, Mr. Perez reports that his pain is becoming intolerable, despite previous attempts at pain management. Practitioner Ross informs Mr. Perez that his prognosis is poor and that he may have only 1 or 2 weeks to live. She also mentions the possibility of CSU to manage his refractory symptoms. Mr. Perez expresses interest in the possibility of CSU, but he is concerned about the effects it may have on his ability to interact with his family during his final days.

   What is the ethical issue(s)?

   How should the ethical issue(s) be addressed?

2. Practitioner Olivia Taylor, a 45-year-old primary care practitioner, is infuriated with her colleague, Practitioner James Thompson, for putting a patient in CSU. The patient is an 80-year-old man with end-stage cancer suffering from refractory symptoms despite maximal treatment. Practitioner Taylor argues that CSU is no different than active euthanasia, which is illegal and unprofessional. She believes that the patient should have been provided with palliative care instead.

   What is the ethical issue(s)?

   How should the ethical issue(s) be addressed?

* * *

WEB LINKS

1. *Code of Medical Ethics*. American Medical Association. https://code-medical-ethics.ama-assn.org
   The AMA *Code of Medical Ethics*, established by the American Medical Association, is a comprehensive guide for healthcare practitioners. It addresses issues and challenges,

promotes adherence to standards of care, and is continuously updated to reflect contemporary practices and challenges.
2. "Is the Clock Ticking for Terminally Ill Patients in Israel? Preliminary Comment on a Proposal for a Bill of Rights for the Terminally Ill." Barilan, Y. Michael. 2004. https://doi.org/10.1136/jme.2002.000885
This paper discusses a proposal to legislate on the rights of the dying patient in Israel, including the use of terminal sedation.
3. "Handling End-of-Life Care in Medical Decision Making: On a Bioethical View." Maribel Bont-Paredes, Carmen Malpica-Gracian, and Carlos Rojas-Malpica. 2011. https://doi.org/10.5772/18866
This book chapter explores the ethical considerations of end-of-life care, including terminal and palliative sedation.
4. "Ethical Considerations at the End-of-Life Care." Melahat Akdeniz, Bülent Yardimci, and Ethem Kavukcu. 2021. https://www.ncbi.nlm.nih.gov/pmc/articles/PMC7958189
This article discusses the ethical difficulties that healthcare professionals face when making decisions about terminal sedation.
5. "The Role of End-of-Life Palliative Sedation: Medical and Ethical Aspects—Review." Miriam S. Menezesa and Maria das Graças Mota da Cruz de Assis Figueiredo. 2019. https://www.ncbi.nlm.nih.gov/pmc/articles/PMC9391748
This article defines palliative sedation therapy and discusses its ethical implications.
6. "AMA *Code of Medical Ethics*' Opinions on Sedation at the End of Life." AMA Council on Ethical and Judicial Affairs. https://journalofethics.ama-assn.org/article/ama-code-medical-ethics-opinions-sedation-end-life/2013-05
This article provides ethical guidelines for the use of palliative sedation.
7. "Terminal Suffering and the Ethics of Palliative Sedation." Ben A. Rich. 2012. https://www.cambridge.org/core/journals/cambridge-quarterly-of-healthcare-ethics/article/terminal-suffering-and-the-ethics-of-palliative-sedation/A50E3D79E50696F3C1DC3A781F88AC17
This article discusses the ethical implications of using total sedation to relieve suffering in advanced terminal illness.
8. "Palliative Sedation: Ethical Aspects." Guido Miccinesi, Augusto Caraceni, and Marco Maltoni. 2017. https://pubmed.ncbi.nlm.nih.gov/28707846
This article discusses the ethical aspects of palliative sedation.
9. "Ethics and Clinical Aspects of Palliative Sedation in the Terminally Ill Child." Gail A. Van Norman. 2014. https://link.springer.com/chapter/10.1007/978-1-4939-1390-9_37
This book chapter discusses the ethical and clinical aspects of palliative sedation in terminally ill patients.
10. "The Hard Case of Palliative Sedation." Eran Klein. 2007. https://journalofethics.ama-assn.org/article/hard-case-palliative-sedation/2007-05
This article discusses a case study of palliative sedation.
11. "Ethical Challenges in Palliative Sedation of Adults: Protocol for a Systematic Review of Current Clinical Practice Guidelines." Martyna Tomczyk, Cécile Jaques, and Ralf J. Jox. 2022. https://bmjopen.bmj.com/content/12/7/e059189
This article discusses the ethical challenges in palliative sedation of adults.
12. "Deep and Continuous Palliative Sedation (Terminal Sedation): Clinical–Ethical and Philosophical Aspects." Lars Johan Materstvedt and Georg Bosshard. 2009. https://www.thelancet.com/journals/lanonc/article/PIIS1470-2045(09)70032-4/fulltext
This article discusses the clinical–ethical issues of terminal sedation.
13. "Expanded Terminal Sedation in End-of-Life Care." Laura Gilbertson, Julian Savulescu, Justin Oakley, and Dominic Wilkerson. 2023. https://jme.bmj.com/content/49/4/252
This article discusses the use of terminal sedation in end-of-life care.

TESTIMONIALS AND QUACKERY

61

Any practitioner who advertises a positive cure for any disease, who issues nostrum testimonials, who sells their services to a secret remedy, or who diagnoses and treats by mail patients that they have never seen is a quack.
—Samuel Hopkins Adams

* * *

MYSTERY STORY

TESTIMONIAL DECEPTION: THE SEDUCTION OF QUACKERY AND THE PERILOUS ROAD OF FALSE PROMISES

In a quaint town, where reputation often overshadowed reason, lived Practitioner Kimberly Quacksen. A charismatic figure in her community, Practitioner Quacksen had a devoted following of patients who admired her unorthodox but seemingly effective treatments. Her practice was thriving, due in part to a collection of glowing testimonials prominently featured on her website.

Practitioner Quacksen's life took a fateful turn when she encountered Mr. James Gullibley, a man tormented by chronic pain and disillusioned by conventional medicine. He had found solace in the words of Practitioner Quacksen's satisfied patients and was drawn to her promise of relief. After a series of treatments and a prescription for a mysterious medication, Mr. Gullibley found himself pain-free for the first time in years. Ecstatic, he added his praise to Practitioner Quacksen's collection of testimonials, proclaiming her a "miracle worker."

However, Practitioner Quacksen's methods had long been the subject of whispers among her more skeptical colleagues. They viewed her reliance on testimonials and her tendency to prescribe unproven medications as a dangerous drift toward quackery. Yet their concerns were dismissed, overshadowed by Practitioner Quacksen's charm and the vocal satisfaction of her patients.

The equilibrium of Practitioner Quacksen's world shattered when news broke of Mr. Gullibley's sudden death. The police discovered that the medication Practitioner Quacksen prescribed—a medication not recognized by reputable pharmaceutical authorities—was the culprit.

An investigation ensued, unraveling the dark truth behind Practitioner Quacksen's practice. Her "miracle" treatments were found to be nothing more than elaborate placebos bolstered by manipulated testimonials and exaggerated claims. Practitioner Quacksen's compassionate facade was shattered, revealing a manipulative quack exploiting the desperation of those seeking relief.

The repercussions were swift and severe. Practitioner Quacksen's license was revoked, her reputation was ruined, and she faced criminal charges for her negligence. The community was left to grapple with the betrayal of trust and the loss of a once-beloved figure.

Her colleagues, though vindicated, took no joy in Practitioner Quacksen's downfall. Instead, they used her cautionary tale to emphasize the importance of evidence-based practice, ethical integrity, and patient-centered care. They spoke out against the dangerous allure of quackery and the seductive power of false promises.

Practitioner Kimberly Quacksen's story became a grim reminder of the fragile boundary between ethical practice and deception. It underscored the vital importance of maintaining professional standards, even in the face of temptation, and reinforced the timeless principle that the patient's well-being must always be the guiding star. The alluring road of false promises had led to a dead end, marked by loss, disillusionment, and a shattered trust that would haunt the community for years to come.

* * *

THINK

Testimonials can be an effective marketing tool for educating the public about available services, describing the patient's level of satisfaction, and expressing the quality of the patient–practitioner relationship. This is especially true as social media expands in its influence and ability to help individuals connect with the right service provider for health concerns.

Testimonials are what marketers call social proof. Academics, employers, and the healthcare professions use various testimonials, such as letters of recommendation and attestation of a colleague's expertise, skills, and professionalism.

However, testimonials by patients about a practitioner's unique skills, treatments, and success can also be deceptive and unprofessional, especially if those unique skills and treatments are contraindicative of standards of care and the treatment successes have not been evidentially established and peered reviewed.

Quackery makes exaggerated claims about a practitioner's ability to heal diseases, generally for financial gain. The quack will often claim a formula, method, device, or product unknown to other practitioners or scientists. People are prone to the seduction of quacks, especially in times of stress, pain, and sorrow when any source of hope will be pursued. Educating the public in the basic sciences has had no impact on dispelling public gullibility.

Testimonials from patients have the risk of creating a conflict of interest with the patient–practitioner relationship that is supposed to be patient-centered, not practitioner-centered.

ASSESS

Patient: Autonomy

Autonomy or self-rule is typically understood to exist when a patient with decisional capacity decides after being informed of their diagnosis, prognosis, risks, and benefits of each legitimate treatment option; having all questions addressed; and not being under compulsion or manipulation. The patient–practitioner relationship is essential to trust that the practitioner will keep all the patient's information confidential and private. Without that assurance, the patient might not provide the protected health information (PHI) necessary for the practitioner to assess the information necessary for informed consent, and it is a violation of the Health Insurance Portability and Accountability Act (HIPAA), a federal statute with punishable fines of up to $250,000 and a jail term of up to 5 years.

Asking a patient to provide a testimonial on social media or any other venue may not be a HIPAA violation, but it is asking a patient, who may be vulnerable and dependent on the practitioner, to divulge their PHI. Such a request from a vulnerable patient is, at the very minimum, a type of coercion invalidating the patient's freely given informed consent.

Practitioner: Beneficence and Nonmaleficence

The practitioner's professional obligation is to provide patient-centered healthcare that maximizes the patient's best interests in accordance with the patient's reasonable goals, values, and priorities. To ask a patient to provide testimony as to the practitioner's effectiveness in providing patient-centered healthcare for the practitioner's financial gain is self-contradictory because it is practitioner-centered rather than patient-centered. It is also never professionally permissible for a practitioner to ask "favors" from patients who are currently or who may in the future be dependent on the practitioner for their healthcare needs. This dependent state makes the patient vulnerable and compelled to involuntarily comply with the practitioner's requests, violating the patient–practitioner relationship.

Public Policy: Justice

The Pure Food and Drug Act of 1906 required the labeling of certain ingredients and eliminated numerous quack cures for a variety of serious diseases. However, many of these quack drugs were then repackaged and sold as cures for the common cold and other generalizable ailments. The Federal Food, Drug, and Cosmetic Act of 1938 and the Wheeler–Lea Act established safety and purity standards. However, only the Federal Food, Drug, and Cosmetic Act was amended in 1963 to include proof of the effectiveness of all drugs. It granted the U.S. Food and Drug Administration control over medication advertising.

These laws have curbed many quack remedies, but the public demand for needed remedies, the availability of a gullible population, and the enormous economic profit associated with the business of quackery have only increased its market demand.

CONCLUDE

Practitioners must stay focused on patient-centered care and never compromise the patient–practitioner relationship for pecuniary gains. Making exaggerated claims about one's ability and claiming to have a healthcare cure that is not evidence-based medicine are unprofessional, illegal, and unethical. Asking a patient, who is vulnerable and dependent on the practitioner, to provide a testimonial violates the patient–practitioner relationship.

In summary, practitioners must be vigilant in maintaining their focus on patient-centered care and ethical practices. Requesting testimonials from vulnerable patients can be coercive and unethical, and engaging in quackery for financial gain is both unprofessional and illegal. By upholding the principles of autonomy, beneficence, and nonmaleficence, practitioners can provide the highest quality care for their patients while preserving the integrity of the patient–practitioner relationship.

* * *

REVIEW QUESTIONS

1. Quackery makes exaggerated claims about a practitioner's ability to heal disease, generally for financial gain. The quack will often claim a formula, method, device, or product unknown to other practitioners or scientists.
   A. True
   B. False

2. People are prone to the seduction of quacks, especially in times of stress, pain, and sorrow when any source of hope will be pursued.
   A. True
   B. False

3. Educating the public in the basic sciences has had a tremendous impact on dispelling public gullibility.
   A. True
   B. False

4. Asking a patient to provide a testimonial on social media or any other venue is asking the patient, who may be vulnerable and dependent on the practitioner, to divulge their private information.
   A. True
   B. False

5. It is never professionally permissible for a practitioner to ask "favors" from patients who are currently or who may in the future be dependent on the practitioner for their healthcare needs.
   A. True
   B. False

6. Public demand for needed remedies, the availability of a gullible population, and the enormous economic profit associated with the business of quackery have increased its market demand.
   A. True
   B. False

Answers: 1A, 2A, 3B, 4A, 5A, 6A

* * *

CLINICAL VIGNETTES

1. Ms. Emily Hall is a 50-year-old accountant who presented to her primary care practitioner with complaints of fatigue, weight gain, and hair loss. She reports feeling depressed and anxious and has noticed that her skin is dry and itchy. Her practitioner orders blood tests, which reveal an elevated level of thyroid-stimulating hormone and a decreased level of free thyroxine. The differential diagnosis for Ms. Hall includes hypothyroidism, depression, and anxiety. Which of the following is professionally permissible?
   A. Ms. Hall's practitioner asks her to provide a testimonial on social media about the effectiveness of her treatment.
   B. Ms. Hall's practitioner offers her a discount on her healthcare bill in exchange for a positive review on a healthcare review site.
   C. Ms. Hall's practitioner provides her with information about hypothyroidism, including treatment options and potential risks and benefits, and encourages her to make an informed decision about her care.
   D. Ms. Hall's practitioner recommends an unproven thyroid treatment and offers to provide the treatment at a discounted rate if Ms. Hall provides a testimonial.

   Explanation: Providing Ms. Hall with information about hypothyroidism, including treatment options and potential risks and benefits, and encouraging her to make an informed decision about her care is the most ethical option because it respects Ms. Hall's autonomy and supports patient-centered care. The other options are incorrect because asking a patient to provide a testimonial on social media about their healthcare treatment is unethical and violates the patient–practitioner relationship, offering a discount on healthcare bills in exchange for a positive review on a healthcare review website is similarly unprofessional and puts the practitioner's financial gain ahead of patient-centered care, and recommending an unproven treatment and offering a discount in exchange for a testimonial is illegal and dangerous because it puts the patient's health at risk and is considered quackery. *Answer*: C

2. Mr. Robert Brown is a 35-year-old computer programmer who has been experiencing chronic low back pain for several months. He has tried over-the-counter pain relievers and stretching exercises, but nothing has provided long-term relief. He searches online for alternative treatments and finds a website advertising a "revolutionary" new treatment for back pain called "spinal manipulation therapy" that uses a special device to manipulate the spine. The website claims that this treatment is more effective than traditional healthcare treatments and has no side effects. Mr. Brown decides to make an appointment with the practitioner offering this treatment. Which of the following quack claims or practices is represented in this clinical vignette?
   A. The practitioner claims to have a unique method of treating back pain that has never been published or evaluated by other healthcare professionals but uses evidence-based treatment methods.
   B. The practitioner uses a special device to manipulate the patient's spine and claims that it is more effective than traditional healthcare treatments, implying that it has not been approved by the U.S. Food and Drug Administration (FDA).

C. The practitioner claims that the patient's back pain is caused by "subluxations" in their spine that can only be corrected through the practitioner's treatment.
D. The practitioner encourages the patient to continue using over-the-counter pain relievers in conjunction with their new treatment without evaluating potential drug interactions.

Explanation: Using a device that the FDA has not approved to manipulate the spine and claiming that it is more effective than traditional healthcare treatments is an example of quackery. Using unapproved healthcare devices is illegal and can be dangerous, and making exaggerated claims about the effectiveness of a treatment that has not been supported by scientific evidence is quackery. The key characteristic of quackery is making exaggerated claims about the effectiveness of a treatment that are unsupported by scientific evidence. If the practitioner uses evidence-based treatment methods, those methods must have undergone peer review and been published in the healthcare literature to be considered a standard of care. If the practitioner's unique method has not undergone the same level of scrutiny and evaluation, then claiming that it is effective without scientific evidence is a hallmark of quackery. The other options are incorrect because evidenced-based treatment methods must be peer-reviewed and published to a standard of care, because the concept of "subluxations" is not recognized by mainstream medicine and is not supported by scientific evidence, and because encouraging the use of over-the-counter pain relievers without evaluating potential drug interactions is unprofessional and potentially dangerous. *Answer*: B

3. Mr. Nathan Cook is a 60-year-old retiree who has been experiencing severe back pain for several months. He decides to visit a practitioner who claims to have a unique method of treating back pain. During the consultation, the practitioner tells Mr. Cook that his back pain is caused by "energy blockages" and that the practitioner's treatment can unblock the energy and provide relief. Which of the following is true about this quack practitioner?
   A. The practitioner's method has undergone scientific scrutiny and is effective.
   B. The practitioner uses an FDA-approved device to manipulate Mr. Cook's energy fields and claims that it is more effective than traditional healthcare treatments.
   C. The practitioner's diagnosis of Mr. Cook's back pain as being caused by energy blockages is an unsupported and exaggerated claim.
   D. The practitioner encourages Mr. Cook to continue using over-the-counter pain relievers in conjunction with the new treatment without evaluating potential drug interactions.

Explanation: The practitioner's diagnosis of Mr. Cook's back pain as being caused by energy blockages is not recognized by mainstream medicine and is not supported by scientific evidence. Making exaggerated claims about the effectiveness of a treatment that is unsupported by scientific evidence is the definition of quackery. The other options are incorrect because the practitioner's method has not undergone scientific scrutiny and is not supported by the information provided in the vignette; the practitioner is not using an FDA-approved device to manipulate energy fields and the treatment is not supported by scientific evidence; and although encouraging the use of over-the-counter pain relievers without evaluating potential drug interactions is unprofessional and potentially dangerous, there is no information in this scenario that indicates that this is happening. *Answer*: C

4. Ms. Stella Long is a 40-year-old woman who has been diagnosed with breast cancer. She is receiving standard healthcare treatment, but she is also interested in alternative therapies that might help with her symptoms and overall well-being. She comes across an advertisement for a practitioner who claims to have a treatment that can cure cancer. The practitioner claims that chemotherapy and radiation therapy are ineffective and even harmful. Ms. Long is tempted to try the treatment but is unsure if it is legitimate. Which of the following is true about the quack practitioner's claims?
   A. The practitioner's claims are supported by scientific evidence and are a legitimate alternative to standard healthcare treatment.

B. The practitioner's claims are unproven and unsupported by scientific evidence, but they do not pose any significant risk to Ms. Long's health.
C. The practitioner's claims are unproven and unsupported by scientific evidence, and they pose a significant risk to Ms. Long's health.
D. The practitioner's claims are unproven and unsupported by scientific evidence, but they are worth trying in conjunction with standard healthcare treatment.

Explanation: The practitioner's claims that their treatment can cure cancer and that chemotherapy and radiation therapy are ineffective and even harmful are unsupported by scientific evidence and are potentially dangerous. This is an example of quackery, which makes exaggerated claims about a practitioner's ability to heal diseases, generally for financial gain. The other options are incorrect because the practitioner's claims are not supported by scientific evidence, the practitioner's claims pose a significant risk to Ms. Long's health by discouraging her from receiving effective standard healthcare treatment for her cancer, there is no evidence to support the effectiveness of the practitioner's treatment, and it is potentially dangerous to forgo standard healthcare treatment for cancer. *Answer*: C

5. Ms. Alice Cruz is a 30-year-old woman who has been seeing a chiropractor for several months for chronic back pain. The chiropractor has helped alleviate her symptoms, and Ms. Cruz is very happy with the treatment she has received. The chiropractor asks Ms. Cruz if she would be willing to provide a testimonial for their website and social media accounts, describing the treatment she has received and how it has helped her. Ms. Cruz is initially hesitant but eventually agrees. Which of the following is true about the chiropractor's request for a testimonial from Ms. Cruz?
   A. The chiropractor's request is an acceptable way to promote their services and provide social proof to potential patients.
   B. The chiropractor's request is a violation of HIPAA because it involves the disclosure of Ms. Cruz's PHI.
   C. The chiropractor's request is a form of coercion and violates the patient-centered nature of the patient–practitioner relationship.
   D. The chiropractor's request is acceptable as long as Ms. Cruz is not financially compensated for providing the testimonial.

Explanation: Asking a patient to provide a testimonial is a form of coercion and violates the patient-centered nature of the patient–practitioner relationship. Patients may feel obligated to comply with the request due to their dependence on the practitioner for their healthcare needs. Therefore, the request is unacceptable, even if the patient agrees to provide the testimonial. The other options are incorrect because asking for a testimonial from a vulnerable patient is never acceptable, asking for a testimonial does not necessarily involve the disclosure of PHI and is not a violation of HIPAA unless the testimonial includes PHI, and financial compensation does not make the request for a testimonial acceptable if it is still a violation of the patient–practitioner relationship. *Answer*: C

REFLECTION VIGNETTES

1. Practitioner Hannah Bennett, a 35-year-old family practitioner, communicates with one of her patients on social media and asks if they would provide an exemplary review about the success of the cures and the personableness of her care.

   What is the ethical issue(s)?

   How should the ethical issue(s) be addressed?

2. Practitioner Vivian Baily, a 40-year-old naturopathic practitioner and owner of an online business, has been marketing a unique blend of vitamins as a cure for various ailments and

improved sports performance. One of her products, with undisclosed proprietary ingredients, claims to reduce respiratory excretions and improve the body's cardiovascular system's overall efficiency in absorbing and transferring oxygen. The online site claims that this "breakthrough new product, not yet recognized by the healthcare community, can be used for respiratory ailments and increase aerobic sports performance."

What is the ethical issue(s)?

How should the ethical issue(s) be addressed?

* * *

WEB LINKS

1. *Code of Medical Ethics*. American Medical Association. https://code-medical-ethics.ama-assn.org
   The AMA *Code of Medical Ethics*, established by the American Medical Association, is a comprehensive guide for healthcare practitioners. It addresses issues and challenges, promotes adherence to standards of care, and is continuously updated to reflect contemporary practices and challenges.
2. "Fraud and Deceit in Medical Research." Frideriki Poutoglidou, Marios Stavrakas, Nikolaos Tsetsos, et al. 2022. https://journals.library.columbia.edu/index.php/bioethics/article/view/8940
   This article discusses the prevalence of scientific fraud, particularly data falsification and fabrication, in healthcare research.
3. "Medical Quackery: The Pseudo-Science of Health and Well-Being." Audiey Kao. 2000. https://journalofethics.ama-assn.org/article/medical-quackery-pseudo-science-health-and-well-being/2000-04
   This article explores how quack physicians exploit public ignorance and desperation through false patient testimonials.
4. "That Glowing Patient Testimonial May Not Be What It Seems." Lindsay Gellman. 2021. https://www.wired.com/story/ban-patient-testimonials-ethics
   This article argues that patient testimonials can exacerbate the imbalance in the provider–patient relationship and potentially mislead others.
5. "The Case of Practitioner Oz: Ethics, Evidence, and Does Professional Self-Regulation Work?" Jon C. Tilburt, Megan Allyse, and Frederic W. Hafferty. 2017. https://journalofethics.ama-assn.org/article/case-dr-oz-ethics-evidence-and-does-professional-self-regulation-work/2017-02
   This article discusses the ethical implications of promoting unproven treatments, as exemplified by practitioner Oz.
6. "The Ethics of Truth-Telling in Health-Care Settings." Yusrita Zolkefli. 2018. https://www.ncbi.nlm.nih.gov/pmc/articles/PMC6422557
   This article emphasizes the importance of truth-telling in healthcare, particularly with respect to patient autonomy.
7. "Principles of Clinical Ethics and Their Application to Practice." Basil Varkey. 2021. https://pubmed.ncbi.nlm.nih.gov/32498071
   This review provides an overview of the main ethical principles in clinical practice, including autonomy (informed consent), beneficence (do good), nonmaleficence (do no harm), and justice (be fair).
8. "The Top 10 Most-Read Bioethics Articles in 2021." Kevin B. O'Reilly. 2021. https://www.ama-assn.org/delivering-care/ethics/top-10-most-read-medical-ethics-articles-2021
   This article lists the most popular bioethics articles of 2021, covering a range of topics.
9. "Bioethical Issues in Healthcare Management." CSP Global. https://online.csp.edu/resources/article/bioethical-issues-health-care-management
   This article discusses the bioethical issues arising from advances in healthcare technology.

10. "What Is Bioethics?" Johns Hopkins Berman Institute of Bioethics. https://bioethics.jhu.edu/about/what-is-bioethics
This article provides a broad overview of the field of bioethics.
11. "Ethical Issues in Biomedical Research: Perceptions and Practices of Postdoctoral Research Fellows Responding to a Survey." Susan Eastwood, Pamela Derish, Evangeline Leash, and Stephen Ordway. 1996. https://pubmed.ncbi.nlm.nih.gov/11657788
This survey explores the perceptions and practices of postdoctoral research fellows regarding ethical issues in biomedical research.

TORTURE

62

[As a practitioner] I will follow that system of regimen which, according to my ability and judgment, I consider for the benefit of my patients and abstain from whatever is deleterious and mischievous.
—Hippocrates

* * *

MYSTERY STORY

TORTURE'S BETRAYAL: THE PROHIBITION OF TORTURE AND HEALTHCARE PROVIDERS' RESPONSIBILITIES

The hospital room was a stark testament to humanity's darkest impulses. Practitioner Olive Ford's eyes widened as they fell upon the patient in room 305, a man covered in unmistakable signs of torture. The cuts, bruises, and burns told a horrifying story, one that pulled her into a profound moral crisis.

As the head of the hospital's ethics committee, Practitioner Ford knew that she was standing at a crossroad that had been strictly defined by the healthcare profession. The American College of Physicians' (ACP) *Ethics Manual* and the American Medical Association's (AMA) *Code of Medical Ethics* had clear and unambiguous guidelines: no association with torture or enhanced interrogation techniques. No participation. No presence. The principles of autonomy, beneficence, nonmaleficence, and justice dictated a clear path.

But as she reviewed the hospital's surveillance footage, a path of betrayal unfolded before her. The police, after arresting the patient on suspicion of terrorism, had engaged in enhanced interrogation techniques, with the hospital staff playing a grim supporting role. They had provided healthcare only to ensure that the tortured could endure more suffering.

The shock that reverberated through Practitioner Ford's being was not just moral; it was professional and legal as well. Torture was incompatible with patient autonomy and informed consent. This man's best interests had been violated, and the hospital staff had crossed a boundary they were never meant to approach.

A dark cloud of realization settled over Practitioner Ford as she understood the gravity of the situation. The prohibition against torture and enhanced interrogation was not only a moral boundary but also a legal obligation. The failure to oppose, expose, and report such practices was legally liable, falling into the same category as the failure to report child or elder abuse.

An emergency meeting of the ethics committee was convened, and the implications were clear. The healthcare profession had several boundaries, and this hospital had breached the most sacred of them. A commitment to patient-centered healthcare, informed consent, and the principles of biomedical ethics had been broken.

With a heavy heart but unwavering resolve, Practitioner Ford guided the ethics committee in determining the course of action. The guilty staff members would be reported to the proper authorities, and the hospital would implement a new policy. There would be training on the prohibition of torture, stricter protocols for reporting, and mandatory reporting to both the ethics committee and outside authorities.

Practitioner Ford knew that this incident had to serve as a turning point, a reaffirmation of the profession's core values and boundaries. Torture and enhanced interrogation were beyond the boundary, and this was a line that must never be crossed again.

* * *

THINK

Practitioners must never participate to any degree in torture or enhanced interrogations, and the practitioner has a professional, legal, and moral duty to oppose, expose, and report torture or enhanced interrogations.

The ACP *Ethics Manual* states, "Physicians must not be a party to and must speak out against torture or other abuses of human rights." The AMA *Code of Medical Ethics* 9.7.5 states, "Physicians must oppose and must not participate in torture for any reason. . . . Physicians must not be present when torture is used or threatened."

Torture and enhanced interrogation are in total contradiction with the healthcare profession's directive of maximizing the patient's best interests, using the biomedical principles of autonomy (informed consent), beneficence (do good), nonmaleficence (do no harm), and justice (be fair).

The healthcare profession has several boundaries that the practitioner must never cross because of the nature of the patient–practitioner relationship, patient-centered healthcare, and the principles of biomedical ethics. Torture and enhanced interrogation are beyond this boundary.

ASSESS

Patient: Autonomy

Torture, by definition, is incompatible with autonomy and informed consent. Torture or enhanced interrogation techniques are not for the tortured person's best interests, and therefore the autonomous informed consent process cannot commence, nor have the victim and practitioner established a patient–practitioner relationship.

Practitioner: Beneficence and Nonmaleficence

Torture and enhanced interrogation practices are prohibited as a violation of the professional principle of beneficence (do good) and the professional principle of nonmaleficence (do no harm).

Practitioners may use the professional principle of beneficence as the basis for treating those who have been injured by torture. However, no treatment should be provided that would allow a person to recover only to endure further torture, as that would be a violation of the professional principle of nonmaleficence.

These prohibitions and duties to report and oppose torture are applicable to practitioners in legal war zones and under the direct orders of military superiors. "I was only following orders" is no justification for participation. As a matter of nonmaleficence, it is always professionally unethical for a practitioner to purposefully participate in a person's physical or psychological harm, regardless of who orders it or what court of law legally mandates it.

Not only is nonmaleficence a professional boundary that must not be crossed but also the practitioner is obligated to oppose, expose, and report any torture or enhanced interrogations as a matter of beneficence.

Public Policy: Justice

Torture or enhanced interrogation practices must always be opposed, exposed, and reported as a matter of justice or being fair. As mandatory reporters, practitioners have a legal obligation to oppose, expose, and report any signs of possible torture as a violation of human rights to those who have the authority to intervene, investigate, and adjudicate.

Prisoners are, by federal law (Common Rule 45CFR46), a protected vulnerable population. Failure to report torture or enhanced interrogation is in the same category as failure to report child and elder abuse, making the practitioner legally liable for any subsequent injury or harm resulting from the failure to oppose, expose, and report.

CONCLUDE

The healthcare profession will never deviate from the prohibition of any practitioner association with torture or enhanced interrogation techniques. The practitioner has a professional, legal, and moral duty to oppose, expose, and report torture and enhanced interrogations to those who have the authority to intervene, investigate, and adjudicate. Because prisoners are federally categorized as a vulnerable population, failure to report prisoner abuse is in the same category as failure to report child and elder abuse, making the practitioner legally liable for any subsequent injury or harm that befalls the victim.

In summary, the healthcare profession's stance against torture and enhanced interrogation techniques remains unwavering. Practitioners have a duty to oppose, expose, and report these practices, fulfilling their professional, legal, and moral obligations. Any association with torture or enhanced interrogation techniques is a violation of the principles of biomedical ethics and the patient–practitioner relationship. As protectors of human rights and providers of patient-centered healthcare, practitioners must always work to ensure the well-being of those in their care, including those who have suffered from torture. (See Chapters 9, 10, 15, 21, and 31.)

* * *

REVIEW QUESTIONS

1. Practitioners must never participate in any degree of torture or enhanced interrogations, and they have a professional, legal, and moral duty to oppose, expose, and report torture and enhanced interrogations.
   A. True
   B. False

2. Torture or enhanced interrogation techniques are not for the tortured person's best interests, and therefore the autonomous informed consent process cannot commence.
   A. True
   B. False

3. Torture and enhanced interrogation practices are sometimes permissible for the practitioner if, by the professional principle of beneficence (do good), others benefit and, by the principle of justice (be fair), burdens are more fairly distributed.
   A. True
   B. False

4. No treatment should be provided that would allow a person to recover only to endure further torture, as that would violate the professional principle of nonmaleficence (do no harm).
   A. True
   B. False

5. It is professionally, legally, and ethically permissible for a practitioner to purposefully participate in a person's physical or psychological harm if a court of law legally mandates it.
   A. True
   B. False

6. Prisoners are, by federal law (Common Rule 45CFR46), a protected vulnerable population. Failure to report torture or enhanced interrogation is in the same category as failure to

report child and elder abuse, making the practitioner legally liable for any subsequent injury or harm resulting from the failure to oppose, expose, and report.
   A. True
   B. False

Answers: 1A, 2A, 3B, 4A, 5B, 6A

* * *

CLINICAL VIGNETTES

1. Mr. Adam Harris is a 34-year-old veteran who served in Iraq and Afghanistan as a military intelligence analyst. He presents to the clinic with symptoms of depression, anxiety, and post-traumatic stress disorder (PTSD). During the consultation, the practitioner learns that Mr. Harris had been subjected to enhanced interrogation techniques during his time in the military. The differential diagnosis includes PTSD related to combat, PTSD related to torture, depression, and anxiety disorders. What ethical principle is violated if the practitioner fails to report possible torture or enhanced interrogation techniques used on Mr. Harris?
   A. Autonomy
   B. Beneficence
   C. Nonmaleficence
   D. Justice

   Explanation: According to the healthcare profession's ethical guidelines, practitioners have a legal obligation to oppose, expose, and report any signs of possible torture as a violation of human rights to those who have the authority to intervene, investigate, and adjudicate. As mandatory reporters, practitioners must report torture or enhanced interrogation as a matter of justice or fairness. Failure to report torture or enhanced interrogation is considered the same as failure to report child and elder abuse, making the practitioner legally liable for any subsequent injury or harm resulting from the failure to oppose, expose, and report. Therefore, the correct answer is justice because failing to report possible torture or enhanced interrogation violates the principle of the public policy. *Answer*: D

2. Practitioner Samantha Washington is a 28-year-old licensed psychologist who works at a detention center for undocumented immigrants. She has been approached by a government official to provide psychological evaluations for detainees who have been subjected to enhanced interrogation techniques. The official offers her a substantial amount of money in exchange for her cooperation. Practitioner Washington is conflicted about what to do and seeks advice from a colleague. What is the ethical obligation of Practitioner Washington as a healthcare professional regarding participation in torture or enhanced interrogation techniques?
   A. She should comply with the request to perform psychological evaluations as long as the detainees' welfare is her top priority.
   B. She should refuse to participate in any degree of torture or enhanced interrogation techniques and report the official's request to higher authorities.
   C. She should accept the offer for monetary gain and perform the evaluations only if she believes the detainees' welfare will not be negatively impacted.
   D. She should comply with the request to perform psychological evaluations as long as the official ensures her protection from legal liability.

   Explanation: As a healthcare professional, Practitioner Washington must never participate in any degree of torture or enhanced interrogation techniques, and she has a professional, legal, and moral duty to oppose, expose, and report torture or enhanced interrogations. She should refuse to participate in any degree of torture or enhanced interrogation techniques and report the official's request to higher authorities. Practitioners must not be a party to and must speak out against torture or other abuses of human rights. The principles of autonomy (informed consent), beneficence (do good), nonmaleficence (do no harm), and justice (be fair) dictate that practitioners must never cross the boundary of participating in torture or enhanced interrogation. Practitioners have a legal obligation to report possible torture as a violation of human rights, and failure to report abuse is considered the same as failure to report child or elder abuse and can result in legal liability. Practitioner Washington has a professional, legal, and moral duty to refuse to participate in any degree of torture or enhanced interrogation techniques and to report the official's request to higher authorities. *Answer*: B

3. Practitioner John Kelly is a 42-year-old practitioner working in a military hospital in a war zone. He is approached by his commanding officer, who orders him to perform healthcare procedures on detainees who have been subjected to enhanced interrogation techniques so that the enhanced interrogation can continue. The officer threatened Mr. Kelly with disciplinary action and court-martial if he refused to comply. What is Practitioner Kelly's ethical obligation as a healthcare professional regarding participation in torture or enhanced interrogation techniques?
   A. He should comply with the orders and perform the healthcare procedures as long as the detainees' welfare is his top priority.
   B. He should refuse to participate in any degree of torture or enhanced interrogation techniques and report the officer's orders to higher authorities.
   C. He should accept the orders and perform the healthcare procedures only if he believes that the information to be attained is important.
   D. He should comply with the orders and perform the healthcare procedures because they are military orders from a superior.

   Explanation: As a healthcare professional, Practitioner Kelly must never participate in any degree of torture or enhanced interrogation techniques, and he has a professional, legal, and moral duty to oppose, expose, and report torture or enhanced interrogations. He should refuse to participate in any degree of torture or enhanced interrogation techniques and report the officer's orders to higher authorities. The healthcare profession holds practitioners accountable for reporting abuse, including in legal war zones or under the orders of superiors. Torture and enhanced interrogation are in total contradiction with the healthcare profession's directive of maximizing the patient's best interests, using the biomedical principles of autonomy (informed consent), beneficence (do good), nonmaleficence (do no harm), and justice (be fair). "I was only following orders" is no justification for participation in physical or psychological harm, regardless of who orders it or what court of law legally mandates it. Practitioners have a legal obligation to report possible torture as a violation of human rights, and failure to report abuse is considered the same as failure to report child or elder abuse and can result in legal liability. Therefore, the correct answer is that Practitioner Kelly has a professional, legal, and moral duty to refuse to participate in any degree of torture or enhanced interrogation techniques and to report the officer's orders to higher authorities. *Answer*: B

4. Practitioner David Martin is a 50-year-old practitioner working in a correctional facility. One of his patients, a detainee subjected to enhanced interrogation techniques, requests that Practitioner Martin not report any injuries he sustained during the interrogation to the authorities. The detainee states that he fears retaliation from the interrogators if they discover he reported the injuries. What is Practitioner Martin's ethical obligation as a healthcare professional regarding confidentiality and the duty to report torture or enhanced interrogation?
   A. He should prioritize the patient's confidentiality and not report any injuries sustained during the interrogation, even if it puts the patient at risk of further harm.

B. He should report the injuries sustained during the interrogation to the authorities, as he has a legal and ethical obligation to report torture or enhanced interrogation.
C. He should negotiate with the patient and attempt to convince him to allow the injuries sustained during the interrogation to be reported to the authorities while ensuring the patient's safety.
D. He should seek legal advice before deciding whether to report the injuries sustained during the interrogation to the authorities, as the situation presents a conflict between the patient's confidentiality and the duty to report.

Explanation: Practitioner Martin's duty to report torture or enhanced interrogation precedes the patient's confidentiality. As a healthcare provider, he has a legal and ethical duty to report any instances of torture or abuse, even if it is against the patient's wishes. Failure to report torture or enhanced interrogation could lead to legal liability and jeopardize the safety and well-being of the patient. Therefore, Practitioner Martin must prioritize reporting the incident and ensuring the patient's safety. Although negotiating with the patient and attempting to convince him to allow the injuries sustained during the interrogation to be reported to the authorities while ensuring the patient's safety is a valid ethical consideration, ultimately, Practitioner Martin is mandated to report the incident of torture. Seeking legal advice may also be helpful, but it should not delay or override Practitioner Martin's duty to report the incident. Therefore, Practitioner Martin should report the injuries sustained during the interrogation to the authorities, as he has a legal and ethical obligation to report torture or enhanced interrogation. *Answer*: B

5. Practitioner Samantha Brown is a 28-year-old healthcare professional working in a refugee camp in a country with a history of human rights abuses. She is approached by a government official who requests that she conduct healthcare examinations on political prisoners subjected to enhanced interrogation techniques. The official tells Practitioner Brown that the examinations are necessary to ensure the health and safety of the prisoners. What is Practitioner Brown's ethical obligation as a healthcare professional regarding the duty to provide healthcare, the duty to report abuse, and the potential risks of collaborating with an oppressive government?
    A. She should comply with the government official's request and conduct healthcare examinations on the political prisoners, as it is her duty to provide healthcare to those in need.
    B. She should refuse to comply with the government official's request and report the incident to higher authorities, as she has a legal and ethical obligation to report torture or enhanced interrogation.
    C. She should negotiate with government officials and attempt to find alternative ways to ensure the health and safety of the political prisoners while avoiding collaborating with an oppressive government.
    D. She should seek guidance from her colleagues, healthcare associations, or human rights organizations before deciding whether to comply with the government official's request or report the incident.

Explanation: Practitioner Brown's duty to provide healthcare to those in need must be balanced with her duty to report abuse and her potential risks of collaborating with an oppressive government. Compliance with the government official's request may indirectly contribute to the continuation of torture or enhanced interrogation, which is against Practitioner Brown's professional obligations. Refusing to comply with the government official's request and reporting the incident to higher authorities is a valid ethical consideration, but it may also expose Practitioner Brown to risks such as retaliation from the government or expulsion from the country. In some situations, negotiating with government officials and attempting to find alternative ways to ensure the health and safety of the political prisoners while avoiding collaborating with an oppressive government may also be appropriate. Still, it may not be feasible or effective in all cases. Seeking guidance from her colleagues, healthcare associations, or human rights organizations may help Ms. Brown make a more informed decision and reduce her potential risks of collaborating with an oppressive government. Therefore, in order to fulfill her ethical obligation, Practitioner Brown should seek guidance from her colleagues, healthcare associations, or human rights organizations before deciding whether to comply with the government official's request or report the incident. In addition, if torture or enhanced interrogation occurs, she must report the abuse, as failure to do so would make her legally liable for any subsequent injury or harm, similar to the failure to report child and elder abuse. *Answer*: D

REFLECTION VIGNETTES

1. The commanding officer orders Practitioner Bella Gray, a 35-year-old military practitioner, to provide blood to a prisoner being tortured so the enhanced interrogation techniques can be continued. The prisoner is experiencing severe physical trauma and blood loss due to the torture, and the officer wants to use the practitioner's healthcare expertise to keep the prisoner alive and conscious during the interrogation.

   What is the ethical issue(s)?

   How should the ethical issue(s) be addressed?

2. Practitioner Leah Evans is a 38-year-old practitioner and former military healthcare officer who has recently transitioned to private practice. She receives a request from a former colleague still serving in the military seeking her advice on making interrogation techniques more effective and humane. The colleague is specifically seeking advice on administering medication to enhance interrogation techniques and is asking Practitioner Evans to provide her expertise in pharmacology.

   What is the ethical issue(s)?

   How should the ethical issue(s) be addressed?

* * *

WEB LINKS

1. *Code of Medical Ethics*. American Medical Association. https://code-medical-ethics.ama-assn.org
   The AMA *Code of Medical Ethics*, established by the American Medical Association, is a comprehensive guide for healthcare practitioners. It addresses issues and challenges, promotes adherence to standards of care, and is continuously updated to reflect contemporary practices and challenges.
2. "Bioethics, Health, and Inequality." Giovanni Berlinguer. 2004. https://doi.org/10.1016/s0140-6736(04)17066-9
   This article discusses the intersection of bioethics and health inequality, touching on the ethical implications of torture.
3. "'Do Not Attempt Resuscitation' Orders in the Peri-Operative Period." Michael E. McBrien and Gary Heyburn. 2006. https://doi.org/10.1111/j.1365-2044.2006.04702.x
   This article explores the ethical dilemmas surrounding "do not attempt resuscitation" orders, which can be relevant in situations of torture.
4. "From Healthcare to Warfare and Reverse: How Should We Regulate Dual-Use Neurotechnology?" Marcello Ienca, Fabrice Jotterand, and Bernice S. Elger. 2018. https://doi.org/10.1016/j.neuron.2017.12.017
   This article discusses the regulation of dual-use neurotechnology, which can be used for both beneficial and harmful purposes, including torture.

5. "Book Review: *Principles of Biomedical Ethics*, 5th edn." S. Holm. 2002. https://doi.org/10.1136/jme.28.5.332-a
   This review of a classic text in biomedical ethics provides a broad overview of the field, including the ethical considerations of torture.
6. "Unspeakably Cruel—Torture, Medical Ethics, and the Law." George J. Annas. 2005. https://www.nejm.org/doi/full/10.1056/NEJMlim044131
   This article reviews international and U.S. laws and U.S. Supreme Court decisions relevant to torture in the context of bioethics.
7. "Torture: The Bioethics Perspective." Steven H. Miles. 2015. https://www.thehastingscenter.org/briefingbook/torture-the-bioethics-perspective
   This article discusses the role of healthcare personnel in torture, a universally illegal yet widely practiced act.
8. "Medical Ethics and Torture: Revising the Declaration of Tokyo." Steven H. Miles and Alfred M. Freedman. 2009. https://www.thelancet.com/journals/lancet/article/PIIS0140-6736(09)60097-0/fulltext
   This article suggests revisions to the Declaration of Tokyo to harmonize bioethics codes with international law regarding torture.
9. "Should Bioethics Justify Violence?" M. H. Kottow. 2006. https://www.ncbi.nlm.nih.gov/pmc/articles/PMC2563389
   This article discusses the role of biomedical ethics in justifying or opposing violence, including torture.
10. "Medical Involvement in Torture Today?" Kenneth Boyd. 2016. https://jme.bmj.com/content/42/7/411
    This article discusses the moral dilemmas faced by healthcare professionals involved in torture.
11. "Torture and Human Rights." Timothy F. Murphy and Peter J. Johnson. 2004. https://journalofethics.ama-assn.org/article/torture-and-human-rights/2004-09
    This article discusses the ethical implications of torture and the responsibility of physicians to oppose it.
12. "Torture." American Medical Association. https://code-medical-ethics.ama-assn.org/ethics-opinions/torture
    This article emphasizes that physicians must oppose and not participate in torture.

TRIAGE

63

Patients pay the practitioner for their trouble; for the practitioner's kindness, patients remain in debt.
—Seneca

* * *

MYSTERY STORY

> **TRIAGE DILEMMA: BALANCING NEEDS AND LIMITED RESOURCES**
>
> The emergency room was a battlefield, as paramedics rushed in with multiple casualties from a disastrous car accident. Practitioner Grace Queue, the lead practitioner on duty, found herself in the epicenter of chaos and uncertainty. Her responsibility was not just healthcare but ethical, moral, and bound by the principles of patient-centered healthcare.
>
> With precision and care, Practitioner Queue began to triage the patients, sorting them based on the severity of their injuries. Those most in need of life-saving measures were prioritized, and those with a low probability of surviving were curtailed.
>
> As the night wore on, the complexities of triage began to weigh heavily on Practitioner Queue. One dilemma became particularly agonizing: An elderly man with a history of heart disease required urgent attention, but the hospital's only available bed was needed for a young woman severely injured in the car crash.
>
> With both the ethical framework of beneficence and nonmaleficence and the hospital's triage policy in mind, Practitioner Queue faced a decision that transcended mere healthcare judgment. The policy, shaped by federal, state, and institutional guidelines, was clear: It prioritized patients based on the likelihood of a successful outcome and protected the integrity of the patient–practitioner relationship.
>
> After a thorough consultation with colleagues and a review of the patient's healthcare history, Practitioner Queue decided to give the bed to the young woman, who had a higher chance of survival with prompt treatment. The decision was not hers alone but one guided by transparent and agreeable policies that sought to ensure justice and protect vulnerable populations.
>
> The next morning, the sobering news reached Practitioner Queue: The elderly man had passed away in the waiting area. His family was inconsolable and demanded answers.
>
> With empathy and professional integrity, Practitioner Queue explained her decision-making process. She described the principles that guided her actions, emphasizing that the triage decisions were made by public policy, not by individual practitioners, to ensure fair allocation of resources. But the family's grief turned to anger, and they filed a lawsuit against the hospital and Practitioner Queue.
>
> The case went to trial, with Practitioner Queue testifying to her adherence to the principles of autonomy, beneficence, nonmaleficence, and justice. She highlighted the importance of separating public policy from the patient–practitioner relationship to prevent conflicts of interest and uphold the patient's best interests.
>
> Despite her best efforts, the jury found Practitioner Queue and the hospital liable for the elderly man's death. The case became a somber reminder of the complex realities of triage, where decisions made in the best interest of one patient could lead to the tragic loss of another.
>
> Triage, a term derived from the French word *trier*, meaning to sort or select, proved to be more than a healthcare practice: It was a mirror to the human condition, reflecting the virtues and vulnerabilities of those who dedicate themselves to the art of healing.

* * *

THINK

The word "triage" is etymologically from the French word *trier*, which means to separate, sort, or select. Triage was implemented when the patient demand for healthcare facilities, treatment, medications, or equipment was greater than what was available. How triage has been implemented has varied widely.[1]

In war, it was not unusual to provide treatment first to those patients who would most quickly recover so that they could get back to fulfilling their responsibilities. At other times, treatment was given to patients most in need of life-saving treatment, while more minor ailments would not be treated. However, treatment for those with a low probability of surviving would be curtailed or not provided in either scenario.

Triage or not, the practitioner has a professional duty to pursue their patient's healthcare best interests before those of any other person with whom they do not have a patient–practitioner relationship. This is why triage decision-making policies need to be socially determined by federal, state, and institutional policy, not by the practitioner, as that would violate fair and impartial resource allocation or the patient–practitioner relationship.

ASSESS

Patient: Autonomy

A patient's autonomous informed consent is supposed to be coherent with the maximization of the patient's best interests as determined by the patient's reasonable goals, values, and priorities. The practitioner provides the patient with a diagnosis, prognosis, treatment options, risks, and benefits and answers questions. Then the patient chooses a treatment option and authorizes the practitioner to provide the treatment. However, sometimes the treatment will be limited because of an excess volume of patients, demand, or scarce availability.

Practitioner: Beneficence and Nonmaleficence

The practitioner always has the obligation of maximizing the patient's best interests in accordance with the patient's reasonable goals, values, and priorities. This is based on the patient–practitioner relationship, in which the patient trusts that the practitioner will always champion the patient's best interests, even when treatments are limited. This is why practitioners should not be involved in triage decision-making with regard to their own patients, as doing so would oblige the practitioner to either choose their own patient to be treated at the expense of other practitioners' patients or violate the patient–practitioner relationship by choosing another practitioner's patient to be treated before one's own patient.

For triage, there needs to be a clear separation between public policy and the patient–practitioner relationship to prevent either of those scenarios. This distinction makes it possible for the practitioner to honestly inform their patients that they will do everything humanly possible to get them the treatment they need within the parameters and resources available.

Public Policy: Justice

Triage decisions must be established by federal, state, or institutional policies and by those who can make objective and fair allocation decisions independent of but still respecting the patient–practitioner relationship. Triage policies must be transparent and agreeable to the public, promoting understanding, trust, and acceptance while protecting vulnerable populations such as the elderly, the disabled, and the socially disadvantaged.

CONCLUDE

It is imperative that society can know and trust that the practitioner will always act and make decisions that will promote each patient's best interests as determined by the patient's reasonable

---

1. Triage. Wikipedia. https://en.wikipedia.org/wiki/Triage.

goals, values, and priorities. Triage needs to be made by federal, state, or institutional policies to avoid practitioner violation of the patient–practitioner relationship. Patient-centered healthcare for the patient's best interests should always be the practitioner's prime directive in the practice of the art of healing.

In summary, triage decisions must be made by public policy to protect the patient–practitioner relationship and uphold the practitioner's responsibility to prioritize their patient's best interests. By separating public policy from the patient–practitioner relationship, practitioners can maintain their commitment to providing patient-centered healthcare within the constraints of available resources. Transparent and fair triage policies also promote understanding, trust, and acceptance while ensuring the protection of vulnerable populations. Practitioners must always strive to act in the best interests of their patients, guided by their patients' reasonable goals, values, and priorities.

* * *

REVIEW QUESTIONS

1. Triage or not, the practitioner has a professional duty to be fair and impartial when distributing healthcare resources, even if that means providing treatment for another practitioner's patient before one's own patient.
   A. True
   B. False

2. Triage decision-making policies must be socially determined by federal, state, and institutional policy, not by the patient's practitioner, as that would violate fair and impartial resource allocation or the patient–practitioner relationship.
   A. True
   B. False

3. Practitioners should not be involved in triage decision-making with regard to their own patients, as doing so would oblige the practitioner to either choose their own patient to be treated at the expense of other practitioners' patients or violate the patient–practitioner relationship by choosing another practitioner's patient to be treated before one's own patient.
   A. True
   B. False

4. Triage decision-making needs to be established by federal, state, or institutional policies and by those who can make objective and fair allocation decisions independent of, but still respecting, the patient–practitioner relationship.
   A. True
   B. False

5. Triage policies must be transparent and agreeable to the public, promoting understanding, trust, and acceptance while protecting vulnerable populations such as the elderly, the disabled, and the socially disadvantaged.
   A. True
   B. False

Answers: 1B, 2A, 3A, 4A, 5A

* * *

CLINICAL VIGNETTES

1. Ms. Michelle Cook is a 35-year-old teacher who presents to the emergency department with a severe headache and loss of vision in her right eye. The healthcare team suspects that she may have a brain tumor and orders a computed tomography (CT) scan to confirm the diagnosis. However, there are only two CT scanners in the hospital, and there are multiple critical patients waiting for the same test. Which of the following triage options is most ethical?
   A. Order a CT scan for Ms. Cook immediately, as her condition is potentially life-threatening and requires timely diagnosis and treatment
   B. Place Ms. Cook on a waiting list for the CT scan, as there are other patients waiting for the same test and her condition is stable for the time being
   C. Prioritize Ms. Cook's CT scan over other patients, as she is a young teacher with a potentially serious condition that requires immediate healthcare attention
   D. Follow institutional policies for triage and promote Ms. Cook's best interests

   Explanation: The most ethical position is to adhere to institutional policies for triage while promoting the patient's best interests. Following institutional policies for triage and allocating resources based on healthcare urgency aligns with the principles of fairness and impartiality. The other options are not correct because ordering a CT scan for Ms. Cook immediately without considering institutional policies would violate the principle of fair and impartial resource allocation, placing Ms. Cook on a waiting list could lead to her condition deteriorating and causing preventable harm, and prioritizing Ms. Cook's CT scan over institutional policies would violate the principle of fair and impartial resource allocation. In summary, following institutional policies for triage and allocating resources based on healthcare urgency promotes fair and impartial resource allocation and aligns with socially determined triage policies. *Answer*: D

2. Mr. Michael Russell is a 50-year-old construction worker who presents to the emergency department with severe lower back pain and difficulty walking. The healthcare team suspects that he may have a herniated disc and orders a magnetic resonance imaging (MRI) scan to confirm the diagnosis. However, there is only one MRI machine in the hospital, and there are multiple patients waiting for the same test. Which of the following triage options is most ethical?
   A. Order an MRI scan for Mr. Russell immediately, as his condition is potentially life-altering and requires timely diagnosis and treatment
   B. Place Mr. Russell on a waiting list for the MRI scan, as there are other patients waiting for the same test and his condition is stable for the time being
   C. Prioritize the MRI scan for patients who are younger, as they have a longer life expectancy and will benefit more from timely diagnosis and treatment
   D. Follow institutional policies for triage and allocate resources based on healthcare urgency, taking into account the severity of Mr. Russell's condition

   Explanation: The most ethical option is to follow institutional policies for triage and allocate resources based on healthcare urgency, taking into account the severity of Mr. Russell's condition. This approach aligns with the principles of fairness and impartiality, and it ensures that resources are allocated to those who need them the most based on healthcare urgency. Although Mr. Russell's condition may be potentially life-altering, the other options are not correct because ordering an MRI scan for him immediately without considering institutional policies would violate the principle of fair and impartial resource allocation; placing Mr. Russell on a waiting list may not be the best option, as his condition could worsen and cause preventable harm; and prioritizing the MRI scan for patients who are younger may not be the best option, as age should not be a determining factor in triage decisions. The most ethical option is to follow institutional policies for triage and allocate resources based on healthcare urgency, taking into account the severity of Mr. Russell's condition. *Answer*: D

3. Ms. Emma Rankin is a 65-year-old retiree who is brought to the emergency department with symptoms of a heart attack. She is taken to the triage area and assessed by the healthcare staff. The healthcare team determines that Ms. Rankin is in need of immediate treatment and that she should be taken to the cardiac catheterization lab for urgent intervention. However, the hospital is currently experiencing a shortage of beds and resources due to an outbreak of a highly contagious virus. The triage team has to make a difficult decision on which patients will receive the limited resources available. The team determines that Ms. Rankin's condition is severe but not immediately life-threatening and that other patients with more severe conditions should receive the limited resources first. The healthcare team informs Ms. Rankin of the situation and advises her that there may be a delay in receiving treatment. Ms. Rankin expresses her concerns about the delay and the potential consequences of waiting for treatment. The healthcare team explains the situation to her and reassures her that it will do everything possible to provide her with the necessary treatment as soon as possible. How should the healthcare team balance Ms. Rankin's urgent healthcare needs with the limited resources available in the hospital during an outbreak?
   A. Prioritize Ms. Rankin's urgent healthcare needs and provide her with immediate treatment, regardless of the limited resources available
   B. Follow the triage protocol established by the hospital and prioritize patients with more severe conditions for the limited resources available, even if it means delaying treatment for Ms. Rankin
   C. Consult with Ms. Rankin's family and loved ones to determine whether or not she should receive immediate treatment
   D. Discharge Ms. Rankin and advise her to seek healthcare attention at another hospital with more resources available

   Explanation: In situations in which there are limited resources, triage protocols are established to ensure that the resources are allocated fairly and impartially based on the severity of the patient's healthcare condition. The healthcare team must follow these protocols to ensure that the allocation of resources is fair and just and that the needs of all patients are met. In this case, Ms. Rankin's condition is severe but not immediately life-threatening, and other patients with more severe conditions require immediate treatment. The healthcare team must follow the established triage protocol to ensure the limited resources are allocated fairly and impartially. Consultation with Ms. Rankin's family and loved ones is not necessary because it is the responsibility of the healthcare team to make treatment decisions based on the patient's healthcare condition. Discharging Ms. Rankin and advising her to seek healthcare attention elsewhere is not ethical because it may put her at risk and not provide her with the necessary care she requires. *Answer*: B

4. Mr. Wesley James is a 55-year-old man with a history of heart disease, hypertension, and diabetes. He presents to the emergency department with symptoms of chest pain and shortness of breath. After initial evaluation, it is determined that he requires urgent cardiac catheterization to assess for possible coronary artery disease. However, the hospital's catheterization lab is currently at capacity, and there are two other patients who also require urgent catheterizations. Upon further evaluation, it is discovered that Mr. James is a prominent local politician and community leader, and there is public pressure from his supporters to prioritize his care. In addition, the hospital's administration is concerned about negative publicity if Mr. James is not given priority. What should be the triage decision in this case?
   A. Prioritize Mr. James for the catheterization procedure based on his prominent status in the community
   B. Follow the hospital's triage protocol and prioritize the patients based on healthcare need, regardless of social status
   C. Delay the procedure for all three patients until the catheterization lab has more availability
   D. Transfer Mr. James to another hospital with available catheterization resources

Explanation: The ethical principle of justice requires that triage decisions be based on healthcare needs rather than social status. Although Mr. James may be a prominent figure in the community, his status should not be a factor in the triage decision. The hospital's triage protocol should be followed, and patients should be prioritized based on healthcare needs, regardless of social status or other external factors. Delaying the procedure for all three patients and transferring Mr. James to another hospital are not optimal solutions because these actions may harm the patients and not address the underlying issue of resource allocation. It is essential that the hospital's administration prioritize patient care based on healthcare needs and ensure that resource allocation is fair and equitable. *Answer*: B

5. Ms. Lily Chen is a 45-year-old teacher who has been diagnosed with a respiratory tract infection (RTI) and admitted to a hospital. Despite her worsening condition, she insists on being transferred to a private hospital that offers better amenities and services than the public hospital where she is currently receiving treatment. The private hospital is known to have better healthcare resources and a higher success rate in treating RTI patients. However, the hospital has limited beds and is currently only accepting patients with severe RTI. What ethical principle should guide the hospital's triage decision-making in this scenario?
   A. Autonomy of the patient
   B. Beneficence of the healthcare provider
   C. Justice in resource allocation
   D. Nonmaleficence of the healthcare provider

Explanation: In this scenario, the principle of justice in resource allocation should guide the hospital's triage decision-making. Triage decisions must prioritize allocating limited resources to those with the greatest need and likelihood of survival. Although Ms. Chen can choose where she receives treatment, the hospital's resources must be distributed equitably to all patients. Similarly, although beneficence requires healthcare providers to act in the best interests of their patients, the limited availability of resources means that difficult choices must be made. Finally, although nonmaleficence obliges healthcare providers to do no harm to their patients, in this case, the harm to Ms. Chen is due to the limitations of resources and not a result of any intentional or negligent action on the part of the healthcare provider. The principle of justice in resource allocation must guide triage decision-making to ensure that resources are distributed fairly and equitably to all patients. Hospitals must establish transparent and agreed-upon triage policies prioritizing allocating limited resources to those with the greatest need and likelihood of survival while protecting vulnerable populations. In this scenario, the hospital must weigh Ms. Chen's desire to be transferred to a private hospital against the need to allocate the limited resources available to other patients with more severe cases of RTI. *Answer*: C

REFLECTION VIGNETTES

1. Two patients with COVID-19 need the only available extracorporeal membrane oxygenation (ECMO) machine. Both patients came in simultaneously and are equal in allocation determinants. The only difference is that Patient A is unvaccinated, and Patient B is vaccinated. It is not possible to share the ECMO machine.
   What is the ethical issue(s)?
   How should the ethical issue(s) be addressed?

2. A practitioner has a Patient A with COVID-19 and requires the only ECMO machine. Patient B, transferred from another hospital, also needs the ECMO machine. The only difference between the two is that Patient A is unvaccinated and Patient B is vaccinated. It is not possible to share the ECMO machine, and there are no federal, state, or institutional triage policies in place, so the practitioner must be the one who determines which patient gets the ECMO machine.
   What is the ethical issue(s)?
   How should the ethical issue(s) be addressed?

3. Practitioner Sarah Carter, a 40-year-old attending practitioner, admits a 55-year-old patient with quadriplegia who has tested positive for COVID-19. The patient's condition deteriorates rapidly, and admission to the intensive care unit is required. However, Practitioner Carter informs the patient's family that the patient's quality of life is already poor and that aggressive treatment would only prolong the patient's suffering. The patient's family members, who are the patient's healthcare surrogates, disagree with Practitioner Carter's assessment and request that the patient receive all possible treatment.

What is the ethical issue(s)?

How should the ethical issue(s) be addressed?

* * *

WEB LINKS

1. *Code of Medical Ethics*. American Medical Association. https://code-medical-ethics.ama-assn.org
   The AMA *Code of Medical Ethics*, established by the American Medical Association, is a comprehensive guide for healthcare practitioners. It addresses issues and challenges, promotes adherence to standards of care, and is continuously updated to reflect contemporary practices and challenges.
2. "Fair Allocation of Scarce Medical Resources in the Time of COVID-19." E. J. Emanuel, G. Persad, R. Upshur, et al. 2020. https://www.nejm.org/doi/full/10.1056/NEJMsb2005114
   This article discusses the principles for allocating healthcare resources in the context of the COVID-19 pandemic, including triage.
3. "The Toughest Triage—Allocating Ventilators in a Pandemic." R. D. Truog, C. Mitchell, and G. Q. Daley. 2020. https://www.nejm.org/doi/full/10.1056/NEJMp2005689
   This article discusses the ethical challenges of triage in the allocation of ventilators during a pandemic.
4. "A Framework for Rationing Ventilators and Critical Care Beds During the COVID-19 Pandemic." D. B. White and B. Lo. 2020. https://jamanetwork.com/journals/jama/fullarticle/2763953
   This article provides a framework for rationing ventilators and critical care beds during the COVID-19 pandemic.
5. "Facing COVID-19 in Italy—Ethics, Logistics, and Therapeutics on the Epidemic's Front Line." L. Rosenbaum. 2020. https://www.nejm.org/doi/full/10.1056/NEJMp2005492
   This article discusses the ethical and logistical challenges faced by Italy during the COVID-19 pandemic, including issues related to triage.
6. "Triage of Scarce Critical Care Resources in COVID-19: An Implementation Guide for Regional Allocation: An Expert Panel Report of the Task Force for Mass Critical Care and the American College of Chest Physicians." R. C. Maves, J. Downar, J. R. Dichter, et al. 2020. https://journal.chestnet.org/article/S0012-3692(20)30691-7/pdf
   This article provides a guide for the regional allocation of scarce critical care resources during the COVID-19 pandemic.
7. "Triage: Care of the Critically Ill and Injured During Pandemics and Disasters: CHEST Consensus Statement." M. D. Christian, C. L. Sprung, M. A. King, et al. 2014. https://journal.chestnet.org/article/S0012-3692(15)51990-9/fulltext
   This article provides a consensus statement on the care of the critically ill and injured, including triage, during pandemics and disasters.

WITHHOLDING AND WITHDRAWING TREATMENT

64

Conscientious and careful practitioners assign disease causes to natural laws, while the best scientists go back to medicine for their first principles.
—Aristotle

* * *

MYSTERY STORY

THE RIGHT TO CHOOSE: A STORY OF AUTONOMY AND CARE

Practitioner Samantha Lewis, a compassionate physician at Autonomy Health & Wellness Center, had always lived by the mantra *primum non nocere*—first do no harm. She often reminded herself of the words from the Hippocratic Corpus, ὠφελέειν, ἢ μὴ βλάπτειν—to do good or to do no harm.

Her patient, Ms. Emily Thompson, a strong-willed woman in her late 60s, had been battling a relentless illness. Ms. Thompson had the spirit of a fighter, and she had clear ideas about her treatment. Despite the fear and uncertainty that the illness brought, Ms. Thompson had full autonomy in deciding her healthcare, guided by informed consent.

Ms. Thompson's choices were only sometimes aligned with Dr. Lewis' professional opinion. But Dr. Lewis understood that she had the moral obligation to pursue Ms. Thompson's best interests in accordance with her reasonable goals, values, and priorities. This respect for autonomy allowed Ms. Thompson to feel confident in trying different treatment options, knowing she could withdraw at any time.

As Ms. Thompson's illness progressed, her decision-making capacity began to wane. Fortunately, she had prepared a living will and had designated a durable power of attorney for healthcare decisions to her eldest daughter, Sarah.

Disagreements arose among Ms. Thompson's family members regarding the withholding and withdrawal of healthcare treatment. Sarah, guided by the living will, advocated for withholding specific treatments that Ms. Thompson had previously objected to, whereas Ms. Thompson's husband insisted on continuing those interventions, fearing the worst.

Practitioner Lewis found herself caught in a professional and ethical dilemma. She reached out to the hospital's ethics committee to seek an ethics consultation, aiming to bring the family to a consensus.

The ethics consultation was a rigorous process that took into account the hierarchical structure of authority as outlined in Ms. Thompson's documents and the law. Careful attention was given to Ms. Thompson's reasonable goals, values, and priorities, as well as the legal rights and common law precedents in the United States.

The committee helped the family understand that there was no legal, professional, or moral difference between withholding and withdrawing healthcare treatment. Ms. Thompson's autonomy and the principle of *primum non nocere* were the guiding lights.

Through careful discussion, empathy, and a clear understanding of the law and healthcare ethics, the family reached an agreement that honored Ms. Thompson's autonomy and Dr. Lewis' moral obligation to do no harm.

In the end, the principles of autonomy, beneficence, nonmaleficence, and justice were upheld, illustrating the intricate dance between the patient's right to choose and the practitioner's commitment to care—a dance that respects the human right to dignity, control, and, above all, the pursuit of the best possible life, even in the face of uncertainty and fear.

* * *

THINK

Professionally, legally, and morally, there is no difference between refusing a healthcare treatment provided and withdrawing from a healthcare treatment already begun. Making a false distinction between the two increases the patient's fear that once a treatment is started, the treatment can never be stopped, no matter how harmful to the patient or against the patient's later judgment.

Patient fear resulted in the mantra of the healthcare profession, *primum non nocere*, which is Latin for "first do no harm." The closest approximation of the mantra is from the Epidemics in the Hippocratic Corpus, ἀσκέειν, περὶ τὰ νουσήματα, δύο, ὠφελέειν, ἢ μὴ βλάπτειν:[1] "[The practitioner must] ... have two special objects in view concerning disease, namely, to do good or to do no harm" (Book 1, Sect. 11).

The Hippocratic Oath states, "I will keep them from harm and injustice."

The patient's right to withhold and to withdraw from treatment is a recognition of the patient's autonomous right to be in control of what is done to their body and a foundational professional obligation of the practitioner: *first do no harm*.

ASSESS

Patient: Autonomy

Based on the individual moral principle of autonomy, all competent adults with decisional capacity have the right to exercise informed consent, meaning the informed determination of what treatments they are to receive, withhold, or withdraw. There is no professional, legal, or moral distinction between withholding and withdrawing healthcare treatment, even though there may be an emotional or experiential distinction.[2] The patient always has the right to stop a treatment that has been started. This autonomous moral right to withdraw treatment provides the patient the freedom and opportunity to experiment and try out various treatment options knowing that, at any time and for any reason, the patient can exercise their autonomous authority and withdraw from the treatment, regardless and independently of other people's opinions, beliefs, and objections.

Practitioner: Beneficence and Nonmaleficence

Professionally, the practitioner has the moral obligation to pursue the patient's best interests in accordance with the patient's reasonable goals, values, and priorities. If the patient loses their decisional capacity, then in hierarchical order, the practitioner should use the following:

1. The most recent oral or written documents indicating patient consent for the withholding or withdrawing treatment
2. A living will that expresses what treatment options the patient would wish to have or not have
3. A durable power of attorney for healthcare indicating who is to make treatment decisions for the patient. (*Durable* means that the proxy's authority to make treatment decisions becomes effective once the patient loses decisional capacity or consciousness.)
4. Next of kin, which may vary slightly from state to state, but in general the order of surrogate authority is the patient's
   a. Guardian
   b. Spouse
   c. Adult offspring(s)
   d. Either parent
   e. Any adult sibling
   f. Any adult grandchild

---

1. *primum non nocere*. Wikipedia. https://en.wikipedia.org/wiki/Primum_non_nocere.
2. "Critical Care. Withdrawing May Be Preferable to Withholding." https://ccforum.biomedcentral.com/articles/10.1186/cc3486.

g. Close friend
h. Guardian of the estate
i. Temporary custodian

The assumption is that the surrogate decision-making list is in the hierarchical order of which individual(s) would have the best knowledge of the patient's reasonable goals, values, and priorities so that the surrogate(s), using substituted judgment, can make treatment decisions the way the patient would have. This hierarchical structure of authority exemplifies the high degree of importance of individual patient autonomy (informed consent) by oral or written document, or a living will, to a patient's proxy whose goal is to make treatment decisions using substituted judgment—that is, using the patient's reasonable goals, values, and priorities when making treatment decisions. This high degree of patient-centered decision-making is believed to maximize the patient's best interests.

If surrogates are unable to determine whether or not the patient would wish to withdraw life-sustaining medical treatment (LSMT) and there is disagreement regarding whether or not that would be in the patient's best interests, then this would be a case in which an "ethics consultation" would help bring the surrogates into consensus.

Public Policy: Justice

In the United States, common law has legally established that patients have the autonomous right to withhold and withdraw from healthcare treatment. Although there is state controversy about what evidence is accepted as proof of a patient's desire to withdraw LSMT after losing decisional capacity, there is no legal controversy that after the patient does lose decisional capacity, the surrogate has the legal right to withhold and withdraw healthcare treatment.

CONCLUDE

Patients have the legal, professional, and moral right to withhold and withdraw healthcare treatment. However, issues arise when surrogates cannot agree on whether or not to withdraw LSMT for a patient when the patient's desires are not determinable. In such situations, the practitioner should encourage surrogate discussion. If a consensus cannot be reached, an ethics consultation can be sought to help guide the decision-making process. Ultimately, maintaining the patient's autonomy, interests, and well-being should be the primary focus of both patients and practitioners.

In summary, patients have the right to withhold and withdraw from healthcare treatment based on the principle of autonomy and the moral obligation of practitioners to pursue the patient's best interests. The patient's autonomy is protected through informed consent, living wills, durable power of attorney for healthcare, and surrogate decision-making. Legally, in the United States, common law recognizes the patient's right to withhold and withdraw from healthcare treatment. The practitioner's moral obligation is to first do no harm and to pursue the patient's best interests. When surrogates cannot agree on whether or not to withdraw LSMT for a patient when the patient's desires are not determinable, an ethics consultation can be sought to help guide the decision-making process.

* * *

REVIEW QUESTIONS

1. The mantra of the healthcare profession is *primum non nocere*, which is Latin for "first do no harm."
    A. True
    B. False

2. There is no professional, legal, or moral distinction between withholding and withdrawing healthcare treatment, even though there may be an emotional or experiential distinction.
   A. True
   B. False

3. The autonomous moral right to withdraw treatment provides the patient the freedom and opportunity to experiment and try various treatment options, knowing that at any time and for any reason, the patient can exercise their autonomous authority and withdraw from the treatment.
   A. True
   B. False

4. It is assumed that the surrogate decision-making list is in the hierarchical order of which individual(s) would have the best knowledge of the patient's reasonable goals, values, and priorities so that the surrogate(s), using substituted judgment, can make treatment decisions the way the patient would have.
   A. True
   B. False

5. If surrogates are unable to determine whether or not the patient would wish to withdraw LSMT, and there is disagreement regarding whether or not that would be in the patient's best interests, then that would be a case in which an "ethics consultation" would help bring the surrogates into consensus.
   A. True
   B. False

6. After the patient loses decisional capacity, there are various legal controversies regarding whether or not a surrogate has the legal right to withhold and withdraw LSMT.
   A. True
   B. False

Answers: 1A, 2A, 3A, 4A, 5A, 6B

* * *

CLINICAL VIGNETTES

1. Ms. Maria Roberts is a 50-year-old accountant who has been diagnosed with early onset Alzheimer's disease. She has been living with her husband and children, but her condition has worsened, and she has been admitted to the hospital due to complications from the disease. Ms. Roberts has lost decisional capacity, and her husband has been appointed as her surrogate decision-maker. The healthcare team has recommended that Ms. Roberts be placed on a ventilator due to respiratory failure, but her husband is conflicted about what decision to make. Ms. Roberts' advance directive did not include any instructions about life-sustaining treatment, and her husband is unsure whether to proceed with the ventilation or to withhold the treatment. What ethical principle is at stake in this case?
   A. Autonomy
   B. Beneficence
   C. Nonmaleficence
   D. Justice

Explanation: The ethical principle at stake in this case is beneficence. Ms. Roberts' husband, as her surrogate decision-maker, has the responsibility to make decisions that are in her best interests and align with her reasonable goals, values, and priorities. If Ms. Roberts had previously expressed her wishes about healthcare treatment, her husband should have followed those wishes to the best of his ability. If her wishes are not known, then her husband should make decisions that are in her best interests based on the healthcare team's recommendations and other relevant information. In this case, the healthcare team has recommended placing Ms. Roberts on a ventilator due to respiratory failure, and her husband must weigh the potential benefits of the treatment against the potential risks and harms. Although the healthcare team can provide guidance and recommendations, Ms. Roberts' surrogate decision-maker should make the final decision, and it should be based on what is in her best interests. Autonomy is not directly at stake in this case because Ms. Roberts has lost decisional capacity, but it is still an important consideration, as Ms. Roberts' husband should make decisions that align with her reasonable goals, values, and priorities and that respect her autonomy as much as possible. Nonmaleficence and justice are also relevant considerations in this case. However, the primary ethical principle at stake is beneficence, as Ms. Roberts' husband must decide in her best interests. *Answer*: B

2. Ms. Mia Garcia is a 72-year-old retired teacher who has been diagnosed with terminal cancer. She has been admitted to the hospital with severe pain and difficulty breathing. Ms. Garcia is still able to communicate and has decisional capacity. She has expressed to her healthcare team that she does not want any further healthcare interventions and prefers to focus on palliative care to manage her symptoms. Ms. Garcia has designated her daughter, Maria, as her healthcare proxy in case she loses decisional capacity. However, Maria disagrees with her mother's decision and wants the healthcare team to do everything possible to prolong her life. What ethical principle is at play in this situation?
   A. Autonomy
   B. Beneficence
   C. Nonmaleficence
   D. Justice

   Explanation: Ms. Garcia has the right to make her own treatment decisions based on her values and preferences as long as she has decisional capacity. While dealing with her emotions, her daughter struggles with a decision that aligns with her mother's previously expressed wishes. The healthcare team should support Ms. Garcia's autonomy and work with her daughter to address her concerns or questions. In situations in which surrogates cannot determine what the patient wishes and there is disagreement about what would be in the patient's best interests, an ethics consultation can help reach a consensus. It is important for the healthcare team to respect Ms. Garcia's autonomy and support her in her decision to focus on palliative care while also addressing her daughter's concerns. *Answer*: A

3. Ms. Susan Jenkins is a 75-year-old retired teacher who was recently diagnosed with advanced ovarian cancer. She has been receiving chemotherapy for several months, but her cancer has continued to progress. Ms. Jenkins has now lost decisional capacity, and her husband has been appointed as her surrogate decision-maker. The healthcare team is recommending a new treatment option that could potentially extend Ms. Jenkins's life, but it also carries significant risks and could decrease her quality of life. What action should be taken?
   A. The healthcare team should proceed with the new treatment option without discussing the risks with Ms. Jenkins' husband, as it is the team's duty to pursue treatment that may benefit the patient's health.
   B. Ms. Jenkins' husband should refuse the new treatment option, as it carries significant risks and could decrease her quality of life.
   C. The healthcare team should discuss the risks and benefits of the new treatment option with Ms. Jenkins' husband, and they should make a joint decision based on what is in Ms. Jenkins' best interests.
   D. Ms. Jenkins' husband should make the decision on his own, without input from the healthcare team, because he has been appointed as her surrogate decision-maker.

Explanation: The healthcare team should discuss the risks and benefits of the new treatment option with Ms. Jenkins' husband, and they should make a joint decision based on what is in Ms. Jenkins' best interests. Although the healthcare team has a duty to pursue treatment that may benefit the patient's health, the team must also consider the patient's well-being and quality of life. Because Ms. Jenkins has lost decisional capacity, her husband has been appointed as her surrogate decision-maker. However, the healthcare team should still involve him in discussing the risks and benefits of the treatment option and make a joint decision based on what is in Ms. Jenkins' best interests. This approach honors the principles of patient autonomy and beneficence while also ensuring that the patient's best interests are prioritized. *Answer*: C

4. Ms. Sophia Baker is a 78-year-old woman with a history of heart disease and dementia. She has been living in a nursing home for the past year and has recently been admitted to the hospital with pneumonia. While in the hospital, she experiences respiratory failure and requires mechanical ventilation. Ms. Baker no longer has decisional capacity. The healthcare team informs her daughter, who is her designated surrogate, that Ms. Baker may require a tracheostomy to continue the mechanical ventilation for an extended period of time. The daughter is hesitant to consent to the tracheostomy because she knows her mother would not have wanted to be on prolonged mechanical ventilation. However, she is also hesitant to withdraw life-sustaining treatment because she feels guilty and wants to do everything possible to keep her mother alive. The daughter asks the healthcare team to provide guidance on what is the best course of action. Which action should be taken?
   A. The daughter should consent to the tracheostomy because it is the best option to keep her mother alive.
   B. The daughter should withdraw life-sustaining treatment because it is what her mother would have wanted.
   C. The healthcare team should seek an ethics consultation to help the daughter make a decision.
   D. The healthcare team should make the decision on behalf of the patient, as the daughter is too emotionally invested to make an objective decision.

   Explanation: The daughter is in a difficult position because she wants to honor her mother's wishes but is also struggling with the guilt of potentially withdrawing life-sustaining treatment. The healthcare team should seek an ethics consultation to help the daughter decide. An ethics consultation can help guide honoring the patient's autonomy while considering the surrogate's emotional and psychological state. It is important to remember that the surrogate's role is to make decisions based on what the patient would have wanted, not what the surrogate personally wants. The healthcare team should support the surrogate in making an informed decision that aligns with the patient's values and beliefs. *Answer*: C

5. Mr. Michael Ward is a 68-year-old man with advanced Parkinson's disease. He has been living in a nursing home for the past year and is frequently admitted to the hospital for treatment of infections. He has recently been admitted to the hospital with pneumonia and is having difficulty breathing. While in the hospital, he experiences respiratory failure and requires mechanical ventilation. Mr. Ward has decisional capacity and expresses to the healthcare team that he does not want to be on prolonged mechanical ventilation. However, his daughter, who is his designated surrogate, insists that he should receive all available treatment options to prolong his life, regardless of his wishes. What is the professional obligation of the healthcare team?
   A. The healthcare team should honor the patient's wishes and not provide prolonged mechanical ventilation, regardless of the surrogate's wishes.
   B. The healthcare team should provide all available treatment options, as requested by the surrogate.

C. The healthcare team should seek an ethics consultation to help resolve the conflict between the patient's wishes and the surrogate's wishes.
D. The healthcare team should make the decision on behalf of the patient, as the surrogate is making an emotionally driven decision.

Explanation: Mr. Ward has decisional capacity and has expressed his wishes regarding his healthcare. The healthcare team should respect his autonomy and not provide prolonged mechanical ventilation. It is important to remember that the designated surrogate only becomes a surrogate after the patient has lost their decisional capacity. After that has happened, then the surrogate's role is to make decisions based on what the patient would have wanted, not what the surrogate personally wants. The healthcare team should support the surrogate in understanding the patient's wishes and provide education on the benefits of honoring the patient's autonomy. If there is a conflict between the patient's and the surrogate's wishes, seeking an ethics consultation may help resolve the conflict and find a solution that aligns with the patient's values and beliefs. *Answer*: A

REFLECTION VIGNETTES

1. Practitioner Emma Lopez, a 50-year-old critical care practitioner, admits a patient named Mr. Charles Davis, who is 70 years old, to the intensive care unit (ICU) due to multiple organ failure from sepsis. Mr. Davis is alert and oriented with decisional capacity, and he tells practitioner Lopez that he no longer wishes to undergo any aggressive LSMT, including mechanical ventilation or vasopressors. Mr. Davis explains that his quality of life is poor, and he wishes to die peacefully.

   What is the ethical issue(s)?

   How should the ethical issue(s) be addressed?

2. Ms. Evelyn Watson is a 65-year-old retired teacher who is now in the ICU with a severe case of pneumonia. Ms. Watson has three siblings who are making treatment decisions for her. Two siblings, Mr. Michael Johnson and Ms. Patricia Brown, believe that their sister's quality of life is so poor that all LSMT should be forgone. Mr. Johnson and Ms. Brown are concerned that their sister's suffering is unbearable and believe that continuing aggressive treatment would be futile. However, the third sibling, Mr. Robert Watson, is adamantly against such a decision. Mr. Watson believes that his sister would want all possible healthcare interventions, no matter what the outcome may be. The family is in disagreement about what treatment decisions to make, and the attending practitioner must navigate this complicated situation while upholding ethical principles and the patient's wishes.

   What is the ethical issue(s)?

   How should the ethical issue(s) be addressed?

3. An adult parent and their 13-year-old child are presented to the emergency department (ED) unconscious and in critical condition. Death is imminent unless they both get blood. There is no living will or durable power of attorney on record. The patient's spouse, who is also the other biological parent of the child, informs the ED practitioner that both of the patients are Jehovah's Witnesses and that it is against their religion to receive any blood and therefore requests that no blood be given to both the spouse and child. The spouse also states they had numerous discussions about such a situation occurring, and not getting blood was what they wanted. Both patients are also wearing No Blood healthcare alert wristbands.

   What is the ethical issue(s)?

   How should the ethical issue(s) be addressed?

4. Practitioner Anne Wilson, a 43-year-old ED practitioner, evaluates a 36-year-old patient named Ronald Turner who presents with acute rectal bleeding. The patient has a history of schizophrenia and is experiencing severe paranoia. A colonoscopy is recommended as the preferred diagnostic and treatment option to address the bleeding. However, Mr. Turner refuses to undergo the procedure, stating that he believes the colonoscopy will transmit HIV to him. As a result of his paranoia, he is resistant to discussing the issue further.

What is the ethical issue(s)?

How should the ethical issue(s) be addressed?

* * *

WEB LINKS

1. *Code of Medical Ethics*. American Medical Association. https://code-medical-ethics.ama-assn.org
   The AMA *Code of Medical Ethics*, established by the American Medical Association, is a comprehensive guide for healthcare practitioners. It addresses issues and challenges, promotes adherence to standards of care, and is continuously updated to reflect contemporary practices and challenges.
2. "Medical Murder by Omission? The Law and Ethics of Withholding and Withdrawing Treatment and Tube Feeding." John Keown. 2003. https://www.ncbi.nlm.nih.gov/pmc/articles/PMC4953644
   This article discusses the legal and ethical aspects of withholding and withdrawing treatment.
3. "Overview of Medical Ethics." Vermont Ethics Network. https://vtethicsnetwork.org/medical-ethics/overview
   This article provides an overview of bioethics, including the topic of withholding and withdrawing treatment.
4. "Frequently Asked Questions: Health Care Ethics." Vermont Ethics Network. https://vtethicsnetwork.org/medical-ethics/frequently-asked-questions
   This article answers frequently asked questions about healthcare ethics, including the topic of withholding and withdrawing treatment.
5. "Withholding and Withdrawing Treatment, Ethical and Legal Aspects." Reidar Pedersen, Marianne Klungland Bahus, and Erik Martinsen Kvisle. 2007. https://pubmed.ncbi.nlm.nih.gov/17571104
   This article discusses the ethical and legal aspects of withholding and withdrawing treatment.
6. "Are Medical Ethicists out of Touch? Practitioner Attitudes in the US and UK Toward Decisions About Medical Futility." Donna L. Dickenson. 2000. https://jme.bmj.com/content/26/4/254
   This article discusses the attitudes of practitioners toward decisions about healthcare futility, including withholding and withdrawing treatment.
7. "Intentionally Hastening Death by Withholding or Withdrawing Treatment." Georg Bosshard, Susanne Fischer, Agnes van der Heide, Guido Miccinesi, and Karin Faisst. 2006. https://pubmed.ncbi.nlm.nih.gov/16855920
   This article discusses the intentional hastening of death by withholding or withdrawing treatment.
8. "361 Ethics Week 9: Withholding and Withdrawing Treatment. Ch. 9." Quizlet. https://quizlet.com/235629635

APPENDIX

Step Exams

IN 1994, THE STEP 1 EXAM WAS established with an average score of 200 out of 300, marking the first step toward healthcare licensure eligibility. Residency directors subsequently adopted Step 1 scores as a primary criterion for granting residency interviews, perceiving it as an "objective" method to assess a diverse pool of applicants. This reliance resulted in a supplementary healthcare school curriculum that utilized third-party materials to boost Step scores, consequently raising the average Step 1 score from 200 to 230 since its inception.

The transition to a pass/fail system for Step 1 was endorsed by healthcare school course directors and deans, anticipating that students would prioritize the healthcare school curriculum over third-party study programs. Contrarily, residency directors and students opposed this change, recognizing the necessity for a differentiating factor in the residency matching process. Consequently, directors now employ more subjective criteria, such as healthcare school prestige and extracurricular activities, as well as new objective criteria, including the number of Step 1 failures and Step 2 CK (Clinical Knowledge) scores. Currently, examinees are limited to four attempts per Step exam, as opposed to the previous six, and United States Medical Licensing Examination (USMLE) transcripts now detail a complete exam history, including any Step 1 failures. A failure to pass Step 1 on the first attempt is now considered a negative factor by residency directors. Moreover, if Step 2 CK scores are utilized (which are not pass/fail), examinees must complete Step 2 CK in time to receive their scores before applying for residency, considering that the exam is taken early in the fourth year and can only be retaken after 6 months.

Licensure eligibility exams are pivotal in ensuring a minimum standard of healthcare competency. Annually, the USMLE, sponsored by the Federation of State Medical Boards and the National Board of Medical Examiners (NBME), has increased the number of ethics-related board questions in the Step exams. The format and ethics content for each Step exam are as follows:

Step 1 exam: Format—7 blocks, each 60 minutes with 40 questions, totaling 280
  Ethics content: 10–15% of questions, amounting to 28–42 questions
Step 2 CK: Format—8 blocks, each 60 minutes with 40 questions, totaling 320
  Ethics content: 10–15% of questions, amounting to 32–48 questions
Step 3: Day 1 format—6 blocks, each 60 minutes with 38 or 39 questions, totaling 240; Day 2 format—6 blocks, each 45 minutes with 30 questions, totaling 180 questions, plus 13 case simulations
  Ethics content: 10–15% of the 420 questions, amounting to 42–63 questions

Considering that ethics questions comprise more than twice the number of any specialty area and that NBME data reveal students typically score lower in ethics than in specialty areas, a concerted focus on ethics is essential for success in the Step exams.

To effectively prepare for the ethics questions in the USMLE Step exams, learners should do the following:

1. Study the four principles of bioethics: Patient autonomy (informed consent), practitioner beneficence (do good), practitioner nonmaleficence (do no harm), and public justice (be fair)
2. Engage with clinical vignettes that illustrate these bioethics principles
3. Stay informed about pertinent laws, regulations, and standards of care
4. Practice responding to bioethics questions under timed conditions
5. Utilize exam-specific materials in preparation

6. Engage in discussions about bioethical concepts with family, friends, and peers to deepen understanding
7. Employ bioethics practice exams to identify areas requiring further review

Implementing these strategies will ensure comprehensive preparation to address the bioethics questions on national board exams, equipping learners to skillfully navigate ethical dilemmas in their future healthcare practice.

GLOSSARY

ABORTION The legality and regulations surrounding abortion vary significantly by state and are often subject to rapid changes influenced by partisan politics. Regardless of the legal landscape, healthcare practitioners have a moral and legal responsibility to provide safe and appropriate healthcare referrals for patients seeking abortion services. In addition, practitioners are socially recognized as having the responsibility to encourage minors who are pregnant to discuss their pregnancy with their parents, while also respecting minors' confidentiality and autonomy.

ABUSE—CHILD, ELDER, AND INTIMATE PARTNER Practitioners have a mandatory responsibility to report any suspected child or elder abuse to the appropriate Child or Adult Protective Services. In cases of intimate partner abuse, practitioners should provide support and encouragement to the victim to report the abuse, while also respecting their autonomy and decision-making process.

ADDRESSING ONESELF AND PATIENT Healthcare students are required to transparently communicate their student status to patients to avoid any misunderstandings about their role in the patients' care. The use of titles and respectful forms of address helps establish clear social boundaries and clarify decision-making authority within the healthcare setting, ensuring professional and ethical interactions.

ADVANCE DIRECTIVE Advance directives are crucial legal documents that empower a patient to outline their preferred healthcare treatments and interventions in case they become unable to communicate or make decisions for themself. They typically consist of two primary components: a living will, which specifies accepted treatment options, and a durable power of attorney, which appoints a surrogate decision-maker. The process of creating an advance directive should ideally be initiated during outpatient visits with the primary care provider, ensuring clear communication and understanding of the patient's wishes. It is essential to revisit and readdress advance directives during each hospital admission to confirm that the patient's current wishes are accurately reflected and respected. In situations requiring healthcare decisions, the directives outlined in the advance directive take precedence over the wishes of family members, thereby honoring the patient's autonomy and previously stated preferences.

ALTERNATIVE THERAPY A treatment or therapeutic approach that falls outside conventional healthcare practices. These may include acupuncture, herbal medicine, or other nontraditional methods. When a patient expresses interest in an alternative therapy that diverges from standards of care, it is important for the practitioner first to seek to understand the patient's reasons for this interest. This allows for open communication and helps the practitioner provide informed and patient-centered care.

ASSISTED SUICIDE Assisted suicide, which involves a healthcare practitioner providing assistance to a patient in ending their life, is strictly prohibited by the healthcare profession. This practice is considered unethical and is not aligned with the primary responsibility of healthcare providers to protect and preserve human life.

BELMONT REPORT The Belmont Report is a foundational document in the field of healthcare ethics, setting forth the principles of principlism as a guide for ethical decision-making in medicine and research. Principlism is rooted in the four core principles of patient autonomy (informed consent), practitioner beneficence (do good), practitioner nonmaleficence (do no harm), and public justice (be fair). These principles collectively provide a comprehensive and morally sound framework for navigating complex ethical dilemmas in healthcare practice and research, as established by federal law and widely accepted ethical standards.

BEST INTERESTS A principle in healthcare that emphasizes the importance of prioritizing and maximizing the patient's reasonable goals, values, and priorities within the framework of standards of care. The patient's best interests should guide all healthcare decisions, ensuring that their values and priorities are respected and upheld throughout the course of their care.

BOARD EXAMS AND ETHICS Licensure eligibility exams are pivotal in ensuring a minimum standard of healthcare competency. Annually, the United States Medical Licensing Examination, sponsored by the Federation of State Medical Boards and the National Board of Medical Examiners (NBME), has increased the number of ethics-related board questions in the Step exams. The format and ethics content for each Step exam are as follows:

Step 1 exam: Format—7 blocks, each 60 minutes with 40 questions, totaling 280 questions
  Ethics content: 10–15% of questions, amounting to 28–42 questions
Step 2 CK: Format—8 blocks, each 60 minutes with 40 questions, totaling 320 questions
  Ethics content: 10–15% of questions, amounting to 32–48 questions
Step 3: Day 1 format—6 blocks, each 60 minutes with 38 or 39 questions, totaling 240 questions; Day 2 format—6 blocks, each 45 minutes with 30 questions, totaling 180 questions, plus 13 case simulations
  Ethics content: 10–15% of the 420 questions, amounting to 42–63 questions

Considering that ethics questions comprise more than twice the number of any specialty area and that NBME data reveal students typically score lower in ethics than in specialty areas, a concerted focus on ethics is essential for success in the Step exams.

CAPITAL PUNISHMENT–EXECUTIONS Participation in executions is deemed impermissible for healthcare practitioners. Upholding social trust, maintaining a reputable professional standing, and adhering to professional obligations are considered paramount. Engaging in capital punishment activities compromises the ethical standards and integrity of the healthcare profession.

CHAPERONES AND PERSONAL PRIVACY In accordance with Health Insurance Portability Accountability Act confidentiality regulations, patients must be provided the opportunity for privacy and confidentiality when a chaperone is present during healthcare examinations or procedures. It is essential to maintain the patient's comfort and trust by respecting their personal privacy while also adhering to professional standards and legal requirements.

CHECKLISTS A systematic tool utilized in various fields, including healthcare, as a risk mitigation procedure. By outlining specific steps or necessary components in a process, checklists help prevent negligent acts, omissions, and undesired outcomes, ensuring that all essential tasks are completed accurately and efficiently.

CHILD ABUSE Refers to physical or emotional neglect or exploitation directed toward a child. Clinicians hold a legal and ethical obligation to report instances of child abuse, neglect, and exploitation to the appropriate authorities. When assessing a child, healthcare professionals should conduct interviews in a private setting, away from potential abusers, to create a safe environment that minimizes the risk of intimidation. Empathetic interviewing techniques should be employed to facilitate open and trusting communication. In cases in which child abuse is suspected, Child Protective Services should be contacted immediately to initiate necessary protective measures.

COMPETENT PATIENTS Individuals with decisional capacity can understand, process, and make informed choices regarding their healthcare treatments. Competent patients typically have the legal right to accept or refuse any healthcare intervention or treatment based on their understanding of the potential risks and benefits and their personal goals, values, and priorities.

CONFIDENTIALITY A fundamental principle in healthcare that emphasizes the importance of protecting patients' private health information. Healthcare providers must take stringent measures to ensure that patients' healthcare records and personal details are kept confidential and secure. Exceptions to this rule occur when maintaining confidentiality would pose a risk to the health and welfare of the patient or others. In such cases, it is advisable to weigh the principles of practitioner beneficence (do good), practitioner nonmaleficence (do no harm), and public justice (be fair) more than patient autonomy (informed consent) to prevent harm to the patient and others. In addition, patients should be encouraged to openly communicate with loved ones or individuals impacted by their healthcare condition, such as those who may have been exposed to a sexually transmitted disease through contact with the patient.

CONFIDENTIALITY, FAMILY A principle that upholds the patient's right to privacy in their healthcare affairs. Healthcare providers must respect the patient's wishes regarding disclosing their health information. Healthcare professionals must not disclose healthcare information to family members or unauthorized individuals if a patient objects.

CONFLICT OF BELIEFS In a diverse and multicultural society, healthcare practitioners often encounter differences in beliefs and values between themselves and their patients. Regardless of these differences, the patient's best interests must always be paramount. When the practitioner's personal beliefs conflict with their professional responsibilities, the duty to the public and the patient's needs may supersede the practitioner's convictions. In such cases, treatment should be administered professionally, respectfully, and nonjudgmentally. However, in non-emergency situations, healthcare providers are not obligated to perform healthcare procedures or services contrary to their beliefs. In these instances, the practitioner should offer a referral to another provider willing to conduct the requested procedure while maintaining a professional and nonjudgmental demeanor.

CONFLICT OF INTERESTS A conflict of interest arises when an individual's personal, professional, or financial interests may potentially influence their decision-making or actions in a situation. In such cases, the individual must recuse themself from the activity in question and disclose the conflict to all parties involved to maintain transparency, integrity, and trust.

CONSENT A patient gives a voluntary agreement based on adequate knowledge and understanding to undergo a specific healthcare procedure or intervention. Consent must be obtained from the patient, and the involvement of a spouse or significant other is not required. This applies to all types of procedures, including sterilization. The patient's autonomy and right to make informed decisions about their own healthcare should be respected and upheld.

CONSENT, MINORS Refers to situations in which individuals younger than age 18 years who have not been legally emancipated can provide informed consent for specific healthcare treatments or interventions without the need for parental consent or notification. These particular situations include:

Prenatal care: Minors can consent to receiving healthcare related to pregnancy.
Contraception: Minors can consent to birth control and other contraceptive services.
Diagnosis or treatment of sexually transmitted diseases (STDs): Minors can consent to receiving healthcare attention to diagnose or treat STDs.
Drug or alcohol rehabilitation: Minors can consent to receiving treatment for drug or alcohol addiction.

CONSENT, PARENTAL Refers to the legal requirement for healthcare providers to obtain informed consent from a parent or legal guardian before administering treatment to a minor. When a parent or guardian refuses to consent a child to receive necessary healthcare treatment for a non-emergency yet potentially life-threatening condition, healthcare providers are responsible for seeking a court order to proceed with the treatment, ensuring the best interests and well-being of the child are prioritized.

CONTRACEPTION Contraception refers to the intentional use of various methods or devices to prevent pregnancy. It is a fundamental right of all individuals, regardless of their age, to have access to safe and effective contraceptive options, enabling them to make informed decisions about their reproductive health and family planning.

DEATH WITH DIGNITY; ASSISTED DYING "Death with dignity" or "aid in dying" refers to providing assistance to terminally ill patients seeking to end their lives in a dignified and peaceful manner. This practice is legally permitted in 11 jurisdictions. Despite its legal status, healthcare boards hold that practitioner-assisted suicide, euthanasia, and assisted death are unacceptable practices within the healthcare profession, emphasizing the importance of preserving life and adhering to ethical standards.

DECLARATION OF GENEVA As a cornerstone of modern healthcare ethics, the Declaration of Geneva captures the core principles that guide healthcare professionals in their practice. By pledging to prioritize patient well-being, uphold confidentiality, respect colleagues, and continue life-long learning, healthcare professionals demonstrate their commitment to ethical conduct. The Declaration serves as a reminder of the values and principles that underpin the healthcare profession, inspiring generations of practitioners to provide compassionate, patient-centered care while respecting human rights and civil liberties.

DIFFICULT AND NONCOMPLIANT PATIENTS The terms "difficult" and "noncompliant" are now considered inappropriate and judgmental because they can imply a negative moral or character assessment and assume the decision-making authority of the practitioner or healthcare team over the patient. Instead, healthcare providers must maintain a professional demeanor and approach each patient with empathy and respect. They should consider the patient's unique healthcare and psychological needs and work collaboratively to achieve the best possible health outcomes using the patient's reasonable goals, values, and priorities.

DIFFICULT NEWS A more appropriate term than "bad news" because the latter can imply moral judgment or negative connotations. Difficult news refers to information that may be distressing or upsetting to receive, such as a healthcare diagnosis or prognosis. When disclosing difficult news, providing the necessary details clearly and comprehensively while being empathetic and supportive of the patient's emotional response is essential. Healthcare providers should strive to be attuned to the patient's perception and reactions to the situation, ensuring open and sensitive communication. Ideally, difficult news should be delivered in person, accompanied by open-ended questions to gauge the patient's understanding and emotional state.

DISABILITY BENEFITS Refers to the financial assistance provided to individuals unable to work due to a disability. Healthcare providers often play a crucial role in evaluating patients' eligibility for disability benefits. In cases in which patients may have urgent or demanding expectations, healthcare practitioners must clarify the importance of conducting a comprehensive assessment before making any determinations. Furthermore, practitioners must emphasize their ethical obligation to report the findings honestly and accurately based on the assessment results.

DISAGREEMENTS, HIERARCHICAL In the healthcare field, hierarchical disagreements, such as those between a resident and an attending practitioner, should be handled with professionalism and respect for the established chain of command. It is inappropriate for a resident to discuss their disagreements or conflicts with an attending practitioner in front of or with the patient. Instead, the resident should address their concerns or differing opinions directly with the attending practitioner in a private and respectful manner to resolve the issue and maintain a united front in the care of the patient.

DOCTRINE OF DOUBLE EFFECT The doctrine of double effect, rooted in medieval natural law tradition, posits that actions with good intentions are morally permissible even if they result in unintended harm. However, this doctrine is based on the practitioner's intention and neglects patient autonomy, reflecting a paternalistic approach to medicine that is no longer acceptable. In contrast, modern bioethics relies on four principles—patient autonomy (informed consent), practitioner beneficence (do good), practitioner nonmaleficence (do no harm), and public justice (be fair)—which are specified and balanced to evaluate the morality of an action and hold practitioners accountable for the consequences of their actions, independent of their intentions. This approach, which focuses on standards of care, patient autonomy, and treatment results, renders the doctrine of double effect obsolete in today's healthcare practice, in which professionalism, patient rights, and moral accountability are paramount.

DONOR BODIES Refers to the bodies of individuals who have generously donated their remains for educational, research, or clinical purposes. It is of utmost importance that all interactions and contact with donor bodies are conducted with the highest degree of dignity and respect, acknowledging the invaluable contribution made by the donor in advancing scientific knowledge and healthcare practice.

DO NOT RESUSCITATE ORDER (DNR) A do not resuscitate (DNR) order is a healthcare directive specifying that cardiopulmonary resuscitation (CPR) should not be performed in the event of cardiac or respiratory arrest. It is important to note that a DNR order solely pertains to CPR and does not impact or diminish any other aspect of healthcare or patient management. All other appropriate healthcare interventions and treatments will continue to be provided to the patient as indicated.

ELDER ABUSE Refers to the harmful treatment, neglect, or exploitation of older individuals. Healthcare practitioners have a legal and ethical duty to report any instances of elder abuse. To ensure the safety and well-being of the patient, it is recommended that patients are interviewed in private, away from potential abusers, to prevent any form of intimidation. Practitioners should employ empathetic interviewing techniques when addressing cases of abuse to create a supportive and nonthreatening environment that encourages open communication.

EMAIL AND ELECTRONIC COMMUNICATION Electronic communications, including email, between patients and practitioners should be conducted with the utmost security, privacy, and confidentiality to protect patient information. Furthermore, professional boundaries must be maintained in all electronic communications to uphold the integrity of the patient–practitioner relationship.

EMERGENCY, ADULT In emergencies, when an adult patient cannot make decisions due to impaired capacity, healthcare providers can administer necessary lifesaving treatment without obtaining prior consent. This principle prioritizes the immediate health and well-being of the patient in critical and time-sensitive circumstances.

EMERGENCY MEDICAL TREATMENT AND LABOR ACT (EMTALA) A federal law enacted by Congress to ensure that all individuals, regardless of their financial status or ability to pay, receive appropriate healthcare screening, examination, and stabilization in emergency situations. EMTALA prohibits hospitals from refusing treatment to indigent patients or from transferring or discharging them inappropriately. This law is crucial in safeguarding the rights and health of patients seeking emergency healthcare assistance.

EMERGENCY, MINOR In emergency situations, healthcare providers are obligated to provide necessary and potentially lifesaving therapy to minors, regardless of parental consent or presence. This principle is based on the ethical duty to prioritize the health and well-being of the patient in critical, time-sensitive circumstances.

ERRORS In the event of a healthcare error, practitioners have an ethical responsibility to fully disclose the mistake to the patient, regardless of the potential harm or outcomes resulting from the error. This transparent communication is vital in maintaining trust and integrity in the patient–practitioner relationship.

EUTHANASIA Regardless of its legal status, practitioners are ethically prohibited from participating in euthanasia because it is contrary to the core principles and responsibilities inherent in the healthcare profession.

FINANCIAL DISCLOSURES Practitioners are required to transparently disclose any financial conflicts of interest to their patients, ensuring honesty and integrity in the patient–practitioner relationship.

FOUR PRINCIPLES–PRINCIPLISM The four principles of biomedical ethics are a foundational framework that guides ethical decision-making in healthcare practice and research. The principles are as follows:

Patient autonomy (informed consent): Emphasizes the importance of informed consent, respecting patients' rights to make decisions about their own healthcare

Practitioner beneficence (do good): Encourages practitioners to act in the patient's best interests, promoting their well-being and advocating for positive health outcomes in accordance with the patient's reasonable goals, values, and priorities

Practitioner nonmaleficence (do no harm): Adheres to the principle of "do no harm," ensuring that the potential benefits of any intervention outweigh the risks

Public policy justice (be fair): Focuses on fairness and equity, addressing healthcare disparities and striving to create a just and ethical healthcare system that serves the needs of all individuals, regardless of their background or circumstances

Together, these principles prioritize patient interests, guide professional conduct, and promote a just and equitable approach to healthcare.

FUTILE TREATMENT When a patient desires a treatment deemed futile by healthcare standards, the practitioner must communicate openly and honestly with the patient. This discussion should correct any misinformation and clarify why the desired treatment is futile, ultimately ensuring that the patient is fully informed before making healthcare decisions.

GENETIC INFORMATION NONDISCRIMINATION ACT (GINA) Enacted in 2008, GINA protects individuals from discrimination based on their genetic information in both employment and health insurance contexts. The act prohibits employers and healthcare professionals from requiring or using genetic information and family healthcare history to make employment-related decisions. This safeguards individuals from potential bias and ensures the confidentiality and non-misuse of their genetic data.

GIFTS In healthcare, it is crucial to maintain professional boundaries to ensure unbiased and ethical care. Accepting expensive gifts from patients or their families can potentially influence or give the appearance of influencing a healthcare provider's decisions or actions. Therefore, although expressing appreciation for the gesture is important, healthcare professionals should politely decline such gifts to uphold the integrity of the patient–provider relationship.

GUNSHOT WOUNDS In accordance with legal requirements, healthcare practitioners must report all gunshot wounds to the appropriate authorities, regardless of the patient's wishes or objections. This mandatory reporting is crucial for public safety and potential criminal investigations.

HEALTHCARE PROVIDER DISAGREEMENTS In cases in which healthcare providers have differing opinions regarding patient care, the attending practitioner holds the ultimate responsibility and decision-making authority for the patient's care and management. This responsibility entails ensuring that the chosen course of action is in the best interest of the patient's health and well-being.

HEALTH INSURANCE PORTABILITY ACCOUNTABILITY ACT (HIPAA), HEALTHCARE RECORDS HIPAA is a federal statute that serves multiple functions in protecting patients' healthcare information. First, it grants patients the legal right to access and obtain copies of their healthcare records within a specified time frame, ensuring transparency in their healthcare information. Second, HIPAA safeguards patients' privacy and confidentiality by requiring verbal or written authorization from the patient before releasing health information, even to family members. Disclosure of protected health information (PHI) to friends and family is permissible only if the patient provides explicit consent or does not object when reasonably given the opportunity. Hospitals and practitioners' offices may establish additional policies and procedures in compliance with HIPAA. In situations in which healthcare providers are obligated to disclose information, they must limit the disclosure to only the necessary information required. Noncompliance with HIPAA regulations can result in significant penalties, including fines of up to $250,000 and imprisonment for up to 5 years.

HIPPOCRATIC OATH The Hippocratic Oath (Oath) is a historical oath of ethics traditionally taken by practitioners, believed to have originated between the 6th century BCE and 1st century CE. The Oath emphasizes a practitioner's ethical responsibilities, including prioritizing patient well-being, committing to lifelong learning, and maintaining patient confidentiality. Although it is considered archaic and not used in its original form today, the Oath's core principles continue to significantly influence modern healthcare ethics and professional conduct.

HOSPICE Hospice is a palliative, interprofessional model of care designed to provide comprehensive and compassionate support for patients with a prognosis of 6 months or less to live. The focus is on improving the quality of life, managing pain and symptoms, and providing emotional and spiritual support to the patients and their families during this challenging time.

IMMUNIZATION, VACCINE HESITANCY Healthcare practitioners are ethically and professionally obliged to maintain current immunizations to prevent the spread of infectious diseases, particularly to vulnerable patient populations. Failure to adhere to recommended vaccination schedules is considered professional misconduct and compromises patient safety.

IMPAIRED COLLEAGUES Healthcare providers have both an ethical and a legal obligation to report colleagues who are impaired on the job or on-call in a timely manner. This responsibility is crucial for maintaining patient safety and upholding the integrity of the healthcare profession. Reporting impaired colleagues not only helps prevent potential harm to patients but also supports the impaired individual in receiving the necessary evaluation and treatment to address their impairment.

IMPAIRED DRIVER Patients are responsible for self-reporting seizure disorders and other conditions that may impair their driving abilities. In certain states, practitioners are mandated to report such impairments to the department of motor vehicles, which then determines if restrictions, suspension, or revocation of the driver's license is necessary. It is important to note that healthcare providers do not possess the authority to directly restrict, suspend, or revoke a patient's driving privileges.

INFORMAL TREATMENT Providing informal healthcare treatment to family and friends, except in emergencies, is generally considered problematic and not recommended. This is due to the potential lack of comprehensive understanding of the individual's healthcare history, the absence of essential risk factors that may not be disclosed or available during informal treatment, and the inappropriateness of conducting an intimate examination under such conditions, each of which can negatively affect the quality and safety of the care provided.

INFORMATION, DISCLOSURE Patients have a fundamental right to be informed of their healthcare diagnosis. In cases in which family members request the withholding of a diagnosis to the patient, healthcare providers should explore the underlying reasons for such requests while also considering the patient's wishes and best interests in accordance with the patient's reasonable goals, values, and priorities.

INFORMATION, WITHHOLDING Patients possess the autonomy and right to decide if they want to be informed about their healthcare condition. If a patient chooses not to receive information about their health, their wishes should be respected and honored by healthcare providers in accordance with ethical guidelines and patient-centered care.

INFORMED CONSENT Patient informed consent is a fundamental ethical and legal requirement in healthcare, in which a patient freely and voluntarily agrees to undergo a specific healthcare intervention. The healthcare professional obtaining the informed consent must have the competence and knowledge to provide the patient with (a) a comprehensive explanation of the various treatment options, including no treatment; (b) an accurate description of the interventions; (c) a clear outline of the potential risks and benefits associated with the various treatment options, including no treatment; and (d) answers to any questions posed by the patient.

INTERROGATION Practitioners are ethically bound to prioritize the patient's best interests in all interactions. They must refrain from participating in any form of interrogation or activity that could compromise the patient's well-being.

INTIMATE PARTNER VIOLENCE Intimate partner violence refers to physical, emotional, or psychological harm caused by a current or former partner or spouse. Patients experiencing such violence should be assessed for safety in a supportive, nonjudgmental manner and using open-ended questions. In addition, it is crucial to help these patients develop an emergency safety plan to protect themselves from further harm.

JEHOVAH'S WITNESS Jehovah's Witnesses may refuse blood transfusions due to their religious beliefs. In a life-threatening emergency, if a Jehovah's Witness patient has not clearly refused blood transfusions, and there is no advance directive or other documentary evidence of their wishes, the healthcare providers are not to withhold blood, even if family and friends suggest otherwise. The patient's safety and well-being should always be the top priority.

LANGUAGE Clear and effective communication is crucial in healthcare. Trained healthcare foreign language interpreters should be utilized to facilitate accurate understanding and ensure informed consent. This includes the provision of sign language interpreters for patients who are deaf or hard of hearing, ensuring that all patients receive optimal and equitable care.

LIFELONG LEARNING States require practitioners to earn a minimum of lifelong learning and continuing professional development credits per year. This includes, but is not limited to, continuing healthcare education credits, continuing nursing education credits, continuing pharmacy education credits, continuing dental education credits, continuing education units, maintenance of certification points, performance improvement continuing healthcare education, continuing professional development points, and quality improvement points. Lifelong learning is a healthcare and legal requirement; failure to comply can result in revocation of a healthcare license.

LITERACY Recognizing and addressing low literacy levels is essential in providing effective healthcare and ensuring patient adherence to treatment plans. Healthcare providers should employ various communication strategies, including auditory and visual aids, to ensure that healthcare information is accessible and understandable to all patients, regardless of their literacy levels.

LIVING WILL A living will is a legal document that outlines a patient's wishes regarding healthcare treatment if they lose the capacity to decide for themself. In such situations, the directives stated in the living will precede the wishes of family members or other individuals. It is an essential tool that helps uphold the patient's autonomy and preferences, ensuring their goals, values, and priorities are respected even if the patient cannot communicate them.

MALPRACTICE In healthcare practice, malpractice refers to professional negligence by a healthcare provider that results in harm or injury to the patient. The two essential components for a malpractice claim are (a) the practitioner's negligence, where they failed to provide the standard of care expected of them; and (b) demonstrable harm or injury suffered by the patient as a direct result of the practitioner's negligence.

MEDIA COMMUNICATIONS In the realm of media communications, it is imperative for practitioners to uphold patient confidentiality and privacy. This means that without explicit authorization from the patient, a healthcare provider must not disclose the patient's PHI to the media or any other external entities. This is in compliance with HIPAA regulations and guidelines that safeguard patient privacy rights. Noncompliance with HIPAA regulations can result in significant penalties, including fines of up to $250,000 and imprisonment for up to 5 years.

HEALTHCARE ERRORS Healthcare providers have a professional and ethical obligation to disclose healthcare errors to patients or their designated proxies in a timely manner, regardless of the presence or absence of harm. This disclosure should include a clear and transparent explanation of the error; the steps taken to rectify the situation; and, when appropriate, an apology. Open and honest communication in the event of healthcare errors is crucial to maintaining trust between healthcare providers and their patients.

HEALTHCARE INTERPRETERS To ensure effective communication, comprehension, and valid informed consent, trained healthcare interpreters should be employed for patients who speak a different language or use sign language. This facilitates a clear understanding of the healthcare information, procedures, and potential risks and benefits, empowering patients to make informed decisions about their healthcare.

HEALTHCARE RECORDS Although healthcare institutions are the owners of both electronic and paper healthcare records, it is a fundamental right of patients to have access to the information contained within these records. This ensures transparency and fosters a collaborative partnership between patients and healthcare providers, empowering patients to participate actively in their healthcare journey.

MINOR PATIENTS Practitioners are generally required to obtain parental consent for the healthcare treatment of minor patients. Exceptions to this rule include emergency care and treatment for sexually transmitted infections, contraception, and substance abuse in most states. Furthermore, parental consent may not be required for prenatal care or if the minor is emancipated, which can occur due to homelessness, parenthood, marriage, military service, or financial independence. It is important to note that parents cannot deny their child lifesaving standards of care based on their personal beliefs or preferences.

NAME ADDRESSING Healthcare providers should inquire about patients' preferred form of address and use a formal mode of address unless instructed otherwise by the patient. This practice fosters respect and helps establish a positive patient–provider relationship.

NURSES AND ALLIED HEALTH PROFESSIONALS Physicians are primarily responsible for ensuring patient management is conducted competently and professionally. This encompasses overseeing the activities and contributions of nurses and allied health professionals to guarantee that patient care meets established standards.

ORGAN DONATIONS Organ donation consent that is solicited by a practitioner is categorically inappropriate and should always be avoided to maintain professional and ethical boundaries.

PALLIATIVE CARE Palliative care, which may be provided with or separate from hospice care, enhances the quality and duration of life for patients facing serious illnesses, emphasizing comfort, symptom management, and holistic well-being.

PARENT, MINORS Parents who are minors have the legal right to provide consent for the healthcare treatment of their children, ensuring that the health and well-being of the children are prioritized and appropriately addressed.

PATIENT ADVANCES Inappropriate patient requests or advances should be addressed by healthcare providers with a polite yet firm response, ensuring that the boundaries of professionalism and respect are maintained at all times.

PATIENT CONFIDENTIALITY A fundamental principle in healthcare, patient confidentiality ensures that patients can freely disclose personal and sensitive information, confident that their privacy will be respected and protected. This confidentiality is crucial for building trust between patients and practitioners, thereby facilitating the provision of optimal care. However, there are exceptions to this rule, such as instances of child or elder abuse, knife or gunshot wounds, diagnosis of reportable infectious diseases, and situations in which the patient poses a risk of harm to themselves or others, which must be reported to the appropriate authorities.

PATIENT DISCLOSURE In accordance with privacy laws and professional and ethical standards, healthcare providers must not confirm or deny the identity or presence of an individual as their patient to anyone, including other healthcare providers, unless they are directly involved in the patient's care and have the necessary authorizations to access such information. This protocol protects patient confidentiality and safeguards their personal and healthcare information from unauthorized access or disclosure.

PATIENT INFORMATION Confidential PHI is crucial to the patient's privacy and should be shared exclusively with healthcare professionals directly involved in the patient's care. Healthcare providers must exercise caution to prevent unintentional disclosure of PHI by avoiding discussing a patient's healthcare condition in public spaces or any area where unauthorized individuals could overhear conversations.

PATIENT-PRACTITIONER INTIMACY Engaging in romantic or sexual relationships with current patients is considered unethical due to the inherent power imbalance and the potential for compromising patient care. In the case of non-psychiatric patients, a relationship with a former patient may be permissible under very specific and limited conditions. However, any form of relationship with current or former psychiatric patients is unethical and illegal, given the vulnerable nature of these patients and the need to maintain a strictly professional and supportive therapeutic environment.

PATIENT-PRACTITIONER RELATIONSHIP The relationship between a patient and a practitioner is foundational to effective healthcare. Patient abandonment constitutes a breach of this relationship and is considered a form of healthcare malpractice. If, for any reason, the patient–practitioner relationship is to be terminated, it is essential to provide sufficient time and resources for the patient to secure alternative care, ensuring continuity of treatment and safeguarding patient welfare.

PATIENT REFUSAL In cases in which a patient declines a potentially lifesaving treatment, the healthcare provider must engage in a comprehensive dialogue with the patient. The discussion should cover the patient's reasons for refusing the treatment and the potential consequences and outcomes of this decision. The provider must ensure the patient is fully informed before respecting and honoring their choice.

PHYSICAL ABUSE When encountering suspected cases of physical abuse, healthcare providers should employ empathic interviewing techniques, utilizing open-ended questions that allow patients to recount their experiences in their own words. The primary objective is to gather accurate and comprehensive details about the abuse while prioritizing patient safety. Subsequent actions may include addressing any resulting psychological symptoms and fulfilling legal obligations to report the abuse.

PRACTITIONER ACCESS A principle emphasizing the importance of private communication between patients and their healthcare practitioners. Regardless of age, patients should be provided the opportunity to speak with their practitioner in a confidential setting, without the presence of family members or other individuals, to ensure open and honest communication and to respect the patient's autonomy and privacy.

PRAYER In acute settings, if a patient desires prayer or spiritual support, healthcare providers should offer compassion and empathy while respecting their and the patient's beliefs. Adhering to the principle of nonmaleficence (do no harm), providers should avoid disagreements or religious debates. Upholding the principle of beneficence (do good), a respectful and inclusive response, such as "I will be keeping you in my thoughts," is appropriate.

PRECISION MEDICINE Precision medicine is a healthcare approach that utilizes genetic testing to tailor healthcare treatment to the individual characteristics of each patient. However, it is important to note that this complex technology has the potential to exacerbate healthcare disparities if not applied equitably across diverse populations.

PREGNANT PATIENT A pregnant patient retains her autonomy and has the right to refuse healthcare treatment, even if such refusal may pose risks to the unborn fetus. Healthcare providers must respect the patient's decision while ensuring the patient is fully informed of the potential consequences for the mother and the fetus. In addition, standards of care for pregnant patients vary by state, subject to specific legal regulations that dictate the treatments and interventions not permitted.

PRODUCT SALES Combining medicine with product sales for capitalistic gain is generally impermissible because it compromises the integrity of patient-centered healthcare. This practice can create conflicts of interest and potentially jeopardize the primary focus on the patient's wellbeing. It is essential to uphold ethical standards and prioritize patient care over economic benefits.

PROTECTED HEALTH INFORMATION (PHI) Healthcare providers are responsible for safeguarding patient PHI by taking precautions to prevent accidental disclosure, especially in public places, including general settings within a hospital. This includes avoiding discussions about patients in areas where conversations may be overheard and ensuring that patient records and other forms of PHI are kept confidential and secure.

PROXY A proxy is an individual who has been legally appointed to make healthcare treatment decisions on behalf of a patient if the patient loses the capacity to make their own healthcare decisions. The appointment of a proxy takes precedence over the standard hierarchy of surrogate decision-makers, including family members, ensuring that the patient's healthcare choices are respected and upheld according to the patient's reasonable goals, values, and priorities as interpreted by the designated proxy.

PSYCHIATRIC PATIENTS Individuals receiving psychiatric care retain the right to provide informed consent for their treatment as long as they possess the necessary decision-making capacity. This right ensures that psychiatric patients are actively involved in their healthcare choices and receive treatments aligned with their goals, values, and priorities, contributing to a more patient-centered approach to psychiatric care.

PUBLIC RISK Although patients generally have the right to refuse healthcare treatment, this right is not absolute when the refusal poses a serious threat to public health. When an individual's decision to forgo treatment could harm others, healthcare providers must balance the patient's autonomy with the greater need to protect the community from public health risks.

RACIAL CONCORDANCE Racial concordance refers to the alignment of the patient's and healthcare provider's race or ethnicity. Although treating patients equally and not violating civil rights and liberties is a fundamental requirement in healthcare, studies have shown that racial concordance can positively impact patient outcomes. However, it is crucial to note that unjust discrimination based on race or ethnicity is illegal, unprofessional, and unethical. Healthcare providers must strive to treat all patients with the same high standard of care, regardless of their racial or ethnic background.

REFERRALS AND FEE SPLITTING Referrals should be made based on the best interests of the patient and the appropriate expertise required for their care. Fee splitting, in which a practitioner receives compensation for referring a patient to a specific provider or service, can create a conflict of interest and potentially compromise patient care. Practitioners must mitigate any conflict of interest and communicate openly and honestly with patients on any activity perceived as a conflict of interest. In many jurisdictions, fee splitting is considered unethical and is prohibited by law and professional guidelines.

REPORTABLE ILLNESSES Practitioners are ethically and legally mandated to report certain infectious diseases and illnesses to the appropriate public health agencies because these conditions may pose a significant risk to public health. This requirement is crucial in preventing the spread of infectious diseases and protecting the community's health. Although practitioners must uphold the principles of patient autonomy (informed consent) and confidentiality, protecting public health may necessitate disclosing certain patient information to relevant authorities. In such cases, practitioners should communicate with their patients and provide the necessary support and information to help them understand the importance of reporting and the measures taken to protect public health.

RESEARCH AND CLINICAL EQUIPOISE Clinical equipoise refers to genuine uncertainty within the healthcare community regarding the comparative therapeutic merits of each arm in a clinical trial. Although clinical equipoise and the null hypothesis are fundamental concepts in the design and conduct of randomized clinical trials (RCTs), they are not explicitly mentioned in the Belmont Report or the Common Rule (45CFR46), which are federal regulations governing the ethical conduct of research involving human subjects. Nonetheless, the principle of clinical equipoise is crucial in ensuring the ethical validity of RCTs by justifying the random assignment of participants to different treatment arms when there is no clear evidence favoring one intervention over another.

RESEARCH ON HUMAN SUBJECTS The National Commission for the Protection of Human Subjects of Biomedical and Behavioral Research established the Belmont Report, which outlines ethical principles and guidelines for research involving human subjects. These principles include (a) respect for persons, (b) beneficence, and (c) justice. The Belmont Report's guidelines were subsequently incorporated into federal regulations by the U.S. Department of Health, Education, and Welfare in 1979, resulting in the Common Rule (45CFR46), which governs the ethical conduct of research involving human subjects. The Common Rule sets forth requirements for ensuring informed consent, assessing risks and benefits, and selecting subjects equitably, among other protections.

ROOT CAUSE ANALYSIS Root cause analysis is a systematic process used in quality improvement to identify the fundamental factors contributing to a preventable adverse outcome or event. The goal is to understand what happened, how it happened, and why it happened to prevent recurrence. The process often begins with interviews of individuals involved in the event to gather comprehensive information.

SELF-TREATMENT AND FAMILY TREATMENT The American Medical Association's *Code of Medical Ethics* advises against practitioners treating themselves or members of their own families (Opinion 1.2.1). This is because self-treatment and family treatment can compromise the professional objectivity and judgment necessary for effective healthcare, potentially violating the four principles of biomedical and behavioral ethics: patient autonomy (informed consent), practitioner beneficence (do good), practitioner nonmaleficence (do no harm), and public justice (be fair). With self-treatment and family treatment, personal relationships can cloud clinical objectivity, there is a potential lack of comprehensive understanding of an individual's healthcare history, essential risk factors may not be disclosed to family members, conducting an intimate examination on family members is inappropriate, and there is the potential to disrupt family hierarchy, each of which can negatively affect the quality and safety of the care provided.

SEXUAL BOUNDARIES Maintaining professional boundaries is fundamental to the patient–practitioner relationship. Any sexual, romantic, or familial involvement between a patient and a practitioner is strictly prohibited because it can compromise the integrity of the therapeutic relationship, potentially harm the patient, and is considered unethical and unprofessional. Establishing clear and appropriate boundaries is crucial to preserving the trust and respect necessary for effective patient care.

SEXUAL HISTORY A crucial aspect of patient assessment is that the sexual history should be conducted in a neutral, open, and nonjudgmental manner by healthcare providers. Practitioners should not assume a patient's sexual orientation and must inquire about all sexual partners to gather comprehensive and accurate information necessary for patient care.

SEXUALLY TRANSMITTED INFECTIONS (STIS) The management of STIs requires careful consideration of both legal regulations and ethical principles. Expedited partner therapy, which involves treating a patient's sexual partner(s) without an examination, is not legally accepted in all states. Furthermore, in dealing with STIs, the ethical principles of nonmaleficence (do no harm) and justice (be fair) can often take precedence over autonomy (informed consent). This approach aims to prevent harm and ensure fairness in public health while respecting the patient's autonomy to the greatest extent possible.

SOCIAL MEDIA BOUNDARIES In social media, healthcare professionals must prioritize their professional and institutional obligations over personal freedoms. This includes adhering to patient confidentiality, maintaining professional conduct, and upholding their respective institutions' standards and policies. Personal expressions and interactions on social media must be carefully managed to avoid any breach of professional ethics or institutional guidelines.

STAGES OF GRIEF In facing terminal illness, patients often go through a series of emotional responses known as the stages of grief, which typically include denial, anger, bargaining, depression, and acceptance. It is important to note that the order and number of these stages may differ among individuals. When patients experience grief in ways that do not impair their relationships or interfere with their care, practitioners should respect their emotional process without confrontation.

STANDARDS OF CARE Established norms and protocols are derived from peer-reviewed, evidence-based research recognized and upheld by the healthcare community. Standards of care are formulated by professional opinions and councils, setting the expectations for acceptable and appropriate clinical practice to ensure patient safety and optimal health outcomes.

STERILIZATION The patient's decision-making authority is paramount with regard to sterilization. A patient with decisional capacity who has provided informed consent has the right to be sterilized, regardless of the objections or opinions of a spouse, partner, parents, or anyone else. The healthcare provider's responsibility is to respect and uphold the patient's autonomous decision, ensuring that all necessary information has been provided and understood for true informed consent.

STRIKES–UNIONIZATION For strikes and unionization, the primary focus should always be advancing the patient's best interests, including access to more affordable and innovative patient services and technologies. Self-interest justifications, such as increased practitioner pay and benefits, should not be the driving force behind unionization efforts in the healthcare setting. The patient's well-being and quality of care must remain the top priority.

STRUCTURAL INJUSTICE Structural injustice in healthcare refers to the systemic and inequitable distribution of healthcare benefits and burdens, resulting in disparities in access to and quality of care. This injustice often disproportionately affects marginalized and vulnerable populations, contributing to significant health disparities and barriers to achieving optimal health outcomes.

STUDENT PATIENT CARE Healthcare students must identify themselves as students when interacting with patients. They are required to obtain informed consent before conducting any procedure, ensuring that patients are fully aware of the students' status and the nature of the procedure. In addition, a qualified attending practitioner must supervise all student interactions with patients to maintain the highest patient care and safety standards.

SURROGATE DECISION-MAKING When patients cannot make healthcare decisions, a surrogate decision-maker is appointed to make healthcare decisions. The healthcare practitioner must provide the surrogate with comprehensive and understandable information necessary for informed consent, ensuring they are well informed about the patient's healthcare condition, prognosis, treatment options, associated risks, and benefits. The surrogate's role is to make healthcare decisions based on the patient's reasonable goals, values, and priorities. All their questions should be answered clearly to assist them in making informed decisions that align with the patient's wishes.

TELEMEDICINE Telemedicine refers to the use of electronic communication to provide healthcare services remotely. Despite the physical distance, the fiduciary patient–practitioner relationship remains foundational to medicine. Practitioners must uphold this relationship by ensuring high-quality care and adhering to all ethical guidelines and standards. In addition, telemedicine is subject to federal mandates that require full compliance with HIPAA regulations, ensuring the confidentiality and privacy of patient information are maintained throughout the process.

TERMINAL/PALLIATIVE SEDATION Terminal or palliative sedation involves administering medication to reduce a patient's consciousness to alleviate refractory symptoms, making the final stages of dying more tolerable. Importantly, terminal sedation is distinct from euthanasia. Unlike euthanasia, terminal sedation does not aim to hasten death but, rather, focuses on providing comfort and relief from distressing symptoms during the natural dying process. Most patients who undergo terminal sedation are typically only days or hours away from death. Therefore, terminal sedation has minimal effect on the length of life and is not considered euthanasia.

TESTIMONIALS AND QUACKERY Engaging in quackery by making exaggerated claims about one's healthcare abilities or promoting cures that lack a scientific basis is unprofessional and unethical, as well as potentially illegal. Furthermore, soliciting testimonials from patients who may be vulnerable and dependent undermines the integrity of the patient–practitioner relationship. Such practices compromise the trust necessary for effective healthcare and should be strictly avoided.

TORTURE Healthcare providers must strictly refrain from participating in or being associated with torture or harsh interrogation techniques. It is a moral and ethical obligation to uphold all individuals' dignity and human rights. Furthermore, any knowledge or suspicion of such inhumane practices must be reported immediately, per mandatory reporting requirements.

TRIAGE To uphold and maintain patients' and society's trust, practitioners must always act in their patients' best interest. Triage decisions, which can significantly impact patient outcomes, should be guided by policies established by federal, state, or institutional bodies. This approach ensures that practitioners do not compromise the foundational patient–practitioner relationship by having to make triage decisions that may conflict with their own patient's best interests.

UNEMANCIPATED MINOR An unemancipated minor typically lacks the legal capacity to provide informed consent for healthcare treatments. In such cases, consent must be obtained from parents or legal guardians, with informed consent from just one parent or guardian being legally sufficient. Although practitioners generally require parental consent for treating minor patients, there are exceptions, such as emergency care and treatment for STIs, contraception, and substance abuse in most states. In addition, prenatal care and situations in which the minor is emancipated—due to factors such as homelessness, parenthood, marriage, military service, or financial independence—may not require parental consent. Parents cannot deny their children lifesaving access to standards of care treatments based on personal beliefs or preferences.

WITHHOLDING AND WITHDRAWING TREATMENT Patients have legal, professional, and moral rights to withhold or withdraw healthcare treatment. However, complexities may arise when surrogates involved in the decision-making process disagree. In such cases, the practitioner must foster open communication and encourage discussion among the surrogates to reach a consensus that aligns with the patient's best interests as determined by the patient's reasonable goals, values, and priorities.